SYSTEMS, NETWORKS,
AND COMPUTATION

**McGRAW-HILL
BOOK COMPANY**

New York
St. Louis
San Francisco
Düsseldorf
Johannesburg
Kuala Lumpur
London
Mexico
Montreal
New Delhi
Panama
Paris
São Paulo
Singapore
Sydney
Tokyo
Toronto

MICHAEL ATHANS
MICHAEL L. DERTOUZOS
RICHARD N. SPANN
SAMUEL J. MASON
Department of Electrical Engineering
Massachusetts Institute of Technology

Systems, Networks, and Computation

MULTIVARIABLE METHODS

This book was set in Times Roman by Composition Technology, Inc.
The editors were Kenneth J. Bowman and Madelaine Eichberg;
the cover designer was Joseph Gillians;
the production supervisor was Thomas J. LoPinto.
The drawings were done by ANCO Technical Services.
The Maple Press Company was printer and binder.

Library of Congress Cataloging in Publication Data
Main entry under title:

Systems, networks, and computation: multivariable
methods.

    1. System analysis. 2. Electric networks:
I. Athans, Michael.
QA402.S98   003   74-2021
ISBN 0-07-002430-8

**SYSTEMS, NETWORKS,
AND COMPUTATION**
Multivariable Methods

1 2 3 4 5 6 7 8 9 0 MA MM 7 9 8 7 6 5 4

To the memory of

SAMUEL J. MASON

who left us prematurely
in March 1974

CONTENTS

This book is the outgrowth of a one-semester course intended for undergraduates in electrical engineering and computer sciences. This subject has been taught in an evolving manner during the past 6 years at M.I.T. Although the subject has been elected primarily by sophomores and juniors in electrical engineering, the character of the class has changed lately by attracting students from many other engineering disciplines—mathematics, economics, and management.

The prerequisites assumed are a sophomore level course in linear algebra and an introductory course in network analysis. The linear-algebra prerequisite is essential to covering the material in a single semester. The knowledge of circuits required is relatively minimal; we assume that the student is familiar with setting up differential equations for simple networks.

The basic objective of this book is to introduce certain topics in systems analysis. During the past decade complex systems have become increasingly important, not only in traditional engineering applications but also in fields with strong interdisciplinary overtones. Apart from technological and socioeconomic motivating forces, the rapid rise in the popularity of complex-systems analysis has been greatly influenced by the increasing use of digital computers for simulation.

In spite of the systems revolution, the teaching of systems at the undergraduate level has been somewhat nebulous. Students have been introduced to systems by studying subjects that are relatively narrow from an applications viewpoint. Thus one can take excellent courses in network systems, electromechanical systems, chemical process control systems, transportation systems, aerospace systems, economic systems, urban systems, and so forth. The fact that similar conceptual, analytical, and algorithmic tools can be used to analyze and design such systems had to be learned more or less by osmosis.

The increasing quest for relevance among the students of today cannot be ignored. The greater opportunities for undergraduate research and the mushrooming interdisciplinary laboratories have also contributed to the need for a more unified treatment of systems earlier in the undergraduate curriculum. It is this need that this book is intended to fill.

One may argue that the analysis of complex dynamical systems does not belong to the sophomore or junior curriculum.We disagree. The power of the digital computer for system simulation and algorithmic solutions surely cannot be appreciated by dealing with second-order RLC networks. It is important to let the student know, as early as possible, that systems with different physical characteristics can be analyzed and designed with a common analytical and algorithmic methodology. It is important to learn that large interconnected systems can exhibit counterintuitive response, which cannot be predicted by intuitive back-of-the-envelope analysis. Finally, it is important to present to the new electrical engineering student relevant applications of general systems concepts. The students of today are vitally interested in pollution systems, social problems, economic systems, transportation systems, and so on. They know and expect that electrical engineers will be intrumental in solving pressing national problems. It therefore follows that one should motivate the need for general system theoretic ideas by considering such relevant examples. We have found that the students respond well to this multidisciplinary flavor.

Before one can motivate the mathematical analysis and design tolls for dynamical systems, one must motivate the need for such a development by treading on familiar ground. The general approach we have taken is to proceed from the familiar to the novel.

In Chapter 1 we present some general remarks about the concepts of systems analysis and dynamical systems. This chapter essentially lays the foundation for the material that follows.

A summary of the sequence of chapters, the basic motivation, the key concepts presented, and the main outcome is tabulated on page xix. We now elaborate on the reason for this sequence and, in particular, on the motivational interaction between the chapters.

Table I   CHAPTER INTERRELATIONSHIPS

Chapter number	Motivation	Key contents	Outcome
2	How to set up equilibrium equations for multivariable static systems	Systematic analysis methods for linear and nonlinear resistive networks; node method; cut-set method; linearization	Need to solve vector equations of the form $\mathbf{Ax} = \mathbf{b}$ or $\mathbf{x} = \mathbf{f}(\mathbf{x})$
3	How to solve vector equations of form $\mathbf{x} = \mathbf{f}(\mathbf{x})$ (from Chap. 2)	Iterative methods: Picard's and Newton's methods; optimization methods: steepest-descent,	Vector difference equations of the form $\mathbf{x}(k + 1) = \mathbf{f}(\mathbf{x}(k))$, where k = iteration index
	Formulation of static optimization problems and equations that must be solved	conjugate-gradient, Fletcher-Powell methods	Conditions for convergence; speed of convergence
4	Additional properties of equations $\mathbf{x}(k + 1) = \mathbf{f}(\mathbf{x}(k))$ (from Chap. 3); analysis of socio-economic systems	Theory of linear and non-linear vector difference equations; introduction to stability; the role of eigen-values; linearization concepts; use of feedback	General analysis and design methods for discrete-time dynamical systems
5	To present a nontrivial example of a discrete-time system	Description of macroeconomic models for the United States economy	General under-standing of use of dynamic systems
6	To set up equations of continuous-time networks	Systematic derivation of state-variable equations for linear and nonlinear RLC networks using cut-set method	Vector differential equations of the form $\dot{\mathbf{x}}(t) = \mathbf{f}(\mathbf{x}(t), \mathbf{u}(t))$ and $\dot{\mathbf{x}}(t) = \mathbf{Ax}(t) + \mathbf{Bu}(t)$
7	Properties of vector differential equations (from Chap. 6); equations of nonelectrical dynamical systems	Theory of linear and nonlinear vector differential equations; introduction to stability; role of eigenvalues; linearization and use of feedback	General analysis and design methods for continuous-time dynamical systems
8	Reconciliation of input-output and state-variable models for linear systems	Going back and forth between state-variable models and input-output system or transfer-function models	Relations between the two main approaches to linear-system modeling
9	How to obtain numerical solutions to vector differential equations (from Chaps. 6 to 8)	Numerical integration algorithms (Euler, trapezoidal, Simpson's rule); numerical solution of differential equations (Euler, Heun, fourth-order Runge-Kutta, implicit iterative methods)	Accuracy sensitivity, and complexity trade-offs of different numerical methods

Chapter 2 deals with *large memoryless systems*. It stresses the systematic nature of obtaining network equations, which contain sources and linear or nonlinear resistors, using the node and cut-set methods. From a pragmatic point of view, this chapter serves as a vehicle for the student's familiarizing himself with actual use of matrices and vectors to equations. From a motivational point of view, we use resistive networks to show the need for solving linear vector equations (of the form $\mathbf{Ax} + \mathbf{b} = \mathbf{0}$) and nonlinear vector equations [of the form $\mathbf{x} = \mathbf{f(x)}$ or $\mathbf{g(x)} = \mathbf{0}$]. From a utilitarian point of view, the student enhances his knowledge of resistive networks and is also introduced to the idea of linearization (a concept that will be used again and again), a tool that will enable him to analyze approximately the behavior of nonlinear systems using linear systems tools. Finally, we present a brief introduction to fluid networks as additional motivation.

Chapter 3 deals with iterative methods and algorithmic solutions. The basic motivation for algorithmic solution of sets of simultaneous algebraic equations has been presented in Chap. 2. We immediately stress that an iterative algorithm generates a sequence $\{\mathbf{x}_k\}$ of educated-guess vectors, where k is an iterative index, and that iterative algorithms have the general structure $\mathbf{x}_{k+1} = \mathbf{f(x}_k)$. At this point, we wish to communicate to the student that there are some other types of systems problems that lead to the same general structure $\mathbf{x}_{k+1} = \mathbf{f(x}_k)$. For this reason, we spend some time discussing unconstrained optimization problems, i.e., problems in which we wish to find the minimum or maximum of a scalar function of many variables. In this manner, we try to use economy of concept, i.e., the fact that the analysis and solution of a specific class of mathematical problems have many areas of applicability. Next, the student is introduced to Picard's and Newton's method for solving nonlinear vector equations and to the steepest-descent, conjugate-gradient, and Fletcher-Powell methods for solving unconstrained optimization problems in an algorithmic manner. From a utilitarian point of view the student now is in possession of some of the best-known algorithms for solving static and memoryless problems. From an intellectual point of view, questions of existence, uniqueness, convergence, and rates of convergence have come to the surface. Finally, the structure of the algorithm $\mathbf{x}_{k+1} = \mathbf{f(x}_k)$ provides and excellent vehicle for motivating the study of discrete-time dynamical systems described by vector difference equations and using the state-variable method.

Chapter 4 deals with analysis and design methods for linear and nonlinear time-invariant discrete-time systems described by vector difference equations. Motivational examples, apart from iterative algorithms, include socioeconomic problems. Linear discrete-time systems are considered in detail, and the usual superposition properties and tests for stability in terms of the eigenvalues of the system matrix are derived. The steady-state and equilibrium analysis of nonlinear discrete-time systems is used to motivate the need for perturbation analysis to understand the stability or instability properties of an equilibrium point. Once more linearization techniques (first introduced in Chap. 2) are used to show how one can obtain results for nonlinear systems using tools developed for linear systems. In addition to these analysis tools, the student is now exposed to design techniques using state-variable feedback, which can be used to stabilize or improve the performance of a discrete-time system.

Chapter 5 contains a discussion of simple and complex macroeconomic models. This material is included because mathematical representations of economic systems are almost always presented in a discrete-time form. It also illustrates how one converts the usual input-output difference equations of econometric models into a state-variable form. Finally, by presenting the detailed equations of a 28-state-variable, 3-input-variable linear econometric model of the United States economy, we illustrate vividly that in socioeconomic applications, the relative accuracy of a mathematical model is of paramount importance when this model is to be used for forecasting and to help determine policy.

At this point, the student has been exposed to the basic bread-and-butter concepts and techniques dealing with systems analysis and design for discrete-time dynamical systems. Since difference equations lack mathematical abstraction, most of the intellectual effort has been devoted to understanding the key concepts of linearity, superposition, role of the eigenvalues and eigenvectors of the system matrix, stability, linearization, and feedback. From a pedagogical viewpoint we felt that another exposure to this key set of systems concepts would help the student appreciate their universal importance. From a practical viewpoint, the great majority of engineering systems are most naturally described by differential, rather than difference, equations. Hence, the emphasis now shifts into the analysis of continuous-time systems.

Chapter 6 once more starts from ground familiar to the electrical engineer. The motivational purpose of this chapter is to show how the analysis of linear and nonlinear RLC networks gives rise to linear vector differential equations [of the form $\dot{\mathbf{x}}(t) = \mathbf{A}\mathbf{x}(t) + \mathbf{B}\mathbf{u}(t)$] and nonlinear vector differential equations [of the form $\dot{\mathbf{x}}(t) = \mathbf{f}(\mathbf{x}(t), \mathbf{u}(t))$]. This provides motivation for the general system concepts associated with continuous-time dynamical systems presented in Chap. 7, as well as for the numerical-integration algorithms presented in Chap. 9. From a pragmatic point of view, the development of the network differential equations is based upon the cut-set method (discussed in Chap. 2) and the state-variable approach. The state-variable approach to the development of the network equations depends upon seeking the network variables that most naturally describe the energy in the network. Thus, for linear networks we always use as state variables the capacitor voltages and the inductor currents. For nonlinear networks, we use as state variables the capacitor charges and inductor flux linkages.

Chapter 7 contains the basic systems concepts associated with continuous-time dynamical systems. The system concepts follow a development parallel to the ideas presented in Chap. 4, where the same concepts were exposed in the context of discrete-time systems. This is how we achieve the double exposure we feel is necessary. Although the concepts are really the same, the results take on a somewhat different analytical character; and since vector differential equations are somewhat more sophisticated than difference equations, this double exposure still serves a useful intellectual and practical purpose. Additional examples drawn from aerospace, transportation, and ecological applications are used to stress further that many diverse systems do obey vector differential equations. Linear-systems analysis includes the matrix exponential and superposition properties, the role of eigenvalues and eigenvectors in the solution of vector differential equations, the stability of linear systems, and the role of dominant eigenvalues. For dynamical systems described by nonlinear differential equations we present the basic proofs for existence and uniqueness of solutions, equilibrium states, linearization methods about the equilibrium, and local stability properties of the equilibrium states using linear-system stability tests. Finally, we present a brief introduction to system design using state and output feedback concepts (the pole-placement problem).

Chapter 8 is a bridge between input-output analysis and state-variable analysis of single-input–single-output linear time-invariant systems. The student has probably already been exposed to the concept of system functions (using complex exponential excitations) and/or to the concept of transfer functions (using Laplace transforms). Thus, he has been exposed to poles and zeros and their key importance to the system transient-response characteristics and system stability. Since our development in Chap. 6 and 7 for continuous-time dynamical systems has followed the state-variable route, the purpose of Chap. 8 is to reconcile these two methods of system representation, which may hitherto appear unrelated to the student. Thus, the material of Chap. 8 deals with ways of going from a state-variable representation to an input-output representation and vice versa, using the standard controllable and standard observable representations. In the process we show that the eigenvalues of the system matrix are identical to the poles of the system or transfer function, thus bringing these two methods of analysis under a common roof.

Chapter 9 deals with the development of numerical algorithms that can be used to solve linear and nonlinear vector differential equations on a digital computer. For numerical evaluation of integrals we present the Euler, trapezoidal, and Simpson methods. For numerical solution of differential equations we present the Euler, Heun, fourth-order Runge-Kutta, and implicit methods. Apart from its obvious practical content, this chapter provides a bridge between the continuous-time world and the discrete-time world.

The three appendixes provide useful facts about vectors, matrices, linear algebra, Taylor series and a brief introduction to Laplace transforms. We strongly urge the student to review the definitions, notations, and manipulative aspects of vectors and matrices by reading Appendix A before Chap. 2, because vector-matrix notation is used right from the start.

Sections that can be omitted without loss of continuity at first reading are marked with a star.

The importance of doing homework cannot be stressed enough. In this book, the exercises are collected at the end of each chapter and are arranged by sections. Exercises that need most of the concepts of a chapter are given at the end.

Three types of exercises are included. The first type is more or less illustrative of the material presented. The second type asks the student to extend the theoretical material presented; many of these are quite difficult and will present a challenge to even the best student. We remark that many important system-theoretic notions, such as controllability, observability, and state reconstructors and observers, appear in the exercises. The third type of exercise requires a digital computer. We have given each student sufficient computer time to do batch computation. We suggest that during a semester, a minimum of four nontrivial computer homework assignments be given. This sequence of computer assignments can start when Chap. 3 is concluded.

This text, together with our first volume,† is the result of an educational experiment started in 1966. The lecture notes that developed into these two texts were used to teach a two-semester sequence at the sophomore level, planned for electrical engineering students whose primary interests were in the area of communications, computers, and control.

In "Basic Concepts" we concentrated upon the exposition of fundamental ideas and techniques using very simple types of networks and logic systems; we also illustrated some key concepts in computation by concentrating upon the solution of scalar equations. We have found that the sophomore student can grasp very general ideas, concepts, and techniques provided that they are accompanied by simple examples drawn from different areas.

In this volume, one of our goals is to communicate to the student that, given the appropriate tools for mathematical representation, e.g., linear algebra, simple concepts, simple algorithms, and simple systems-analysis tools can *readily* be generalized to analyze systems of great complexity, both with respect to the number

---

† M. L. Dertouzos, M. Athans, R. N. Spann, and S. J. Mason, "Systems, Networks and Computation: Basic Concepts," McGraw-Hill, New York, 1972; this is referred to simply as "Basic Concepts."

of variables and the number of equations that describe the system. Thus, we have attempted to stress the economy of concept, i.e., that one does not have to learn new systems-analysis concepts or new algorithmic concepts just because one is studying more complex systems.

Thus, although each text can be used independently of each other, a certain unified philosophy emerges from their sequential use. The students who have studied "Basic Concepts" will find strong parallelism between the development of the topics. The majority of the concepts developed in scalar form in "Basic Concepts" are generalized in this text. The treatment of the logic networks and finite-state machines in "Basic Concepts" is replaced by the study of discrete-time systems in this book. Otherwise, the notions of iteration, state variables, dynamic networks, integration, etc., are in complete simple-to-complex parallelism.

Both texts have involved the joint effort of all four coauthors. The preparation of the material for this book has been the primary responsibility of M. Athans.

We wish to extend our sincere gratitude to all our past students who have helped us in formulating and presenting ideas. We wish to thank our M.I.T. colleagues R. G. Gallagher, P. L. Penfield, F. C. Schweppe, G. C. Newton, Jr., D. C. White, and especially L. A. Gould, who have taught from the material of this book and have given us valuable suggestions, ideas, and criticism. We also wish to thank Professors C. A. Desoer, R. A. Rohrer, and L. A. Zadeh, of the University of California at Berkeley; J. G. Truxal, of the Polytechnic Institute of Brooklyn; W. S. Levine, of the University of Maryland; and M. E. Valkenburg, of Princeton University, for manuscript reviews, encouragement, and support. P. C. Houpt has greatly contributed to this volume and has been responsible for the preparation of the solutions manual. We are also indebted to our department chairmen, first Professor P. Elias and then Professor L. D. Smullin, for providing the environment and resources that made our project possible. Special thanks are due to Dr. Leslie C. Kramer, now with the M.I.T. Lincoln Laboratory, who from his sophomore year until his Ph.D. was so intimately involved with this project, as a student, consultant, teacher, and friend. Finally, we wish to thank Maureen Stanton, Ionia Lewis, Judi Bucci, and especially Elyse Wolf, for typing and retyping all the versions of this manuscript.

MICHAEL ATHANS
MICHAEL L. DERTOUZOS
RICHARD N. SPANN
SAMUEL J. MASON

1
INTRODUCTION TO DYNAMICAL SYSTEMS

SUMMARY

This chapter presents a qualitative overview of concepts and methods associated with the study of multivariable dynamical systems.

1.1. THE CONCEPT OF A DYNAMICAL SYSTEM

Of key importance in several branches of engineering, economics, and management science is the notion of a dynamical system. Since we shall spend quite a bit of time analyzing several types of dynamical systems, it is important to examine their intuitive properties briefly; these in turn will be used to make a transition into the appropriate mathematical framework that can be used to describe such systems.

1.1.1 Cause and Effect

The notion of a dynamical system cannot be described independently of the variables used to describe its behavior. In general, and at a very abstract conceptual level, we distinguish two types of variables:

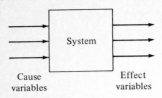

Cause variables   Effect variables

FIGURE 1.1.1
Abstract representation.

1 The *cause variables*, i.e., the variables that are externally applied or act upon a system

2 The *effect variables*, i.e., the variables which are intimately related to the way the system behaves

Figure 1.1.1 gives an abstract visualization of this cause-and-effect relationship. To be sure, this type of description is too abstract to be really useful. Nonetheless, it represents an acceptable starting point. It is interesting to note in the diagram of Fig. 1.1.1 the following:

1 There may be more than one cause variable. Each cause variable is represented by a directed arrow; the arrow is directed *into* the system to indicate that each cause variable directly influences the system under consideration and that the system has no choice but somehow to respond to this cause.

Table 1.1   EXAMPLES OF CAUSE-AND-EFFECT RELATIONSHIPS

Cause variables	System	Effect variables
Voltage across	Resistor	Current through
Current through	Resistor	Voltage across
Position of accelerator Position of brake pedal Position of gear lever Position of steering wheel	Automobile	Forward speed Lateral speed Position on highway
Elevator deflection Aileron deflection Rudder deflection	Aircraft	Pitch angle Roll angle Yaw angle
Air temperature Humidity Walking speed	Human body	Perspiration level
Money supply Government expenditures Tax rates	United States economy	Gross national product Unemployment rate Inflation level Interest rate

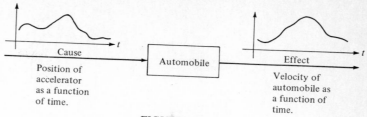

FIGURE 1.1.2

Example of a continuous-time dynamical system.

2 There may be more than one effect variable. The number of effect variables may be larger or smaller than the number of cause variables. Each effect variable is also represented by a directed arrow; note that the arrow is directed *away* from the system, which means that the effect variables are in some sense generated by the system itself.

Before we become too entangled in abstraction, let us give some examples of what we mean. The examples are summarized in Table 1.1.

A little reflection should convince the reader that there may be some element of choice in determining what constitutes a cause and what constitutes an effect for the same system (compare the resistor example in Table 1.1). Often causes and effects are selected as a function of the specific application at hand; however, we are dealing with a physical system, and it is understood that a certain degree of care must be exercised in the definition of these notions. For example, the triplet "voltage across," "aircraft," "perspiration level" constitutes a nonsensical cause-and-effect system relationship.

1.1.2 The Role of Time

Of key importance in analyzing systems is the notion of time. In this book we examine two different ways that time enters into the picture. In a *continuous-time* description we visualize each cause variable and each effect variable as being described by a time function. This is the most physically consistent level for most of the physical systems we are familiar with. In such systems we can visualize a set of time functions being the cause variables and a set of time functions being the effect variables; see, for example, Fig. 1.1.2.

However, there are many types of systems in which a continuous-time description is too detailed for the purposes of system analysis. In such problems one is satisfied with a description of the causes and effects only at discrete instants of time. One then has the so-called *discrete-time* description of a system. Such

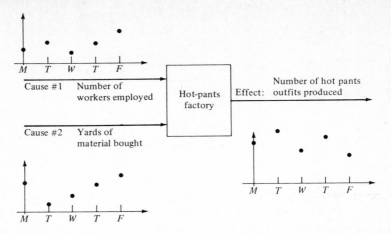

FIGURE 1.1.3
A discrete-time dynamical system.

descriptions are extremely common in economic and business systems because in such problems there may exist a natural time unit (day, week, month, etc.), in which the system response can be adequately analyzed. In discrete systems one deals with time sequences rather than time functions. An example is illustrated in Fig. 1.1.3.

We can now start to see what types of questions we can ask. A typical *systems analysis* question follows.

> Given the system and given the cause variables as time functions (or time sequences), can we find what the effect variables will be as time functions (or time sequences)?

It should be self-evident that there are myriads of applications for which such a question is typical. We would like to examine what steps we must follow to answer such a question in quantitative terms.

Before we start our mathematical analysis, it is useful to narrow down the class of systems under consideration. The class of systems we shall study is the class of *causal* systems. Intuitively speaking, a causal system is one that "does not run before being kicked." In other words, in a causal system the effect does not take place before the cause has been applied. The reader should note that the notion of a causal system requires a time description of the cause and effect variables. Clearly our restriction to causal systems does not place any limitations on the class of physical systems, since by their very nature they are causal.

1.1.3 Static and Dynamical Systems

We are now in a position to distinguish between *static systems* (or systems with no memory) and *dynamical systems* (or systems with memory). Both of these types belong to the class of causal systems.

Intuitively speaking, a *static system* has the property that its effect variables at any time t depend only on the values of its cause variables at the same instant of time t. By causality, the effect variables cannot depend on future values of the cause variables. In a static system the effect variables also cannot depend on past values of the causes; hence, they cannot remember the past causes, and that is why they are called systems without memory.

A familiar example of a static system is a linear constant resistor whose cause variable is the voltage across $v(t)$ and whose effect variable is the current through $i(t)$. Such a system is described by Ohm's law

$$i(t) = \frac{1}{R} v(t)$$

where R is the resistance. Note that the value of the current at any time t depends only on the instantaneous value of the voltage at the same instant of time.

Static systems are not too exciting because there is no time lag between the application of a cause and the appearance of an effect. The systems that exhibit more interesting behavior are the so-called *dynamical systems*.

In a *dynamical system* the values of the effect variables at any t depend in general on past values of the cause variables; in this manner, the system remembers the past, and hence it has memory. Such systems can exhibit extremely interesting behavior as a function of time (instability, cyclic oscillations, etc.) which are often quite the opposite of what one would intuitively expect. Dynamical systems claim most of our attention in this book.

1.2 INFLUENCING A DYNAMICAL SYSTEM

Just why dynamical systems are both interesting and challenging cannot be appreciated until one asks some synthesis-oriented questions. Because of the time lags and memory, the effect of any current decisions (acting as inputs) will not be fully seen until some time in the future. Hence, one can ask the following type of questions.

1 Suppose that I establish a certain tentative set of inputs to be applied from now until some time in the future. How do I predict what the system will do?

2 Suppose that the system behavior is unsatisfactory. What can I do to make the system respond in a more satisfactory way?

One can then see that in dealing with dynamical systems the fundamental problem is that at each instant of time one has to make decisions and apply inputs to the system such that the future system behavior is satisfactory in some sense.

This type of problem is of fundamental importance in many areas of science and engineering. We outline several examples.

1 In an economic system the government has to make decisions pertaining to level of government spending, changes in money supply, setting up tariffs, etc. Such current and past decisions will affect future values of key economic variables, e.g., gross national product, unemployment rate, inflation rate. Contemplated actions may have to be changed if in fact the economy is not behaving in a satisfactory manner due to factors that could not be fully predicted.

2 In complex ecological systems a large number of species may compete for the same resources and depend upon each other for survival. Such species may have reached an ecological equilibrium. On the other hand, natural or man-made causes, e.g., insecticides, may affect the death rate and/or growth rate of one or more of the species. Such ecological disturbances may lead to long-term changes in the species involved (and even extinction). Undesirable trends may in turn be influenced by other inputs to the system, e.g., banning of certain chemicals or hunting restrictions. However, such current corrective actions have to be planned carefully because short-term benefits may not necessarily last and eventually (once more due to the memory interrelations) may lead to undesirable outcomes.

3 In air-traffic control systems, the problem is primarily one of keeping the aircraft separated by a safe distance. In the vicinity of the airport, one can imagine many "highways in the sky" that merge progressively until there is a "single lane" left, leading from the outer marker to the runway. At each instant of time aircraft may be safely separated. However, if each plane followed a nominal individual flight plan, eventually, due to the merging of the air routes, aircraft might get too close to each other. Thus the air-traffic control system must predict the future locations of the aircraft and in case of danger command changes in the aircraft accelerations. Such decision and flow-control problems can be extremely complex when the number of aircraft involved is large, when weather conditions exclude certain air space due to turbulence, when changes in wind direction necessitate runway changes, and when emergency conditions necessitate changes in the normal first come, first served landing patterns.

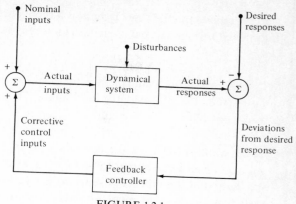

FIGURE 1.2.1
Conceptual diagram of a feedback system.

The problem of constantly changing the input of a dynamical system so that the response of the system is satisfactory is often called the *feedback control problem*, illustrated conceptually in Fig. 1.2.1.

The basic problem in such a feedback system (either at discrete time instants or in continuous time) is to change the actual system inputs to the system directly under our control. One can imagine that the actual inputs to the system are obtained by adding the nominal system inputs and the corrective control inputs. The *nominal system inputs* are those which we believe will lead to a satisfactory system response. They are usually computed before the system is placed in operation and usually involve many assumptions; e.g., no external disturbances are present.

The *corrective control inputs*, on the other hand, are not computed beforehand; their value at each instant of time depends on actual measurements made while the system is in operation. To understand their role, we must examine the system in greater detail.

We assume that measurements can be made on the variables that define the actual system response. The actual system response at each instant of time will depend on the past history of the actual inputs and disturbances. Thus, the actual system response variables may not behave in a satisfactory manner.

The *desired response variables* are time sequences or time functions which reflect an ideal desired response. Like the nominal input variables, they are usually precomputed.

At each instant, we can subtract the desired response variables from the actual response variables and obtain the (undesirable) deviations from the ideal response. If we did nothing (and this corresponds to not using the corrective control variables and continuing to use the nominal inputs as the actual inputs), these currently sensed deviations might grow larger and larger. To prevent this, we send the actually sensed deviations to another system, which may be static or dynamical, called the *feedback controller*.

The basic job of the *feedback controller* is to translate the measured deviations into corrective control signals, so that future deviations of the actual system responses from the desired ones are small. The quantitative specification of the feedback controller is a major and challenging task in system design.

Feedback systems are extremely common. They range from very simple (home temperature control using a thermostat), to complex guidance and control systems (for aircraft, missiles, submarines, etc.), to extremely complex decision-oriented systems in socioeconomic problems. Many feedback systems occur naturally in living systems, e.g., adjustment of pupil size to changes in light intensity, body-temperature control by shivering and perspiration level, or sugar-level control via the insulin cycle.

Man is an essential element of the feedback controller in man-machine systems, e.g., pilot control of an aircraft, aiming antiaircraft guns. In spite of the diversity of systems and tasks, the analysis and design of such systems follows a relatively unified set of conceptual, theoretical, analytical, and algorithmic methods. We shall examine some of these in this text.

1.3. THE ROLE OF QUANTITATIVE METHODS

Clever people can make systems that work in a satisfactory manner without the benefit of mathematics and digital computers. In addition to cleverness, good physical intuition, which is the result of past experience, helps a lot. However, clever aircraft-control-system designers do not necessarily make good chemical-reactor-control designers, and vice versa. For complex systems, intuition may fail and lead to disastrous results.

Since the number of very experienced and very clever people is limited, one needs alternative approaches to systems analysis and design for the ordinary person. Complex systems tend to be multidisciplinary and involve the cooperative effort of workers trained in different disciplines. Communication can be a major obstacle if jargon is used.

Mathematical models and mathematical analysis and design techniques provide the common language of systems. There are obvious advantages in dealing with such a well-developed, logical, and consistent set of mathematical tools. Nonetheless, there are disadvantages, and one must be aware of the dangers involved.

The major danger in dealing with mathematical models is that one has to make an abstraction of reality. To be sure, one has help in modeling a physical system by means of equations; the laws of nature, experiments, and so on. Nonetheless, there are always assumptions and idealizations that have to be made in order to arrive at a quantitative mathematical model of an actual system.

Fortunately, most mathematical models can be simulated on digital computers. By carrying out simulation experiments the engineer can correlate the solution of his equations with what he knows the physical system has done or will do. Since digital computers have taken over the drudgery of equation solving, the use of mathematical system models for analysis, simulation, and design has greatly increased.

Just because computers can solve very complex equations does not mean that there is no value in delving into the analytical (rather than algorithmic) aspects of the mathematical models. Past developments in mathematics offer a rich supply of results that are of great value in the study of systems. Under suitable assumptions, the application of mathematical theorems to a specific system can yield valuable qualitative information on how a system will behave and help us make it behave better. For this reason the study of systems almost always involves a judicial blend of both analytical and algorithmic methods.

The systems we shall study in this book are of the following types:

1 For static or memoryless systems, the mathematical model will be a set of simultaneous algebraic equations. We shall discuss classes of physical systems that give rise to such mathematical models.
2 For dynamical systems, we shall consider only those which are adequately modeled by a simultaneous set of difference equations (discrete-time systems) or ordinary differential equations (continuous-time systems). Once more we shall present several examples of physical systems that are adequately modeled by such mathematical models.

There are many other types of physical systems that do not fall in the above categories, e.g., finite-state machines, automata, distributed-parameter systems, and stochastic systems. Their study is outside the scope of this book.

1.4. LINEAR VERSUS NONLINEAR SYSTEMS

Some classes of physical systems can be adequately modeled by mathematical relationships with "nice" analytical properties. These are often referred to as *linear systems*. On the other hand, physical systems which lead to mathematical relationships with no neat analytical properties are referred to as *nonlinear systems*.

From a pragmatic point of view, one can argue that no systems in nature are truly linear. Even the most linear resistor will change its physical characteristics if the applied voltage exceeds a certain set of values. Hence, one may rightfully ask:

Why should one study linear mathematical models when they represent only an approximation to reality? The key "nice" analytical property of linear systems is that of *superposition*. We shall examine linear systems and superposition properties in great detail. For the time being, the property of superposition means that if we know how a linear system responds to a specific set of inputs, we can readily deduce the system response to other inputs that can be expressed as linear combinations of the specific set. On the other hand, nonlinear systems do not enjoy this property. The superposition property of linear systems results in closed-form types of solutions that can be used to determine the system response without having to resort to algorithmic methods. Thus, the great popularity of linear systems from a mathematical point of view can be traced to the many and varied implications of the superposition property.

From a pragmatic point of view, it turns out that the behavior of many physical systems can be often modeled by a linear mathematical model *with a very high degree of accuracy, provided that the ranges of the physical variables are limited.* To put it another way, a physical system that is best modeled by a nonlinear model in a global sense can often be approximated with a high degree of accuracy by a linear system as long as the variables remain in the neighborhood of certain values. We shall see how this fundamental technique of linearizing a nonlinear system can be carried out in a systematic manner and why this technique is so useful. We shall thus be able to use the tools of linear-systems analysis and design to analyze and synthesize nonlinear systems.

1.5. THE INPUT-OUTPUT VERSUS STATE-VARIABLE METHODS

We conclude this overview with a brief description of the methodology that will be employed to model dynamical systems. Strange as it may appear, there are two (related but different-looking) methods for modeling dynamical systems, the state-variable method and the input-output method.

In the *input-output* method, the behavior of the dynamical system is characterized by mathematical equations that relate the external variables of the system, namely, the inputs (causes) and outputs (effects). The input-output method gained popularity due to its use in network and servomechanism analysis and synthesis. It turns out that for single-input single-output linear time-invariant systems, the input-output method is ideally suited for using Fourier and Laplace transform methods for manipulating and solving the resultant equations. In addition, such input-output models are ideally suited for analyzing the behavior of complex dynamical systems when the variables are undergoing periodic sinusoidal variations; this is the case with many communication and electric-power networks.

In the *state-variable* method, one still deals with the interrelationship of the external inputs and outputs; however, this interrelation involves an additional set of variables, called the system *state variables*. In physical systems, the state variables are associated with the energy-storage variables. Typical state variables are

1 Positions, since they relate to potential energy, e.g., amount that a spring is compressed or distance from a gravitational source
2 Velocities, since they relate to kinetic energy
3 Capacitor voltages, since they relate to stored electric energy
4 Inductor currents, since they relate to stored magnetic energy
5 Temperatures, since they relate to thermal energy

Thus, in the state-variable description the mathematical system model usually splits into two parts:

1 A set of mathematical equations relating the state variables to the input variables
2 A set of mathematical equations relating the state variables and input variables to the output variables

The state-variable method has enjoyed popularity more recently (from 1960 on) than the input-output method (from about 1920 on). The state-variable method can be used to describe linear or nonlinear systems, time-varying or time-invariant, and with many inputs and outputs. As a whole, the state-variable method is better suited for simulation and digital-computation purposes. Also, it is more amenable to general systems analysis and design studies. For this reason, we use the state-variable method almost exclusively in this book.

Specific details on the state-variable method will be given in Chaps. 4, 6, and 7. The interrelationship between the state-variable and input-output methods will be considered in Chap. 8.

2
RESISTIVE NETWORKS AND MEMORYLESS SYSTEMS

SUMMARY

This chapter presents systematic procedures for obtaining the equations that characterize linear and nonlinear resistive networks, using the node-voltage method and the cut-set method. Vector-matrix notation is employed throughout. The key ideas developed in the context of electric networks are extended to fluid networks.

    As we shall see, the analysis of memoryless (or static) systems leads, in general, to a set of n equations in n unknowns. In linear systems, the equations take the form $\mathbf{A}\mathbf{x} = \mathbf{b}$, where the elements of the matrix $\mathbf{A}$ and of the column vector $\mathbf{b}$ are fixed by the system topology and the numerical values of the elements and the components of the vector $\mathbf{x}$ represent the unknowns. In nonlinear systems the sought-for solution vector $\mathbf{x}$ satisfies a nonlinear vector equation of the form $\mathbf{x} = \mathbf{f}(\mathbf{x})$ or $\mathbf{g}(\mathbf{x}) = \mathbf{0}$. Solution methods are described in Chap. 3.

2.1 INTRODUCTION

In this chapter we combine network concepts with the tools of linear algebra† to deduce a general and systematic procedure for obtaining the equations of resistive networks of arbitrary complexity. Thus there are very few new concepts that we

---

† The reader should review Appendix A before starting this chapter.

shall have to develop. Instead we shall emphasize the structure of network equations (linear or nonlinear) by utilizing vector and matrix notation. In addition, we emphasize that this description of networks agrees with current computational practice and capability (in terms of writing, say, FORTRAN programs to solve a problem) as well as with theoretical advances in the analysis and design of other multivariable engineering systems, e.g., fluid systems.

We start with an amplification of the exposition of topological-network concepts and express the implications of Kirchhoff's voltage law and Kirchhoff's current law in matrix-vector equations satisfied by time-invariant resistive networks, both linear and nonlinear.

For nonlinear networks we shall explain the general and powerful linearization method which can be used to deduce the behavior of a nonlinear system for small perturbations about a set of constant operating conditions. This method will be used later to deal with the analysis of nonlinear systems with memory, e.g., nonlinear networks with capacitors and inductors.

From a system-theoretic viewpoint, this chapter deals with the modeling issue—how one interconnects memoryless elements with known characteristics to construct an arbitrarily large system and how the rules of interconnection, coupled with the physical laws of nature, combine with the atomic description of the individual elements to lead to the equations that describe the behavior of the system as a whole. It should be apparent that the characteristics of the individual elements will affect the behavior of the overall system to a certain degree. Nevertheless, the fact that many such elements are collectively used can lead to a large system with interesting and perhaps unexpected global properties. Such questions are at the heart of large-scale system analysis and synthesis. This chapter provides an introduction to such problems.

2.2 ELEMENTARY NETWORK-TOPOLOGICAL DEFINITIONS

The skeleton or *graph* of an electric network consists of branches (line segments) that connect or terminate at nodes (points). Each branch represents a possible network element, an inductance, capacitance, resistance, or a current or voltage source. If the standard symbols for the network elements, e.g., zigzag line for a resistance, are shown, the representation is usually called a *network diagram* or *circuit diagram*. If such symbols are omitted and each element is represented as a simple line segment, the representation is usually called a *network graph* or the *linear graph* of the network. The term *branch network* applies either to the diagram or the graph.

We formalize some of these ideas by offering some precise definitions.

Definition 1   (Graph)   *A graph G is a collection of nodes together with a collection of branches; each branch connects precisely two nodes. G_1 is called a subgraph of G if every node and branch of G_1 is also a node and branch of G. G is called a* connected graph *if there is at least one path between any two nodes of G along the branches of G.*

Definition 2   (Loop)   *A loop L is a subgraph which:*

(a)   *Is connected,*
(b)   *Has precisely two branches incident to each node of L, and*
(c)   *Has precisely two nodes of L connected to each branch of L.*

Definition 3   (Cut Set)   *A set C of branches of a connected graph is called a* cut set *if:*

(a)   *The removal of all the branches of C yields an unconnected graph which consists of two separate parts and*
(b)   *The removal of all but one of the branches of C leaves the graph connected.*

Definition 4   (Tree)   *Given a connected graph G, a* tree T *of G is any subgraph of G which:*

(a)   *Is connected,*
(b)   *Contains all the nodes of G, and*
(c)   *Contains no loops.*

The branches of G which also belong to the tree T are called tree branches *of G. The branches of G that do not belong to the tree T are called* links.

Figure 2.2.1 shows a network with eight nodes and ten branches and with two possible choices of trees. According to our definition, this is a connected graph. The set of nodes 1, 2, 5, and 6 together with the branches a, d, e, h provides an example of a connected subgraph. The nodes 2, 3, 6, and 7 and the branches b, e, f, i define a loop. The branches b and i define a cut set; also the branches d, e, f, g define a cut set. If we choose the tree of Fig. 2.2.1a, the tree branches are c, b, a, d, h, i, j, while the links are e, f, g, that is, the dotted branches; on the other hand, the choice of the tree indicated in Fig. 2.2.1b means that the tree branches are d, e, f, g, h, i, j and the links a, b, c. An additional example of trees is illustrated in Fig. 2.2.2.

It is very easy to establish a relation between the number of tree branches, links, nodes, and branches. Consider a connected graph G such that

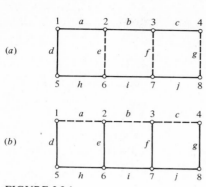

FIGURE 2.2.1
Two possible choices of trees: (a) *abcdhij*
and (b) *defghij*.

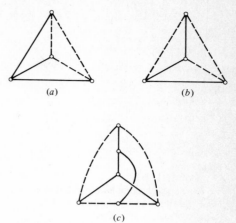

FIGURE 2.2.2
Two different trees in the same network, (a) and
(b), and a tree in a nonplanar network (c).

$$\text{Number of nodes of } G = n + 1$$

$$\text{Number of branches of } G = b$$

Since any tree must contain all the nodes of G, then from Definition 4 it follows that

$$\text{Number of tree branches } = n$$

$$\text{Number of links } = b - n$$

2.3 RELATION BETWEEN NODE VOLTAGES, BRANCH VOLTAGES, AND BRANCH CURRENTS IN AN ARBITRARY NETWORK

In an electric circuit we often assign reference directions to the branch currents and reference polarities for the branch voltages using the passive convention. Let us denote the branch currents by $i_1, i_2, \ldots, i_b$, where b is the total number of branches, and let us denote the branch voltages by $v_1, v_2, \ldots, v_b$. In general, all the branch currents and voltages are functions of time. We can discuss the topology of the network by its *directed graph*; this is simply the graph with directed branches, which represent the assumed positive current directions (see, for example, Fig. 2.3.1).

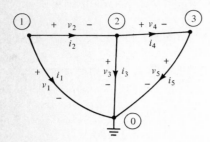

FIGURE 2.3.1
A directed graph containing information about the indexing and polarities of the branch currents and voltages.

FIGURE 2.3.2
The passive sign convention.

Sometimes, the branch-voltage polarities are also indicated; if they are omitted, one assumes that the passive sign convention is used (see Fig. 2.3.2).

As is well known from elementary circuit theory, many factors contributing to the equations must be satisfied by the network variables. Contributing factors are the network topology, i.e., the total number of nodes; how they are interconnected by branches; the physical nature (resistors, capacitors, inductors, sources) of the circuit elements that appear in the branches; their mathematical characteristics (linear, nonlinear); and the quantitative information (value of resistance in ohms, voltage-current characteristics of a diode).

At present we are interested in the implications of network topology, without worrying for the moment about the physical nature of the elements that make up the network branches. In deducing the relations that are satisfied by the voltages and currents we must observe certain laws of nature. These are Kirchhoff's laws:[†]

Kirchhoff current law (KCL)   The algebraic sum of the currents entering a node is zero.

Kirchhoff voltage law (KVL)   The algebraic sum of the voltage differences around any closed path is zero.

---

† See M.L. Dertouzos, M. Athans, R.N. Spann, and S.J. Mason, "Systems, Networks and Computation: Basic Concepts," pp. 84–85, McGraw-Hill, New York, 1972 (referred to henceforth simply as "Basic Concepts").

In this section we examine the general structure of the node equations in a network. We start with an analysis of the graph of the network. We then use KCL and KVL to obtain a general set of vector equations. Our procedure will be to illustrate the idea with a simple example and then assert the generality of the results.

Consider a network with the topology shown in Fig. 2.3.1. This network is described by four nodes (labeled 0, 1, 2, and 3). It has a total of five branches.

We label each branch by establishing an indexed *branch current* through it (thus i_2 is the current in branch 2). The current directions are assigned arbitrarily. With each branch we associate a *branch voltage*. Thus, we have a total of five branch voltages, v_1, v_2, v_3, v_4, and v_5. We *pick* the polarity of the branch voltage using the usual passive convention, illustrated in Fig. 2.3.2, so that the branch-current arrow points from the positive to the negative polarity of the branch voltage.

As a final step in our definitions, we chose a ground (or datum) node; in Fig. 2.3.1 we have selected the node 0 as the datum node. We let e_1, e_2, and e_3 denote the node-to-ground voltages of the nodes 1, 2, and 3, respectively.

We define the following three column vectors:

$$\mathbf{i} = \begin{bmatrix} i_1 \\ i_2 \\ i_3 \\ i_4 \\ i_5 \end{bmatrix} \qquad \text{vector of branch currents} \qquad (1)$$

$$\mathbf{v} = \begin{bmatrix} v_1 \\ v_2 \\ v_3 \\ v_4 \\ v_5 \end{bmatrix} \qquad \text{vector of branch voltages} \qquad (2)$$

$$\mathbf{e} = \begin{bmatrix} e_1 \\ e_2 \\ e_3 \end{bmatrix} \qquad \text{vector of node-to-ground voltages} \qquad (3)$$

Let us now investigate the relation between these vectors $\mathbf{i}$, $\mathbf{v}$, and $\mathbf{e}$. The only relations we can use are KCL and KVL, as they hold independent of the network elements, sources, etc., that may appear in the branches.

We start with the analysis of the network of Fig. 2.3.1 by applying KCL at nodes 1, 2, and 3 (we know that this is enough, as these relations would imply KCL at node 0).

At node 1: $\qquad i_1 + i_2 = 0$

At node 2: $\qquad -i_2 + i_3 + i_4 = 0$ (4)

At node 3: $\qquad -i_4 + i_5 = 0$

Observe that Eq. (4) was written in the form

$$\Sigma \text{ (currents out of node)} = 0 \tag{5}$$

Now Eq. (4) can be written using vector-matrix notation as

$$\underbrace{\begin{bmatrix} 1 & 1 & 0 & 0 & 0 \\ 0 & -1 & 1 & 1 & 0 \\ 0 & 0 & 0 & -1 & 1 \end{bmatrix}}_{\mathbf{A}} \underbrace{\begin{bmatrix} i_1 \\ i_2 \\ i_3 \\ i_4 \\ i_5 \end{bmatrix}}_{\mathbf{i}} = \underbrace{\begin{bmatrix} 0 \\ 0 \\ 0 \end{bmatrix}}_{\mathbf{0}} \tag{6}$$

which is a vector equation of the form

$$\mathbf{Ai} = \mathbf{0} \tag{7}$$

where the **A** matrix is given by

$$\mathbf{A} = \begin{bmatrix} 1 & 1 & 0 & 0 & 0 \\ 0 & -1 & 1 & 1 & 0 \\ 0 & 0 & 0 & -1 & 1 \end{bmatrix} \tag{8}$$

Note that **A** is a 3×5 matrix; its number of rows (three) equals one less than the number of total nodes, i.e., the number of independent nodes, while the number of columns (five) equals the number of branches.

Now we focus our attention on the relationship of the branch voltages v_k to the node-to-ground voltages e_j. A simple application of KVL leads to the relations:

Branch 1: $\qquad v_1 = e_1$

Branch 2: $\qquad v_2 = e_1 - e_2$

Branch 3: $\qquad v_3 = e_2$ (9)

Branch 4: $\qquad v_4 = e_2 - e_3$

Branch 5: $\qquad v_5 = e_3$

These equations can be written

$$
\begin{bmatrix} v_1 \\ v_2 \\ v_3 \\ v_4 \\ v_5 \end{bmatrix} = \begin{bmatrix} 1 & 0 & 0 \\ 1 & -1 & 0 \\ 0 & 1 & 0 \\ 0 & 1 & -1 \\ 0 & 0 & 1 \end{bmatrix} \begin{bmatrix} e_1 \\ e_2 \\ e_3 \end{bmatrix} \tag{10}
$$

$$\underbrace{}_{\mathbf{v}} \qquad \underbrace{}_{\mathbf{B}} \qquad \underbrace{}_{\mathbf{e}}$$

which is a vector equation of the form

$$ \mathbf{v} = \mathbf{Be} \tag{11} $$

$\mathbf{B}$ being the 5×3 matrix

$$
\mathbf{B} = \begin{bmatrix} 1 & 0 & 0 \\ 1 & -1 & 0 \\ 0 & 1 & 0 \\ 0 & 1 & -1 \\ 0 & 0 & 1 \end{bmatrix} \tag{12}
$$

Now, by comparing Eqs. (8) and (12) we note that $\mathbf{B}$ is the transpose of $\mathbf{A}$; that is

$$ \mathbf{B} = \mathbf{A}' \tag{13} $$

Let us recapitulate. For the network of Fig. 2.3.1 we found that KCL yields

$$ \mathbf{Ai} = \mathbf{0} \tag{14} $$

and KVL yields

$$ \mathbf{v} = \mathbf{A}'\mathbf{e} \tag{15} $$

We *assert* that these relations are true for any graph. Thus, we have the following theorem.

Theorem 1   *Consider a network with b branches and $(n + 1)$ nodes. Let $i_1, i_2, \ldots, i_b$ denote the branch currents. Let $v_1, v_2, \ldots, v_b$ denote the branch voltages. Let $e_1, e_2, \ldots, e_n$ denote the node-to-datum voltages. Assume that the passive polarity convention of Fig. 2.3.2 holds. Define*

$$\mathbf{i} = \begin{bmatrix} i_1 \\ i_2 \\ \vdots \\ i_b \end{bmatrix} \qquad \mathbf{v} = \begin{bmatrix} v_1 \\ v_2 \\ \vdots \\ v_b \end{bmatrix} \qquad \mathbf{e} = \begin{bmatrix} e_1 \\ e_2 \\ \vdots \\ e_n \end{bmatrix} \tag{16}$$

Then there is an $n \times b$ matrix $\mathbf{A}$, *called the reduced incidence matrix, such that*

$$\mathbf{Ai} = \mathbf{0} \qquad \text{KCL} \tag{17}$$

$$\mathbf{v} = \mathbf{A'e} \qquad \text{KVL} \tag{18}$$

where the elements $a_{\alpha\beta}$ $(\alpha = 1, 2, \ldots, n; \beta = 1, 2, \ldots, b)$ *of the matrix* $\mathbf{A}$ *are given by*

$$a_{\alpha\beta} = \begin{cases} +1 & \text{if branch current } i_\beta \text{ comes out of node } \alpha \\ -1 & \text{if branch current } i_\beta \text{ enters node } \alpha \\ 0 & \text{if branch current } i_\beta \text{ does not touch node } \alpha \end{cases}$$

The implications of this theorem are many. For example, we shall use it to deduce a form of *Tellegen's theorem*. Let p_k be the power associated with the kth branch

$$p_k \triangleq v_k i_k \qquad k = 1, 2, \ldots, b \tag{19}$$

Then we can establish the following theorem.

Theorem 2 *Independent of the network elements in the branches,*

$$\sum_{k=1}^{b} v_k i_k = 0 \tag{20}$$

PROOF First note that Eq. (20) can be written as the scalar product[†] of the vector $\mathbf{i}$ of the branch currents with the vector $\mathbf{v}$ of the branch voltages. Thus,

$$\sum_{k=1}^{b} v_k i_k = \langle \mathbf{v}, \mathbf{i} \rangle = \langle \mathbf{A'e}, \mathbf{i} \rangle = \langle \mathbf{e}, \mathbf{Ai} \rangle = \langle \mathbf{e}, \mathbf{0} \rangle = 0 \tag{21}$$

where we used Eqs. (18) and (17), the general relationship $\langle \mathbf{x}, \mathbf{My} \rangle = \langle \mathbf{M'x}, \mathbf{y} \rangle$, and the fact that $(\mathbf{M'})' = \mathbf{M}$.

We emphasize that Theorem 2 holds for any type of network (linear, nonlinear, time-varying, etc).

---

† Scalar products are defined in Sec. A.11 of Appendix A.

The basic result of this section is contained in Theorem 1. It simply brings into focus the implications of KVL and KCL for an arbitrary network by interrelating the branch current and voltage vectors and the node-voltage vector through the reduced incidence matrix. It should be clear at this point that no additional results can be obtained unless the elements in the branches are specified; only this can yield a coupling between the branch currents and voltages. In the next section we therefore consider the analysis of linear time-invariant resistive networks in order to examine the use of the branch constraints in the development of the complete network equations.

From a system-theoretic point of view, we have decomposed the problem of complete system specification into well-defined steps. In any physical circuit-analysis problem all the interconnections of physical elements would be completely specified. Although one need not follow a systematic procedure in order to obtain the equations, in the analysis of complex systems it usually pays off to break up the problem of writing equations into well-defined logical steps. Theorem 1 comes in handy, because (independent of the actual devices in the network) certain relations are imposed by the network topology and the laws of nature. Later we shall see how these global interrelations get combined with the local subsystem description, e.g., the value in ohms of a linear resistor in branch 13, to make up the complete mathematical description of the network.

2.4   NODE ANALYSIS OF LINEAR TIME-INVARIANT RESISTIVE NETWORKS

2.4.1   Motivation

In this section we formalize a popular technique for solving network equations, *the node-voltage method*, using vector and matrix representations. As is well known, the node-analysis method for networks that contain linear time-invariant (LTI) resistors and independent voltage and current sources consists in computing the values of all node-to-datum voltages as a function of the independent source voltages and currents and the values (in ohms) of the resistors (see, for example, "Basic Concepts," pp. 98–124). The popularity of the node-voltage method of analysis is due to the fact that many practical networks have few nodes and many branches. Thus, by solving for the node voltages, one solves for only a few variables. We shall demonstrate that once the node-to-datum voltages have been determined, it is a simple matter to obtain from them all other branch currents and voltages.

We proceed as follows. First, we carry out a local description of our network by defining the constraints imposed upon the branch current and branch voltage when the branch consists of a linear resistor, one independent voltage source, and

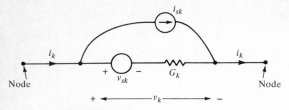

FIGURE 2.4.1
A generalized resistive branch.

one independent current source. The totality of the branch constraints will be expressed by a clean vector equation. Then, we shall combine these local constraints with the global constraints obtained in the previous section in terms of the KVL and KCL equations to arrive at the complete network description. We shall demonstrate the ideas by an example and then indicate how this systematic procedure can be used with a fictitious machine which by interacting with the operator can automatically set up the complete network equations.

2.4.2 Local Network Description

Let us consider a network with $n + 1$ nodes and b branches. We shall assume that every branch contains an LTI resistor, an independent current source, and an independent voltage source. In essence, we are concerned with networks whose kth branch ($k = 1, 2, \ldots, b$) is of the type shown in Fig. 2.4.1. Of course, if there are branches with no sources, one simply sets the source currents and voltages to zero. Thus in reference to Fig. 2.4.1 we have[†]

$$i_k = \text{branch current, A}$$
$$v_k = \text{branch voltage, V}$$
$$G_k = \text{branch conductance, } \Omega^{-1}$$
$$v_{sk} = \text{independent voltage source of } k\text{th branch, V}$$
$$i_{sk} = \text{independent current source of } k\text{th branch, A}$$

The following relation then holds:

$$\boxed{i_k = G_k v_k + i_{sk} - G_k v_{sk} \qquad k = 1, 2, \ldots, b} \tag{1}$$

We can write Eq. (1) in vector form to yield the branch constraints for the entire network. To do this we define the diagonal $b \times b$ *branch-conductance matrix* **G** by

---

[†] In general, i_k, v_k, v_{sk}, i_{sk} are time-varying while G_k is constant. We do not explicitly show the time dependence of the currents and voltages.

$$\mathbf{G} \triangleq \begin{bmatrix} G_1 & 0 & 0 & \cdots & 0 \\ 0 & G_2 & 0 & \cdots & 0 \\ 0 & 0 & G_3 & \cdots & 0 \\ \vdots & \vdots & \vdots & \vdots & \vdots \\ 0 & 0 & 0 & \cdots & G_b \end{bmatrix} \tag{2}$$

the vector of independent current sources $\mathbf{i}_s$ by

$$\mathbf{i}_s \triangleq \begin{bmatrix} i_{s1} \\ i_{s2} \\ \vdots \\ i_{sb} \end{bmatrix} \tag{3}$$

and the vector of independent voltage sources $\mathbf{v}_s$ by

$$\mathbf{v}_s \triangleq \begin{bmatrix} v_{s1} \\ v_{s2} \\ \vdots \\ v_{sb} \end{bmatrix} \tag{4}$$

Then Eq. (1) takes the form of the vector equality

$$\boxed{\mathbf{i} = \mathbf{Gv} + \mathbf{i}_s - \mathbf{Gv}_s} \tag{5}$$

which summarizes the total information about the local description of our system. In a practical problem one has numerical values for the elements of $\mathbf{G}$, $\mathbf{i}_s$, and $\mathbf{v}_s$; the branch-current vector $\mathbf{i}$ and the branch-voltage vector $\mathbf{v}$ are the unknowns. Clearly, there is no way of solving for both $\mathbf{i}$ and $\mathbf{v}$ from Eq. (5) alone. We need to combine the local information provided by Eq. (5) with the global information provided in KVL and KCL to arrive at the complete network description.

2.4.3 Complete Network Description

We now combine Eq. (5) with the two fundamental equations developed in the previous section

$$\boxed{\mathbf{Ai} = \mathbf{0} \qquad \text{KCL}} \tag{6}$$

$$\boxed{\mathbf{v} = \mathbf{A'e} \qquad \text{KVL}} \tag{7}$$

where $\mathbf{e}$ = node-voltage vector
$\qquad \mathbf{A}$ = reduced incidence matrix

From Eqs. (5) and (6) we have

$$\mathbf{A}(\mathbf{G}\mathbf{v} + \mathbf{i}_s - \mathbf{G}\mathbf{v}_s) = 0 \tag{8}$$

or

$$\mathbf{A}\mathbf{G}\mathbf{v} + \mathbf{A}\mathbf{i}_s - \mathbf{A}\mathbf{G}\mathbf{v}_s = 0 \tag{9}$$

Substituting Eq. (7) into Eq. (9), we obtain

$$\mathbf{A}\mathbf{G}\mathbf{A}'\mathbf{e} + \mathbf{A}\mathbf{i}_s - \mathbf{A}\mathbf{G}\mathbf{v}_s = 0 \tag{10}$$

or

$$\mathbf{Y}\mathbf{e} = \mathbf{A}\mathbf{G}\mathbf{v}_s - \mathbf{A}\mathbf{i}_s \tag{11}$$

where

$$\boxed{\mathbf{Y} \triangleq \mathbf{A}\mathbf{G}\mathbf{A}'} \tag{12}$$

$\mathbf{Y}$ is often called the *node-admittance* matrix. Note that

 1 $\mathbf{Y}$ is an $n \times n$ matrix (number of nodes = $n + 1$), since $\mathbf{A}$ is $n \times b$, $\mathbf{G}$ is $b \times b$, and $\mathbf{A}'$ is $b \times n$.
 2 $\mathbf{Y}$ is a symmetric matrix (see Sec. A.6 of Appendix A) because

$$\mathbf{Y}' = (\mathbf{A}\mathbf{G}\mathbf{A}')' = (\mathbf{A}')'\mathbf{G}'\mathbf{A}' = \mathbf{A}\mathbf{G}\mathbf{A}' = \mathbf{Y} \tag{13}$$

Thus $\mathbf{Y}$ has n real eigenvalues (see Appendix A, Sec. A.9). We *assert* that if all the conductances G_k are positive, then

$$\det \mathbf{Y} \neq 0 \tag{14}$$

so that $\mathbf{Y}^{-1}$ exists. Thus, from Eq. (11) we have

$$\boxed{\mathbf{e} = \mathbf{Y}^{-1}(\mathbf{A}\mathbf{G}\mathbf{v}_s - \mathbf{A}\mathbf{i}_s)} \tag{15}$$

The meaning of Eq. (15) is that we can solve for the node-to-datum voltages in terms of the voltage sources and current sources. Thus, if we specify $\mathbf{v}_s$ and $\mathbf{i}_s$, the network topology specifies $\mathbf{A}$ and the network resistors specify $\mathbf{G}$. Computationally, we must find $Y = AGA'$ and invert it. This procedure is ideally suited for digital-computer calculations.

Once the vector e has been computed from Eq. (15), it is a simple matter (without requiring matrix inversion) to compute all other branch currents and voltages. To see this, note that the KVL equation (7)

$$\boxed{v = A'e} \tag{16}$$

yields directly the branch-voltage vector v. From Eqs. (5), (7), and (15) we obtain

$$\boxed{i = GA'e + i_s - Gv_s} \tag{17}$$

which yields the entire branch-current vector i.

We shall now demonstrate, in a somewhat heuristic way, that the matrix $Y = AGA'$ is indeed nonsingular, provided that all the conductances G_k are positive. We have already demonstrated in Eq. (13) that Y is symmetric; if we show that Y is a positive definite matrix, we know that this would in turn imply that Y^{-1} exists.[†] Now from Eq. (2) we note that G is a diagonal matrix; since the G_k's are assumed positive, this means that G is positive definite. Therefore, for any *nonzero* column b-vector q we have the scalar-product inequality

$$\langle q, Gq \rangle > 0 \qquad \text{for all } q \neq 0 \tag{18}$$

Let $q = A'p$; then (18) yields

$$\langle A'p, GA'p \rangle = \langle p, AGA'p \rangle = \langle p, Yp \rangle > 0 \qquad \text{for all } q \neq 0 \tag{19}$$

If we show that

$$\langle p, Yp \rangle > 0 \qquad \text{for all } p \neq 0 \tag{20}$$

this implies that Y is positive definite. So the question is: Can $q = 0$ for $p \neq 0$? The answer is no, because the relation $q = A'p$ is of the same form as the KVL equation $v = A'e$ [see Eq. (7)], and so this situation would occur if all the branch voltages were zero for a nonzero node-voltage vector. This cannot occur, and so Eq. (18) holds; this, in turn, implies the positive definiteness, and hence the nonsingularity, of Y. We emphasize, however, that this plausibility argument does *not* constitute a rigorous proof.

[†] Recall that a positive definite matrix has positive eigenvalues. A singular matrix must have at least one zero eigenvalue. Therefore, a positive definite matrix cannot be singular; see Appendix A, Sec. A.12, Theorem 2.

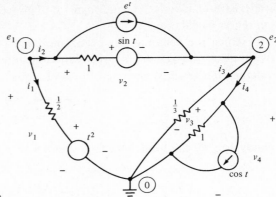

FIGURE 2.4.2
A resistive network.

2.4.4 An Example

Consider the network shown in Fig. 2.4.2. The values are in ohms. This network has
four branches ($b = 4$) and three nodes ($n + 1 = 3$, and so $n = 2$). The branch-
incidence matrix $\mathbf{A}$ is the 2×4 matrix

$$\mathbf{A} = \begin{bmatrix} 1 & 1 & 0 & 0 \\ 0 & -1 & 1 & 1 \end{bmatrix}$$

The branch-conductance matrix $\mathbf{G}$ is the constant 4×4 matrix

$$\mathbf{G} = \begin{bmatrix} 2 & 0 & 0 & 0 \\ 0 & 1 & 0 & 0 \\ 0 & 0 & 3 & 0 \\ 0 & 0 & 0 & 1 \end{bmatrix}$$

The voltage-source vector $\mathbf{v}_s$ is the time-varying vector

$$\mathbf{v}_s = \begin{bmatrix} t^2 \\ \sin t \\ 0 \\ 0 \end{bmatrix}$$

The current-source vector $\mathbf{i}_s$ is the time-varying vector

$$\mathbf{i}_s = \begin{bmatrix} 0 \\ e^t \\ 0 \\ \cos t \end{bmatrix}$$

The 2×2 node-admittance matrix $\mathbf{Y} = \mathbf{AGA}'$ is computed to be

$$Y = \begin{bmatrix} 3 & -1 \\ -1 & 5 \end{bmatrix}$$

Note that det $Y = 14 \neq 0$, so that

$$Y^{-1} = \frac{1}{14}\begin{bmatrix} 5 & 1 \\ 1 & 3 \end{bmatrix}$$

The vector of node-to-datum voltages is given by [see Eq. (15)]

$$\begin{bmatrix} e_1 \\ e_2 \end{bmatrix} = e = Y^{-1}(AGv_s - Ai_s)$$

The vector AGv_s is computed to be

$$AGv_s = \begin{bmatrix} 2t^2 + \sin t \\ -\sin t \end{bmatrix}$$

The vector Ai_s is computed to be

$$Ai_s = \begin{bmatrix} e^t \\ -e^t + \cos t \end{bmatrix}$$

Thus

$$\begin{bmatrix} e_1 \\ e_2 \end{bmatrix} = \frac{1}{14}\begin{bmatrix} 5 & 1 \\ 1 & 3 \end{bmatrix}\begin{bmatrix} 2t^2 + \sin t - e^t \\ -\sin t + e^t - \cos t \end{bmatrix}$$

so that

$$e_1 = \tfrac{1}{14}(10t^2 + 5 \sin t - 5e^t - \sin t + e^t - \cos t)$$

$$e_2 = \tfrac{1}{14}(2t^2 + \sin t - e^t - 3 \sin t + 3e^t - 3 \cos t)$$

or

$$e_1 = \tfrac{1}{14}(10t^2 + 4 \sin t - 4e^t - \cos t)$$

$$e_2 = \tfrac{1}{14}(2t^2 - 2 \sin t + 2e^t - 3 \cos t)$$

★ 2.4.5 Computer-aided Nodal Analysis†

An engineer must live in phase with the technological advances, current and projected, of his age. He must be aware of new devices and utilize them when they are appropriate. Anyone who has witnessed the rapid transition from vacuum tubes

† Starred sections and subsections can be omitted with no loss of continuity.

to transistors and integrated circuits in the relative short span of 15 years fully appreciates the need for an open mind and keeping up with new devices and tools for design.

Technological advances are not limited to devices per se. The advent of the digital computer has already removed much of the drudgery associated with routine but lengthy calculations. On the other hand, utilization of the digital computer and associated peripheral equipment for computer-aided analysis and design has been rather slow. Graphics terminals, which consist of a cathode-ray-tube display tied to a digital computer, can take over the drudgery of obtaining the equations of a large network and solving them. This section shows how a modern graphics terminal can carry out this task.

Anyone who has written a digital-computer program knows that it is necessary to provide the computer with a very precise list of systematic instructions. In this chapter we are presenting techniques for the systematic analysis of large networks. Our objective now is to indicate how this systematic analysis can pay off in delegating the task of analyzing a resistive network to a computer.

Description of the machine We shall assume that our machine consists of a graphics terminal and a keyboard tied to a digital computer. Inputs to the machine are provided by a light pen and from the keyboard. We shall further assume that someone has written the specific programming instructions that enable the engineer to communicate with the machine. We are not interested in the details of programming, stressing instead how we can use available computer hardware to take over tedious equation writing.

General approach The task of analyzing an LTI resistive network using the nodal method can be split up into two parts, implications of KVL and KCL from the graph of the network and implications of the values of the resistors and sources in the network branches. For this reason, our communication with the machine will also be divided into two parts. The first basically duplicates the material in Sec. 2.3, and the second deals with the material in Sec. 2.4.1.

KCL, KVL, and the reduced incidence matrix Let us now imagine that we are sitting in front of the machine ready to analyze our network.

STEP 1 In this step we draw the nodes of our network with the light pen. We have made an agreement with the machine (via the internal program) that the first dot we put in the screen is the datum node. So we turn on the light pen, and the machine draws the datum symbol (see Fig. 2.4.3a). Next, we turn the light pen on at another location to draw the first node, and the machine responds by drawing ① next to

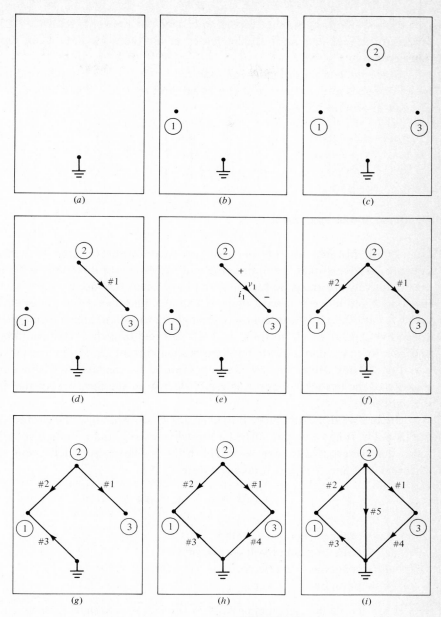

FIGURE 2.4.3
Building a network on the display screen.

that spot; we both know that this is node 1 of our network (see Fig. 2.4.3*b*). This procedure is repeated until all the nodes in our network have been drawn and numbered (see Fig. 2.4.3*c*).

Since the computer now knows the total number $n + 1$ (including the datum node) of nodes in our network, it can assign and reserve in its memory the vector **e** of node-to-ground voltages

$$\mathbf{e} = \begin{bmatrix} e_1 \\ e_2 \\ \vdots \\ e_n \end{bmatrix}$$

Of course, in our example $n = 3$.

STEP 2   In this step we use the light pen to draw the directed branches of the network. We recall that the numbering and the reference direction of the branches are completely arbitrary. So for the sake of argument suppose we start our light pen on node 2 and move it toward node 3. The machine then responds by drawing an arrow (established by the direction of the light-pen motion) and showing the number 1 (see Fig. 2.4.3*d*) which means that this is directed branch 1. One could be fancier and have the machine indicate both the branch current i_1 and the branch voltage v_1 (see Fig. 2.4.3*e*). Note that by preprogramming the passive sign convention, the polarity of the branch voltage v_1 is established from the direction of the motion of the light pen.

Before we draw any more branches, let us see what else the computer can do at this point. It has now established a branch current i_1 and a branch voltage v_1. We know that we eventually want to establish the implications of KCL and KVL (see Theorem 1 of Sec. 2.3) which take the form

$$\mathbf{Ai} = \mathbf{0} \qquad \mathbf{v} = \mathbf{A'e}$$

where **A** = reduced incidence matrix
  e = node-to-ground voltage vector
  i = branch-current vector
  v = branch-voltage vector

We let $a_{\alpha\beta}$ denote the elements of **A**. Now the fact that branch 1 comes out of node 2 immediately fixes the element a_{21} of **A** as

$$a_{21} = +1$$

Also since branch 1 enters node 3, this fixes the element a_{31} as

$$a_{31} = -1$$

Now let us draw another branch. Suppose we start at node 2 and connect it with node 1. The computer then labels it branch 2 (see Fig. 2.4.3f). It reserves a branch current i_2 and branch voltage v_2 and in fact computes

$$a_{22} = +1 \qquad a_{12} = -1$$

Next, suppose we connect the datum node to node 1. The computer labels this branch 3 (see Fig. 2.4.3g) and reserves a branch current i_3 and voltage v_3. The computer computes

$$a_{13} = -1$$

The fact that branch 3 comes out of the datum node is not used to compute anything.

Let us draw two more branches, as indicated in Fig. 2.4.3h and 2.4.3i. When we draw branch 4, the computer deduces

$$a_{34} = +1$$

and when we draw branch 5, the computer deduces

$$a_{25} = +1$$

Now the computer knows the total number of branches b (in our example $b = 5$), and hence it knows the dimension of the branch current and voltage vectors

$$\mathbf{i} = \begin{bmatrix} i_1 \\ i_2 \\ \vdots \\ i_b \end{bmatrix} \qquad \mathbf{v} = \begin{bmatrix} v_1 \\ v_2 \\ \vdots \\ v_b \end{bmatrix}$$

Let us recall that the reduced incidence matrix $\mathbf{A}$ is an $n \times b$ matrix. In our example, $\mathbf{A}$ is a 3×5 matrix, and so it looks like

$$\mathbf{A} = \begin{bmatrix} a_{11} & a_{12} & a_{13} & a_{14} & a_{15} \\ a_{21} & a_{22} & a_{23} & a_{24} & a_{25} \\ a_{31} & a_{32} & a_{33} & a_{34} & a_{35} \end{bmatrix}$$

Some of the elements of the $\mathbf{A}$ matrix have already been computed; the rest of the elements must be zero, and the computer now uses this fact for complete specification of the $\mathbf{A}$ matrix, which is

$$\mathbf{A} = \begin{bmatrix} 0 & -1 & -1 & 0 & 0 \\ +1 & +1 & 0 & 0 & +1 \\ -1 & 0 & 0 & +1 & 0 \end{bmatrix}$$

If we wish, we can ask the computer to type out (or project on the screen) the detailed KCL and KVL equations. Clearly, this can be done since the numbers b and n and the reduced incidence matrix $\mathbf{A}$ have been found. Thus, the KCL equation $\mathbf{Ai} = \mathbf{0}$ can be used to print

$$-i_2 - i_3 = 0 \qquad \text{KCL at node 1}$$
$$i_1 + i_2 + i_5 = 0 \qquad \text{KCL at node 2}$$
$$-i_1 + i_4 = 0 \qquad \text{KCL at node 3}$$

which, of course, are the correct KCL equations.

The detailed KVL equations are obtained from $\mathbf{v} = \mathbf{A'e}$:

$$v_1 = e_2 - e_3$$
$$v_2 = -e_1 + e_2$$
$$v_3 = -e_1$$
$$v_4 = e_3$$
$$v_5 = e_2$$

These are, of course, the correct KVL equations.

The branch constraints   At this stage we and our machine have obtained the full implications of the network topology and of KCL and KVL. The nodes and branches have all been indexed and all reference directions have been established. It remains to establish the values of the conductances and of the independent sources in the branches in order to complete the specification of the network. In this part of the analysis the visual display plays a secondary role and is used for the visual convenience of the engineer.

The information available up to now is sufficient for the machine to know that the engineer must specify the diagonal conductance matrix $\mathbf{G}$ [see Eq. (2)], the vector of independent current sources $\mathbf{i}_s$ [see Eq. (3)], and the vector of independent

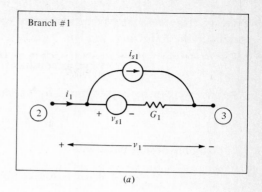

(a)

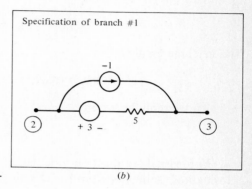

(b)

FIGURE 2.4.4
Display information for branch 1.

voltage sources $\mathbf{v}_s$ [see Eq. (4)]. In our example, this corresponds to fixing the elements of

$$
\mathbf{G} = \begin{bmatrix} G_1 & 0 & 0 & 0 & 0 \\ 0 & G_2 & 0 & 0 & 0 \\ 0 & 0 & G_3 & 0 & 0 \\ 0 & 0 & 0 & G_4 & 0 \\ 0 & 0 & 0 & 0 & G_5 \end{bmatrix} \qquad \mathbf{i}_s = \begin{bmatrix} i_{s1} \\ i_{s2} \\ i_{s3} \\ i_{s4} \\ i_{s5} \end{bmatrix} \qquad \mathbf{v}_s = \begin{bmatrix} v_{s1} \\ v_{s2} \\ v_{s3} \\ v_{s4} \\ v_{s5} \end{bmatrix}
$$

The assumption that all the network branches have a standard form, as illustrated in Fig. 2.4.1, can be used to advantage in the following step.

STEP 3   The machine projects on the screen the information depicted in Fig. 2.4.4a. It identifies the generalized structure of branch 1, including the nodes it connects and the already established branch-current reference direction and branch-voltage polarity. The machine can be instructed to print in conversational language and in sequence the following:

What is conductance G_1 in mhos?

to which we can reply by typing, for example,

$$G_1 = 5$$

The machine comes back with

What is source voltage v_{s1} in volts?

to which we reply by typing

$$v_{s1} = 3$$

The machine finally types

What is source current i_{s1} in amperes?

to which we reply be typing

$$i_{s1} = -1$$

For visual verification one could program the machine to display this information, as illustrated in Fig. 2.4.4b, before going on. Now the machine knows that

$$G_1 = 5 \qquad v_{s1} = 3 \qquad i_{s1} = -1$$

We next proceed with branch 2. The machine projects on the screen Fig. 2.4.5a. It types

What is conductance G_2 in mhos?

we reply

$$G_2 = 3$$

The machine asks

What is source voltage v_{s2} in volts?

we reply

$$v_{s2} = 0$$

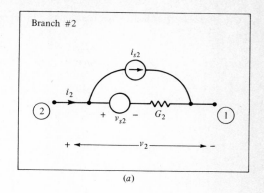

Branch #2

(a)

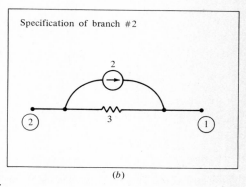

Specification of branch #2

(b)

FIGURE 2.4.5
Display information for branch 2.

The machine asks

What is source current i_{s2} in amperes?

we reply

$$i_{s2} = 2$$

The machine now projects Fig. 2.4.5b; it has eliminated from the picture the voltage source since we have said it was zero (recall that a voltage source with zero source voltage is simply a short circuit), and the values

$$G_2 = 3 \qquad v_{s2} = 0 \qquad i_{s2} = 2$$

have been established.

In a similar way the machine can be instructed to eliminate current sources if

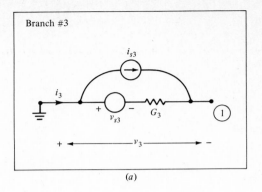

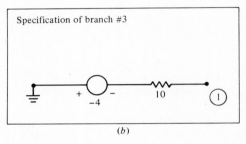

FIGURE 2.4.6
Display information for branch 3.

these are zero. To illustrate, let us go on. The machine projects Fig. 2.4.6a and our conversation continues

What is conductance G_3 in mhos?

$$G_3 = 10$$

What is source voltage v_{s3} in volts?

$$v_{s3} = -4$$

What is source current i_{s3} in amperes?

$$i_{s3} = 0$$

The machine comes back with Fig. 2.4.6b.

This procedure is continued until the remaining two branches have been specified. As a final double check one can ask the machine to project the full network, as shown in Fig. 2.4.7, which is obtained from the available information (we assume that branch 4 and 5 do not contain sources).

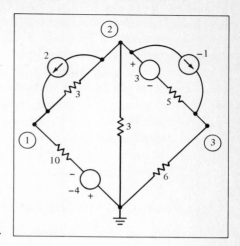

FIGURE 2.4.7
Display picture of complete network.

At this stage, the machine knows that

$$
\mathbf{G} = \begin{bmatrix} 5 & 0 & 0 & 0 & 0 \\ 0 & 3 & 0 & 0 & 0 \\ 0 & 0 & 10 & 0 & 0 \\ 0 & 0 & 0 & 6 & 0 \\ 0 & 0 & 0 & 0 & 3 \end{bmatrix} \qquad \mathbf{i}_s = \begin{bmatrix} -1 \\ 2 \\ 0 \\ 0 \\ 0 \end{bmatrix} \qquad \mathbf{v}_s = \begin{bmatrix} 3 \\ 0 \\ -4 \\ 0 \\ 0 \end{bmatrix}
$$

STEP 4   Now the digital computer takes over. Since $\mathbf{A}$ and $\mathbf{G}$ are known, the computer evaluates the matrix $\mathbf{Y}$ [see Eq. (12)]

$$\mathbf{Y} = \mathbf{AGA}'$$

and its inverse $\mathbf{Y}^{-1}$. Since $\mathbf{v}_s$ and $\mathbf{i}_s$ have been specified, the vector $\mathbf{AGv}_s$ and $\mathbf{Ai}_s$ are computed. From Eq. (15) the computer can now evaluate the node-to-datum voltages by

$$\mathbf{e} = \mathbf{Y}^{-1}(\mathbf{AGv}_s - \mathbf{Ai}_s)$$

All the branch voltages can then be found from the KVL equation [see Eq. (16)]

$$\mathbf{v} = \mathbf{A}'\mathbf{e}$$

and all the branch currents can be found from Eq. (17)

$$\mathbf{i} = \mathbf{GAe} + \mathbf{i}_s - \mathbf{Gv}_s$$

Concluding remarks We hope we have shown how a systematic procedure for analyzing networks of arbitrary complexity can be used to program a computer display system so that the analysis of networks can be fun and done without drudgery. Of course this type of machine is not as readily available as a slide rule, but it is safe to predict that once management appreciates the economic savings arising from freeing the engineer from routine tasks, such machines will become commonplace. The key to such advances is precise problem formulation even under the most general conditions; precise mathematics lead to precise programming and delegation of routine tasks to the machine.

⋆ 2.5 NODE ANALYSIS OF IMPEDANCE NETWORKS

It is well known that a network which contains LTI resistors, capacitors, and inductors and complex excitations of the form Ve^{st} or Ie^{st} can be analyzed at steady state by replacing all network elements with their impedances, provided that the excitation frequency s is dominant with respect to the natural frequencies of the network. At steady state, all branch currents and voltages also contain the e^{st} term, and the complex magnitudes of the steady-state voltages and currents satisfy the KVL and KCL equations. This fact enabled us to analyze networks at steady state by treating the impedances just like resistances.

These considerations, coupled with the material on resistive networks presented in the previous section, provide us with the motivation to extend the techniques of the preceding section to impedance networks. Thus, we shall consider networks whose kth branch, $k = 1, 2, \ldots, b$, is of the type shown in Fig. 2.5.1. Thus, in reference to Fig. 2.5.1, we have

$$I_k e^{st} = \text{steady-state branch current}$$
$$V_k e^{st} = \text{steady-state branch voltage}$$
$$Y_k(s) = \text{admittance of } k\text{th branch}$$
$$V_{sk} e^{st} = \text{independent voltage source}$$
$$I_{sk} e^{st} = \text{independent current source}$$

We know that at steady state the following relation holds

$$I_k e^{st} = Y_k(s)V_k e^{st} + I_{sk} e^{st} - Y_k(s)V_{sk} e^{st} \tag{1}$$

and so

$$I_k = Y_k(s)V_k + I_{sk} - Y_k(s)V_{sk} \qquad k = 1, 2, \ldots, b \tag{2}$$

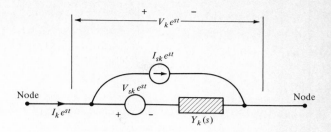

FIGURE 2.5.1
A typical branch containing an admittance $Y_k(s)$, an independent voltage source $V_{sk}e^{st}$, and an independent current source $I_{sk}e^{st}$.

We now define the diagonal $b \times b$ *branch-admittance matrix* $\mathbf{Y}(s)$ by

$$\mathbf{Y}(s) \triangleq \begin{bmatrix} Y_1(s) & 0 & \cdots & 0 \\ 0 & Y_2(s) & \cdots & 0 \\ \vdots & \vdots & \vdots & \vdots \\ 0 & 0 & \cdots & Y_b(s) \end{bmatrix} \tag{3}$$

the current-source vector $\mathbf{I}_s$, and the voltage-source vector $\mathbf{V}_s$,

$$\mathbf{I}_s \triangleq \begin{bmatrix} I_{s1} \\ I_{s2} \\ \vdots \\ I_{sb} \end{bmatrix} \qquad \mathbf{V}_s \triangleq \begin{bmatrix} V_{s1} \\ V_{s2} \\ \vdots \\ V_{sb} \end{bmatrix} \tag{4}$$

the branch-current vector $\mathbf{I}$, and the branch-voltage vector $\mathbf{V}$

$$\mathbf{I} = \begin{bmatrix} I_1 \\ I_2 \\ \vdots \\ I_b \end{bmatrix} \qquad \mathbf{V} = \begin{bmatrix} V_1 \\ V_2 \\ \vdots \\ V_b \end{bmatrix} \tag{5}$$

With these quantities Eq. (2) takes the form

$$\boxed{\mathbf{I} = \mathbf{Y}(s)\mathbf{V} + \mathbf{I}_s - \mathbf{Y}\mathbf{V}_s} \tag{6}$$

Since the amplitudes I_k and V_k of the complex branch voltages and currents satisfy KCL, they must satisfy the vector equations

$$\boxed{\mathbf{AI} = \mathbf{0} \qquad \text{KCL}} \qquad (7)$$

$$\boxed{\mathbf{V} = \mathbf{A}'\mathbf{E} \qquad \text{KVL}} \qquad (8)$$

where $\mathbf{A}$ is the $n \times n$ reduced incidence matrix of the network and $\mathbf{E}$ is the column n-vector of the node-voltage amplitudes $E_1 e^{st}$, $E_2 e^{st}$, ..., $E_n e^{st}$. Combining Eqs. (6) to (8) as in Sec. 2.4, we arrive at

$$\boxed{\mathbf{E} = [\mathbf{AY}(s)\mathbf{A}']^{-1}[\mathbf{AY}(s)\mathbf{V}_s - \mathbf{AI}_s]} \qquad (9)$$

which implies that one can solve for the steady-state node voltages in terms of the network admittances and the source amplitudes.

2.6 CUT-SET ANALYSIS AND NETWORK-TOPOLOGY EQUATIONS

2.6.1 Motivation

In this section we present an alternative approach to node-voltage analysis involving the use of cut sets. The reason for presenting yet another method for analysis of electric circuits is that the analysis of networks which contain nonlinear resistors and the analysis of networks that contain capacitors and inductors (both linear and nonlinear) by the state-variable method are much simpler by the cut-set method. This is not to say that one cannot analyze a network with resistors, capacitors, and inductors using more familiar methods. The cut-set analysis technique is offered because it aids in the *systematic* analysis of networks, which can then be exploited to let the computer do the messy equation setup for any given network.

2.6.2 Definitions

The cut set was defined in Sec. 2.2, Definition 3. We shall rapidly review the concept here.

Cut sets Consider a network consisting of b branches and $n + 1$ nodes. Basically, a *cut set is a set of branches*. A set of branches is called a cut set if the following two properties hold:

1 If *all* the branches of the cut set are removed, i.e., cut, the network is split into two parts which are not connected; i.e., there are no paths from one part of the network to the other.

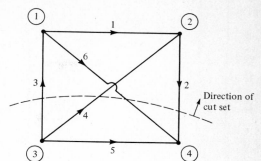

FIGURE 2.6.1
Notation for a cut set.

2 If any one of the cut-set branches is replaced, the network becomes connected again.

To illustrate this idea consider the network shown in Fig. 2.6.1. Cut sets are often denoted by a dashed line crossing the branches of the cut set. Thus the indicated dashed line crosses branches 3, 4, 6, and 2. To see if it defines a cut set we must check whether it satisfies both properties stated above. So let us break these branches; in so doing we obtain the graph of Fig. 2.6.2. Clearly the first property is satisfied since there is no path connecting either node 3 or 4 to node 1 or 2. Now we observe that if we close any one of the open branches in Fig. 2.6.2, we obtain a connected network. Therefore, we conclude that the set consisting of branches 3, 4, 6, and 2 is indeed a cut set.

We note that a cut set provides a natural boundary across which KCL must hold. Let us illustrate this with reference to Fig. 2.6.1, where the arrows indicate current directions for the branch currents $i_1, i_2, \ldots, i_6$ associated with each branch. If we consider that part of the network which lies below the cut line in Fig. 2.6.1, we can write the following relation for the currents leaving that subnetwork:

$$i_3 + i_4 - i_6 - i_2 = 0 \tag{1}$$

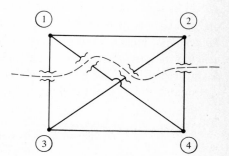

FIGURE 2.6.2
The network of Fig. 2.6.1 once the branches defining the cut set are broken.

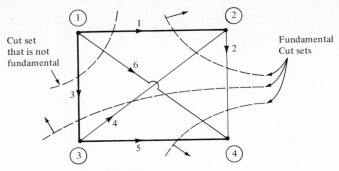

FIGURE 2.6.3
Examples of cut sets; the tree branches are 1, 3, and 5.

To verify that this relation holds let us use KCL at nodes 3 and 4:

$$\text{At node 3:} \qquad i_3 + i_4 + i_5 = 0 \qquad\qquad (2)$$

$$\text{At node 4:} \qquad -i_2 - i_5 - i_6 = 0 \qquad\qquad (3)$$

Adding Eqs. (2) and (3), we obtain Eq. (1).†

In the context of Eq. (1), we can assign a direction to the cut set, as shown in Fig. 2.6.1, such that Eq. (1) represents the sum of the currents flowing out of the cut set.

Fundamental cut sets   We know that in a network with b branches and $n + 1$ nodes we can choose a tree which consists of n branches such that all nodes are connected by some path and there are no loops. Clearly:

1   A tree must contain n tree branches.
2   There are $l = b - n$ links.

Let us consider a cut set that contains *only one* tree branch. Evidently, there are precisely n such cut sets (the same as tree branches). Such a cut set is called a *fundamental cut set. It is customary to assign a direction to the fundamental cut set which is the same as the direction of the current in the associated tree branch.* This idea is illustrated in Fig. 2.6.3, where the tree branches are drawn with heavy lines to distinguish them from the links.

---

† Note that an inductive extension of this procedure will establish an alternative form of KCL: the algebraic sum of the currents into any (sub) network is zero.

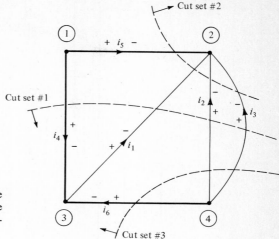

FIGURE 2.6.4
A directed four-node graph. The tree branches are denoted by heavy lines. The three fundamental cut sets and their directions are also indicated.

2.6.3 Cut-Set Analysis of Networks

We shall illustrate the cut-set method of network analysis by considering the specific network of Fig. 2.6.4, in which we have indicated the three fundamental cut sets. We define the currents as follows:

$$\mathbf{i}_l = \begin{bmatrix} i_1 \\ i_2 \\ i_3 \end{bmatrix} \quad \text{link-current vector} \qquad (4)$$

$$\mathbf{i}_t = \begin{bmatrix} i_4 \\ i_5 \\ i_6 \end{bmatrix} \quad \text{tree-branch current vector} \qquad (5)$$

$$\mathbf{i} = \begin{bmatrix} \mathbf{i}_l \\ \text{- - -} \\ \mathbf{i}_t \end{bmatrix} = \begin{bmatrix} i_1 \\ i_2 \\ i_3 \\ \text{- - -} \\ i_4 \\ i_5 \\ i_6 \end{bmatrix} \begin{matrix} \\ \end{matrix} \mathbf{i}_l \\ \begin{matrix} \\ \end{matrix} \mathbf{i}_t \quad \text{branch-current vector} \quad (6)$$

Using our standard passive polarity convention, we define

$$\mathbf{v}_l = \begin{bmatrix} v_1 \\ v_2 \\ v_3 \end{bmatrix} \qquad \text{link-voltage vector} \qquad (7)$$

$$\mathbf{v}_t = \begin{bmatrix} v_4 \\ v_5 \\ v_6 \end{bmatrix} \qquad \text{tree-branch voltage vector} \qquad (8)$$

$$\mathbf{v} = \begin{bmatrix} \mathbf{v}_l \\ --- \\ \mathbf{v}_t \end{bmatrix} = \left.\begin{matrix} \left. \begin{matrix} v_1 \\ v_2 \\ v_3 \end{matrix} \right\} \mathbf{v}_l \\ --- \\ \left. \begin{matrix} v_4 \\ v_5 \\ v_6 \end{matrix} \right\} \mathbf{v}_t \end{matrix}\right. \qquad \text{branch-voltage vector} \qquad (9)$$

Note also that *the cut sets are indexed like the tree-branch currents that define them.* Thus cut set 1 is associated with i_4, cut set 2 is associated with i_5, and so on.

KCL implications We shall now use KCL across the cut sets using the convention

$$\sum \text{(currents into cut set, i.e., same direction as the reference direction of the cut set)}$$
$$= 0 \qquad (10)$$

This leads to

$$\begin{aligned} \text{Cut set 1:} && -i_1 - i_2 - i_3 + i_4 &= 0 \\ \text{Cut set 2:} && i_1 + i_2 + i_3 + i_5 &= 0 \\ \text{Cut set 3:} && i_2 + i_3 + i_6 &= 0 \end{aligned} \qquad (11)$$

These equations can be written in matrix-vector notation

$$\underbrace{\left[\begin{array}{ccc:ccc} -1 & -1 & -1 & 1 & 0 & 0 \\ 1 & 1 & 1 & 0 & 1 & 0 \\ 0 & 1 & 1 & 0 & 0 & 1 \end{array}\right]}_{\mathbf{Q}} \underbrace{\begin{bmatrix} i_1 \\ i_2 \\ i_3 \\ --- \\ i_4 \\ i_5 \\ i_6 \end{bmatrix}}_{\mathbf{i}} = \underbrace{\begin{bmatrix} 0 \\ 0 \\ 0 \end{bmatrix}}_{\mathbf{0}} \qquad (12)$$

with the underbraced labels $\mathbf{F}$ and $\mathbf{I}$ within $\mathbf{Q}$.

Note that Eq. (12) is in the form

$$\boxed{\mathbf{Qi} = \mathbf{0} \qquad \text{from KCL}} \tag{13}$$

The matrix $\mathbf{Q}$ is often called the *fundamental cut-set matrix*. In fact, as indicated in Eq. (12), the last three columns of $\mathbf{Q}$ are the identity matrix $\mathbf{I}$. To stress this property of $\mathbf{Q}$ we write Eq. (13) in the form

$$[\mathbf{F} \,\vdots\, \mathbf{I}] \begin{bmatrix} \mathbf{i}_l \\ \text{- - -} \\ \mathbf{i}_t \end{bmatrix} = \mathbf{0} \qquad \mathbf{I} = \text{identity matrix} \tag{14}$$

All we have done is to partition $\mathbf{Q}$ into its two submatrices $\mathbf{F}$ and $\mathbf{I}$. Equation (14) can also be written

$$\boxed{\mathbf{Fi}_l + \mathbf{Ii}_t = \mathbf{Fi}_l + \mathbf{i}_t = \mathbf{0}} \tag{15a}$$

or

$$\boxed{\mathbf{Fi}_l = -\mathbf{i}_t} \tag{15b}$$

KVL implications   The next step in our analysis is to use KVL to obtain relations involving the branch voltages. In fact, we wish to solve for the entire vector of branch voltages $\mathbf{v}$ [see Eq. (9)] in terms of the tree-branch voltage vector $\mathbf{v}_t$. Thus, we seek a relation of the form

$$\mathbf{v} = \mathbf{B}\mathbf{v}_t$$

We start by using KVL to solve for the link voltages. This can be done by placing each link in the tree *one at a time*; then we have a single loop, and we use KVL so that we relate the single link voltage to the tree-branch voltages. If we do this for the three links, we obtain

$$\begin{aligned}
\text{Link 1:} \quad & v_1 = -v_4 + v_5 \\
\text{Link 2:} \quad & v_2 = -v_4 + v_5 + v_6 \\
\text{Link 3:} \quad & v_3 = -v_4 + v_5 + v_6
\end{aligned} \tag{16}$$

Observe that the tree-branch voltages v_4, v_5, and v_6 appear in the right-hand side of each equation, so that Eq. (16) takes the form

$$\begin{bmatrix} v_1 \\ v_2 \\ v_3 \end{bmatrix} = \begin{bmatrix} -1 & 1 & 0 \\ -1 & 1 & 1 \\ -1 & 1 & 1 \end{bmatrix} \begin{bmatrix} v_4 \\ v_5 \\ v_6 \end{bmatrix} \qquad (17)$$

$$\underbrace{}_{\mathbf{v}_l} \qquad \underbrace{}_{\mathbf{F}'} \quad \underbrace{}_{\mathbf{v}_t}$$

If we compare Eq. (17) with Eq. (12), we see that the matrix appearing in Eq. (17) is the transpose of the **F** matrix. Thus, Eq. (17) takes the form

$$\boxed{\mathbf{v}_l = \mathbf{F}'\mathbf{v}_t} \qquad (18)$$

But

$$\mathbf{v}_t = \mathbf{I}\mathbf{v}_t \qquad \mathbf{I} = \text{identity matrix} \qquad (19)$$

We combine Eqs. (18) and (19) to obtain

$$\begin{bmatrix} \mathbf{v}_l \\ \text{- - -} \\ \mathbf{v}_t \end{bmatrix} = \begin{bmatrix} \mathbf{F}' \\ \text{- - -} \\ \mathbf{I} \end{bmatrix} \mathbf{v}_t \qquad (20)$$

$$\underbrace{}_{\mathbf{v}} \qquad \underbrace{}_{\mathbf{Q}'}$$

or

$$\boxed{\mathbf{v} = \mathbf{Q}'\mathbf{v}_t \qquad \text{from KVL}} \qquad (21)$$

To recapitulate, for this example, the cut-set analysis of the network has yielded the two relations

$$\boxed{\mathbf{Q}\mathbf{i} = \mathbf{0} \qquad \text{or} \qquad \mathbf{F}\mathbf{i}_l = -\mathbf{i}_t \qquad \text{KCL}} \qquad (22)$$

$$\boxed{\mathbf{v} = \mathbf{Q}'\mathbf{v}_t \qquad \text{or} \qquad \mathbf{v}_l = \mathbf{F}'\mathbf{v}_t \qquad \text{KVL}} \qquad (23)$$

where the cut-set fundamental matrix **Q** is partitioned into the submatrices **F** and **I**

$$\mathbf{Q} = \begin{bmatrix} \mathbf{F} & \vdots & \mathbf{I} \end{bmatrix} \qquad (24)$$

2.6.4 Generalization of the Cut-Set Analysis

We have examined a specific example to illustrate how one obtains the KCL and KVL relations using cut-set analysis. This procedure is systematic, and it can be readily generalized to an arbitrary network. We now outline the steps involved in this systematic analysis.

Consider a network with b branches and $n + 1$ nodes. The step-by-step cut-set analysis method proceeds as follows:

STEP 1 Pick a tree (it will contain n branches).

STEP 2 Number the links (there are $b - n$ links) $1, 2, \ldots, b - n$.

STEP 3 Number the tree branches $b - n + 1, b - n + 2, \ldots, b$.

STEP 4 Define the current vectors

$$\mathbf{i}_l = \begin{bmatrix} i_1 \\ i_2 \\ \vdots \\ i_{b-n} \end{bmatrix} \quad \text{vector of link currents} \tag{25}$$

$$\mathbf{i}_t = \begin{bmatrix} i_{b-n+1} \\ i_{b-n+2} \\ \vdots \\ i_b \end{bmatrix} \quad \text{vector of tree-branch currents} \tag{26}$$

$$\mathbf{i} = \begin{bmatrix} \mathbf{i}_l \\ --- \\ \mathbf{i}_t \end{bmatrix} \quad \text{vector of branch currents} \tag{27}$$

STEP 5 Define the voltage vectors

$$\mathbf{v}_l = \begin{bmatrix} v_1 \\ v_2 \\ \vdots \\ v_{b-n} \end{bmatrix} \quad \text{vector of link voltages} \tag{28}$$

$$\mathbf{v}_t = \begin{bmatrix} v_{b-n+1} \\ v_{b-n+2} \\ \vdots \\ v_b \end{bmatrix} \qquad \text{vector of tree-branch voltages} \qquad (29)$$

$$\mathbf{v} = \begin{bmatrix} \mathbf{v}_l \\ - - - \\ \mathbf{v}_t \end{bmatrix} \qquad \text{vector of branch voltages} \qquad (30)$$

STEP 6 Identify the fundamental cut sets associated with the chosen tree. Assign a reference direction to each cut set using the current direction of the tree-branch current defining it. Identify

Cut set 1 with *tree branch $b - n + 1$*

Cut set 2 with *tree branch $b - n + 2$*

. . . .

Cut set n with *tree branch b*

STEP 7 Use KCL across the cut sets. This results in the relation

$$\boxed{\mathbf{Qi} = \mathbf{0} \qquad \text{or} \qquad \mathbf{Fi}_l = -\mathbf{i}_t \quad \text{KCL}} \qquad (31)$$

where $\mathbf{Q}$ is an $n \times b$ matrix, called the *fundamental cut-set matrix*, whose elements $q_{\alpha\beta}$ are defined by

$$q_{\alpha\beta} = \begin{cases} +1 & \text{if branch } \beta \text{ belongs to cut set } \alpha \text{ and has} \\ & \text{\textit{same} reference direction as cut set} \\ -1 & \text{if branch } \beta \text{ belongs to cut set } \alpha \text{ and has} \\ & \text{\textit{opposite} reference direction to that of cut set} \\ 0 & \text{if branch } \beta \text{ does not belong to cut set } \alpha \end{cases} \qquad (32)$$

for $\alpha = 1, 2, \ldots, n$; $\beta = 1, e, \ldots, b$. Moreover, the fundamental cut-set matrix $\mathbf{Q}$ can be partitioned into

$$\boxed{\mathbf{Q} = \begin{bmatrix} \mathbf{F} & \vdots & \mathbf{I} \end{bmatrix}} \qquad (33)$$

where $\mathbf{F} = n \times (b - n)$ matrix
$\mathbf{I} = n \times n$ identity matrix

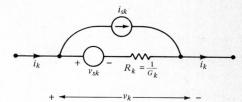

FIGURE 2.6.5
A generalized resistive branch.

STEP 8   Replace the links one at a time and use KVL in the single loop thus formed to obtain

$$\mathbf{v} = \mathbf{Q}'\mathbf{v}_t \quad \text{or} \quad \mathbf{v}_l = \mathbf{F}'\mathbf{v}_t \quad \text{KVL} \qquad (34)$$

The reader should note that the cut-set analysis of networks is somewhat similar to the node analysis. In both cases, KCL yields a relation that must be satisfied by the branch currents. In the node-analysis method, using KVL yielded a relation between the branch voltages and the node voltages. In cut-set analysis from KVL one obtains a relation between the branch voltages and the tree-branch voltages. At this point, one cannot proceed farther unless the network elements in the branches are specified so that it is possible to obtain a relation between the branch voltages and currents.

2.6.5   LTI Resistive Networks

We now indicate how to use cut-set analysis to obtain a general solution for networks in which each branch contains an LTI resistor, an independent voltage source, and an independent current source. Thus we assume that each branch has the structure indicated in Fig. 2.6.5, so that

$$i_k = G_k v_k + i_{sk} - G_k v_{sk} \qquad k = 1, 2, \ldots, b \qquad (35)$$

If we define

$$\mathbf{G} \triangleq \begin{bmatrix} G_1 & 0 & \cdots & 0 \\ 0 & G_2 & \cdots & 0 \\ \vdots & \vdots & \vdots & \vdots \\ 0 & 0 & \cdots & G_b \end{bmatrix} \quad \mathbf{i}_s \triangleq \begin{bmatrix} i_{s1} \\ i_{s2} \\ \vdots \\ i_{sb} \end{bmatrix} \quad \mathbf{v}_s \triangleq \begin{bmatrix} v_{s1} \\ v_{s2} \\ \vdots \\ v_{sb} \end{bmatrix} \qquad (36)$$

then Eq. (35) takes the form

$$\boxed{\mathbf{i} = \mathbf{Gv} + \mathbf{i}_s - \mathbf{Gv}_s} \qquad (37)$$

We now have an equation which relates the branch-current vector $\mathbf{i}$ and the branch-voltage vector $\mathbf{v}$ to the sources and the resistances of the network. We shall show how to solve for all branch currents and voltages using the fundamental results of cut-set analysis.

From the KCL relation $\mathbf{Qi} = \mathbf{0}$ and Eq. (37) we obtain

$$\mathbf{QGv} + \mathbf{Qi}_s - \mathbf{QGv}_s = \mathbf{0} \qquad (38)$$

Substituting the KVL equation $\mathbf{v} = \mathbf{Q}'\mathbf{v}_t$ into Eq. (38), we get

$$(\mathbf{QGQ}')\mathbf{v}_t = \mathbf{QGv}_s - \mathbf{Qi}_s \qquad (39)$$

The $n \times n$ symmetric matrix

$$\boxed{\mathbf{Y}_q \overset{\triangle}{=} \mathbf{QGQ}'} \qquad (40)$$

is often called the *cut-set admittance matrix*. We *assert* that $\mathbf{Y}_q^{-1}$ exists. Thus,

$$\boxed{\mathbf{v}_t = \mathbf{Y}_q^{-1}(\mathbf{QGv}_s - \mathbf{Qi}_s)} \qquad (41)$$

This means that we can solve for the tree-branch voltages in terms of the sources. Once this is done, we can solve for the entire branch-voltage vector $\mathbf{v}$ using the KVL relation $\mathbf{v} = \mathbf{Q}'\mathbf{v}_t$ so that

$$\boxed{\mathbf{v} = \mathbf{Q}'\mathbf{Y}_q^{-1}(\mathbf{QGv}_s - \mathbf{Qi}_s)} \qquad (42)$$

The branch-current vector $\mathbf{i}$ can be found from Eqs. (42) and (37), i.e.,

$$\mathbf{i} = \mathbf{GQ}'\mathbf{Y}_q^{-1}(\mathbf{QGv}_s - \mathbf{Qi}_s) + \mathbf{i}_s - \mathbf{Gv}_s \qquad (43)$$

or

$$\boxed{\mathbf{i} = (\mathbf{GQ}'\mathbf{Y}_q^{-1}\mathbf{Q} - \mathbf{I})(\mathbf{Gv}_s - \mathbf{i}_s)} \qquad (44)$$

The conclusion is that cut-set analysis can be used systematically to solve for all the branch currents and voltages.

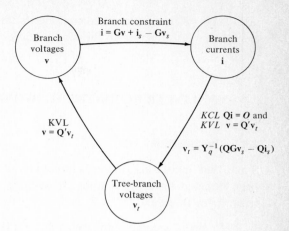

FIGURE 2.6.6
Equation flow graph for cut-set analysis
of LTI resistive networks.

The consistency of these equations is illustrated in the flow chart in Fig. 2.6.6, which shows that the branch voltage and current vectors $\mathbf{v}$ and $\mathbf{i}$ and the tree-branch voltage vector $\mathbf{v}_t$ can be evaluated from KVL, KCL, and the network branch constraint relation (37).

2.7 DISCUSSION

We have introduced two general techniques for the analysis of arbitrary networks and have specialized the results to the case of resistive networks which contain LTI resistors and independent sources. The reader may wonder why we have introduced the method of cut-set analysis in addition to the method of nodal analysis. After all, for the analysis of LTI resistive networks the cut-set method looks more complicated than the nodal method.

The benefits of cut-set analysis will become evident when we consider dynamical networks in Chap. 6; by this we mean networks that besides resistors contain capacitors and/or inductors. For such networks, the method of cut-set analysis is well suited to the development of the state-variable differential equations for the network in a step-by-step systematic way which can be fully utilized for automated writing of equations. As we shall see, by prudently choosing the tree branches to contain all the capacitors and by making all the inductive branches links, the differential equations of the network state variables (which are the capacitor voltages and the inductor currents) are obtained in a very natural way.

The methods of cut-set analysis also lead to a systematic analysis procedure when we deal with networks that contain *nonlinear* resistive elements. This important class of network problems merits serious consideration. We present an elementary introduction in the next section.

2.8 NONLINEAR RESISTIVE NETWORKS

2.8.1 Introduction

Since 1968 increasing interest among circuit theorists and electronic engineers has been directed toward the study of resistive nonlinear networks. There are several reasons for this trend.[†]

1 From a component viewpoint, most new solid-state electronic devices and integrated circuits are of the nonlinear resistive type.
2 From an analysis viewpoint, the combination of systematic analysis methods and the use of the digital computer to provide iterative solutions to large sets of simultaneous algebraic equations has been of practical importance for quantitative analysis.

Nonlinear resistive circuits differ from linear resistive circuits because the mathematical equations that describe them may have no solution, a unique solution, several solutions, or an infinite number of solutions. This multiplicity of solutions to the mathematical problem may be shared by the physical network, or it may be due to oversimplified modeling. Hence, *one must be far more careful in analyzing nonlinear networks*; the experienced designer finds a constant interplay between physical reasoning and mathematical formalism. This is another reason for the intellectual appeal of this subject, in addition to its engineering importance.

Although we provide here only a brief introduction to this fascinating subject, we stress that the important theoretical and practical questions in this area are far from being wrapped up. Very little is known about networks characterized by multiple solutions, for example. Hence, this area represents a fruitful ground for individual research by a well-motivated sophomore or junior with a good probability of obtaining significant results that advance the state of the art.

From a motivational point of view, this section serves as our primary vehicle for setting up simultaneous nonlinear algebraic equations. In Chap. 3 we develop a framework by which we can obtain solutions to these sets of nonlinear equations, using iterative, i.e., educated trial-and-error, digital-computer algorithms.

[†] The interested reader is encouraged to study the survey paper by E. S. Kuh and I. N. Haji, Nonlinear Circuit Theory: Resistive Networks, *Proc. IEEE*, vol. 59, no. 3, pp. 340–355, March 1971, for the state of the art in this area and an extended list of references.

The study of nonlinear networks will also demonstrate a general analysis tool associated with nonlinear systems. Known as *small-signal analysis, linearization, or perturbation analysis,* this technique is of universal value in understanding the *local* behavior of nonlinear systems (not only networks) in the vicinity of an operating or equilibrium set of conditions. We shall encounter this method frequently in later chapters because it can be extended to the analysis of nonlinear dynamic networks and systems.

2.8.2 Nonlinear Resistive Elements

In Sec. 2.6 we used the method of cut-set analysis to determine the branch-current and -voltage vectors for LTI resistive networks. In this section we extend these techniques to obtain the equations of a class of nonlinear resistive networks. In the absence of energy-storage elements we expect that the network equations will be algebraic (memoryless) but nonlinear.

Let us consider networks whose branches contain

1 Voltage-controlled resistors
2 Current-controlled resistors

Definition 1 *A* voltage-controlled *resistor is characterized by a current-voltage relation of the form*

$$i = g(v) \tag{1}$$

and a current-controlled *resistor by*

$$v = r(i) \tag{2}$$

where i = current through resistor
v = voltage across resistor

(see also Fig. 2.8.1). The functions $g(.)$ and $r(.)$ are scalar-valued functions which may not have an inverse.† Thus, our analysis must be conducted in such a way that $g^{-1}(.)$ or $r^{-1}(.)$ never appear in the equations.

We shall then consider two types of branch in the network. Typical current-controlled resistive branches are shown in Fig. 2.8.2a, and typical voltage-controlled resistive branches are shown in Fig. 2.8.2b. Thus, if we define

---

† If $g^{-1}(.)$ or $r^{-1}(.)$ exists, the resistor can be considered as both current- and voltage-controlled.

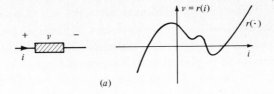

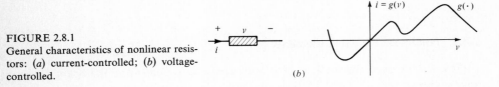

FIGURE 2.8.1
General characteristics of nonlinear resistors: (*a*) current-controlled; (*b*) voltage-controlled.

$$i_k = \text{branch current}$$
$$v_k = \text{branch voltage}$$
$$j_k = \text{independent current source}$$
$$e_k = \text{independent voltage source}$$

the two types of branch are described by

$$\text{Current-controlled:} \qquad v_k = r_k(i_k - j_k) \tag{3}$$

$$\text{Voltage-controlled:} \qquad i_k = g_k(v_k - e_k) \tag{4}$$

The nonlinear functions $r_k(.)$ and $g_k(.)$ are, again, scalar-valued and not necessarily invertible.

2.8.3 Network Analysis

We shall now make the following assumption: *a tree can be defined such that all the current-controlled branches are tree branches and all voltage-controlled branches are links; this tree is called a proper tree.*

To analyze networks that satisfy this assumption we proceed as follows. As before, we let $\mathbf{i}_l$ and $\mathbf{i}_t$ denote the link and tree-branch current vectors, respectively. Similarly, $\mathbf{v}_l$ and $\mathbf{v}_t$ denote the link and tree-branch voltage vectors. We let $\mathbf{j}$ denote the independent current-source vector, and we let $\mathbf{e}$ denote the independent voltage-source vector. Then, all the current-controlled branches are described by the vector relation

$$\boxed{\mathbf{v}_t = \mathbf{r}(\mathbf{i}_t - \mathbf{j})} \tag{5}$$

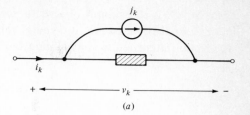

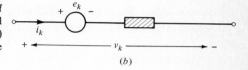

FIGURE 2.8.2
The assumed structure of the branches of
a nonlinear resistive network: (*a*) typical
current-controlled resistive branch; (*b*)
typical voltage-controlled resistive
branch.

where $\mathbf{r}(.)$ is a vector-valued *nonlinear* function whose components are the scalar-valued functions $r_1(.)$, $r_2(.)$, ... Similarly, all the voltage-controlled branches are described by the vector equation

$$\boxed{\mathbf{i}_l = \mathbf{g}(\mathbf{v}_l - \mathbf{e})} \tag{6}$$

where $\mathbf{g}(.)$ is a vector-valued nonlinear function whose components are the scalar-valued functions $g_1(.)$, $g_2(.)$, ...

Since a tree has been chosen, we can readily determine the fundamental cut sets. A standard cut-set analysis (see Sec. 2.6) yields the following pair of relations (note that they are independent of the fact that we have nonlinear resistors):

$$\boxed{\mathbf{Fi}_l = -\mathbf{i}_t \quad \text{KCL}} \tag{7}$$

$$\boxed{\mathbf{v}_l = \mathbf{F}'\mathbf{v}_t \quad \text{KVL}} \tag{8}$$

We now have four vector equations, namely Eqs. (5) to (8), in which the source vectors $\mathbf{j}$ and $\mathbf{e}$ are assumed known, the vector-valued functions $\mathbf{r}(.)$ and $\mathbf{g}(.)$ are assumed known, and the matrix $\mathbf{F}$ is readily evaluated from the network topology. The unknowns are the vectors $\mathbf{i}_l$, $\mathbf{i}_t$, $\mathbf{v}_l$, and $\mathbf{v}_t$. It is easy to eliminate three of these unknowns to obtain an equation for the fourth. To see this we proceed as follows:

$$
\begin{aligned}
\mathbf{v}_l = \mathbf{F}'\mathbf{v}_t &= \mathbf{F}'\mathbf{r}(\mathbf{i}_t - \mathbf{j}) && \text{by Eq. (5)} \\
&= \mathbf{F}'\mathbf{r}(-\mathbf{Fi}_l - \mathbf{j}) && \text{by Eq. (7)} \\
&= \mathbf{F}'\mathbf{r}(-\mathbf{Fg}(\mathbf{v}_l - \mathbf{e}) - \mathbf{j}) && \text{by Eq. (6)}
\end{aligned}
\tag{9}
$$

Equation (9) is an example of a vector implicit equation in the unknown $\mathbf{v}_l$. It should be obvious that we cannot in general solve Eq. (9) analytically. For the time being we are not interested in the solution but in its structure, which is of the form[†]

$$\boxed{\mathbf{v}_l = \mathbf{h}(\mathbf{v}_l)} \tag{10}$$

where $\mathbf{h}$ is a vector-valued function defined by the right-hand side of Eq. (9). Questions about the existence and uniqueness of the solution and iterative numerical techniques for evaluating the solution will be discussed in Chap. 3 (see also Ref. 8).

For our current purposes, let us suppose that Eq. (9) can be solved for $\mathbf{v}_l$. Then, we can immediately determine $\mathbf{i}_l$ by using Eq. (6), then $\mathbf{i}_t$ by using Eq. (7), and finally $\mathbf{v}_t$ by using Eq. (5). Therefore, once Eq. (9) is solved, all the remaining voltages and currents can be evaluated.

2.8.4 Linearization Techniques and Small-Signal Analysis

The equations of the nonlinear resistive network illustrate the technique of small-signal analysis, which involves the linearization of a set of nonlinear equations about a constant operating point.

Let us first suppose that all independent current and voltage sources in the nonlinear network are constant. In other words, the source-voltage vector $\mathbf{e}$ and the source-current vector $\mathbf{j}$ are constant vectors[‡]. To stress this we use the notation $\mathbf{e}^*$ and $\mathbf{j}^*$. Let $\mathbf{v}_l^*$ satisfy Eq. (9):

$$\mathbf{v}_l^* = \mathbf{F}'\mathbf{r}(-\mathbf{Fg}(\mathbf{v}_l^* - \mathbf{e}^*) - \mathbf{j}^*) \tag{10}$$

Now suppose that each current and voltage source has a *small* time variation about its constant value. Thus, a typical voltage source generates a voltage

$$e_k(t) = e_k^* + \delta e_k(t) \tag{11}$$

while a typical current source generates a current

$$j_m(t) = j_m^* + \delta j_m(t) \tag{12}$$

[†] In the scalar case, examples of such functions are $x = \sin x$, $y = e^y$, etc. In the vector case, an example of an equation we have in mind is

$$x_1 = x_1 \sin x_1 e^{x_1 x_2} \qquad x_2 = x_1(x_1 + x_2)^2$$

[‡] It may be useful to think of these constant voltages and currents as biases.

In both cases, $\delta e_k(t)$ is *small* compared to the constant voltage e_k^*, and $\delta j_m(t)$ is small[†] compared to the constant current j_m^*. In vector form we then have

$$\mathbf{e}(t) = \mathbf{e}^* + \boldsymbol{\delta}\mathbf{e}(t) \tag{13}$$

$$\mathbf{j}(t) = \mathbf{j}^* + \boldsymbol{\delta}\mathbf{j}(t) \tag{14}$$

where the vectors $\boldsymbol{\delta}\mathbf{e}(t)$ and $\boldsymbol{\delta}\mathbf{j}(t)$ denote the small time variations of the voltage and current sources, respectively.

It should then be evident that the link-voltage vector $\mathbf{v}_l(t)$ will also be time-varying. Let

$$\mathbf{v}_l(t) = \mathbf{v}_l^* + \boldsymbol{\delta}\mathbf{v}_l(t) \tag{15}$$

Clearly these time-varying vectors satisfy the equation

$$\mathbf{v}_l(t) = \mathbf{F}'\mathbf{r}(-\mathbf{Fg}(\mathbf{v}_l(t) - \mathbf{e}(t)) - \mathbf{j}(t)) \tag{16}$$

By subtracting Eq. (10) from Eq. (16) we obtain

$$\boldsymbol{\delta}\mathbf{v}_l(t) = \mathbf{F}'[\mathbf{r}(-\mathbf{Fg}(\mathbf{v}_l^* + \boldsymbol{\delta}\mathbf{v}_l - \mathbf{e}^* - \boldsymbol{\delta}\mathbf{e}(t)) - \mathbf{j}^* - \boldsymbol{\delta}\mathbf{j}(t)) - \mathbf{r}(-\mathbf{Fg}(\mathbf{v}_l^* - \mathbf{e}^*) - \mathbf{j}^*)] \tag{17}$$

For the sake of notational convenience we define the vector $\mathbf{x}$ by

$$\mathbf{x} \overset{\triangle}{=} \mathbf{v}_l - \mathbf{e} \tag{18}$$

Then

$$\mathbf{x}^* \overset{\triangle}{=} \mathbf{v}_l^* - \mathbf{e}^* \tag{19}$$

and

$$\mathbf{x}(t) \overset{\triangle}{=} \mathbf{v}_l(t) - \mathbf{e}(t) \overset{\triangle}{=} \mathbf{x}^* + \boldsymbol{\delta}\mathbf{x}(t) \tag{20}$$

so that

$$\boldsymbol{\delta}\mathbf{x}(t) = \boldsymbol{\delta}\mathbf{v}_l(t) + \boldsymbol{\delta}\mathbf{e}(t) \tag{21}$$

In this manner,

[†] It may be useful to think of the $\delta e_k(t)$ and $\delta j_m(t)$ as signals. This terminology is consistent with what happens in transistor amplifiers. Certain constant voltages provide the bias or operating point. The small signals are those which must be amplified. Indeed our overall development is a generalization of what one goes through to analyze in a linear manner networks that contain such inherently nonlinear elements as transistors, vacuum tubes, tunnel diodes, etc.

$$g(v_l(t) - e(t)) \triangleq g(x(t)) = g(x^* + \delta x(t)) \tag{22}$$

Let us expand $g(x^* + \delta x(t))$ in a Taylor series (see Appendix B) about x^* and *neglect* the quadratic and higher-order terms. Thus, we obtain

$$g(x^* + \delta x(t)) \approx g(x^*) + \frac{\partial g}{\partial x}\bigg|_* \delta x(t) \tag{23}$$

where $\partial g/\partial x|_*$ is the *Jacobian matrix* of $g(x)$ evaluated at $x = x^*$. For the sake of notational convenience define the constant matrix G^* by

$$G^* \triangleq \frac{\partial g}{\partial x}\bigg|_* \tag{24}$$

Therefore, Eq. (23) reduces to

$$g(x^* + \delta x(t)) \approx g(x^*) + G^* \delta x(t) \tag{25}$$

With this notation, Eq. (17) yields

$$\delta v_l(t) \approx F'[r(-Fg(x^*) - FG^*\delta x(t) - j^* - \delta j(t)) - r(-Fg(x^*) - j^*)] \tag{26}$$

Again for the sake of notational convenience, define the vectors

$$z^* \triangleq -Fg(x^*) - j^* \tag{27}$$

$$\delta z(t) \triangleq -FG^*\delta x(t) - \delta j(t) \tag{28}$$

$$z(t) \triangleq z^* + \delta z(t) \tag{29}$$

Then Eq. (26) takes the form

$$\delta v_l(t) \triangleq F'[r(z^* + \delta z(t)) - r(z^*)] \tag{30}$$

Once again we expand $r(z(t))$ about z^* using a Taylor series; by neglecting the quadratic and higher-order terms we obtain

$$r(z^* + \delta z) \approx r(z^*) + R^*\delta z(t) \tag{31}$$

where R^* is the Jacobian matrix of $r(.)$ evaluated at z^*, that is,

$$R^* \triangleq \frac{\partial r(z)}{\partial z}\bigg|_* \qquad \text{a constant matrix} \tag{32}$$

From Eqs. (30) and (31) we deduce that

$$\delta \mathbf{v}_l(t) \approx \mathbf{F'R^*} \delta \mathbf{z}(t) \approx \mathbf{F'R^*}[-\mathbf{FG^*} \delta \mathbf{x}(t) - \delta \mathbf{j}(t)]$$

$$\approx \mathbf{F'R^*}[-\mathbf{FG^*} \delta \mathbf{v}_l(t) + \mathbf{FG^*} \delta \mathbf{e}(t) - \delta \mathbf{j}(t)] \qquad \text{by Eq. (28)} \tag{33}$$

or

$$(\mathbf{I} + \mathbf{F'R^* FG^*}) \delta \mathbf{v}_l(t) \approx \mathbf{F'R^*}[\mathbf{FG^*} \delta \mathbf{e}(t) - \delta \mathbf{j}(t)] \tag{34}$$

and so

$$\boxed{\delta \mathbf{v}_l(t) \approx (\mathbf{I} + \mathbf{F'R^* FG^*})^{-1} \mathbf{F'R^*}[\mathbf{FG^*} \delta \mathbf{e}(t) - \delta \mathbf{j}(t)]} \tag{35}$$

provided the indicated matrix inverse exists.

The implication of Eq. (35) is that the time variations in the link voltages $\delta \mathbf{v}_l(t)$ can be found from the time variations $\delta \mathbf{e}(t)$ and $\delta \mathbf{j}(t)$ of the source voltages once the matrices $\mathbf{R^*}$ and $\mathbf{G^*}$ are determined. We shall leave it to the reader to show that:

1 $\mathbf{R^*}$ and $\mathbf{G^*}$ are diagonal matrices.
2 The elements of $\mathbf{R^*}$ have the units of resistance (ohms).
3 The elements of $\mathbf{G^*}$ have the units of conductance (mhos).

It should be noted that the matrices $\mathbf{R^*}$ and $\mathbf{G^*}$ depend on the particular values of $\mathbf{e^*}$, $\mathbf{j^*}$, and $\mathbf{v}_l^*$. The analysis is valid only for *small* variations about these constant values.

Once the link-voltage variations $\delta \mathbf{v}_l(t)$ have been found, one can compute the time variations in all other currents and voltages. Thus, if we let

$$\mathbf{i}_l(t) \triangleq \mathbf{i}_l^* + \delta \mathbf{i}_l(t) \tag{36}$$

$$\mathbf{i}_t(t) \triangleq \mathbf{i}_t^* + \delta \mathbf{i}_t(t) \tag{37}$$

$$\mathbf{v}_t(t) \triangleq \mathbf{v}_t^* + \delta \mathbf{v}_t(t) \tag{38}$$

where the starred vectors are constant, then

$$\boxed{\delta \mathbf{i}_l(t) \approx \mathbf{G^*}[\delta \mathbf{v}_l(t) - \delta \mathbf{e}(t)]} \tag{39}$$

$$\boxed{\delta \mathbf{i}_t(t) \approx -\mathbf{FG^*}[\delta \mathbf{v}_l(t) - \delta \mathbf{e}(t)]} \tag{40}$$

$$\boxed{\delta \mathbf{v}_t(t) \approx \mathbf{R^*}[-\mathbf{FG^*}[\delta \mathbf{v}_l(t) - \delta \mathbf{e}(t)] - \delta \mathbf{j}(t)]} \tag{41}$$

We shall leave it to the reader to verify Eqs. (39) to (41).

2.9 ANALYSIS OF A CLASS OF FLUID SYSTEMS

2.9.1 Introduction

One of the purposes of this book is to illustrate that the general system analysis techniques motivated by electric-network examples are also applicable in the analysis of other types of engineering systems. In this section we present such an example. Our treatment is brief since our purpose is not to provide a thorough discussion but to illustrate the general applicability of the ideas. The interested reader should consult Refs. 10 to 13.

2.9.2 The Basic Elements

Let us consider the flow of fluids in networks made up of components such as pumps, pipes, valves, and other such devices. This is a special subclass of the types of systems studied in fluid mechanics. Several complications make the study of such systems, which involve the flow of gases and fluids, much more complex than the study of electric networks. One may deal with compressible or incompressible fluids. The flow may be laminar or turbulent. As a result the basic relations right at the start are extremely nonlinear and complex.

We shall deal with some simple components that can be interconnected and present an idealized mathematical description of these devices. The two key physical variables we shall be concerned with are pressure p, usually measured in pounds per square inch, and flow w, usually measured in units of mass per unit time. There is strong analogy between these two variables and the electrical variables. Thus

1 Pressure is analogous to voltage.
2 Flow is analogous to current.

As in the case of electric networks, it is often convenient to think of a pressure drop across an element; this is analogous to voltage drop across an electric element. Continuing our analogies, long, large pipes have negligible pressure drop and hence are analogous to wires in electric networks. On the other hand, short devices that contain some constrictions (orifices, nozzles, valves) do have a pressure drop and as such are analogous to electric resistors; however, they are often described by nonlinear equations relating flow to pressure drop. Devices that store mass (tanks, reservoirs, and the like) are analogous to electric capacitors. Finally, pumps are analogous to independent voltage sources.

We now give some ideal mathematical descriptions of fluid-system elements.

Definition 1 *An ideal pump is a device that produces a pressure differential Δp*

FIGURE 2.9.1
A common symbol for an ideal pump. An ideal pump produces a pressure differential $\Delta p = p_2 - p_1$ independent of the flow through it. The pressure differential is either constant or time-varying; it is regulated externally by varying the pump speed.

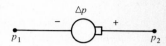

across its terminals

$$\Delta p = p_2 - p_1 \qquad (1)$$

independent of the flow through it.

Figure 2.9.1 shows a common symbol for an ideal pump.

Definition 2 *A constriction is a device (orifice, nozzle, valve, etc.) which relates a pressure drop Δp across it, to the flow w through it by means of a nonlinear relation*

$$\Delta p = r(w) \qquad (2)$$

Remarks In most physical constrictions, the relationship between the pressure drop p and the flow w is *nonlinear*. Often a good approximation can be obtained by modeling such constrictions by a quadratic law

$$\Delta p = Rw|w| \qquad (3)$$

where R is a constant which is related to the "resistance" offered by the constriction to flow.

Physically the flow in a constriction is from the high-pressure end to the low-pressure end. We shall use the valve symbol shown in Fig. 2.9.2 to model such

FIGURE 2.9.2
The valve symbol. Physically, if $p_1 > p_2$, the flow is along the direction of the arrow and the pressure drop $\Delta p = p_1 - p_2$ is positive. A good approximation is then $\Delta p = Rw|w|$, where the value of the constant R depends on the geometry of the constriction.

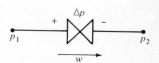

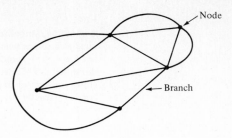

FIGURE 2.9.3
The graph of a fluid network.

constrictions. The validity of Eq. (3) hinges on the polarity assumption of Fig. 2.9.2. A positive pressure drop yields positive flow and vice versa.

We also note that a real pump can be modeled as an ideal pump in series with a constriction.

2.9.3 Fluid-System Topology and the Analogs to KVL and KCL

From a topological point of view, fluid networks resemble electric networks. They are represented by branches connecting nodes. Each branch may contain many pumps, valves, or storage devices.

Figure 2.9.3 shows the graph of a fluid system. A branch interconnects two nodes; typically each branch carries fluid through pipes, pumps, valves, etc. A node is a place where at least two branches meet.

As with electric networks, we can deduce certain equations (completely analogous to KCL and KVL) yielding a set of constraining equations relating pressures and flows, which are the consequences of the network topology and the laws of physics. To establish this we need the notion of a directed branch, as shown in Fig. 2.9.4, in which we use the passive sign convention. The arrow in the directed branch corresponds to the assumed direction of flow, plus and minus signs corresponding to the assumed direction of the pressure drop (recall that in a valve or constriction the actual fluid flow is from the point of high pressure toward the point of low pressure).

FIGURE 2.9.4
A directed branch; $\Delta p \triangleq p_1 - p_2$; w is the branch flow, p_1 and p_2 are node pressures measured with respect to a common standard, and Δp is the pressure drop. The passive convention implies that the arrow goes from + to −.

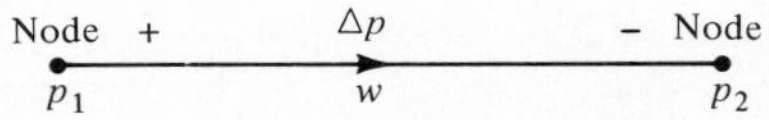

Node + Δp − Node
p_1 w p_2

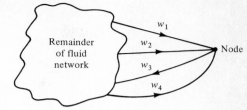

FIGURE 2.9.5
Illustration of the flow law $w_1 + w_2 - w_3 + w_4 = 0$.

The flow law Conservation-of-mass arguments for an incompressible fluid lead to the analog of KCL:

$$\text{Algebraic sum of flows into a node} = 0$$

Figure 2.9.5 shows the implications of the flow law for a particular node.

As with electric networks, the law holds not only for an individual node but also for a collection of nodes. Figure 2.9.6 illustrates these concepts.

The pressure law The equivalent law to KVL states that the algebraic sum of the pressure drops around any closed path is zero.

Figure 2.9.7 illustrates the implications of the pressure law.

2.9.4 Discussion

We can see that the topological constraints of fluid systems are exactly analogous to those of electric networks. Hence both the node-voltage methods (Sec. 2.3) and the cut-set method (Sec. 2.6) carry through without change, leading to the same type of linear equations as those obtained from KCL and KVL for electric networks.

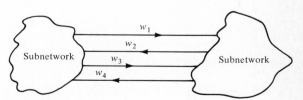

FIGURE 2.9.6
Alternate illustration of the fluid law $w_1 - w_2 + w_3 - w_4 = 0$.

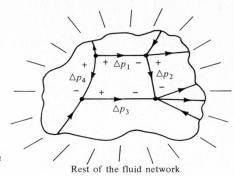

FIGURE 2.9.7
Illustration of the pressure law $\Delta p_1 + \Delta p_2$
$- \Delta p_3 - \Delta p_4 = 0$.

Rest of the fluid network

2.9.5 Introducing the Branch Constraints

Once the constraining equations from the fluid-system topology have been obtained, the next step is to introduce the exact specification of the pumps and constrictions that appear in the network branches. This procedure is completely analogous to the way in which we have analyzed nonlinear resistive networks in Sec. 2.8. For each branch of the fluid network, we can obtain from the pumps and constrictions that appear a branch-constraint equation of the form

$$\Delta p_i = r_i(w_i) \tag{4}$$

which relates the flow w_i through the ith branch to the pressure differential Δp_i across it. This set of equations, together with the equations obtained by application of the pressure and flow laws, provides us with a complete set of nonlinear algebraic equations. Their solution, obtained by some iterative algorithm of the type to be described in Chap. 3, will specify the pressures and flows in each part of the network. Since there is a complete one-to-one correspondence between these types of fluid networks and nonlinear resistive networks, repetition of all the formal equations is not worthwhile. At any rate, our purpose is not to provide expertise in fluid dynamics but to illustrate that concepts and techniques developed in the context of electric networks are applicable to other classes of physical problems.

2.10 CONCLUDING REMARKS

This chapter has presented systematic techniques for setting up the equations of static networks. This class of problem invariably leads to a set of simultaneous equations of the form

Linear case: $\qquad$ $\mathbf{A}\mathbf{x} = \mathbf{b}$

Nonlinear case: $\qquad$ $\mathbf{x} = \mathbf{f}(\mathbf{x})$

and we seek the solution vector $\mathbf{x}$.

We have illustrated the ideas by means of electric and fluid networks. Many other types of problems have the same type of characteristics: power systems (see, for example, Ref. 14), general flow problems (see, for example, Refs. 5 and 6), and even some static economic problems (see, for example, Ref. 7).

All these problems are characterized by

1 A topological-network interconnection of memoryless systems
2 A set of equilibrium conditions, e.g., KCL and KVL
3 A set of branch constraints relating the branch variables

The analysis of such systems proceeds generally along the lines outlined in this chapter. Of course, the physics of the problem may be different, but the mathematical structure is the same.

REFERENCES

1 DESOER, C. A., and E. S. KUH: "Basic Circuit Theory," McGraw-Hill, New York, 1969.
2 STERN, T. E.: "Theory of Nonlinear Networks and Systems," Addison-Wesley, Reading, Mass., 1965.
3 CHUA, L. O.: "Introduction to Nonlinear Network Theory," McGraw-Hill, New York, 1969.
4 DENNIS, J. B.: "Mathematical Programming and Electrical Networks," Wiley, New York, 1959.
5 IRI, M.: "Network Flow, Transportation and Scheduling Theory and Algorithms," Academic, New York, 1969.
6 FRANK, H., and I. FRISCH: "Transmission and Communication Networks," Addison-Wesley, Reading, Mass., 1971.
7 NIKAIDO, H.: "Convex Structures and Economic Theory," Academic, New York, 1968.
8 KUH, E. S., and I. N. HAJI: Nonlinear Circuit Theory: Resistive Networks, *Proc. IEEE*, vol. 59, no. 3, pp. 340–355, March 1971.
9 KUO, F. F., and W. G. MAGNUSON, JR.: "Computer Oriented Circuit Design," Prentice-Hall, Englewood Cliffs, N. J., 1969.
10 CANNON, R. H., JR.: "Dynamics of Physical Systems," McGraw-Hill, New York, 1967.
11 MOODY, L. F.: Friction Factors in Pipe Flow, *Trans. ASME*, vol. 66, pp. 671–684, 1944.
12 HUNSAKER, J. C., and B. G. RIGHTMIRE: "Engineering Applications of Fluid Mechanics," McGraw-Hill, New York, 1947.
13 VENNARD, J. K.: "Elementary Fluid Mechanics," 4th ed., Wiley, New York, 1961.
14 ELGERD, O. I.: "Electric Energy System Theory," McGraw-Hill, New York, 1971.

EXERCISES

Section 2.2

2.2.1   Given a connected graph *G*. Select any tree *T*. Prove that every tree branch of *T* together with some links defines a *unique* cut set of *G*.

Section 2.3

2.3.1   Prove Theorem 1.

2.3.2   Determine the reduced incidence matrix for the network shown. Verify that Eqs. (17) and (18) indeed hold. Label the branch variables by using the branch number as index.

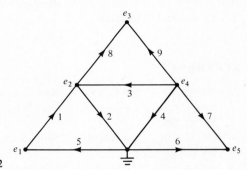

Fig. Ex. 2.3.2

2.3.3   Develop a procedure for constructing the directed graph of a network given its reduced incidence matrix **A**. Use your procedure to generate the directed graphs for the following matrix.

$$\mathbf{A} = \begin{bmatrix} 1 & 1 & 0 & 0 & 0 & 0 \\ 0 & -1 & 1 & 0 & 0 & 1 \\ 0 & 0 & 0 & 1 & -1 & -1 \end{bmatrix}$$

2.3.4   The reduced incidence matrix **A** (obtained from a node analysis) for a network is

$$\mathbf{A} = \begin{bmatrix} 1 & 1 & 0 & 0 & 0 & 0 & 0 & 0 \\ 0 & -1 & 1 & 1 & 0 & 0 & 0 & 0 \\ 0 & 0 & 0 & -1 & 1 & 1 & 0 & 0 \\ 0 & 0 & 0 & 0 & 0 & -1 & 1 & -1 \end{bmatrix}$$

Generate the *directed graph* of the network.

Section 2.4

2.4.1   Consider the network shown. Resistors are specified by their *conductance* in mhos.

(*a*)   Redraw the network to indicate clearly how the various elements might be lumped together to form generalized branches. Define voltage and current variables for your network of generalized branches.

(*b*)   Determine the reduced incidence matrix of your network.

(*c*)   Determine the conductance matrix **G**, the voltage-source vectors $\mathbf{v}_s$, and the current-source vector $\mathbf{i}_s$ in terms of the sources and conductances above.

(*d*)   Define and determine a vector of node-to-datum voltages in terms of the source vectors, as explained in Sec. 2.4.

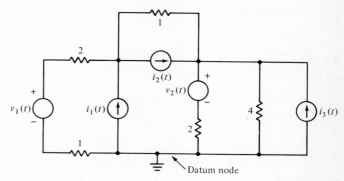

Fig. Ex. 2.4.1

2.4.2   This problem deals with rearranging sources within a network to generate generalized branches. If the current through a voltage source or the voltage across a current source is not of particular interest, we can relocate these sources as follows.

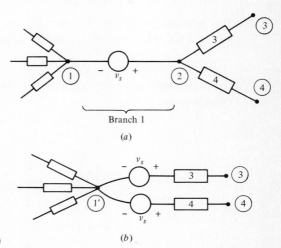

Fig. Ex. 2.4.2 (*a*) and (*b*)

CASE 1   Note that in the circuit in Fig. Ex. 2.4.2*b* branch 1 has been eliminated while a new node 1' has been introduced which results from the merger of nodes 1 and 2 of the circuit in Fig. Ex. 2.4.2*a*. Demonstrate that this transformation does not effect the solution of the network by showing that:

(*a*)   The KCL equation at node 1' is consistent with the old equations at nodes 1 and 2.

(*b*)   All KVL equations involving elements 3 and 4 are unchanged. The above procedure allows us to eliminate a branch consisting of a voltage source alone.

CASE 2   By a development analogous to that of case 1 show that the networks in Fig. Ex. 2.4.2*c* and *d* are equivalent. This transformation allows us to eliminate a branch consisting of a current source alone.

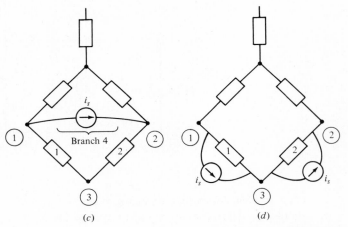

(*c*)   (*d*)

Fig. Ex. 2.4.2 (*c*) and (*d*)

2.4.3   This problem generalizes the results of Sec. 2.4 to include dependent sources. One way of doing this is sketched. Consider the generalized branch. Suppose

$$i_{sk} = k_c v_j \qquad v_{sk} = k_v v_l$$

i.e., the voltage and current sources in the *k*th branch depend upon the branch voltages v_l and v_j, respectively.

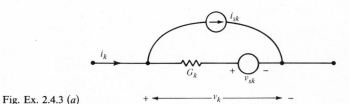

Fig. Ex. 2.4.3 (*a*)

(a)   Write i_k in terms of G_k, v_k, v_j, v_l.

(b)   Show that this relation for the kth branch in the network can be incorporated into the *vector* equations by

(1)   Interpreting our previous source vectors $\mathbf{v}_s$ and $\mathbf{i}_s$ as vectors of *independent* sources; thus the v_{sk} and i_{sk} components are zero.

(2)   Introducing off-diagonal terms in the conductance matrix **G**. (How is this done?)

(c)   How would the above be modified (for a network of LTI resistors) if instead of depending on branch voltages, the dependent sources were controlled by branch currents?

(d)   Consider the example sketched in Fig. Ex. 2.4.3b.

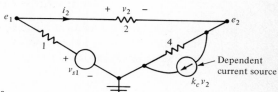

Fig. Ex. 2.4.3 (b). Values are in mhos.

(1)   What is the reduced incidence matrix **A**?

(2)   What is the conductance matrix **G**?

(3)   What is the node-admittance matrix **Y**?

(4)   Let $k_c = 10$. Let $v_{s1}(t) = \text{const} = 1$. Solve for e_1 and e_2.

(5)   Let $k_c = 14$. Let $v_{s1}(t) = 1$. Solve for e_1 and e_2.

Try to explain physically any strange behavior you discover in (4) or (5).

2.4.4

(a)   Redraw the network illustrated to indicate clearly how the various elements might be lumped together to form generalized branches. Define voltage and current variables for your network of generalized branches.

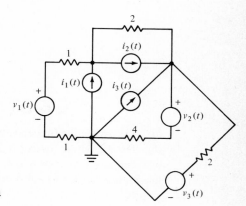

Fig. Ex. 2.4.4

(*b*) Determine the reduced incidence matrix of your network.

(*c*) Determine the conductance matrix **G**, the voltage-source vector **v**$_s$, and the current-source vector **i**$_s$ in terms of the sources and conductances above.

(*d*) Define and determine a vector of node-to-datum voltages in terms of the source vectors as explained in Sec. 2.4.

2.4.5 The LTI resistive network shown is excited by an independent voltage source and an independent current source. It is desired to use the node-voltage method to analyze this network. We remind you that in this method the equations take the form

$$\mathbf{Ai} = 0 \qquad \text{KCL}$$

$$\mathbf{v} = \mathbf{A}'\mathbf{e} \qquad \text{KVL}$$

$$\mathbf{i} = \mathbf{i}_s + \mathbf{G}(\mathbf{v} - \mathbf{v}_s) \qquad \text{branch constraints}$$

(*a*) Determine the matrices **A** and **G** and the vectors **i**$_s$ and **v**$_s$. Clearly identify the branch and node indices and the reference directions.

(*b*) Solve for the node-voltage vector **e**.

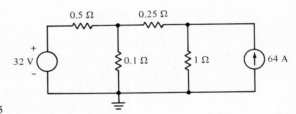

Fig. Ex. 2.4.5

Section 2.6

2.6.1 Consider the directed graph of an LTI resistive network. The links are indicated by dashed lines and the tree branches by solid lines.

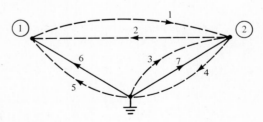

Fig. Ex. 2.6.1

(*a*) Determine the fundamental cut sets.

(*b*) Determine the fundamental cut-set matrix **Q** for this network.

(*c*) Suppose that each tree branch $k = 1, 2, \ldots, 7$ is a generalized conductive branch of the form illustrated in Fig. 2.6.6. For each branch indexed by k assume

$$G_k = k \, \Omega^{-1} \qquad v_{sk} = k \text{ V} \qquad i_{sk} = k \text{ A}$$

Thus, for example, for branch 3 we have

$$G_3 = 3 \, \Omega^{-1} \qquad v_{s3} = 3 \text{ V} \qquad i_{s3} = 3 \text{ A}$$

Determine the tree-branch voltages v_6 and v_7 in volts.

2.6.2   Outline a procedure which uses cut-set analysis for an LTI resistive network in which a branch, say j, contains a dependent voltage source whose voltage is a constant times the branch voltage across branch $k \, (k \neq j)$.

2.6.3   Give a heuristic or formal proof that the cut-set admittance matrix $\mathbf{Y}_q = \mathbf{QGQ'}$ is invertible provided that $\mathbf{G}$ is a diagonal positive definite matrix.

2.6.4   Given below is a matrix $\mathbf{F}$ resulting from the cut-set analysis of a directed graph.

$$\mathbf{F} = \begin{bmatrix} -1 & -1 & 0 & 0 \\ 0 & 1 & 0 & 0 \\ 0 & -1 & -1 & 1 \\ 0 & -1 & -1 & 0 \\ 0 & 0 & 0 & 1 \end{bmatrix}$$

(*a*)   How many tree branches does the network have?

(*b*)   How many links?

(*c*)   Find a directed graph and indicate a tree such that a cut-set analysis yields the above $\mathbf{F}$. (Number links from 1 to $i - 1$ and tree branches from i to n.)

2.6.5   (*a*)   An LTI resistor network with two sources is shown in Fig. Ex. 2.6.5a. The branches and nodes are numbered according to Fig. Ex. 2.6.5b, and the reference directions for the branch currents are shown. It is desired to use the node-voltage approach to formulate the equations of this network. In this approach, the equations have the form

$$\mathbf{Ai} = \mathbf{0} \qquad \mathbf{v} = \mathbf{A'e} \qquad \mathbf{i} = \mathbf{i}_s + \mathbf{G(v} - \mathbf{v}_s)$$

so that elimination of $\mathbf{v}$ and $\mathbf{i}$ yields equations to be solved for the node-voltage vector $\mathbf{e}$ of the form

$$\mathbf{Ye} + \mathbf{Ai}_s - \mathbf{AGv}_s = \mathbf{0}$$

(*1*)   Find the matrices $\mathbf{A}$, $\mathbf{G}$, $\mathbf{Y}$, $\mathbf{i}_s$, and $\mathbf{v}_s$ for this network.

(*2*)   Solve for $\mathbf{e}$.

(*b*)   It is now desired to use the cut-set-analysis approach to formulate the equations of the network of Fig. Ex. 2.6.5a and b. In this approach, the equations are of the form

$$\mathbf{Qi} = \mathbf{0} \qquad \mathbf{v} = \mathbf{Q'v}_t \qquad \mathbf{i} = \mathbf{i}_s + \mathbf{G(v} - \mathbf{v}_s)$$

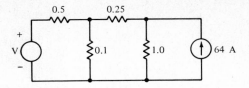

Fig. Ex. 2.6.5 (*a*) Network with resistances in ohms, voltage in volts, current in amperes.

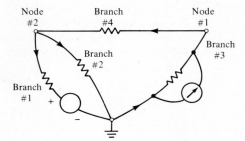

Fig. 2.6.5 (*b*) Network redrawn to emphasize topology.

so that elimination of **v** and **i** yields equations to be solved for the tree-branch voltage vector $\mathbf{v}_t$ of the form

$$\mathbf{Y}\mathbf{v}_t + \mathbf{Q}\mathbf{i}_s - \mathbf{Q}\mathbf{G}\mathbf{v}_s = \mathbf{0}$$

(*1*) Find the matrices **Q**, **G**, **Y**, $\mathbf{i}_s$, and $\mathbf{v}_s$ using branches 3 and 4 as a tree. In particular, state whether **G**, **Y**, $\mathbf{i}_s$, and $\mathbf{v}_s$ are the same as those in part (*a*).

(*2*) Solve for $\mathbf{v}_t$.

2.6.6 Consider the directed graph of an LTI resistive network. The links are indicated by dashed lines and the tree branches by solid lines.

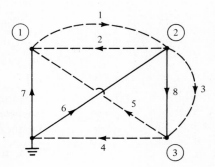

Fig. Ex. 2.6.6

(a) Determine the fundamental cut sets.

(b) Determine the fundamental cut-set matrix **Q** for this network.

(c) Suppose that each branch $k = 1, 2, 3, \ldots, 8$ is a generalized conductive branch of the form illustrated in Fig. 2.6.6. For each branch indexed by k assume

$$G_k = k \; \Omega^{-1} \qquad v_{sk} = k \; \text{V} \qquad i_{sk} = k \; \text{A}$$

Thus, for example, for branch 3 we have

$$G_3 = 3 \; \Omega^{-1} \qquad v_{s3} = 3 \; \text{V} \qquad i_{s3} = 3 \; \text{A}$$

Determine the tree-branch voltages v_6, v_7, and v_8 in volts.

Section 2.8

2.8.1 Prove that the matrices **G*** [see Eq. (24)] and **R*** [see Eq. (32)] are diagonal matrices.

2.8.2 Verify Eqs. (39) to (41).

2.8.3 Repeat the development of this entire section by finding an equation of the form $\mathbf{v}_t = \mathbf{q}(\mathbf{v}_t)$ for the tree-branch voltage vector. Carry out a small-signal analysis in this case and show that the resultant equations are identical to (30), (40), (41), and (35).

2.8.4 Consider a resistive network which contains one nonlinear resistor (all values are in ohms). The nonlinear resistor is characterized by $v = 4e^{i-3}$.

(a) Determine v and i.

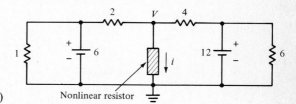

Fig. Ex. 2.8.4 (a)

(b) Suppose that the 6-V battery is replaced by a battery with voltage $6 + \epsilon$, where ϵ is a very small number. An engineer claims that the linear network illustrated in Fig. 2.8.4b can be used to compute the currents and voltages

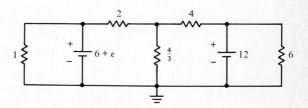

Fig. Ex. 2.8.4 (b)

when ϵ is small. Is the engineer right or wrong? If he is right, verify the fact. If he is wrong, give the right answer.

2.8.5 Consider an electric network with the graph shown. Assume that each branch contains a nonlinear resistive device. The characteristics of these resistors in terms of the branch voltages v_k and branch currents i_k are as follow:

Branch 1: $v_1 = \exp 4i_1$ Branch 2: $v_2 = 6i_2$ Branch 3: $v_3 = 4i_3{}^3$

Branch 4: $v_4 = 8i_4$ Branch 5: $v_5 = i_5$ Branch 6: $v_6 = -2i_6$

Set up the network equations using cut-set analysis. Specify the exact (nonlinear) equations you must solve in order to find all currents and voltages in the network. Can you solve this set of equations? At any rate, suppose that you could obtain the solution. Outline precisely how you would linearize this nonlinear network about the operating point.

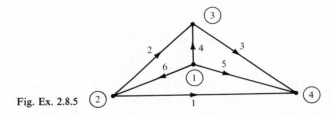

Fig. Ex. 2.8.5

2.8.6 Consider the nonlinear resistive network shown. The nonlinear resistors shown are characterized by

$$v_3 = \tfrac{1}{2}i_3{}^3 \qquad v_4 = 3i_4{}^2 \tag{1}$$

Show that the following equations hold:

$$i_3 = -\tfrac{1}{2}(v_3 - E) - (v_3 - v_4) \tag{2}$$

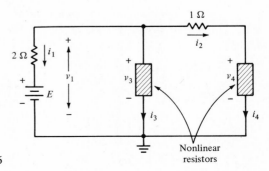

Fig. Ex. 2.8.6

$$i_4 = v_3 - v_4 \tag{3}$$

Let $E = 10$ V. Then it can be shown that

$$v_3 = 4 \text{ V} \qquad v_4 = 3 \text{ V} \tag{4}$$

You are asked, as an engineer, to determine *a good but approximate* value for the current i_2 for the following values of the battery voltage E:

(a) $E = 9.7$ V
(b) $E = 10.0$ V
(c) $E = 10.3$ V

Note: Exercises involving fluid networks will be given in Chap. 3.

3
ITERATIVE METHODS
FOR THE SOLUTION
OF SIMULTANEOUS EQUATIONS
AND OPTIMIZATION PROBLEMS

SUMMARY

This chapter introduces techniques suitable for numerical evaluation of solutions of vector equations of the form $\mathbf{x} = \mathbf{f(x)}$ or $\mathbf{g(x)} = \mathbf{0}$, which can be implemented via iterative digital-computer algorithms. Such equations were encountered in Chap. 2. In addition, the problem of parameter optimization, i.e., finding the minimum or maximum of a function of several variables, is considered; it is shown that the solution of such optimization problems requires us to be able to solve, once more, a set of simultaneous equations.

      Several methods for the numerical solution of such problems will be given. We shall see that these techniques lead to equations of the form $\mathbf{x}_{k+1} = \mathbf{h(x}_k)$, where $\mathbf{x}_k$ is the current guess for the solution and $\mathbf{x}_{k+1}$ is the next guess. The structure of such difference equations leads us to broaden our understanding of discrete-time dynamical systems, the subject of Chap. 4.

3.1 INTRODUCTION

We have seen in Chap. 2 that if we are dealing with the problem of determining the branch currents and voltages in a nonlinear resistive network, we are invariably faced with the task of solving a set of, say, n simultaneous equations in n unknowns $x_1, x_2, \ldots, x_n$ of the form

$$x_1 = f_1(x_1, x_2, \ldots, x_n)$$
$$x_2 = f_2(x_1, x_2, \ldots, x_n)$$
$$\vdots$$
$$x_n = f_n(x_1, x_2, \ldots, x_n)$$

(1)

or, in vector form

$$\boxed{\mathbf{x} = \mathbf{f}(\mathbf{x})}$$

(2)

Even in the scalar case (see, for example, chap. 5 of "Basic Concepts") such nonlinear equations can seldom be solved analytically. Hence, one must develop alternate methods in order to obtain solutions. For scalar problems, i.e., to find the solution to a scalar equation $x = f(x)$, one can often use graphical methods. Although the geometric ideas can be extended to arbitrary dimensions, they cease to be practical.

Fortunately, digital computers are ideally suited for carrying out repetitive calculations, and so they can be used to solve such sets of nonlinear equations. The art of constructing an algorithm that can be programmed on a digital computer is the subject of this chapter.

We stress two methods for solving nonlinear vector equations, *Picard's method* (or the method of successive approximations) and *Newton's method*. These two algorithms span the range of practical methods actually used. Picard's method is the simplest to program and requires the least amount of calculation at each step. Newton's method is quite sophisticated and uses a great deal of computation at each step (inversion of a matrix). We discuss later the advantages and disadvantages of each.

Equations such as (1) arise not only in the solution of nonlinear memoryless systems and networks but also in a completely different class of problems, which are often called *unconstrained static optimization problems*. In such problems one has a single function of many variables which can be thought of as a *profit function* or a *cost function*. Usually, if one has a profit function, one would like to find the values of variables which *maximize* the profit. On the other hand, if one has a cost (or loss) function, one is interested in *minimizing* it. In either case, one is interested in *optimization*.

The solution methods for a wide class of optimization problems are *identical* to those employed to finding solution(s) for a nonlinear vector equation. We shall see why in this chapter. This close interplay is another manifestation of the general nature of system-theoretic tools and of their great practical value in solving many different problems. Since optimization problems are so important, we shall also include some algorithms tailored to their solution.

The study of numerical iterative algorithms for the solution of nonlinear vector equations and for optimization is an important and large discipline. Needless to say, we can only scratch the surface. The references at the end of this chapter should provide ample opportunity for the interested reader to pursue this subject further. It is a fascinating area full of contributions from researchers of diverse backgrounds, and yet there is room for creative contributions at all levels.

The material in this chapter will serve not only as a set of tools that can be used to solve problems but also as a backdoor introduction to a class of dynamical systems, namely, *discrete-time dynamical systems*. Such systems arise from many other disciplines and modeling of diverse physical processes. For this reason they are considered in subsequent chapters. However, we remark at this point that many of the issues associated with discrete-time dynamical systems (such as stability) are intimately related to concepts arising from the study of iterative algorithms (such as convergence). We shall exploit this interplay quite a bit, in the true spirit of modern systems analysis.

3.2 UNCONSTRAINED STATIC OPTIMIZATION PROBLEMS

In this section we shall define precisely what we mean by an optimization problem and illustrate why the solution of an optimization problem requires us to be able to solve vector equations of the form

$$g_1(x_1, x_2, \ldots, x_n) = 0$$
$$g_2(x_1, x_2, \ldots, x_n) = 0$$
$$\vdots$$
$$g_n(x_1, x_2, \ldots, x_n) = 0$$

(1)

or, in compact vector form,

$$\boxed{\mathbf{g}(\mathbf{x}) = \mathbf{0}}$$

(2)

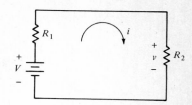

FIGURE 3.2.1
A simple linear resistive network.

3.2.1 A Simple Analytical Example

Let us start with an extremely simple analytical example that arises in network design. Consider the simple network shown in Fig. 3.2.1, containing a battery and two linear resistors R_1 and R_2. The value of R_1 is given, and we cannot change it. We are interested in selecting the value of the load resistor R_2 such that *the power dissipated in R_2 is maximum* (this is an extremely common problem in electrical engineering; e.g., it arises in hi-fi equipment in which R_2 represents the speakers and we wish maximum power transfer to the speakers).

We note that in this problem R_2 represents a parameter that is to be selected so that a scalar quantity, the power, is maximized. To solve the problem, we compute the power p for any value of R_2

$$p = \frac{v^2}{R_2} = \frac{R_2}{(R_1 + R_2)^2} V^2 \tag{3}$$

To find the maximum of the p-vs.-R_2 curve we must compute the root of

$$\frac{\partial p}{\partial R_2} = \frac{(R_1 - R_2)V^2}{(R_1 + R_2)^3} = 0 \tag{4}$$

which implies that the optimum value of R_2, denoted by R_2^*, is given by

$$R_2^* = R_1 \tag{5}$$

In this problem, we can see that the maximization of the power in Eq. (3) involves the solution of Eq. (4), which is of the form

$$g(R_2) = 0 \tag{6}$$

In this problem, we were able to compute the solution analytically; this is not the issue. What really matters is that by setting the derivative of the power function equal to zero we reduced the optimization problem to a root-finding problem.

Let us attempt now to generalize the problem conceptually to see how multivariable optimization problems can arise. Imagine, as illustrated in Fig. 3.2.2, that we have an electric network with linear resistors and sources (not shown) and n load resistors $R_1, R_2, \ldots, R_n$.

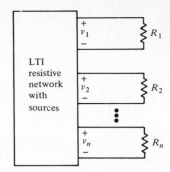

FIGURE 3.2.2
An LTI resistive network with n load resistors.

The optimization problem is now to select the values of the load resistors $R_1, R_2, \ldots, R_n$ such that the total power to these load resistors is maximized. The total power is given by

$$p = \frac{v_1{}^2}{R_1} + \frac{v_2{}^2}{R_2} + \cdots + \frac{v_n{}^2}{R_n} \tag{7}$$

If we know the source and resistor values inside the network, then for any given values of $R_1, R_2, \ldots, R_n$ we can compute (using the tools of Chap. 2) the branch voltages $v_1, v_2, \ldots, v_n$. Since only the values of $R_1, R_2, \ldots, R_n$ are to be found, the total power must be a function of the R_i's

$$p(R_1, R_2, \ldots, R_n) \tag{8}$$

and other parameters that we suppress. The optimization problem is then to find the specific values (optimal) resistor values

$$R_1^*, R_2^*, \ldots, R_n^* \tag{9}$$

such that

$$p(R_1^*, R_2^*, \ldots, R_n^*) \geq p(R_1, R_2, \ldots, R_n) \tag{10}$$

for all values of $R_1, R_2, \ldots, R_n$.

This is an example of a multidimensional optimization problem. What we need is a procedure for characterizing the values of the optimal resistances by a procedure which is analogous to setting the derivative equal to zero in the scalar example.

3.2.2 Definition of Optimization Problem

Let $x_1, x_2, \ldots, x_n$ denote scalar real variables. Let $\mathbf{x}$ be a column n-vector whose components are the x_i, that is,

$$\mathbf{x} = \begin{bmatrix} x_1 \\ x_2 \\ \vdots \\ x_n \end{bmatrix} \tag{11}$$

Let

$$h(\mathbf{x}) = h(x_1, x_2, \ldots, x_n) \tag{12}$$

denote a real scalar-valued function of $\mathbf{x}$.

Definition 1 *The vector $\mathbf{x}^*$ is said to be a global minimum of $h(\mathbf{x})$ if*

$$h(\mathbf{x}^*) \leq h(\mathbf{x}) \qquad \textit{for all} \quad \mathbf{x} \in R_n \tag{13}$$

If

$$h(\mathbf{x}^*) < h(\mathbf{x}) \qquad \textit{for all} \quad \mathbf{x} \neq \mathbf{x}^* \tag{14}$$

then the global minimum is unique. If

$$h(\mathbf{x}^*) \leq h(\mathbf{x}) \qquad \textit{for all} \quad \mathbf{x} \textit{ near } \mathbf{x}^* \tag{15}$$

then $\mathbf{x}^*$ is a local (or relative) minimum.

Remark Completely analogous definitions can be given for a maximum by reversing the inequalities in Definition 1. The problem of finding the maximum of $h(\mathbf{x})$ is equivalent to finding the minimum of $-h(\mathbf{x})$.

The problem of computing the miuimum of a function can be facilitated by deriving certain properties that must be satisfied by the minimizing vector $\mathbf{x}^*$. These properties may be satisfied by other points as well, but in general the set of points that share with $\mathbf{x}^*$ this common property is much smaller than all possible ones.

We shall give a sequence of theorems that state these properties precisely, and they help us in the computation of the minimizing vector.

Theorem 1 *Suppose that*

(a) *$h(\mathbf{x})$ is continuous for all $\mathbf{x}$.*
(b) *The gradient vector[†] $\partial h/\partial \mathbf{x}$ is continuous for all $\mathbf{x}$.*
(c) *The second-derivative matrix $(n \times n)\partial^2 h/\partial \mathbf{x}^2$ is continuous for all $\mathbf{x}$.*

---

† See Sec. A.14 of Appendix A.

Then if $\mathbf{x}^*$ *is a (local or global) minimum of* $h(\mathbf{x})$, *the following properties hold. The gradient vector* $\partial h / \partial \mathbf{x}$ *must be zero at the minimum, i.e.,*

$$\left. \frac{\partial h}{\partial \mathbf{x}} \right|_{\mathbf{x} = \mathbf{x}^*} = \mathbf{0} \tag{16}$$

The second-derivative matrix $\partial^2 h / \partial \mathbf{x}^2$ *must be at least positive semidefinite[†] at the minimum, i.e.,*

$$\left. \frac{\partial^2 h}{\partial \mathbf{x}^2} \right|_{\mathbf{x} = \mathbf{x}^*} \quad \text{is positive semidefinite} \tag{17}$$

PROOF   First, we shall prove Eq. (16). This type of proof is standard for many optimization problems and should therefore be studied with care.

Let us consider a vector $\mathbf{x}$ near $\mathbf{x}^*$. We construct the vector $\mathbf{x}$ as follows:

$$\mathbf{x} = \mathbf{x}^* + \epsilon \mathbf{y} \tag{18}$$

where ϵ is a small, positive or negative, scalar and $\mathbf{y}$ is an arbitrary vector (n-dimensional).

Since, by definition, $\mathbf{x}^*$ is at least a local minimum,

$$h(\mathbf{x}^*) \leq h(\mathbf{x}) = h(\mathbf{x}^* + \epsilon \mathbf{y}) \tag{19}$$

Now let us expand $h(\mathbf{x})$ in a Taylor series (see Appendix B) about $\mathbf{x}^*$

$$h(\mathbf{x}) = h(\mathbf{x}^*) + \left(\left. \frac{\partial h}{\partial \mathbf{x}} \right|_{\mathbf{x} = \mathbf{x}^*} \right)' \epsilon \mathbf{y} + O(\epsilon) \tag{20}$$

The $O(\epsilon)$ term indicates the remainder of the terms in the series. They contain quadratic (ϵ^2), cubic (ϵ^3), etc., terms. This is made concise by stating the property

$$\lim_{\epsilon \to 0} \frac{O(\epsilon)}{\epsilon} = 0 \tag{21}$$

From Eqs. (19) and (20) we then deduce the inequality

$$\left(\left. \frac{\partial h}{\partial \mathbf{x}} \right|_{\mathbf{x} = \mathbf{x}^*} \right)' \epsilon \mathbf{y} + O(\epsilon) \geq 0 \tag{22}$$

Dividing both sides by ϵ yields

† See Sec. A.12 of Appendix A.

$$\left(\left.\frac{\partial h}{\partial \mathbf{x}}\right|_{\mathbf{x}=\mathbf{x}^*}\right)' \mathbf{y} + \frac{O(\epsilon)}{\epsilon} \geq 0 \qquad \text{for } \epsilon > 0 \tag{23}$$

$$\left(\left.\frac{\partial h}{\partial \mathbf{x}}\right|_{\mathbf{x}=\mathbf{x}^*}\right)' \mathbf{y} + \frac{O(\epsilon)}{\epsilon} \leq 0 \qquad \text{for } \epsilon < 0 \tag{24}$$

Taking the limit as $\epsilon \to 0$ and using Eq. (21), we obtain

$$\left(\left.\frac{\partial h}{\partial \mathbf{x}}\right|_{\mathbf{x}=\mathbf{x}^*}\right)' \mathbf{y} \geq 0 \tag{25}$$

$$\left(\left.\frac{\partial h}{\partial \mathbf{x}}\right|_{\mathbf{x}=\mathbf{x}^*}\right)' \mathbf{y} \leq 0 \tag{26}$$

The only way both (25) and (26) can be true is to have

$$\left(\left.\frac{\partial h}{\partial \mathbf{x}}\right|_{\mathbf{x}=\mathbf{x}^*}\right)' \mathbf{y} = 0 \tag{27}$$

Since $\mathbf{y}$ is a completely arbitrary vector, Eq. (27) implies

$$\left.\frac{\partial h}{\partial \mathbf{x}}\right|_{\mathbf{x}=\mathbf{x}^*} = \mathbf{0} \tag{28}$$

which establishes Eq. (16).

Next we shall prove that (17) must hold. To do this, once more we expand $h(\mathbf{x})$ in a Taylor series, but now we retain the quadratic terms, i.e.,

$$h(\mathbf{x}) = h(\mathbf{x}^*) + \epsilon\left(\left.\frac{\partial h}{\partial \mathbf{x}}\right|_{\mathbf{x}=\mathbf{x}^*}\right)' \mathbf{y} + \tfrac{1}{2}\epsilon^2 \mathbf{y}'\left(\left.\frac{\partial^2 h}{\partial \mathbf{x}^2}\right|_{\mathbf{x}=\mathbf{x}^*}\right)\mathbf{y} + O(\epsilon^2) \tag{29}$$

where $O(\epsilon^2)$ has the property that

$$\lim_{\epsilon \to 0} \frac{O(\epsilon^2)}{\epsilon^2} = 0 \tag{30}$$

In view of Eq. (28) the second term in the right-hand side of Eq. (29) is zero; therefore Eqs. (19) and (29) imply the inequality

$$\tfrac{1}{2}\epsilon^2 \mathbf{y}'\left(\left.\frac{\partial^2 h}{\partial \mathbf{x}^2}\right|_{\mathbf{x}=\mathbf{x}^*}\right)\mathbf{y} + O(\epsilon^2) \geq 0 \tag{31}$$

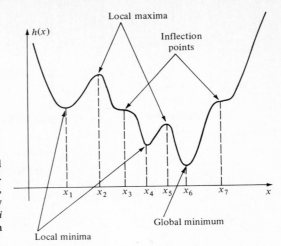

FIGURE 3.2.3
Stationary points $x_1, x_2, \ldots, x_7$ (local maxima, minima, and inflection points). Note that the slope $\partial h/\partial x|_{x=x_i}$, $i = 1, 2, \ldots, 7$, equals zero at all stationary points. Also at minima $\partial^2 h/\partial x^2|_{x=x_i}$, $i = 2, 5$, is nonpositive; and at inflection $\partial^2 h/\partial x^2|_{x=x_i}$, $i = 3, 7$, is zero.

When we divide by ϵ^2 and take $\epsilon \to 0$, Eqs. (30) and (31) yield

$$\tfrac{1}{2}\mathbf{y}'\left(\left.\frac{\partial^2 h}{\partial \mathbf{x}^2}\right|_{\mathbf{x}=\mathbf{x}^*}\right)\mathbf{y} \geq 0 \tag{32}$$

Since $\mathbf{y}$ is completely arbitrary, this means (by definition, see Appendix A, Sec. A.12) that the matrix

$$\left.\frac{\partial^2 h}{\partial \mathbf{x}^2}\right|_{\mathbf{x}=\mathbf{x}^*} \qquad \text{is positive semidefinite} \tag{33}$$

which establishes Eq. (17).

Remark   In a completely analogous manner we can prove that if $\mathbf{x}^*$ is a (local or global) maximum, the following relations must hold:

$$\left.\frac{\partial h}{\partial \mathbf{x}}\right|_{\mathbf{x}=\mathbf{x}^*} = \mathbf{0} \tag{34}$$

$$\left.\frac{\partial^2 h}{\partial \mathbf{x}^2}\right|_{\mathbf{x}=\mathbf{x}^*} \qquad \text{is negative semidefinite} \tag{35}$$

We can see that the condition that the gradient vector is equal to zero is a common property for both (local or global) minima and maxima; this property is also shared by inflection or saddle points, as illustrated in Figs. 3.2.3 and 3.2.4.

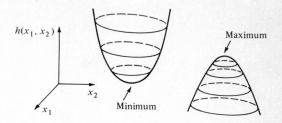

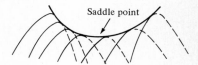

FIGURE 3.2.4
Shape of minima, maxima and saddle points for a function $h(x_1, x_2)$ of two variables.

Minima, maxima, inflection, and saddle points are often called *stationary points* of the function $h(\mathbf{x})$.

Definition 2 *A vector $\hat{\mathbf{x}}$ is said to be a stationary point of $h(\mathbf{x})$ if the gradient vector of $h(\,.\,)$ vanishes at $\hat{\mathbf{x}}$; that is,*

$$\left.\frac{\partial h}{\partial \mathbf{x}}\right|_{\mathbf{x}=\hat{\mathbf{x}}} = \mathbf{0} \tag{36}$$

3.2.3 Toward the Solution of Optimization Problems

Suppose that we are interested in determining the global minimum $\mathbf{x}^*$, assuming that it exists, of a function $h(\mathbf{x})$. In the absence of the theory described above, we would have to revert to a brute-force method, i.e., evaluate $h(\mathbf{x})$ for all $\mathbf{x}$ and find the minimum $\mathbf{x}^*$ numerically, a very time-consuming process.

From the theory developed above, we know that the minimum must be a stationary point. How do we find the stationary points? We start with our scalar function

$$h(\mathbf{x}) \triangleq h(x_1, x_2, \ldots, x_n) \tag{37}$$

and we compute the n partial derivatives

$$g_1(\mathbf{x}) \overset{\triangle}{=} g_1(x_1, x_2, \ldots, x_n) \overset{\triangle}{=} \frac{\partial h}{\partial x_1}(x_1, x_2, \ldots, x_n)$$

$$g_2(\mathbf{x}) \overset{\triangle}{=} g_2(x_1, x_2, \ldots, x_n) \overset{\triangle}{=} \frac{\partial h}{\partial x_2}(x_1, x_2, \ldots, x_n) \tag{38}$$

$$\vdots$$

$$g_n(\mathbf{x}) \overset{\triangle}{=} g_n(x_1, x_2, \ldots, x_n) \overset{\triangle}{=} \frac{\partial h}{\partial x_n}(x_1, x_2, \ldots, x_n)$$

Next we form the system of n equations in n unknowns

$$g_1(x_1, x_2, \ldots, x_n) = 0$$

$$g_2(x_1, x_2, \ldots, x_n) = 0 \tag{39}$$

$$\vdots$$

$$g_n(x_1, x_2, \ldots, x_n) = 0$$

or, in vector form,

$$\mathbf{g}(\mathbf{x}) = \mathbf{0} \tag{40}$$

where
$$\mathbf{g}(\mathbf{x}) \overset{\triangle}{=} \begin{bmatrix} g_1(\mathbf{x}) \\ g_2(\mathbf{x}) \\ \vdots \\ g_n(\mathbf{x}) \end{bmatrix} \overset{\triangle}{=} \frac{\partial h}{\partial \mathbf{x}} \quad \text{gradient vector of } h \tag{41}$$

If we find *all solutions* of Eq. (39), then, according to Definition 2, we shall have found *all the stationary points* (local and global maxima, minima, inflection and saddle points) of $h(\mathbf{x})$.

In almost all practical problems there are a finite number of stationary points. Once we find them all, by solving Eq. (39), we can revert to direct numerical evaluation of $h(\mathbf{x})$ *only* at the stationary points to determine which one is the global minimum.

We have thus demonstrated another class of problems of great practical importance which require solutions of sets of simultaneous equations.

3.2.4   The Use of Second-Derivative Information

We reiterate that the solution of the equation

$$\frac{\partial h}{\partial \mathbf{x}} = \mathbf{0} \tag{42}$$

i.e., the computation of all points at which the gradient vector is equal to zero, yields all stationary points, and it is not sharp enough to distinguish between, say, local maxima and minima. On the other hand, we have seen that there are certain conditions on the second-derivative matrix that must also hold. It is reasonable then to ask whether the additional information

At a minimum $\mathbf{x}^*$: $\qquad \dfrac{\partial^2 h}{\partial \mathbf{x}^2}\bigg|_{\mathbf{x}=\mathbf{x}^*} = \text{positive semidefinite}$ $\qquad\qquad$ (43)

At a maximum $\mathbf{x}^*$: $\qquad \dfrac{\partial^2 h}{\partial \mathbf{x}^2}\bigg|_{\mathbf{x}=\mathbf{x}^*} = \text{negative semidefinite}$ $\qquad\qquad$ (44)

can be always used to distinguish between local minima and maxima.

In general, the answer is no. This can be shown be a very simple scalar example. Consider the function

$$h_1(x) = x^4 \qquad\qquad (45)$$

Clearly it has a global minimum at $x = 0$. It is trivial to verify that

$$\frac{\partial h_1}{\partial x}\bigg|_{x=0} = 0 \qquad \frac{\partial^2 h_1}{\partial x^2}\bigg|_{x=0} = 0 \qquad\qquad (46)$$

On the other hand, consider the function

$$h_2(x) = -x^4 \qquad\qquad (47)$$

This has a global maximum at $x = 0$. Once more

$$\frac{\partial h_2}{\partial x}\bigg|_{x=0} = 0 \qquad \frac{\partial^2 h_2}{\partial x^2}\bigg|_{x=0} = 0 \qquad\qquad (48)$$

Finally, consider the function

$$h_3(x) = x|x|^3 \qquad\qquad (49)$$

which has neither a maximum nor a minimum for finite values of x. However, it has an inflection point at $x = 0$. Note that

$$\frac{\partial h_3}{\partial x}\bigg|_{x=0} = 0 \qquad \frac{\partial^2 h_3}{\partial x^2}\bigg|_{x=0} = 0 \qquad\qquad (50)$$

We can then see on the basis of these three simple cases that the information conveyed by the first- and second-derivative behavior at the stationary point is *not sufficient* to distinguish minima, maxima, and inflection points. This is why the imformation conveyed by Theorem 1 is called a *necessary* (but not sufficient) *condition* for a (local or global) minimum.

★ 3.2.5 Sufficient Conditions

We shall now state sufficient conditions for a local minimum or maximum.

> **Theorem 2**   *Consider the function $h(\mathbf{x})$. Assume that the assumptions of Theorem 1 hold.*
>
> *Suppose that $\mathbf{x}^*$ is a stationary point of $h(\mathbf{x})$, that is,*
>
> $$\frac{\partial h}{\partial \mathbf{x}}\bigg|_{\mathbf{x}=\mathbf{x}^*} = \mathbf{0} \tag{51}$$
>
> *Then*
>
> *(a) If*
>
> $$\frac{\partial^2 h}{\partial \mathbf{x}^2}\bigg|_{\mathbf{x}=\mathbf{x}^*} = positive\ definite \tag{52}$$

then $\mathbf{x}^*$ is a (local or global) minimum.

> *(b) If*
>
> $$\frac{\partial^2 h}{\partial \mathbf{x}^2}\bigg|_{\mathbf{x}=\mathbf{x}^*} = negative\ definite \tag{53}$$

then $\mathbf{x}^*$ is a (local or global) maximum.

> PROOF   The first part of the proof duplicates the proof of Theorem 1 by expanding $h(\mathbf{x})$ in a Taylor series about $\mathbf{x}^*$ [see Eq. (29)]:
>
> $$h(\mathbf{x}) = h(\mathbf{x}^*) + \epsilon\left(\frac{\partial h}{\partial \mathbf{x}}\bigg|_{\mathbf{x}=\mathbf{x}^*}\right)' \mathbf{y} + \tfrac{1}{2}\epsilon^2 \mathbf{y}'\left(\frac{\partial^2 h}{\partial \mathbf{x}^2}\bigg|_{\mathbf{x}=\mathbf{x}^*}\right)\mathbf{y} + O(\epsilon^2) \tag{54}$$
>
> By the hypothesis (51) of Theorem 2, since $\mathbf{x}^*$ is stationary, we have
>
> $$h(\mathbf{x}) = h(\mathbf{x}^*) + \tfrac{1}{2}\epsilon^2 \mathbf{y}'\left(\frac{\partial^2 h}{\partial \mathbf{x}^2}\bigg|_{\mathbf{x}=\mathbf{x}^*}\right)\mathbf{y} + O(\epsilon^2) \tag{55}$$
>
> We now note that
>
> $$\tfrac{1}{2}\epsilon^2 \mathbf{y}'\left(\frac{\partial^2 h}{\partial \mathbf{x}^2}\bigg|_{\mathbf{x}=\mathbf{x}^*}\right)\mathbf{y} + O(\epsilon^2) = \epsilon^2\left[\tfrac{1}{2}\mathbf{y}'\left(\frac{\partial^2 h}{\partial \mathbf{x}^2}\bigg|_{\mathbf{x}=\mathbf{x}^*}\right)\mathbf{y} + \frac{O(\epsilon^2)}{\epsilon^2}\right] \tag{56}$$

Since, by assumption (52) the second-derivative matrix is positive definite, the first term in the bracket in the right-hand side of Eq. (56) is strictly positive for all $\mathbf{y} \neq \mathbf{0}$ · (or equivalently for all $\mathbf{x} \neq \mathbf{x}^*$). Hence, in view of Eq. (30), we can conclude that the bracketed term in the right-hand side of (56) is strictly positive for ϵ sufficiently small. Hence, from (55) we deduce that

$$h(\mathbf{x}) > h(\mathbf{x}^*) \qquad \text{for all } \mathbf{x} \neq \mathbf{x}^* \text{ near } \mathbf{x}^*, \text{ that is, } \epsilon \text{ small} \qquad (57)$$

which implies that $\mathbf{x}^*$ is at least a local minimum.

The same type of argument is used for a local maximum.

We remark that $\mathbf{x}^*$ can be a minimum and violate the sufficient condition of Theorem 2; for example, $h(x) = x^4$. On the other hand, if $\mathbf{x}^*$ is a stationary point, and if it turns out that $\partial^2 h/\partial \mathbf{x}^2|_{\mathbf{x}=\mathbf{x}^*}$ is positive definite, then $\mathbf{x}^*$ *must* be a minimum, i.e., it cannot be a maximum, inflection point, or saddle point.

3.2.6 Analytical Solutions

Very few general optimization problems can be solved analytically. In the theorem below we summarize the most general class.

Theorem 3   *Suppose that*

$$h(\mathbf{x}) = \tfrac{1}{2}\mathbf{x}'\mathbf{A}\mathbf{x} + \mathbf{b}'\mathbf{x} + c \qquad (58)$$

where $\mathbf{x} = $ *n-dimensional column vector*
   $\mathbf{A} = n \times n$ *symmetric, positive definite matrix (given)*
   $\mathbf{b} = $ *n-dimensional column vector (given)*
   $c = $ *real scalar (given)*

Then, $h(\mathbf{x})$ has a unique global minimum at $\mathbf{x}^*$ given by

$$\mathbf{x}^* = -\mathbf{A}^{-1}\mathbf{b} \qquad (59)$$

PROOF   We start by computing

$$\frac{\partial h}{\partial \mathbf{x}} = \mathbf{A}\mathbf{x} + \mathbf{b} \qquad (60)$$

$$\frac{\partial^2 h}{\partial \mathbf{x}^2} = \mathbf{A} \qquad (61)$$

Setting $\partial h/\partial \mathbf{x}\,|_{\mathbf{x}=\mathbf{x}^*} = \mathbf{0}$, we obtain the vector equation

$$\mathbf{A}\mathbf{x}^* + \mathbf{b} = \mathbf{0} \qquad (62)$$

Since **A** is positive definite, it is nonsingular; hence

$$\mathbf{x}^* = -\mathbf{A}^{-1}\mathbf{b} \tag{63}$$

Up to now all we know is that $\mathbf{x}^*$ is a stationary point. But since **A** is assumed positive definite, Eq. (61) implies that $\mathbf{x}^*$ is at least a local minimum. But since (62) has a unique solution, this means that there is one and only one stationary point. Hence, we have only one local minimum, and hence it must be the global one.

3.3   THE PHILOSOPHY OF ALGORITHMIC SOLUTIONS

We have indicated that the solution of a set of nonlinear algebraic equations in general cannot be obtained by closed-form formulas or graphically. It thus becomes important to discuss methods by which one can utilize *trial-and-error* methods to compute the desired solution numerically.

Needless to say, one can devise arbitrary ad hoc trial-and-error methods to solve such equations. Let us examine the essential aspects of such methods and see what they consist of.

The basic problem is to find a vector $\mathbf{x}^*$ such that it is a solution (there may be many) to the equation $\mathbf{g}(\mathbf{x}) = \mathbf{0}$, that is,

$$\mathbf{g}(\mathbf{x}^*) = \mathbf{0} \tag{1}$$

By the very nature of trial and error one makes a first-guess vector, denoted by $\mathbf{x}_0$, plugs it in the equation, and computes the vector $\mathbf{g}(\mathbf{x}_0)$. If $\mathbf{g}(\mathbf{x}_0) = \mathbf{0}$ (by lucky accident), then $\mathbf{x}_0$ is indeed a solution. If, however, $\mathbf{g}(\mathbf{x}_0) \neq \mathbf{0}$, one knows that the first guess was not correct and one tries a new guess vector $\mathbf{x}_1$. The two ways of selecting $\mathbf{x}_1$ are at random, i.e., independent of what $\mathbf{x}_0$ and $\mathbf{g}(\mathbf{x}_0)$ were, or dependent on $\mathbf{x}_0$ and/or $\mathbf{g}(\mathbf{x}_0)$. If one picks the guess vectors $\mathbf{x}_0, \mathbf{x}_1, \mathbf{x}_2, \ldots$ at random, chances are that only by a lucky accident does one make a guess such that, say at the Nth trial, $\mathbf{g}(\mathbf{x}_N) = \mathbf{0}$. On the other hand, if one uses an *educated trial-and-error* method, which means that $\mathbf{x}_1$ depends on $\mathbf{x}_0$, $\mathbf{x}_2$ depends on $\mathbf{x}_1$, etc., one may have a better chance in finding a solution.

At any rate, a trial-and-error scheme generates a sequence of vectors

$$\mathbf{x}_0, \mathbf{x}_1, \mathbf{x}_2, \ldots, \mathbf{x}_k, \mathbf{x}_{k+1}, \ldots \tag{2}$$

Let us suppose that this sequence has a limit, i.e., that

$$\lim_{k \to \infty} \mathbf{x}_k = \hat{\mathbf{x}} \tag{3}$$

If the limit vector $\hat{\mathbf{x}}$ has in addition the property

$$\mathbf{g}(\hat{\mathbf{x}}) = \mathbf{0} \tag{4}$$

then we say that sequence (2) converges to a solution of the equation $\mathbf{g}(\mathbf{x}) = \mathbf{0}$. This convergence is fine from a mathematical point of view; however, it still requires an infinite number of steps and hence an infinite amount of time even in the fastest computer. To circumvent this difficulty the engineer must compromise and stop the convergent sequence at some finite step, say $k = N$, when he feels that the guess at that step, $\mathbf{x}_N$, is close enough to the solution $\mathbf{x}^*$. Hence, he must decide upon some accuracy level which mathematically can be stated as

$$\mathbf{x}_N \approx \mathbf{x}^* \qquad \text{when } \|\mathbf{g}(\mathbf{x}_N)\| < \epsilon \tag{5}$$

where $\|.\|$ is a suitable norm (see Appendix A, Sec. A.15) and ϵ is a positive constant that reflects the accuracy requirements and depends on the problem at hand.

Thus the procedure one must follow in deducing the approximate solution of a vector equation is to construct a convergent sequence of vectors which get close enough to a solution. The means of generating such a sequence is called an *algorithm*. Significant questions that must be answered depend on the programming complexity required to implement the algorithm on a digital computer, the amount of computing time required to generate each new guess $\mathbf{x}_k$, and the total number of guesses required until an approximate solution is obtained.

3.3.1 Some Fundamental Questions That Must Be Answered

The usefulness of the digital computer in engineering analysis and design is enhanced if the engineer can trust the computer printout (assuming that there are no bugs in the computer program). Invariably, if the engineer asks the computer to solve a silly problem, he gets a silly answer. Let us illustrate what we mean. Suppose we wish to solve an equation of the form $\mathbf{g}(\mathbf{x}) = \mathbf{0}$. If such an equation arises from a physical problem, then in general we can confine our search in some region,† say Ω of the euclidean space R_n. Hence, we can ask first of all an *existence question*.

Is there a vector $\mathbf{x}^* \in \Omega \subset R_n$ such that $\mathbf{g}(\mathbf{x}^*) = \mathbf{0}$?

If such a vector does not exist, it would be futile to program the computer to find it; if we feel that there must be a solution somewhere, we shift our search in some other region Ω' and ask the *existence question* once more.

---

† Once more we stress that a given mathematical modeling of a physical problem is valid only for some range of values of its variables.

Let us now suppose that the existence question has been answered in the affirmative. Since we know that there is *a* solution, it is natural to ask if there are many. This is the *uniqueness question*:

> Given the region Ω in R_n and given that there exists an $\mathbf{x}^* \in \Omega$ such that $\mathbf{g}(\mathbf{x}^*) = \mathbf{0}$, is $\mathbf{x}^*$ unique?

Once more it is important to know, if possible, the answer to this question before we start our computer. If more than one solution exists and we wish to find all solutions, we must expect to try our algorithm many times, hoping it is convergent, and construct many sequences of guess vectors, each converging to a different solution. Alternatively, we can narrow our region of interest Ω down into a smaller one Ω' ($\Omega' \subset \Omega$), until Ω' contains only one solution, which we attempt to find.

Suppose now that there is one and only one solution $\mathbf{x}^*$. It is not enough to guarantee that our sequence approaches a limit; this limit must be the sought solution.

These three topics, *existence, uniqueness, and convergence* are of fundamental importance. They must be dealt with before any computer runs so that confidence in the computer results can be established.

★ 3.4 SOME MATHEMATICAL FACTS

Since we shall be dealing with the notions of convergence of sequences of vectors to a limit vector, it is important to review some pertinent mathematical facts. The reader has probably met these notions in the context of scalar sequences and functions. The extensions of these ideas to the vector case is conceptually very simple since one can use the norm of a vector to establish the notion of distance.

Throughout this section we shall assume that $\mathbf{x}_0, \mathbf{x}_1, \ldots, \mathbf{x}_k, \ldots$ is a sequence of vectors in R_n.

> **Definition 1** *The sequence of vectors* $\mathbf{x}_0, \mathbf{x}_1, \ldots, \mathbf{x}_k, \ldots$ *is a convergent sequence and is said to converge to the limit vector* $\hat{\mathbf{x}}$ *if for each* $\epsilon > 0$ *there exists a positive integer* N *such that for* $k \geq N$, $\|\mathbf{x}_k - \hat{\mathbf{x}}\| < \epsilon$. *We write*
> $$\lim_{k \to \infty} \mathbf{x}_k = \hat{\mathbf{x}} \tag{1}$$

Intuitively speaking, one expects the vectors of a convergent sequence to get closer and closer as it approaches the limit. This intuitive notion is formalized as follows.

Lemma 1 *Every convergent sequence of vectors in R_n, $\mathbf{x}_0, \mathbf{x}_1, \ldots, \mathbf{x}_k, \ldots$ has the following property: for every $\epsilon > 0$ there exists a positive integer N such that for $m \geq N$ and $k \geq N$*

$$\|\mathbf{x}_m - \mathbf{x}_k\| < \epsilon \tag{2}$$

(A sequence with such a property is called a Cauchy sequence.)

This lemma is very useful because it allows us to look at the distance $\|\mathbf{x}_m - \mathbf{x}_n\|$ between members of the sequence and deduce that it converges if this distance tends to zero. For this reason, it is important to look at the proof.

PROOF OF LEMMA 1   Since the sequence converges by assumption, we let $\hat{\mathbf{x}}$ be the limit vector. From Definition 1 we know that there exists an N such that for $k \geq N$

$$\|\mathbf{x}_k - \hat{\mathbf{x}}\| < \epsilon' = \frac{\epsilon}{2} \tag{3}$$

Similarly for $m \geq N$

$$\|\mathbf{x}_m - \hat{\mathbf{x}}\| < \epsilon' = \frac{\epsilon}{2} \tag{4}$$

But use of the triangle inequality yields (see Appendix A, Sec. A.15)

$$\|\mathbf{x}_m - \mathbf{x}_k\| = \|\mathbf{x}_m - \hat{\mathbf{x}} + \hat{\mathbf{x}} - \mathbf{x}_k\| < \|\mathbf{x}_m - \hat{\mathbf{x}}\| + \|\mathbf{x}_k - \hat{\mathbf{x}}\|$$
$$< \frac{\epsilon}{2} + \frac{\epsilon}{2} = \epsilon \tag{5}$$

Every convergent sequence is a Cauchy sequence. In the euclidean space R_n every Cauchy sequence converges to a limit in R_n; thus every Cauchy sequence is a convergent sequence (this is not true in arbitrary vector spaces).

The final set of mathematical facts we shall be needing relates to the following question. Suppose that $\mathbf{f}(.)$ is a function from R_n to R_n. Suppose that a sequence $\mathbf{x}_0, \mathbf{x}_1, \ldots, \mathbf{x}_k, \ldots$ of vectors in R_n converges to a limit vector $\hat{\mathbf{x}}$ in R_n. Is it true that

$$\lim_{k \to \infty} \mathbf{f}(\mathbf{x}_k) = \mathbf{f}(\hat{\mathbf{x}}) \tag{6}$$

To answer this question we must recall what we mean by a continuous function.

Definition 2 *Let $\mathbf{f}(.)\colon R_n \to R_n$; $\mathbf{f}(.)$ is said to be continuous at a point $\hat{\mathbf{x}} \in R_n$ if for each $\epsilon > 0$ there exists a $\delta > 0$ such that*

$$\|\hat{\mathbf{x}} - \mathbf{x}\| < \delta$$

implies that

$$\|\mathbf{f}(\hat{\mathbf{x}}) - \mathbf{f}(\mathbf{x})\| < \epsilon$$

If $\mathbf{f}(.)$ is continuous for all $\hat{\mathbf{x}} \in R_n$, it is called a continuous function.

We now shall establish the second important lemma.

Lemma 2   *Suppose that $\mathbf{f}(.): R_n \to R_n$ is a continuous function. Suppose that the sequence of vectors in R_n, $\mathbf{x}_0, \mathbf{x}_1, \ldots, \mathbf{x}_k, \ldots$ converge to the limit vector $\hat{\mathbf{x}}$. Then the sequence of vectors in R_n*

$$\mathbf{f}(\mathbf{x}_0), \mathbf{f}(\mathbf{x}_1), \ldots, \mathbf{f}(\mathbf{x}_k), \ldots$$

converges to the vector $\mathbf{f}(\hat{\mathbf{x}})$.

> PROOF   Since $\mathbf{x}_k$ converges to $\hat{\mathbf{x}}$, by Definition 1 there exists N such that for $k \geq N$
>
> $$\|\mathbf{x}_k - \hat{\mathbf{x}}\| < \delta$$
>
> By assumption $\mathbf{f}(.)$ is a continuous function. In particular it is continuous at $\hat{\mathbf{x}}$. Hence, by Definition 2, $\|\mathbf{x}_k - \hat{\mathbf{x}}\| < \delta$ implies that $\|\mathbf{f}(\hat{\mathbf{x}}) - \mathbf{f}(\mathbf{x}_k)\| < \epsilon$. Therefore, by Definition 1, $\mathbf{f}(\mathbf{x}_k)$ is a convergent sequence and its limit is $\mathbf{f}(\hat{\mathbf{x}})$.

3.5   THE PICARD ALGORITHM

We are now ready to discuss the first algorithm for solving sets of simultaneous equations in an educated trial-and-error manner. This algorithm is often called Picard's method or the method of successive approximations.

First we explain the basic idea of the algorithm in the scalar case, stressing its geometric nature. Then generalize the basic idea behind the algorithm to the vector case.

3.5.1   The Scalar Case†

Consider the problem of finding the solution x^* of the equation

$$g(x) = x - f(x) = 0 \qquad (1)$$

† For an extensive treatment see "Basic Concepts," chap. 5.

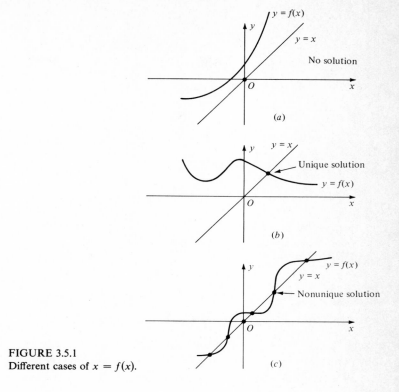

FIGURE 3.5.1
Different cases of $x = f(x)$.

or, equivalently, of

$$x = f(x) \tag{2}$$

when x is a real-valued scalar and $f(x)$ is a scalar-valued function of x.

The fact that Eq. (2) may have no solution, one solution, or many solutions is illustrated in Fig. 3.5.1.

In the scalar case, the Picard algorithm is characterized as follows. One starts with an initial guess x_0. Then, one constructs the sequence $x_1, x_2, \ldots, x_n, \ldots$ as follows:

$$
\begin{aligned}
x_1 &= f(x_0) \\
x_2 &= f(x_1) \\
&\vdots \\
x_{n+1} &= f(x_n)
\end{aligned}
\tag{3}
$$

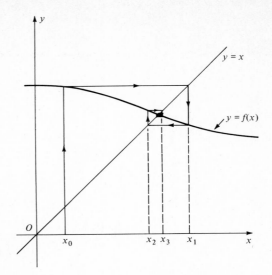

FIGURE 3.5.2
Geometric interpretation of convergent
Picard algorithm.

or, in shorthand way,

$$x_{k+1} = f(x_k) \qquad k = 0, 1, 2, \ldots \tag{4}$$

Figure 3.5.2 shows a convergent case, i.e., the sequence $\{x_k\}$ converges to the solution x^* of $x = f(x)$. Figure 3.5.3 shows a nonconvergent case.

We turn our attention now to the vector case to understand the method in its full generality.

3.5.2 The Vector Case

We are concerned here with solving the set of equations†

$$x_1 = f_1(x_1, x_2, \ldots, x_n)$$
$$x_2 = f_2(x_1, x_2, \ldots, x_n)$$
$$\vdots \tag{5}$$
$$x_n = f_n(x_1, x_2, \ldots, x_n)$$

or, in vector form,

$$\mathbf{x} = \mathbf{f}(\mathbf{x}) \tag{6}$$

† Remember that this was exactly the type of equation that had to be solved for nonlinear resistive networks [Eq. (10) of Sec. 2.8].

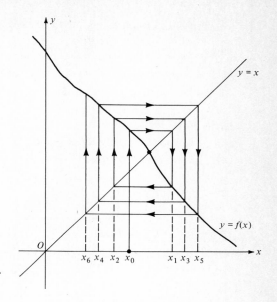

FIGURE 3.5.3
Geometric interpretation of divergent Picard algorithm.

We seek a solution vector $\mathbf{x}^*$ (if it exists), i.e., a vector such that

$$\mathbf{x}^* = \mathbf{f}(\mathbf{x}^*) \tag{7}$$

It is convenient to confine our search to a region Ω contained in R_n. Ω may be a hypercube, i.e.,

$$\Omega = \{\mathbf{x} \in R_n : a \leq x_i \leq b; \quad i = 1, 2, \ldots, n\}$$

hyperparallelepiped

$$\Omega = \{\mathbf{x} \in R_n : a_i \leq x_i \leq b_i; \quad i = 1, 2, \ldots, n\}$$

or some other set.

We summarize our main result in the following theorem.

Theorem 1 (The Picard Algorithm)

(a)   *Suppose that for every* $\mathbf{x} \in \Omega \subset R_n$

$$\mathbf{f}(\mathbf{x}) \in \Omega \subset R_n \tag{8}$$

(b) $\mathbf{f}(\mathbf{x})$ *is continuous for all* $\mathbf{x} \in \Omega$.

(c) $$\|\mathbf{f}(\mathbf{x}_1) - \mathbf{f}(\mathbf{x}_2)\| \leq L\|\mathbf{x}_1 - \mathbf{x}_2\| \qquad L < 1 \tag{9}$$

(*the Lipschitz condition*) *holds for all vectors* $\mathbf{x}_1$ *and* $\mathbf{x}_2$ *in* Ω.

Then

(*1*) *A solution vector* $\mathbf{x}^*$ *to the equation*

$$\mathbf{x} = \mathbf{f}(\mathbf{x}) \tag{10}$$

exists in Ω.

(*2*) *The solution vector* $\mathbf{x}^*$ *is unique.*

(*3*) *The Picard algorithm*

$$\boxed{\mathbf{x}_{k+1} = \mathbf{f}(\mathbf{x}_k)} \qquad k = 0, 1, 2, \ldots \tag{11}$$

converges to $\mathbf{x}^*$, *that is,*

$$\boxed{\lim_{k \to \infty} \mathbf{x}_k = \mathbf{x}^*} \tag{12}$$

PROOF   The proof of the theorem is a constructive one. The existence of $\mathbf{x}^*$ is actually proved by generating the sequence

$$\mathbf{x}_1 = \mathbf{f}(\mathbf{x}_0)$$
$$\mathbf{x}_2 = \mathbf{f}(\mathbf{x}_1)$$
$$\vdots \tag{13}$$
$$\mathbf{x}_{k+1} = \mathbf{f}(\mathbf{x}_k)$$
$$\vdots$$

Note that assumption (*a*) guarantees that if $\mathbf{x}_0 \in \Omega$, then, $\mathbf{x}_1 \in \Omega$, then $\mathbf{x}_2 \in \Omega$, and so on. Thus, every member of the sequence remains in the set Ω.

Let us now examine whether the sequence $\mathbf{x}_0, \mathbf{x}_1, \ldots, \mathbf{x}_k, \ldots$ converges to some limit. Toward this goal let us consider the vectors

$$\mathbf{x}_2 - \mathbf{x}_1 = \mathbf{f}(\mathbf{x}_1) - \mathbf{f}(\mathbf{x}_0)$$
$$\mathbf{x}_3 - \mathbf{x}_2 = \mathbf{f}(\mathbf{x}_2) - \mathbf{f}(\mathbf{x}_1)$$
$$\vdots \tag{14}$$
$$\mathbf{x}_{k+1} - \mathbf{x}_k = \mathbf{f}(\mathbf{x}_k) - \mathbf{f}(\mathbf{x}_{k-1})$$
$$\vdots$$

But successive use of the Lipschitz condition (9) yields

$$\|\mathbf{x}_2 - \mathbf{x}_1\| = \|\mathbf{f}(\mathbf{x}_1) - \mathbf{f}(\mathbf{x}_0)\| \leq L\|\mathbf{x}_1 - \mathbf{x}_0\|$$

$$\|\mathbf{x}_3 - \mathbf{x}_2\| = \|\mathbf{f}(\mathbf{x}_2) - \mathbf{f}(\mathbf{x}_1)\| \leq L\|\mathbf{x}_2 - \mathbf{x}_1\| \leq L^2\|\mathbf{x}_1 - \mathbf{x}_0\|$$

and, proceeding in an inductive manner, we deduce that

$$\|\mathbf{x}_{k+1} - \mathbf{x}_k\| \leq L^k\|\mathbf{x}_1 - \mathbf{x}_0\| = L^k\|\mathbf{f}(\mathbf{x}_0) - \mathbf{x}_0\| \qquad (15)$$

The norm $\|\mathbf{x}_1 - \mathbf{x}_0\|$ is some positive number which is fixed by the initial guess $\mathbf{x}_0$. Since by assumption the Lipschitz constant $L < 1$, it follows that

$$\lim_{k \to \infty} L^k = 0 \qquad (16)$$

and, hence, from Eq. (15)

$$\lim_{k \to \infty} \|\mathbf{x}_{k+1} - \mathbf{x}_k\| = (\lim_{k \to \infty} L^k)\|\mathbf{f}(\mathbf{x}_0) - \mathbf{x}_0\| = 0 \qquad (17)$$

Since we need to show that the sequence $\{\mathbf{x}_k\}$ is a Cauchy sequence, we shall verify that $\|\mathbf{x}_{k+m} - \mathbf{x}_k\| \to 0$ as $k \to \infty$ for any $m \geq 0$.

$$\|\mathbf{x}_{k+m} - \mathbf{x}_k\| = \|\sum_{i=0}^{m-1} (\mathbf{x}_{k+i+1} - \mathbf{x}_{k+i})\|$$

$$\leq \sum_{i=0}^{m-1} \|\mathbf{x}_{k+i+1} - \mathbf{x}_{k+i}\| \qquad \text{by triangle inequality}^\dagger \ m \text{ times}$$

$$\leq \sum_{i=0}^{m-1} L^{k+i}\|\mathbf{x}_1 - \mathbf{x}_0\| \qquad \text{by Eq. (15)}$$

$$\leq L^k\|\mathbf{x}_1 - \mathbf{x}_0\| \sum_{i=0}^{m-1} L^i \leq L^k\|\mathbf{x}_1 - \mathbf{x}_0\| \sum_{i=0}^{\infty} L^i$$

$$= \frac{L^k}{1-L}\|\mathbf{x}_1 - \mathbf{x}_0\| \qquad \text{since } 0 \leq L < 1$$

Therefore

$$\lim_{k \to \infty} \|\mathbf{x}_{k+m} - \mathbf{x}_k\| \to 0 \qquad \text{for any } m \text{ [by Eq. (16)]}$$

Therefore $\{\mathbf{x}_k\}$ is a Cauchy sequence in R_n and hence has a limit point $\hat{\mathbf{x}}$,

$$\lim_{k \to \infty} \mathbf{x}_k = \hat{\mathbf{x}} \in \Omega \qquad (18)$$

---

† See Appendix A, Sec. A.15.

The next step is to investigate whether $\hat{\mathbf{x}}$ is indeed a solution to $\mathbf{x} = \mathbf{f}(\mathbf{x})$. Note that

$$\mathbf{x}_{k+1} = \mathbf{f}(\mathbf{x}_k) \tag{19}$$

and so

$$\lim_{k \to \infty} \mathbf{x}_{k+1} = \lim_{k \to \infty} \mathbf{f}(\mathbf{x}_k) \tag{20}$$

But $\mathbf{x}_{k+1}$ converges to a limit vector $\hat{\mathbf{x}}$

$$\lim_{k \to \infty} \mathbf{x}_{k+1} = \hat{\mathbf{x}} \tag{21}$$

Since by assumption (b), $\mathbf{f}(.)$ is a continuous function and the sequence converges to $\hat{\mathbf{x}}$, we have

$$\lim_{k \to \infty} \mathbf{f}(\mathbf{x}_k) = \mathbf{f}(\hat{\mathbf{x}}) \tag{22}$$

Hence,

$$\hat{\mathbf{x}} = \mathbf{f}(\hat{\mathbf{x}}) \tag{23}$$

and this establishes the *existence* of a solution vector $\mathbf{x}^* = \hat{\mathbf{x}}$ in Ω.

We next investigate the *uniqueness* of the solution $\mathbf{x}^*$. As with most uniqueness proofs, we shall use a contradiction argument. Suppose that there are two distinct solutions $\mathbf{x}_1^*$ and $\mathbf{x}_2^*$ in Ω; this means that

$$\begin{aligned} \mathbf{x}_1^* &= \mathbf{f}(\mathbf{x}_1^*) \\ \mathbf{x}_2^* &= \mathbf{f}(\mathbf{x}_2^*) \end{aligned} \tag{24}$$

Hence, subtraction of the vectors in Eqs. (24) yields

$$\mathbf{f}(\mathbf{x}_1^*) - \mathbf{f}(\mathbf{x}_2^*) = \mathbf{x}_1^* - \mathbf{x}_2^* \tag{25}$$

and this implies that

$$\|\mathbf{f}(\mathbf{x}_1^*) - \mathbf{f}(\mathbf{x}_2^*)\| = \|\mathbf{x}_1^* - \mathbf{x}_2^*\| \tag{26}$$

But the Lipschitz condition (9) requires that

$$\|\mathbf{f}(\mathbf{x}_1^*) - \mathbf{f}(\mathbf{x}_2^*)\| < \|\mathbf{x}_1^* - \mathbf{x}_2^*\| \tag{27}$$

Relations (26) and (27) are *contradictory*; hence, $\mathbf{x}_1^*$ and $\mathbf{x}_2^*$ cannot be distinct, and the uniqueness of solutions has been established.

Since the solution $\mathbf{x}^*$ exists and is unique, and since we have shown that the Picard algorithm converges to $\hat{\mathbf{x}} = \mathbf{x}^*$, the statements of the theorem have been verified.

3.5.3 Speed of Convergence

We now turn our attention to the speed of convergence of the Picard algorithm. Intuitively speaking, we are interested in how fast our initial error $\mathbf{x}_0 - \mathbf{x}^*$ decreases. Since we do not know the solution $\mathbf{x}^*$, and since our first guess $\mathbf{x}_0$ is completely arbitrary, the magnitude of our initial error

$$\|\mathbf{x}_0 - \mathbf{x}^*\| \tag{28}$$

cannot be regulated. The speed of convergence is related to the way the error magnitude at the kth step

$$\|\mathbf{x}_k - \mathbf{x}^*\| \tag{29}$$

is related to the initial error magnitude (28). To obtain such a relation we start by examining

$$\mathbf{x}_1 - \mathbf{x}^* = \mathbf{f}(\mathbf{x}_0) - \mathbf{f}(\mathbf{x}^*)$$

Use of the Lipschitz condition (9) yields

$$\|\mathbf{x}_1 - \mathbf{x}^*\| = \|\mathbf{f}(\mathbf{x}_0) - \mathbf{f}(\mathbf{x}^*)\| \leq L\|\mathbf{x}_0 - \mathbf{x}^*\| \tag{30}$$

Similarly from the equality

$$\mathbf{x}_2 - \mathbf{x}^* = \mathbf{f}(\mathbf{x}_1) - \mathbf{f}(\mathbf{x}^*) \tag{31}$$

we obtain, using the Lipschitz condition and (30),

$$\|\mathbf{x}_2 - \mathbf{x}^*\| \leq L\|\mathbf{x}_1 - \mathbf{x}^*\| \leq L^2\|\mathbf{x}_0 - \mathbf{x}^*\| \tag{32}$$

and at the kth step

$$\boxed{\|\mathbf{x}_k - \mathbf{x}^*\| \leq L^k\|\mathbf{x}_0 - \mathbf{x}^*\|} \tag{33}$$

Therefore, *at the kth step we have reduced the original error magnitude* $\|\mathbf{x}_0 - \mathbf{x}^*\|$ *by at least a factor of* L^k. The larger the value of L $(0 \leq L < 1)$ the slower the guaranteed rate of convergence; the smaller the value of L the faster the rate of convergence. For example, suppose that we know from the physics of the problem that

$$\|\mathbf{x}_0 - \mathbf{x}^*\| \leq 10$$

and we are interested in obtaining an accuracy level of 10^{-3}, that is, we set

$$\mathbf{x}_N \approx \mathbf{x}^* \qquad \text{provided } \|\mathbf{x}_N - \mathbf{x}^*\| \leq 10^{-3}$$

To determine the maximum number of steps N we need to attain this accuracy we form the relation

$$\|\mathbf{x}_N - \mathbf{x}^*\| \leq L^N \|\mathbf{x}_0 - \mathbf{x}^*\| \leq 10 L^N \leq 10^{-3}$$

Hence,

$$L^N \leq 10^{-4}$$

The worst case is achieved when

$$L^N = 10^{-4}$$

so that

$$N \log L = -4$$

If L is relatively large, for example $L = 0.9$, we find that we need approximately

$$N = -\frac{4}{\log 0.9} \approx \frac{4}{0.04} = 100 \text{ steps}$$

If, on the other hand, the Lipschitz constant is small, for example, $L = 0.1$, we find that we only need

$$N = -\frac{4}{\log 0.1} = \frac{4}{1} = 4 \text{ steps}$$

for the same computational accuracy.

We shall exploit this dependence of the guaranteed rate of convergence upon the value of L in the next section, where we develop the Newton method.

Although Eq. (33) is useful for estimating how the initial error $\|\mathbf{x}_0 - \mathbf{x}^*\|$ shrinks at the kth step of the iteration, it does not provide us with an actual estimate of the error at the kth step $\|\mathbf{x}_k - \mathbf{x}^*\|$. We now present a formula that gives such an estimate, in terms of an upper bound.

We have already shown that

$$\|\mathbf{x}_{k+m} - \mathbf{x}_k\| \leq \frac{L^k}{1 - L} \|\mathbf{x}_1 - \mathbf{x}_0\| \tag{34}$$

In Eq. (34) if we let

$$k = 0 \qquad m \to \infty \tag{35}$$

$$\lim_{m \to \infty} \|\mathbf{x}_m - \mathbf{x}_0\| \leq \frac{L^0}{1 - L} \|\mathbf{x}_1 - \mathbf{x}_0\| \leq \frac{1}{1 - L} \|\mathbf{x}_1 - \mathbf{x}_0\| \tag{36}$$

which implies

$$\|\mathbf{x}^* - \mathbf{x}_0\| \leq \frac{1}{1-L}\|\mathbf{x}_1 - \mathbf{x}_0\| \tag{37}$$

But since

$$\mathbf{x}_1 = \mathbf{f}(\mathbf{x}_0) \tag{38}$$

expressions (37) and (38) yield

$$\|\mathbf{x}_0 - \mathbf{x}^*\| \leq \frac{1}{1-L}\|\mathbf{x}_0 - \mathbf{f}(\mathbf{x}_0)\| \tag{39}$$

Note that Eq. (39) provides us with an upper bound on the initial error magnitude $\|\mathbf{x}_0 - \mathbf{x}^*\|$. This can now be combined with Eq. (33) to obtain the sought for upper bound on the error magnitude $\|\mathbf{x}_k - \mathbf{x}^*\|$ at the kth step. Thus from Eqs. (39) and (33) we deduce that

$$\|\mathbf{x}_k - \mathbf{x}^*\| \leq \frac{L^k}{1-L}\|\mathbf{x}_0 - \mathbf{f}(\mathbf{x}_0)\| \tag{40}$$

3.5.4 Some Comments on the Use of Picard's Method

We shall now comment on the preliminary steps required to verify the assumptions of Theorem 1 before the actual programming itself. Once the assumptions are verified, one can proceed with confidence that the Picard algorithm will indeed converge to the unique solution.

The equation is $\mathbf{x} = \mathbf{f}(\mathbf{x})$. The continuity of $\mathbf{f}(.)$ can be usually deduced by inspection. The determination of the set Ω [in which the relation $\mathbf{f}(\mathbf{x}) \in \Omega$ for all $\mathbf{x}$ in Ω holds] is not straightforward, and the difficulty of isolating the proper set Ω will depend on the problem at hand. However, if such equations arise from an examination of a class of physical or engineering systems, one usually has an idea of the range in which the solution is to be expected.

The most difficult task is to verify that the Lipschitz condition (9) holds at all pairs of vectors $\mathbf{x}_1$ and $\mathbf{x}_2$ in the set Ω. The verification of the Lipschitz condition is often a considerably simpler task *if the function* $\mathbf{f}(\mathbf{x})$ *in addition to being continuous also has continuous partial derivatives with respect to the components of the vector* $\mathbf{x}$. If this is the case, one can express the Lipschitz condition in a somewhat simpler form. This is the topic of the remainder of this section.

In Appendix A we discussed the concept of a *Jacobian matrix*. The Jacobian matrix of $\mathbf{f(x)}$ is

$$
\frac{\partial \mathbf{f}}{\partial \mathbf{x}} \triangleq
\begin{bmatrix}
\dfrac{\partial f_1}{\partial x_1} & \dfrac{\partial f_1}{\partial x_2} & \cdots & \dfrac{\partial f_1}{\partial x_n} \\[2ex]
\dfrac{\partial f_2}{\partial x_1} & \dfrac{\partial f_2}{\partial x_2} & \cdots & \dfrac{\partial f_2}{\partial x_n} \\[2ex]
\vdots & \vdots & \vdots & \vdots \\[2ex]
\dfrac{\partial f_n}{\partial x_1} & \dfrac{\partial f_n}{\partial x_2} & \cdots & \dfrac{\partial f_n}{\partial x_n}
\end{bmatrix}
\tag{41}
$$

Thus the ijth element of the Jacobian matrix $\partial \mathbf{f}/\partial \mathbf{x}$ is $\partial f_i/\partial x_j$, where $i, j = 1, 2, \ldots, n$. If $\mathbf{f(x)}$ has a continuous partial derivative, this means that all the elements $\partial f_i/\partial x_j$ of the Jacobian matrix are continuous functions of $x_1, x_2, \ldots, x_n$.

We shall now state the vector form of the *mean-value theorem*.

Theorem 2 *Suppose that $\mathbf{f(x)}$ and $\partial \mathbf{f}/\partial \mathbf{x}$ are continuous in $\mathbf{x}$. Let $\mathbf{x}_1$ and $\mathbf{x}_2$ be arbitrary vectors in R_n. Then there exists a vector $\hat{\mathbf{x}}$ in the line segment joining $\mathbf{x}_1$ and $\mathbf{x}_2$, that is,*

$$
\hat{\mathbf{x}} = \alpha \mathbf{x}_1 + (1 - \alpha)\mathbf{x}_2 \qquad 0 \le \alpha \le 1
\tag{42}
$$

such that

$$
\mathbf{f}(\mathbf{x}_1) = \mathbf{f}(\mathbf{x}_2) + \left(\frac{\partial \mathbf{f}}{\partial \mathbf{x}} \bigg|_{\mathbf{x}=\hat{\mathbf{x}}} \right)(\mathbf{x}_1 - \mathbf{x}_2)
\tag{43}
$$

where the notation $\partial \mathbf{f}/\partial \mathbf{x}|_{\mathbf{x}=\hat{\mathbf{x}}}$ means that the elements of the Jacobian matrix are evaluated at $\hat{\mathbf{x}}$.

We now examine how one can use the mean-value theorem to obtain an alternate form of the Lipschitz condition (9) which states

$$
\|\mathbf{f}(\mathbf{x}_1) - \mathbf{f}(\mathbf{x}_2)\| < L\|\mathbf{x}_1 - \mathbf{x}_2\| \qquad \begin{array}{c} \mathbf{x}_1, \mathbf{x}_2 \in \Omega \\ L < 1 \end{array}
\tag{44}
$$

If the Jacobian matrix $\partial \mathbf{f}/\partial \mathbf{x}$ is continuous for all $\mathbf{x} \in \Omega$, then Eq. (43) yields

$$
\mathbf{f}(\mathbf{x}_1) - \mathbf{f}(\mathbf{x}_2) = \left(\frac{\partial \mathbf{f}}{\partial \mathbf{x}} \bigg|_{\mathbf{x}=\hat{\mathbf{x}}} \right)(\mathbf{x}_1 - \mathbf{x}_2)
\tag{45}
$$

Using the properties of norms, we deduce that

$$
\|\mathbf{f}(\mathbf{x}_1) - \mathbf{f}(\mathbf{x}_2)\| = \left\| \left(\frac{\partial \mathbf{f}}{\partial \mathbf{x}} \bigg|_{\mathbf{x}=\hat{\mathbf{x}}} \right)(\mathbf{x}_1 - \mathbf{x}_2) \right\| \le \left\| \frac{\partial \mathbf{f}}{\partial \mathbf{x}} \bigg|_{\mathbf{x}=\hat{\mathbf{x}}} \right\| \cdot \|\mathbf{x}_1 - \mathbf{x}_2\|
\tag{46}
$$

By comparing inequalities (44) and (46) we deduce that the function $\mathbf{f}(\mathbf{x})$ will satisfy the Lipschitz condition if

$$\left\| \frac{\partial \mathbf{f}}{\partial \mathbf{x}} \right\| \leq M < 1 \qquad \text{for all } \mathbf{x} \in \Omega \tag{47}$$

This means that we can check the Lipschitz condition by evaluating the norm of the Jacobian matrix in the set of interest Ω. The Lipschitz constant L is then

$$L = \max_{\mathbf{x} \in \Omega} \left\| \frac{\partial \mathbf{f}}{\partial \mathbf{x}} \right\| \tag{48}$$

We summarize these ideas in the following theorem.

Theorem 3 (Alternate form of Picard algorithm) *Consider the vector equation* $\mathbf{x} = \mathbf{f}(\mathbf{x})$. *Let Ω be a subset of R_n.*
Suppose that

(*a*) $\mathbf{f}(\mathbf{x})$ *is continuous for all* $\mathbf{x} \in R_n$. $\tag{49}$

(*b*) $\mathbf{f}(\mathbf{x}) \in \Omega$ *for all* $\mathbf{x} \in \Omega$. $\tag{50}$

(*c*) $\mathbf{f}(\mathbf{x})$ *has continuous derivatives for all* $\mathbf{x} \in \Omega$. $\tag{51}$

(*d*) $\|\partial \mathbf{f}/\partial \mathbf{x}\| < 1$ *for all* $\mathbf{x} \in \Omega$. $\tag{52}$

Then

(*1*) *There exists a unique solution* $\mathbf{x}^* \in \Omega$, *that is,* $\mathbf{x}^* = \mathbf{f}(\mathbf{x}^*)$, *and*
(*2*) *The Picard algorithm*

$$\mathbf{x}_{k+1} = \mathbf{f}(\mathbf{x}_k) \qquad k = 0, 1, 2, \ldots \tag{53}$$

converges to $\mathbf{x}^*$, *that is,*

$$\lim_{k \to \infty} \mathbf{x}_k = \mathbf{x}^* \tag{54}$$

EXAMPLE To illustrate these ideas let us consider a two-dimensional example. The equations are

$$x_1 = f_1(x_1, x_2) = 0.2x_1 + 0.5 \cos x_2$$
$$x_2 = f_2(x_1, x_2) = 0.1e^{-x_1} \sin x_2$$

Consider the set Ω defined by

$$0 \le x_1 \le 1$$
$$0 \le x_2 \le 1$$

which is a square in the $x_1 x_2$ plane.

Clearly $f_1(.)$ and $f_2(.)$ are continuous in Ω. Furthermore, for any values of x_1 and x_2 in the square Ω

$$0 \le 0.2x_1 + 0.5 \cos x_2 \le 1$$
$$0 \le 0.1e^{-x_1} \sin x_2 \le 1$$

so that condition (50) is satisfied.

Let us now compute the partial derivatives

$$\frac{\partial f_1}{\partial x_1} = 0.2 \qquad \frac{\partial f_1}{\partial x_2} = -0.5 \sin x_2$$

$$\frac{\partial f_2}{\partial x_1} = -0.1e^{-x_1} \sin x_2 \qquad \frac{\partial f_2}{\partial x_2} = 0.1e^{-x_1} \cos x_2$$

which are continuous functions of x_1 and x_2 in the square Ω.

The Jacobian matrix (41) for this problem is the 2×2 matrix

$$\frac{\partial \mathbf{f}}{\partial \mathbf{x}} = \begin{bmatrix} 0.2 & -0.5 \sin x_2 \\ -0.1e^{-x_1} \sin x_2 & 0.1e^{-x_1} \cos x_2 \end{bmatrix}$$

If we compute the euclidean norm of the Jacobian matrix, we find

$$\left\| \frac{\partial \mathbf{f}}{\partial \mathbf{x}} \right\| = \sqrt{(0.2)^2 + (-0.5 \sin x_2)^2 + (0.1e^{-x_1} \sin x_2)^2 + (0.1e^{-x_1} \cos x_2)^2}$$

$$= \sqrt{0.04 + 0.25 \sin^2 x_2 + 0.01e^{-2x_1}(\sin^2 x_2 + \cos^2 x_2)}$$

$$= \sqrt{0.04 + 0.25 \sin^2 x_2 + 0.01e^{-2x_1}}$$

The largest possible value of this expression is obtained when $x_1 = 0$ and $x_2 = 1$. Hence

$$L = \max_{\mathbf{x} \in \Omega} \sqrt{0.04 + 0.25 \sin^2 1 + 0.01} = \sqrt{0.04 + 0.21 + 0.01}$$

$$= \sqrt{0.216}$$

$$= 0.465 < 1$$

Hence, condition (52) is satisfied. Therefore we deduce that there exists a unique solution x_1^*, x_2^* in the square Ω.

Let us now use the Picard algorithm to find the solution using an accuracy of the following type

$$\|\mathbf{x}_N - \mathbf{x}^*\| < 10^{-2}$$

which means that we are satisfied if our answer $\mathbf{x}_N$ falls in a circle of radius 0.01 about the point $\mathbf{x}^*$.

Let us start for our initial guess

$$k = 0 \qquad x_1 = 1 \qquad x_2 = 1$$

The worst possible initial-error magnitude $\|\mathbf{x}_0 - \mathbf{x}^*\|$ is then $\sqrt{2}$. We know that at the Nth step we must have

$$\|\mathbf{x}_N - \mathbf{x}^*\| \le L^N \|\mathbf{x}_0 - \mathbf{x}^*\| = (0.465)^N \sqrt{2} \le 10^{-2}$$

From this relation we can conclude that the *maximum* number of required steps is

$$N = 7$$

If we apply the Picard method to this equation, we find:

		x_1	x_2
$k = 0$	initial guess	1.000	1.000
$k = 1$	first iterate	0.209	0.031
$k = 2$	second iterate	0.504	0.000
$k = 3$	third iterate	0.608	0.000
$k = 4$	fourth iterate	0.621	0.000
$k = 5$	fifth iterate	0.622	0.000

The true solution happens to be

$$x_1^* = 0.625 \qquad x_2^* = 0.000$$

We can see that the posed accuracy was attained at $k = 4$, which is less than the predicted maximum number of steps ($N = 7$).

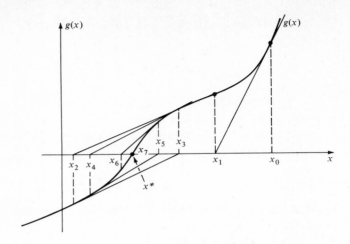

FIGURE 3.6.1
Graphical illustration of the scalar Newton method in a convergent case.

3.6 NEWTON'S METHOD

Another popular algorithm for solving systems of nonlinear algebraic equations is Newton's method. As we shall see, Newton's method contrasted to Picard's method has three main characteristics:

1 The conditions which guarantee the convergence of Newton's method are more stringent than those associated with Picard's method.
2 Newton's method requires more calculations at each step and more programming effort.
3 When Newton's method converges, it converges at a much faster rate than the Picard algorithm.

Thus, the increase in programming complexity is often counterbalanced by the rapid rate of convergence. For this reason the Newton algorithm enjoys a wide popularity.

We shall derive Newton's algorithm through a judicious use of the Picard algorithm.†

3.6.1 The Scalar Case

The reader is probably familiar with the use of Newton's method for finding the root x^* of the equation

$$g(x) = 0 \qquad (1)$$

† The same approach can be used in the scalar case; see, for example, "Basic Concepts," chap. 5.

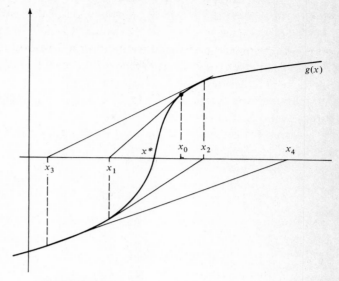

FIGURE 3.6.2
Graphical illustration of Newton's method in a divergent case.

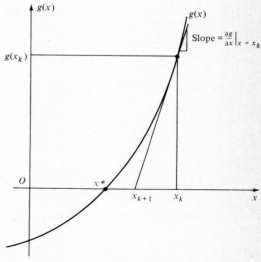

FIGURE 3.6.3
The geometry of Newton's method.

Figure 3.6.1 shows the graphical interpretation of the scalar Newton's method when it converges. Figure 3.6.2 shows that Newton's method can diverge.

The graphical interpretation of Newton's method is as follows (see Fig. 3.6.3 for the geometry). Suppose x_k is the current guess. One computes the slope of the function $g(x)$ at the point $x = x_k$; the slope is $\partial g/\partial x \mid_{x=x_k}$. One draws then a straight line from the point $(x_k, g(x_k))$ with the above slope. Its intersection with the x-axis is the next guess x_{k+1}.

From the geometry of Fig. 3.6.3 we can get the analytical form of the Newton algorithm

$$\text{Slope} = \frac{\partial g}{\partial x}\bigg|_{x=x_k} = \frac{g(x_k)}{x_k - x_{k+1}} \tag{2}$$

Rearranging, we obtain the common form of the Newton algorithm

$$\boxed{x_{k+1} = x_k - \frac{g(x_k)}{\partial g/\partial x|_{x=x_k}} \qquad k = 0, 1, 2, \ldots} \tag{3}$$

3.6.2 Vector Case: Problem Definition

The set of equations we shall attempt to solve is the set of the following n equations with n unknowns:

$$g_1(\mathbf{x}) = g_1(x_1, x_2, \ldots, x_n) = 0$$
$$g_2(\mathbf{x}) = g_2(x_1, x_2, \ldots, x_n) = 0$$
$$\vdots \tag{4}$$
$$g_n(\mathbf{x}) = g_n(x_1, x_2, \ldots, x_n) = 0$$

This set of equations is written in vector form as

$$\boxed{\mathbf{g}(\mathbf{x}) = \mathbf{0}} \tag{5}$$

where

$$\mathbf{x} \in R_n \tag{6}$$

As usual, we denote the *solution vector* (if it exists) by $\mathbf{x}^*$. Clearly, $\mathbf{x}^* \in R_n$, and

$$\mathbf{g}(\mathbf{x}^*) = \mathbf{0} \tag{7}$$

3.6.3 A Philosophy of Approach

The derivation of the Newton algorithm from the Picard algorithm involves two main ideas.

Idea 1 *Starting from an equation of the form*

$$\mathbf{g(x) = 0} \tag{8}$$

find an equation of the form

$$\mathbf{x = f(x)} \tag{9}$$

(with $\mathbf{x} \in R_n$) such that the solution vector $\mathbf{x^*}$ of Eq. (8) is also a solution vector of Eq. (9). In other words, if $\mathbf{x^*}$ has the property $\mathbf{g(x^*) = 0}$ the function $\mathbf{f(\,.\,)}$ must be chosen in such a way that

$$\mathbf{x^* = f(x^*)} \tag{10}$$

Let us elaborate on this concept before proceeding. There are many ways of constructing, given $\mathbf{g(\,.\,)}$, a function $\mathbf{f(\,.\,)}$ with such a property. To see this we start with our equation

$$\mathbf{g(x) = 0} \tag{11}$$

Let $\mathbf{A(x)}$ be an arbitrary $n \times n$ matrix whose elements $a_{ij}(\mathbf{x})$ are functions of the vector $\mathbf{x}$. Premultiplying both sides of (11) by $\mathbf{A(x)}$, we obtain

$$\mathbf{A(x)g(x) = 0} \tag{12}$$

Adding $\mathbf{x}$ to both sides of Eq. (12), we obtain

$$\mathbf{x = x + A(x)g(x)} \tag{13}$$

Now let us evaluate (13) at the solution vector $\mathbf{x^*}$

$$\mathbf{x^* = x^* + A(x^*)g(\overset{0}{\mathbf{x^*}})} \tag{14}$$

which is obviously a true identity. Now if we define the function $\mathbf{f(\,.\,)}$ by

$$\mathbf{f(x) \triangleq x + A(x)g(x)} \tag{15}$$

the equation $\mathbf{x = f(x)}$ has the required property $\mathbf{x^* = f(x^*)}$.

From these considerations and in view of the fact that the matrix $\mathbf{A(x)}$ is completely arbitrary, it should be evident that there are many ways of constructing $\mathbf{f(\,.\,)}$ given $\mathbf{g(\,.\,)}$.

Now let us suppose that we have constructed $\mathbf{f}(\mathbf{x})$ according to Eq. (15) and we apply the Picard algorithm to the equation $\mathbf{x} = \mathbf{f}(\mathbf{x})$. We recall that the speed of convergence of the Picard algorithm

$$\mathbf{x}_{k+1} = \mathbf{f}(\mathbf{x}_k) \qquad k = 0, 1, 2, \ldots \tag{16}$$

hinges on the value of the Lipschitz constant L ($0 \leq L < 1$) associated with $\mathbf{f}(.)$. We indicated that *the smaller the value of L the faster the Picard algorithm converges.* This quest for a fast-converging algorithm leads us to the second main idea.

Idea 2 *Since one has great freedom in selecting* $\mathbf{f}(.)$ *given* $\mathbf{g}(.)$, *let us try to choose an* $\mathbf{f}(.)$ *whose Lipschitz constant L is as small as possible so that if we apply the Picard algorithm to the equation* $\mathbf{x} = \mathbf{f}(\mathbf{x})$, *it will converge rapidly.*

We now utilize this second idea in our selection of $\mathbf{f}(.)$. The main train of development proceeds as follows. In view of Eq. (15), our equation is

$$\mathbf{x} = \mathbf{f}(\mathbf{x}) = \mathbf{x} + \mathbf{A}(\mathbf{x})\mathbf{g}(\mathbf{x}) \tag{17}$$

If we apply the Picard algorithm (16) to Eq. (17), we obtain

$$\mathbf{x}_{k+1} = \mathbf{x}_k + \mathbf{A}(\mathbf{x}_k)\mathbf{g}(\mathbf{x}_k) \tag{18}$$

Let us suppose that the solution vector $\mathbf{x}^*$ is in some subset Ω of R_n. We have indicated in the previous section (see Theorem 3) that we can define

$$L = \max_{\mathbf{x} \in \Omega} \left\| \frac{\partial \mathbf{f}}{\partial \mathbf{x}} \right\| \tag{19}$$

where $\partial \mathbf{f}/\partial \mathbf{x}$ is the $n \times n$ Jacobian matrix of $\mathbf{f}(.)$. Evaluation of the Jacobian matrix $\partial \mathbf{f}/\partial \mathbf{x}$ requires care.

We have indicated that $\mathbf{A}(\mathbf{x})$ is an $n \times n$ matrix whose elements $a_{ik}(\mathbf{x})$ are scalar-valued functions of the vector $\mathbf{x}$. We define the column n-vector $\mathbf{a}_k(\mathbf{x})$ to be the vector whose components are the elements of the kth column of the matrix $\mathbf{A}(\mathbf{x})$. Thus

$$\mathbf{a}_k(\mathbf{x}) \triangleq \begin{bmatrix} a_{1k}(\mathbf{x}) \\ a_{2k}(\mathbf{x}) \\ \vdots \\ a_{nk}(\mathbf{x}) \end{bmatrix} \tag{20}$$

so that the matrix $\mathbf{A}(\mathbf{x})$ looks like

$$\mathbf{A}(\mathbf{x}) = \begin{bmatrix} \uparrow & \uparrow & & \uparrow \\ \mathbf{a}_1(\mathbf{x}) & \mathbf{a}_2(\mathbf{x}) & \cdots & \mathbf{a}_n(\mathbf{x}) \\ \downarrow & \downarrow & & \downarrow \end{bmatrix} \tag{21}$$

Now let us consider the vector equation

$$\mathbf{f}(\mathbf{x}) = \mathbf{x} + \mathbf{A}(\mathbf{x})\mathbf{g}(\mathbf{x}) \tag{22}$$

Let us write this vector equation in terms of components; for $i = 1, 2, \ldots, n$

$$f_i(\mathbf{x}) = x_i + \sum_{k=1}^{n} a_{ik}(\mathbf{x})g_k(\mathbf{x}) \tag{23}$$

But

$$\frac{\partial f_i(\mathbf{x})}{\partial x_j} = \frac{\partial x_i}{\partial x_j} + \sum_{k=1}^{n} a_{ik}(\mathbf{x})\frac{\partial g_k(\mathbf{x})}{\partial x_j} + \frac{\partial a_{i1}(\mathbf{x})}{\partial x_j}g_1(\mathbf{x})$$

$$+ \frac{\partial a_{i2}(\mathbf{x})}{\partial x_j}g_2(\mathbf{x}) + \cdots + \frac{\partial a_{in}(\mathbf{x})}{\partial x_j}g_n(\mathbf{x}) \tag{24}$$

But

$$\frac{\partial f_i(\mathbf{x})}{\partial x_j} = ij\text{th element of Jacobian matrix } \frac{\partial \mathbf{f}}{\partial \mathbf{x}}$$

$$\frac{\partial x_i}{\partial x_j} = \left\{ \begin{array}{ll} 1 & \text{if } i = j \\ 0 & \text{if } i \neq j \end{array} \right\} = ij\text{th element of identity matrix } \mathbf{I}$$

$$\frac{\partial g_k(\mathbf{x})}{\partial x_j} = kj\text{th element of Jacobian matrix } \frac{\partial \mathbf{g}}{\partial \mathbf{x}}$$

$$\sum_{k=1}^{n} a_{ik}(\mathbf{x})\frac{\partial g_k(\mathbf{x})}{\partial x_j} = ij\text{th element of product matrix } \mathbf{A}(\mathbf{x})\frac{\partial \mathbf{g}(\mathbf{x})}{\partial(\mathbf{x})}$$

$$\frac{\partial a_{i1}(\mathbf{x})}{\partial x_j} = ij\text{th element of Jacobian matrix } \frac{\partial \mathbf{a}_1(\mathbf{x})}{\partial \mathbf{x}}$$

$$\vdots$$

$$\frac{\partial a_{in}(\mathbf{x})}{\partial x_j} = ij\text{th element of Jacobian matrix } \frac{\partial \mathbf{a}_n(\mathbf{x})}{\partial \mathbf{x}}$$

In this manner, Eq. (24) can be written in matrix form as

$$\frac{\partial \mathbf{f}(\mathbf{x})}{\partial \mathbf{x}} = \mathbf{I} + \mathbf{A}(\mathbf{x})\frac{\partial \mathbf{g}(\mathbf{x})}{\partial \mathbf{x}} + \sum_{k=1}^{n} \frac{\partial \mathbf{a}_k(\mathbf{x})}{\partial \mathbf{x}}g_k(\mathbf{x}) \tag{25}$$

Our objective is to choose $\mathbf{f}(.)$ such that the Lipschitz constant L as defined by (16), is as small as possible. Since $0 \leq L < 1$, the smallest value of L possible is $L = 0$. We cannot, however, pick $\mathbf{f}(.)$ so that $L = 0$ is true for all $\mathbf{x} \in \Omega$. However, it is possible to pick $\mathbf{A}(\mathbf{x})$ and hence $\mathbf{f}(.)$ so that $L = 0$ when $\mathbf{x} = \mathbf{x}^*$ and have L very small in the immediate neighborhood of $\mathbf{x}^*$. To see how this is possible, let us *select $\mathbf{A}(\mathbf{x})$ to be the inverse of the Jacobian matrix* $\partial \mathbf{g}(\mathbf{x})/\partial \mathbf{x}$, that is,

$$\mathbf{A}(\mathbf{x}) \triangleq -\left[\frac{\partial \mathbf{g}(\mathbf{x})}{\partial \mathbf{x}}\right]^{-1} \tag{26}$$

If we substitute (26) into (25), we see that

$$\frac{\partial \mathbf{f}}{\partial \mathbf{x}} = \mathbf{I} - \left[\frac{\partial \mathbf{g}(\mathbf{x})}{\partial \mathbf{x}}\right]^{-1}\frac{\partial \mathbf{g}(\mathbf{x})}{\partial \mathbf{x}} + \sum_{k=1}^{n} \frac{\partial \mathbf{a}_k(\mathbf{x})}{\partial \mathbf{x}} g_k(\mathbf{x}) \tag{27}$$

or

$$\frac{\partial \mathbf{f}}{\partial \mathbf{x}} = \sum_{k=1}^{n} \frac{\partial \mathbf{a}_k(\mathbf{x})}{\partial \mathbf{x}} g_k(\mathbf{x}) \tag{28}$$

If we evaluate (28) at the solution vector $\mathbf{x} = \mathbf{x}^*$, then [since, by definition, $g_k(\mathbf{x}^*) = 0$] we have

$$\left.\frac{\partial \mathbf{f}(\mathbf{x})}{\partial \mathbf{x}}\right|_{\mathbf{x}=\mathbf{x}^*} = \mathbf{0} \tag{29}$$

If in addition we assume that $g_k(\mathbf{x})$ is continuous, then for $\mathbf{x}$ near $\mathbf{x}^*$, $g_k(\mathbf{x})$ will be very small provided that the Jacobian matrices $\partial \mathbf{a}_k(\mathbf{x})/\partial \mathbf{x}$ are also continuous in $\mathbf{x}$ near $\mathbf{x}^*$.

This chain of arguments leads us to select $\mathbf{A}(\mathbf{x})$ by means of Eq. (26). In this case (22) becomes

$$\mathbf{f}(\mathbf{x}) = \mathbf{x} - \left[\frac{\partial \mathbf{g}(\mathbf{x})}{\partial \mathbf{x}}\right]^{-1}\mathbf{g}(\mathbf{x}) \tag{30}$$

and the Picard algorithm (16) applied to (30) yields

$$\mathbf{x}_{k+1} = \mathbf{x}_k - \left[\left.\frac{\partial \mathbf{g}(\mathbf{x})}{\partial \mathbf{x}}\right|_{\mathbf{x}=\mathbf{x}_k}\right]^{-1}\mathbf{g}(\mathbf{x}_k) \qquad k = 0, 1, 2, \ldots \tag{31}$$

which is precisely the Newton algorithm in terms of the original function $\mathbf{g}(\mathbf{x})$.

3.6.4 Formal Statement

In deriving the Newton algorithm we have implicitly made many assumptions which we have glossed over. The following theorem summarizes all the assumptions which guarantee the convergence of the Newton algorithm.

> **Theorem 1 (Newton's method)** *Consider the vector equation*
>
> $$\mathbf{g}(\mathbf{x}) = \mathbf{0} \tag{32}$$
>
> *Suppose that*
>
> (*a*) *A solution vector to* (32) $\mathbf{x}^*$ *exists and is unique in some subset Ω of R_n.*
> (*b*) $\mathbf{g}(\mathbf{x})$ *is continuous for all* $\mathbf{x} \in \Omega$.
> (*c*) *The Jacobian matrix $\partial \mathbf{g} / \partial \mathbf{x}$ exists and is continuous in $\mathbf{x}$ for all* $\mathbf{x} \in \Omega$, *and its inverse matrix*
>
> $$-\left[\frac{\partial \mathbf{g}(\mathbf{x})}{\partial \mathbf{x}}\right]^{-1} \triangleq \mathbf{A}(\mathbf{x}) \tag{33}$$
>
> *exists for all* $\mathbf{x} \in \Omega$, *and that the elements of $\mathbf{A}(\mathbf{x})$ are differentiable with respect to $\mathbf{x}$, for all* $\mathbf{x} \in \Omega$.
>
> *Then the Newton algorithm*
>
> $$\mathbf{x}_{k+1} = \mathbf{x}_k - \left[\frac{\partial \mathbf{g}(\mathbf{x})}{\partial \mathbf{x}}\bigg|_{\mathbf{x}=\mathbf{x}_k}\right]^{-1} \mathbf{g}(\mathbf{x}_k) \tag{34}$$
>
> *where* $k = 0, 1, 2, \ldots$, *converges to $\mathbf{x}^*$ that is*
>
> $$\lim_{k \to \infty} \mathbf{x}_k = \mathbf{x}^* \tag{35}$$
>
> *provided that*
>
> $$\left\|\sum_{j=1}^{n} \frac{\partial \mathbf{a}_j(\mathbf{x})}{\partial \mathbf{x}} g_j(\mathbf{x})\right\| < 1 \qquad \text{for all } \mathbf{x} \in \Omega \tag{36}$$
>
> *where* $\mathbf{a}_j(\mathbf{x})$, $j = 1, 2, \ldots, n$, *are the column vectors of $\mathbf{A}(\mathbf{x})$ defined by Eq.* (33).

ELEMENTS OF PROOF   Instead of repeating the entire chain of arguments we indicate the salient points. The convergence of Newton's method hinges on the convergence of the associated Picard algorithm, $\mathbf{x}_{k+1} = \mathbf{f}(\mathbf{x}_k)$, where

$$\mathbf{g}(\mathbf{x}) \triangleq \mathbf{x} + \mathbf{A}(\mathbf{x})\mathbf{g}(\mathbf{x})$$

Theorem 3 of Sec. 3.5 contains sufficient conditions for the convergence of the Picard algorithm. Assumptions (*a*) to (*c*) of this theorem guarantee the continuity of $\mathbf{f}(\mathbf{x})$. Equation (36) guarantees that the Lipschitz constant L associated with the Picard method is less than unity, because expressions (36) and (28) imply that $\|\partial \mathbf{f}(\mathbf{x})/\partial \mathbf{x}\| < 1$ for all $\mathbf{x} \in \Omega$. Hence, the convergence of the Newton algorithm is a consequence of the convergence of the associated Picard algorithm.

3.6.5   Some Remarks on the Newton Algorithm

First we outline the steps one must follow to apply the Newton algorithm. We assume that $\mathbf{g}(\mathbf{x})$ is known analytically.

STEP 1   Calculate analytically the Jacobian matrix $\partial \mathbf{g}/\partial \mathbf{x}$, which is an $n \times n$ matrix.

STEP 2   Suppose that $\mathbf{x}_k$ is the kth guess vector; the computer calculates the column n-vector $\mathbf{g}(\mathbf{x}_k)$.

STEP 3   The computer evaluates the Jacobian matrix $\partial \mathbf{g}/\partial \mathbf{x}$ at $\mathbf{x} = \mathbf{x}_k$.

STEP 4   The computer computes the inverse of the matrix $\partial \mathbf{g}/\partial \mathbf{x}|_{\mathbf{x}_k}$.

STEP 5   The next guess, $\mathbf{x}_{k+1}$, is evaluated by the algorithm

$$\mathbf{x}_{k+1} = \mathbf{x}_k - \left(\frac{\partial \mathbf{g}}{\partial \mathbf{x}}\bigg|_{\mathbf{x}_k}\right)^{-1} \mathbf{g}(\mathbf{x}_k)$$

It is evident that the Newton algorithm requires more programming effort and more computation time at each step than the Picard algorithm; a significant portion of the time at each iteration is used to invert the $n \times n$ Jacobian matrix.

The Newton algorithm solves linear equations in a single step while the Picard algorithm does not. To illustrate this let us consider the linear equation

$$\mathbf{A}\mathbf{x} - \mathbf{b} = \mathbf{0} \qquad \begin{array}{c} \mathbf{x} \in R_n \\ \mathbf{b} \in R_n \end{array} \tag{37}$$

and assume that the $n \times n$ matrix $\mathbf{A}$ is nonsingular. Clearly the solution is

$$\mathbf{x}^* = \mathbf{A}^{-1}\mathbf{b} \tag{38}$$

Now let us apply the Newton algorithm to this equation; clearly

$$\mathbf{g}(\mathbf{x}) = \mathbf{A}\mathbf{x} - \mathbf{b} \tag{39}$$

The Jacobian matrix is

$$\frac{\partial \mathbf{g}}{\partial \mathbf{x}} = \mathbf{A} \tag{40}$$

Let $\mathbf{x}_0$ be an arbitrary first guess. Then

$$\mathbf{x}_1 = \mathbf{x}_0 - \mathbf{A}^{-1}(\mathbf{A}\mathbf{x}_0 - \mathbf{b}) \tag{41}$$

which reduces to

$$\mathbf{x}_1 = \mathbf{x}_0 - \mathbf{x}_0 + \mathbf{A}^{-1}\mathbf{b} = \mathbf{A}^{-1}\mathbf{b} = \mathbf{x}^* \tag{42}$$

so that the solution is obtained in a single step.

In general, the more "linear" the functions $g_i(\mathbf{x})$ the faster the convergence of the Newton method.

In general, it is extremely difficult to verify the sufficient conditions which guarantee the convergence of Newton's method at the start. The reason is that the condition (36) is extremely hard to evaluate before execution of the algorithm. Thus, in practice, one is never quite sure whether the algorithm will converge. One can be sure, however, that *the algorithm will converge if the initial guess is near enough to the solution*, i.e., if the norm $\|\mathbf{x}_0 - \mathbf{x}^*\|$ is small; the trouble is that it is difficult to say how small "small" is. The convergence in such cases is assured by the continuity properties of $\mathbf{g}(\mathbf{x})$. Since $\mathbf{g}(\mathbf{x}^*) = \mathbf{0}$ and $\mathbf{g}(\mathbf{x})$ is continuous, then (by the very definition of continuity) for each $\epsilon > 0$ there exists a $\delta > 0$ such that if $\|\mathbf{x} - \mathbf{x}^*\| < \delta$, then $\|\mathbf{g}(\mathbf{x}) - \mathbf{g}(\mathbf{x}^*)\| < \epsilon$; since $\mathbf{g}(\mathbf{x}^*) = \mathbf{0}$, we can see that $\|\mathbf{g}(\mathbf{x})\| < \epsilon$. This implies that $|g_j(\mathbf{x})| < \epsilon_j$, where $g_j(\mathbf{x})$ are the components of the vector $\mathbf{g}(\mathbf{x})$. From Eq. (36) we can see that

$$\left\| \sum_{j=1}^{n} \frac{\partial \mathbf{a}_j(\mathbf{x})}{\partial \mathbf{x}} g_j(\mathbf{x}) \right\| \leq \sum_{j=1}^{n} \left\| \frac{\partial \mathbf{a}_j(\mathbf{x})}{\partial \mathbf{x}} \right\| \cdot |g_j(\mathbf{x})|$$

$$\leq \sum_{j=1}^{n} \left\| \frac{\partial \mathbf{a}_j(\mathbf{x})}{\partial \mathbf{x}} \right\| \epsilon_j < 1 \tag{43}$$

because the ϵ_j can be selected to be very small.

The fact that the convergence of the Newton algorithm is always guaranteed near the solution can be used to justify using it anyway. In practice, one tries it and monitors the norm $\|\mathbf{x}_{k+1} - \mathbf{x}_k\|$; if it does not get small, one stops and tries another initial guess $\mathbf{x}_0$ until, hopefully, it converges. Nonetheless, any information (from the physics of each particular problem) about the range in which $\mathbf{x}^*$ lies is of paramount importance in such iterative algorithms.

3.6.6 The Modified Newton Algorithm

We have indicated that a significant amount of computation time at each step is used to determine the inverse of the Jacobian matrix $\partial\mathbf{g}/\partial\mathbf{x}|_{\mathbf{x}_k}$. The modified Newton algorithm saves time by computing at the first guess $\mathbf{x}_0$ the inverse of the Jacobian matrix $\partial\mathbf{g}/\partial\mathbf{x}|_{\mathbf{x}_0}$ and using the same inverse at each subsequent iteration. Thus, the modified Newton algorithm is

$$\mathbf{x}_{k+1} = \mathbf{x}_k - \left(\frac{\partial\mathbf{g}}{\partial\mathbf{x}}\bigg|_{\mathbf{x}_0}\right)^{-1}\mathbf{g}(\mathbf{x}_k) \qquad k = 0, 1, 2, \ldots \tag{44}$$

We shall leave it to the reader to verify that the modified Newton algorithm will converge if $\|\mathbf{x}_0 - \mathbf{x}^*\|$ is small enough.

EXAMPLE 1 To illustrate some of the numerical aspects of Newton's method we shall use it to solve a set of two simultaneous algebraic equations:

$$g_1(x_1, x_2) = x_1^3 x_2 + 3x_1^2 x_2^2 + 3x_1 x_2^3 + x_2^4 + x_2 - 1 = 0$$

$$g_2(x_1, x_2) = x_1 + x_2^4 - 4x_2^3 + 8x_2^2 - 7x_2 + 3 = 0$$

The reader can verify that the solution is given by

$$x_1^* = -1.0 \qquad x_2^* = 1.0$$

Before implementing the Newton algorithm we must determine the Jacobian matrix. For this purpose we evaluate the partial derivatives

$$\frac{\partial g_1}{\partial x_1} = 3x_1^2 x_2 + 6x_1 x_2^2 + 3x_2^3$$

$$\frac{\partial g_1}{\partial x_2} = x_1^3 + 6x_1^2 x_2 + 9x_1 x_2^2 + 4x_2^3 + 1$$

$$\frac{\partial g_2}{\partial x_1} = 1$$

$$\frac{\partial g_2}{\partial x_2} = 4x_2^3 - 12x_2^2 + 16x_2 - 7$$

Since the Jacobian matrix $\partial\mathbf{g}/\partial\mathbf{x}$ is the 2×2 matrix

$$\frac{\partial\mathbf{g}}{\partial\mathbf{x}} \triangleq \begin{bmatrix} \dfrac{\partial g_1}{\partial x_1} & \dfrac{\partial g_1}{\partial x_2} \\[2mm] \dfrac{\partial g_2}{\partial x_1} & \dfrac{\partial g_2}{\partial x_2} \end{bmatrix}$$

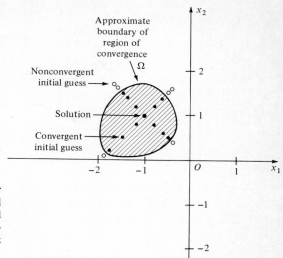

FIGURE 3.6.4
Approximate region of convergence for
Example 1. The dots represent initial
guesses from which Newton's method
converged. The O's represent initial guesses for which Newton's method did not
converge.

its inverse is given by

$$\left(\frac{\partial \mathbf{g}}{\partial \mathbf{x}}\right)^{-1} = \frac{1}{\dfrac{\partial g_1}{\partial x_1}\dfrac{\partial g_2}{\partial x_2} - \dfrac{\partial g_1}{\partial x_2}\dfrac{\partial g_2}{\partial x_1}} \begin{bmatrix} \dfrac{\partial g_2}{\partial x_2} & -\dfrac{\partial g_1}{\partial x_2} \\ -\dfrac{\partial g_2}{\partial x_1} & \dfrac{\partial g_1}{\partial x_1} \end{bmatrix}$$

The Newton algorithm is

$$\begin{bmatrix} x_1^{k+1} \\ x_2^{k+1} \end{bmatrix} = \begin{bmatrix} x_1^{k} \\ x_2^{k} \end{bmatrix} - \left(\frac{\partial \mathbf{g}}{\partial \mathbf{x}}\bigg|_{\mathbf{x}_k}\right)^{-1} \begin{bmatrix} g_1(x_1^{k}, x_2^{k}) \\ g_2(x_1^{k}, x_2^{k}) \end{bmatrix}$$

These equations were simulated on a digital computer. Our objectives were

1 To obtain numerically an estimate of the region of convergence
2 To illustrate that inside the region of convergence the convergence is very
rapid
3 To illustrate the instability outside the region of convergence

An estimate of the region of convergence† Ω is shown in Fig. 3.6.4. If the initial
guess was inside Ω, convergence to the solution occurred. If the initial guess was
outside Ω, convergence was not guaranteed.

† The region of convergence was estimated by trial-and-error numerical computation.

Run 1   CONVERGES

Guess		x_1	x_2
Initial	$(k = 0)$	−1.50000	1.50000
First	$(k = 1)$	−0.31250	1.00000
Second	$(k = 2)$	−1.49051	1.49051
Third	$(k = 3)$	−0.34510	0.99999
Fourth	$(k = 4)$	−1.43855	1.43856
Fifth	$(k = 5)$	−0.50435	1.00000
Sixth	$(k = 6)$	−1.21709	1.21709
Seventh	$(k = 7)$	−0.89908	1.00000
Eighth	$(k = 8)$	−1.00205	1.00205
Ninth	$(k = 9)$	−0.99999	1.00000

We shall now present some typical runs.

In run 1 the method had trouble settling down, and it oscillated until the fifth step. From then on, it zoomed to the solution. This is an indication that the initial guess was near the boundary of Ω (see Fig. 3.6.4). To see this let us examine what happens when the initial guess is placed farther out from the solution.

In run 2, although the method often found the correct value of x_2 ($= 1.0$), it could not determine the correct value of x_1.

Run 2   DIVERGES

Guess		x_1	x_2
Initial	$(k = 0)$	−1.70000	1.70000
First	$(k = 1)$	0.70051	0.99996
Second	$(k = 2)$	−2.66122	2.66150
Third	$(k = 3)$	27.36043	1.00089
Fourth	$(k = 4)$	−3.01039	3.00332
Fifth	$(k = 5)$	56.31958	0.97635
Sixth	$(k = 6)$	−2.73682	2.91968
Seventh	$(k = 7)$	−7.37419	2.47354
Eighth	$(k = 8)$	−5.64575	2.17469

When the initial guess is closer to the solution, there is fast convergence to the solution; this is illustrated in run 3.

Run 3   CONVERGES FAST

Guess		x_1	x_2
Initial	$(k = 0)$	−1.20000	0.80000
First	$(k = 1)$	−0.89885	0.90270
Second	$(k = 2)$	−0.98080	1.00000
Third	$(k = 3)$	−1.00002	1.00001

Sometimes an initial guess which is very far converges by a lucky accident. This is illustrated by run 4.

Run 4 CONVERGES BY LUCK

Guess		x_1	x_2
Initial	$(k = 0)$	1,000.00000	1,000.00000
First	$(k = 1)$	749.58350	750.24976
. . . .		. . .	. . .
Fifth	$(k = 5)$	236.03714	238.06558
. . . .		. . .	. . .
Tenth	$(k = 10)$	54.75928	57.24953
. . . .		. . .	. . .
Fifteenth	$(k = 15)$	8.48112	10.96630
. . . .		. . .	. . .
Twenty-third	$(k = 23)$	0.70513	1.16612
Twenty-fourth	$(k = 24)$	31.51408	-18.28690
. . . .		. . .	. . .
Thirty-fourth	$(k = 34)$	2.44237	0.60983
Thirth-fifth	$(k = 35)$	-0.39734	1.28648
Thirty-sixth	$(k = 36)$	0.74290	0.30420
Thirty-seventh	$(k = 37)$	-0.21214	0.71776
Thirty-eighth	$(k = 38)$	-0.81546	1.02827
Thirty-ninth	$(k = 39)$	-1.02066	1.01999
Fortieth	$(k = 40)$	-1.00002	1.00000

In this run the errors kept decreasing until $k = 23$. Then the iteration overshot its mark at $k = 24$. These errors were decreased, and the iterations gave numbers near the boundary of Ω ($k = 35$ and $k = 37$). At $k = 38$ the point was inside Ω, and convergence from then on was rapid.

EXAMPLE 2   To reinforce some of the conclusions that one can reach about Newton's method we consider another numerical example. The equations to be solved are

$$g_1(x_1, x_2) = x_1 + x_2 = 0 \quad \text{and} \quad g_2(x_1, x_2) = x_1^2 - (x_2 - 1)^2 = 0$$

The solution to these equations is

$$x_1^* = -0.5 \qquad x_2^* = 0.5$$

The Jacobian matrix is

$$\frac{\partial \mathbf{g}}{\partial \mathbf{x}} = \begin{bmatrix} 1 & 1 \\ 2x_1 & -2(x_2 - 1) \end{bmatrix}$$

so that

$$\left(\frac{\partial \mathbf{g}}{\partial \mathbf{x}}\right)^{-1} = \frac{1}{-2(x_1 + x_2 - 1)} \begin{bmatrix} -2(x_2 - 1) & -1 \\ -2x_1 & 1 \end{bmatrix}$$

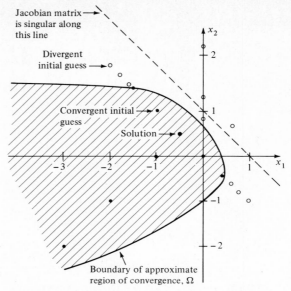

FIGURE 3.6.5

Approximate region of convergence near the solution for Example 2.

The Newton algorithm is

$$\begin{bmatrix} x_1^{k+1} \\ x_2^{k+1} \end{bmatrix} = \begin{bmatrix} x_1^{k} \\ x_2^{k} \end{bmatrix} - \left(\frac{\partial \mathbf{g}}{\partial \mathbf{x}} \Big|_{\mathbf{x}_k} \right)^{-1} \begin{bmatrix} g_1(x_1^{k}, x_2^{k}) \\ g_2(x_1^{k}, x_2^{k}) \end{bmatrix}$$

As in Example 1, an estimate of the approximate region of convergence Ω was obtained by numerical means; the set Ω is illustrated in Fig. 3.6.5 for points near the solution and in Fig. 3.6.6 for points far from the solution. This illustrates that the region of convergence is by no means symmetric about the solution. The convergent and divergent aspects of Newton's method can be illustrated by reproducing some typical computer runs for this example.

The oscillatory nature of the iterates in run 1 suggests that the initial guess is near the boundary of Ω. This is true in fact because if we increase the initial error a bit, the algorithm diverges, as illustrated in run 2.

Run 1   CONVERGENT; NEAR SOLUTION

Guess		x_1	x_2
Initial	$(k = 0)$	−1.55000	1.55000
First	$(k = 1)$	−0.50000	1.05125
Second	$(k = 2)$	−0.16142	0.70506
Third	$(k = 3)$	−0.57953	0.27300
Fourth	$(k = 4)$	−0.48270	0.46357
Fifth	$(k = 5)$	−0.49950	0.49950

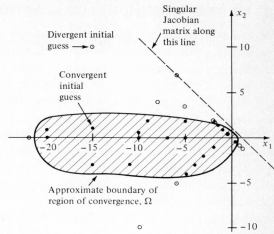

FIGURE 3.6.6
Approximate convergence region for Example 2.

Note that runs 1 and 2 give almost the same answer for $k = 1$. Nevertheless run 2 diverges because for $k = 3$ the guess is to the right of the line of singularity for the Jacobian matrix and very fast divergence follows.

Run 3 shows how fast a convergent Newton algorithm zooms to the solution. Initially the error in x_1 is cut by about one-half at each iteration; then the convergence becomes more rapid.

Run 2   DIVERGENT; NEAR SOLUTION

Guess		x_1	x_2
Initial	$(k = 0)$	−1.60000	1.60000
First	$(k = 1)$	−0.50000	1.10500
Second	$(k = 2)$	−0.03668	1.01803
Third	$(k = 3)$	0.93934	75.94717

Run 3   CONVERGENT; FAR FROM SOLUTION

Guess		x_1	x_2
Initial	$(k = 0)$	−20.00000	0.00000
First	$(k = 1)$	−9.54762	2.19421
Second	$(k = 2)$	−5.22795	2.54228
Third	$(k = 3)$	−2.96667	2.01260
Fourth	$(k = 4)$	−1.47143	1.31346
Fifth	$(k = 5)$	−0.62175	0.81757
Sixth	$(k = 6)$	−0.44651	0.50828
Seventh	$(k = 7)$	−0.50149	0.49948
Eighth	$(k = 8)$	−0.50000	0.50000

3.7 ITERATIVE METHODS FOR FUNCTION MINIMIZATION

We remarked in Sec. 3.2 that the unconstrained optimization problem (finding the minimum or maximum of a scalar-valued function of many variables) involves determining all solutions to a vector equation, since these correspond to the stationary points of the function (maxima, minima, saddle points, etc.). Hence, the techniques of the preceding two sections clearly can be applied to this class of problems.

On the other hand, optimization problems tend to be more structured than problems in which we simply seek the solutions to a set of equations. For this reason, several extremely powerful iterative techniques are available.

Because optimization problems are common in diverse engineering and socioeconomic applications, we present a collection of the most popular available methods. We shall illustrate the basic idea of these methods without discussing how they are derived or their convergence properties. The interested reader can consult the references for more details and examples.

3.7.1 The Basic Problem

For the sake of completeness we shall now state the basic optimization problem. We formulate it as a minimization problem, remembering that the problem of finding the maximum of a function is equivalent to finding the minimum of the negative of that function.

The basic problem is as follows. Given the scalar-valued function

$$h(\mathbf{x}) = h(x_1, x_2, \dots, x_n) \tag{1}$$

find a vector $\mathbf{x}^* \in R_n$ (if it exists) such that

$$h(\mathbf{x}^*) \leq h(\mathbf{x}) \qquad \text{for all } \mathbf{x} \tag{2}$$

3.7.2 Function Minimization by Newton's Method

It is a straightforward task to adapt Newton's method to the basic minimization problem. From Sec. 3.2 we know that the minimum x* is characterized by the property

$$\left. \frac{\partial h}{\partial \mathbf{x}} \right|_{\mathbf{x}=\mathbf{x}^*} = \mathbf{0} \tag{3}$$

Hence, we seek the solution of the vector equation

$$\frac{\partial h}{\partial \mathbf{x}} = \mathbf{0} \tag{4}$$

Applying Newton's method to Eq. (4), we obtain the algorithm for $k = 0, 1, 2, \ldots$

$$\mathbf{x}_{k+1} = \mathbf{x}_k - \left(\frac{\partial^2 h}{\partial \mathbf{x}^2} \Bigg|_{\mathbf{x}=\mathbf{x}_k} \right)^{-1} \frac{\partial h}{\partial \mathbf{x}} \Bigg|_{\mathbf{x}=\mathbf{x}_k} \tag{5}$$

Note that in the algorithm we must analytically compute the gradient vector

$$\frac{\partial h}{\partial \mathbf{x}} = \begin{bmatrix} \dfrac{\partial h}{\partial x_1} \\[6pt] \dfrac{\partial h}{\partial x_2} \\[6pt] \vdots \\[6pt] \dfrac{\partial h}{\partial x_n} \end{bmatrix} \tag{6}$$

and evaluate it at each step at the current value of $\mathbf{x}_k$. Also we must compute the $n \times n$ second-derivative matrix (symmetric)

$$\frac{\partial^2 h}{\partial \mathbf{x}^2} = \begin{bmatrix} \dfrac{\partial^2 h}{\partial x_1{}^2} & \dfrac{\partial^2 h}{\partial x_1\, \partial x_2} & \cdots & \dfrac{\partial^2 h}{\partial x_1\, \partial x_n} \\[10pt] \dfrac{\partial^2 h}{\partial x_1\, \partial x_2} & \dfrac{\partial^2 h}{\partial x_2{}^2} & \cdots & \dfrac{\partial^2 h}{\partial x_2\, \partial x_n} \\[10pt] \vdots & \vdots & \vdots & \vdots \\[10pt] \dfrac{\partial^2 h}{\partial x_1\, \partial x_n} & \dfrac{\partial^2 h}{\partial x_2\, \partial x_n} & \cdots & \dfrac{\partial^2 h}{\partial x_n{}^2} \end{bmatrix} \tag{7}$$

evaluate its elements at the current guess $\mathbf{x}_k$, and then invert it.

It is the inversion of this second-derivative matrix that makes Newton's method expensive to use when the number n of variables is large. In many applications, e.g., optimization of power flow in electric-power systems or economic systems, the number of variables can easily exceed 500. Inversion of 500×500 matrices in real time is not easy.

When we apply Newton's method to minimization problems, we have no guarantee of convergence. Although there are sufficient conditions for the convergence of Newton's method, they are difficult to check and often overconservative. On the other hand, for the same reasons covered in Sec. 3.6, Newton's method will converge if the initial guess x_0 happens to be near the solution x^*. When Newton's method converges, it really zooms to the solution. Usually, only four or five iterations are necessary to achieve excellent accuracy.

For the type of problem that Newton's method works well, the rule of thumb is the more quadratic the function $h(x)$ the faster the convergence. Indeed, if $h(x)$ is quadratic, Newton's method converges in a single step. To illustrate this, consider the quadratic function

$$h(\mathbf{x}) = \tfrac{1}{2}\mathbf{x}'\mathbf{A}\mathbf{x} + \mathbf{b}'\mathbf{x} + c \qquad \mathbf{x} \in R_n \tag{8}$$

where $\mathbf{A} = n \times n$ positive definite matrix
$\quad \mathbf{b} =$ arbitrary n vector
$\quad c =$ arbitrary scalar

We compute

$$\frac{\partial h}{\partial \mathbf{x}} = \mathbf{A}\mathbf{x} + \mathbf{b} \tag{9}$$

$$\frac{\partial^2 h}{\partial \mathbf{x}^2} = \mathbf{A} \tag{10}$$

Let $\mathbf{x}_0 \in R_n$ be an arbitrary initial guess. Then for $k = 1$, the Newton algorithm (5) yields

$$\mathbf{x}_1 = \mathbf{x}_0 - \mathbf{A}^{-1}(\mathbf{A}\mathbf{x}_0 + \mathbf{b}) = -\mathbf{A}^{-1}\mathbf{b} \tag{11}$$

For $k = 2$ we have

$$\mathbf{x}_2 = \mathbf{x}_1 - \mathbf{A}^{-1}(\mathbf{A}\mathbf{x}_1 + \mathbf{b}) = -\mathbf{A}^{-1}\mathbf{b} - \mathbf{A}^{-1}(-\mathbf{A}\mathbf{A}^{-1}\mathbf{b} + \mathbf{b}) = -\mathbf{A}^{-1}\mathbf{b} = \mathbf{x}_1 \tag{12}$$

So $\mathbf{x}_2 = \mathbf{x}_1$. Similarly $\mathbf{x}_3 = \mathbf{x}_2$, and so on, and this means that the minimum is

$$\mathbf{x}^* = \mathbf{x}_1 = -\mathbf{A}^{-1}\mathbf{b} \tag{13}$$

which is of course true, as can be verified by direct substitution.

3.7.3 Function Minimization by the Method of Steepest Descent

One of the simplest and most intuitive methods and one extremely popular for finding the minimum of a function is the so-called *method of steepest descent* or the *gradient method*. This technique has strong geometric overtones that can be exploited to increase its rate of convergence. Before we present the geometric and analytical aspects of the steepest-descent method, we shall intuitively explain it using an example.

Suppose that you have parachuted onto the side of a valley filled with fog and you wish to reach the bottom of the valley, where there is a shelter. Your problem is to try to select the direction you should walk in.† Obviously, you always want to walk downhill, but certain downhill directions are preferable to others. If you happen to have a marble and you drop it, the marble will move in the direction of *local steepest descent* (most downhill). So now you know the direction you should walk in, at least for a while. You take a few steps, stop, drop the marble again, and then change direction. You can then repeat the procedure again and again. Unless you get trapped in a local bottom with no shelter, this procedure is obviously guaranteed to bring you to the bottom of the valley.

The above procedure (and all descent algorithms) have the following two characteristics at each stage:

Question 1 What is a good direction?

Question 2 How far should I move along this good direction before I ask question 1 again?

These intuitive ideas can be formalized once one gains a certain geometric feel for analytical minimization problems. It is instructive to develop such a feel.

In Fig. 3.7.1 we plot the equal-value contours of a function $h(x_1, x_2)$ of two variables (like a topographic map). This is the type of valley that we have discussed above. The vector

$$\mathbf{x}^* = \begin{bmatrix} x_1^* \\ x_2^* \end{bmatrix}$$

is the minimum of the function, i.e., the bottom of the valley. Now suppose that at some stage k of our descent we are at the point (vector) $\mathbf{x}_k$. We seek the direction of steepest descent.

If we had a mathematical description of our valley, we could find the direction of steepest descent without dropping the marble. To do this we compute the gradient vector at $\mathbf{x}_k$

$$\text{Gradient vector at } \mathbf{x}_k \triangleq \frac{\partial h}{\partial \mathbf{x}}\bigg|_{\mathbf{x}_k}$$

† If there were no fog, you could see the shelter and head straight for it.

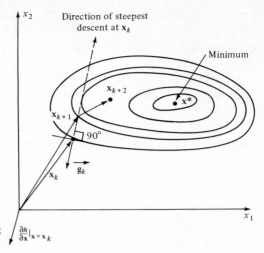

FIGURE 3.7.1
The geometric significance of the gradient vector.

and we plot it as an arrow vector in our coordinate system. Then we move its origin to the point $\mathbf{x}_k$ to obtain the situation illustrated in Fig. 3.7.1 by the vector $\mathbf{g}_k$.

The vector $\mathbf{g}_k$ has the following properties:

1 It is perpendicular to the tangent line to the contour of $h(\mathbf{x})$ at the point $\mathbf{x}_k$.
2 It points to the direction of *steepest ascent* (most uphill).
3 The direction of steepest descent at $\mathbf{x}_k$ is opposite to that of $\mathbf{g}_k$.

Thus, *the direction defined by the gradient vector $\partial h/\partial\mathbf{x}|_{\mathbf{x}=\mathbf{x}_k}$ is opposite to that of the direction of steepest descent. This is true in minimization problems independent of the dimension n of the vector* $\mathbf{x}$.

Now that we have established which way to go, it remains to quantify how far we have decided to go before we reevaluate the direction of steepest descent. Suppose that we have decided to move to the point $\mathbf{x}_{k+1}$ (see Fig. 3.7.1). How do we characterize the point $\mathbf{x}_{k+1}$? Simple vector addition yields

$$\mathbf{x}_{k+1} = \mathbf{x}_k - \alpha_k \frac{\partial h}{\partial \mathbf{x}}\bigg|_{\mathbf{x}=\mathbf{x}_k} \tag{14}$$

where the positive scalar

$$\alpha_k > 0 \tag{15}$$

is called the *step size at the kth step* since it tells us what fraction of the negative gradient vector $-\partial h/\partial \mathbf{x}|_{\mathbf{x}=\mathbf{x}_k}$ to take to arrive at the next point.

Equation (14) is the basic equation of the steepest-descent algorithm with arbitrary step size. From point $\mathbf{x}_{k+1}$ one generates $\mathbf{x}_{k+2}$ by

$$\mathbf{x}_{k+2} = \mathbf{x}_{k+1} - \alpha_{k+1}\frac{\partial h}{\partial \mathbf{x}}\bigg|_{\mathbf{x}=\mathbf{x}_{k+1}} \qquad \alpha_{k+1} > 0$$

We can now see that the steepest-descent algorithm has a certain degree of freedom, which is associated with the selection of the scalar step-size parameter at each and every step. From the geometry of the problem we can see that the speed of convergence of the algorithm will depend on the step size. Small step sizes represent very conservative strategies, and one crawls down the mountain. On the other hand, large step sizes may bounce us from one wall of the valley to the other. There should be some way of selecting the step size at each step so as to minimize (on the average) the number of steps. We remark that the minimization of the number of steps is important because it takes time to evaluate $\partial h/\partial \mathbf{x}$ at any step (sometimes in the absence of analyticity we may have to compute the gradient vector numerically).

If we return to our descent through the fog, and if we imagine that our marbles are made out of gold and get lost at each try, we can develop a good intuitive strategy that ensures our getting to the bottom with a minimum of gold marbles lost. The basic idea is as follows:

> Once the direction of steepest descent has been found at the kth step, continue along that direction as long as we go downhill; step when the next step will be uphill, and reevaluate the direction of the steepest descent.

Figure 3.7.2 illustrates the geometric significance of this policy. For any starting point $\mathbf{x}_k$ the next point is at a tangency of the direction of the steepest descent from $\mathbf{x}_k$ with a contour of $h(\mathbf{x})$. Such a method of deducing the step size is often called the *steepest-descent method with optimum step size*.

To describe this method of selecting an optimum step size mathematically we start at $\mathbf{x}_k$. Let α_k be any positive scalar. Then any vector $\mathbf{x}$ along the direction of steepest descent (dotted line in Fig. 3.7.1) is characterized by

$$\mathbf{x} = \mathbf{x}_k - \alpha_k\frac{\partial h}{\partial \mathbf{x}}\bigg|_{\mathbf{x}=\mathbf{x}_k} \qquad \alpha_k > 0 \qquad (16)$$

We claim that the optimal value of α_k, denoted by α_k^*, is the one that satisfies the inequality

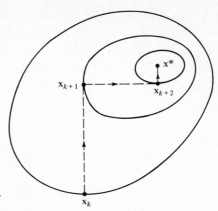

FIGURE 3.7.2
Illustration of the steepest-descent algorithm with optimum step-size control.

$$h\left(\mathbf{x}_k - \alpha_k^* \frac{\partial h}{\partial \mathbf{x}}\bigg|_{\mathbf{x}=\mathbf{x}_k}\right) \leq h\left(\mathbf{x}_k - \alpha_k \frac{\partial h}{\partial \mathbf{x}}\bigg|_{\mathbf{x}=\mathbf{x}_k}\right) \tag{17}$$

because the value of α_k^* finds the minimum of $h(\mathbf{x})$ along the direction of steepest descent from $\mathbf{x}_k$.

Once the value of α_k^* is found from (17), we have

$$\mathbf{x}_{k+1} = \mathbf{x}_k - \alpha_k^* \frac{\partial h}{\partial \mathbf{x}}\bigg|_{\mathbf{x}=\mathbf{x}_k} \tag{18}$$

The reader may now ask: What have we gained? We have simply replaced one minimization problem $[h(\mathbf{x}^*) \leq h(\mathbf{x})]$ with a sequence of others defined by (17).

The essential difference is that at each step (17) defines a minimization problem with respect to a scalar. Scalar minimization problems are easy to solve numerically, and a host of one-dimensional minimization methods can be employed, e.g., golden search or Fibonacci search, to find the optimal step size. We give such an algorithm in Sec. 3.7.6.

Steepest-descent algorithms are easy to implement on a digital computer with a minimum of programming effort. In most well-behaved problems, steepest-descent algorithms do a good job of quickly bringing us near the solution; however, they are very slow in improving the accuracy when we are near the solution.

For this reason one often designs a hybrid algorithm which consists of starting

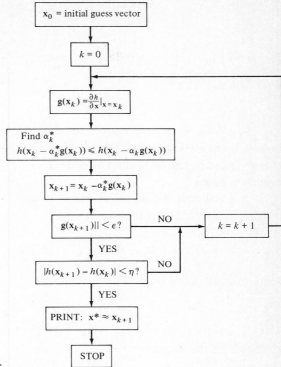

FIGURE 3.7.3
The steepest-descent algorithm.

with the steepest-descent algorithm to bring us near the solution and then switching to Newton's method to zoom in. The progress of Newton's method is monitored so that if divergence is detected, one switches back to the steepest-descent algorithm.

The flow chart of the steepest-descent algorithm is shown in Fig. 3.7.3. The flow chart is complete, including a stopping rule, except for the subroutine that performs the optimum-step-size determination. As we have remarked, this one-parameter minimization can be carried out in several ways. We present a specific method, the *quadratic interpolation*, later in this section.

In spite of the simplicity of the steepest-descent algorithm and the intuitive appeal of moving along the most downhill direction, it is obvious that this policy, even with optimal step control, will not work well in situations in which the function $h(\mathbf{x})$ looks like an elongated valley. Figure 3.7.4 shows the zigzag effect that can result. Because of this effect, many new methods have been developed; they still involve the determination of a direction at each step and an optimum step size, but the direction is *not* that of steepest descent but a modified direction.

FIGURE 3.7.4
The zigzag response of the steepest-descent algorithm.

Initial point

We present two methods, the conjugate-gradient method and the Fletcher-Powell method, without elaborating on how they are derived beyond the remark that they are based on finding a minimum of a quadratic function of n variables in exactly n steps.

A good iterative algorithm must possess a desirable property for quadratics because a Taylor series expansion of a function $h(\mathbf{x})$ about the minimum $\mathbf{x}^*$ yields

$$h(\mathbf{x}) = h(\mathbf{x}^*) + (\mathbf{x} - \mathbf{x}^*)' \frac{\partial h}{\partial \mathbf{x}}\bigg|_{\mathbf{x}=\mathbf{x}^*} + \tfrac{1}{2}(\mathbf{x} - \mathbf{x}^*)' \frac{\partial^2 h}{\partial \mathbf{x}^2}\bigg|_{\mathbf{x}=\mathbf{x}^*}(\mathbf{x} - \mathbf{x}^*) + \cdots \quad (19)$$

Since a minimum is a stationary point, then

$$\frac{\partial h}{\partial \mathbf{x}}\bigg|_{\mathbf{x}=\mathbf{x}^*} = \mathbf{0} \quad (20)$$

so that near the minimum

$$h(\mathbf{x}) \approx h(\mathbf{x}^*) + \tfrac{1}{2}(\mathbf{x} - \mathbf{x}^*)' \frac{\partial^2 h}{\partial \mathbf{x}^2}\bigg|_{\mathbf{x}=\mathbf{x}^*}(\mathbf{x} - \mathbf{x}^*) \quad (21)$$

that is, $h(\mathbf{x})$ looks like a quadratic. Hence, a good algorithm must work well for quadratics. Of course, Newton's method works the best (convergence in one step), but the matrix inversion may be costly. Hence these modified gradient algorithms are faster then the steepest-descent algorithm but not as fast as a convergent Newton's method. On the other hand, at each step they require more computation than the steepest-descent method but less than Newton's method.

⋆ 3.7.4 The Conjugate-Gradient Method

This method is best described in some detail for the first few iterations before presenting the general formula.

1 Start with initial guess vector $\mathbf{x}_0$.
2 Compute

$$\mathbf{s}_0 = -\frac{\partial h}{\partial \mathbf{x}}\bigg|_{\mathbf{x}=\mathbf{x}_0} \quad (22)$$

3 Determine the optimum step size α_0^* such that

$$h(\mathbf{x}_0 + \alpha_0^* \mathbf{s}_0) \le h(\mathbf{x}_0 + \alpha_0 \mathbf{s}_0) \tag{23}$$

4 Set

$$\mathbf{x}_1 = \mathbf{x}_0 + \alpha_0^* \mathbf{s}_0 \tag{24}$$

Up to this point this is identical to the steepest-descent method with optimum step size.

5 Compute the scalar β_0:

$$\beta_0 = \frac{\left(\left.\dfrac{\partial h}{\partial \mathbf{x}}\right|_{\mathbf{x}=\mathbf{x}_1}\right)' \left(\left.\dfrac{\partial h}{\partial \mathbf{x}}\right|_{\mathbf{x}=\mathbf{x}_1}\right)}{\left(\left.\dfrac{\partial h}{\partial \mathbf{x}}\right|_{\mathbf{x}=\mathbf{x}_0}\right)' \left(\left.\dfrac{\partial h}{\partial \mathbf{x}}\right|_{\mathbf{x}=\mathbf{x}_0}\right)} \tag{25}$$

6 Compute

$$\mathbf{s}_1 = -\left.\frac{\partial h}{\partial \mathbf{x}}\right|_{\mathbf{x}=\mathbf{x}_1} + \beta_0 \mathbf{s}_0 \tag{26}$$

7 Compute α_1^* such that

$$h(\mathbf{x}_1 + \alpha_1^* \mathbf{s}_1) \le h(\mathbf{x}_1 + \alpha_1 \mathbf{s}_1) \tag{27}$$

8 Set

$$\mathbf{x}_2 = \mathbf{x}_1 + \alpha_1^* \mathbf{s}_1 \tag{28}$$

9 Compute

$$\beta_1 = \frac{\left(\left.\dfrac{\partial h}{\partial \mathbf{x}}\right|_{\mathbf{x}=\mathbf{x}_2}\right)' \left(\left.\dfrac{\partial h}{\partial \mathbf{x}}\right|_{\mathbf{x}=\mathbf{x}_2}\right)}{\left(\left.\dfrac{\partial h}{\partial \mathbf{x}}\right|_{\mathbf{x}=\mathbf{x}_1}\right)' \left(\left.\dfrac{\partial h}{\partial \mathbf{x}}\right|_{\mathbf{x}=\mathbf{x}_1}\right)} \tag{29}$$

10 Compute

$$\mathbf{s}_2 = -\left.\frac{\partial h}{\partial \mathbf{x}}\right|_{\mathbf{x}=\mathbf{x}_2} + \beta_1 \mathbf{s}_1 \tag{30}$$

and so on.

To write the general format of the algorithm it will be convenient to use the notation

$$\mathbf{g}(\mathbf{x}) = \frac{\partial h}{\partial \mathbf{x}} = \text{gradient vector of } h(\mathbf{x}) \tag{31}$$

$$\mathbf{g}(\mathbf{x}_k) = \frac{\partial h}{\partial \mathbf{x}}\bigg|_{\mathbf{x}=\mathbf{x}_k} = \text{gradient vector of } h(\mathbf{x}) \text{ evaluated at } \mathbf{x} = \mathbf{x}_k \tag{32}$$

Then the conjugate-gradient algorithm is the following

STEP 1 Select $\mathbf{x}_0$ (initial guess) arbitrary.

STEP 2 Set

$$\mathbf{s}_0 = -\mathbf{g}(\mathbf{x}_0) \tag{33}$$

STEP 3 Set $k = 0$.

STEP 4 Find α_k^* such that

$$h(\mathbf{x}_k + \alpha_k^* \mathbf{s}_k) \le h(\mathbf{x}_k + \alpha_k \mathbf{s}_k) \tag{34}$$

STEP 5 Set

$$\boxed{\mathbf{x}_{k+1} = \mathbf{x}_k + \alpha_k^* \mathbf{s}_k} \tag{35}$$

STEP 6 Compute

$$\boxed{\beta_k = \frac{\mathbf{g}'(\mathbf{x}_{k+1})\mathbf{g}(\mathbf{x}_{k+1})}{\mathbf{g}'(\mathbf{x}_k)\mathbf{g}(\mathbf{x}_k)}} \tag{36}$$

STEP 7 Set

$$\boxed{\mathbf{s}_{k+1} = -\mathbf{g}(\mathbf{x}_{k+1}) + \beta_k \mathbf{s}_k} \tag{37}$$

STEP 8 Is

$$\|g(\mathbf{x}_{k+1})\| < \epsilon \quad \text{and} \quad |h(\mathbf{x}_{k+1}) - h(\mathbf{x}_k)| < \eta \tag{38}$$

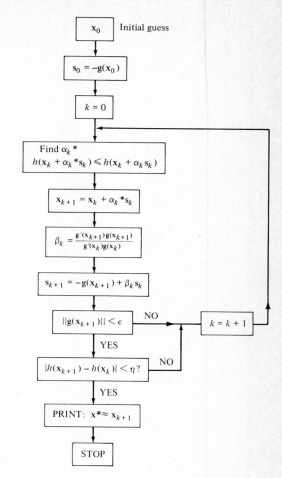

FIGURE 3.7.5
The conjugate-gradient algorithm.

where ϵ and η are tolerance parameters? If yes, stop; $\mathbf{x}^* \approx \mathbf{x}_{k+1}$. If no, set $k = k + 1$ and go to step 4.

The flow chart of this algorithm is shown in Fig. 3.7.5. More details on the derivation of the algorithm can be found in Ref. 17.

⋆ 3.7.5 The Fletcher-Powell Algorithm

The Fletcher-Powell algorithm is generally acknowledged to be one of the best algorithms around. It has been used extensively in many applications. It is characterized by having the freedom, in addition to the initial guess $\mathbf{x}_0$, of selecting

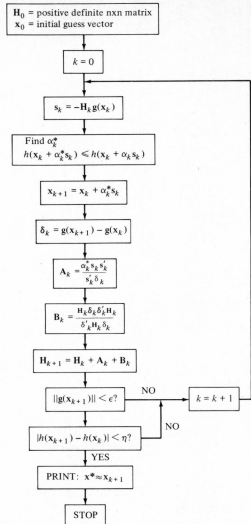

FIGURE 3.7.6
The Fletcher-Powell algorithm.

an $n \times n$ symmetric positive definite matrix $\mathbf{H}_0$.[†] Then the algorithm generates a sequence of directions $\mathbf{s}_k$ (related but not collinear with the direction of steepest descent). These directions $\mathbf{s}_k$ are based on the gradients rotated by a sequence of matrices $\mathbf{H}_k$ that remain positive definite. The flow chart of Fig. 3.7.6 summarizes this popular algorithm. The details of the algorithm are given in Ref. 16.

[†] $\mathbf{H}_0$ often is selected as the identity matrix.

3.7.6   Determination of Optimum Step Size by Quadratic Interpolation

All three direction-based algorithms, namely,

1   The steepest-descent method
2   The conjugate-gradient method
3   The Fletcher-Powell method

involve the determination of an optimum scalar step size. For those who do not feel like experimenting in devising their own, we present here a cookbook subroutine that accomplishes this task.

The basic problem that we are facing is as follows. We are given two vectors **y** and **z** in R_n, a scalar parameter α, $\alpha > 0$, and a scalar-valued function

$$h(\mathbf{y} + \alpha \mathbf{z}) \tag{39}$$

We seek the optimum value of α, denoted by α^*, such that

$$h(\mathbf{y} + \alpha^* \mathbf{z}) \le h(\mathbf{y} + \alpha \mathbf{z}) \tag{40}$$

The quadratic interpolation algorithm proceeds roughly as follows:

1   A reasonable value of α is selected so that the first step produces only a small change in the function, roughly of the order of 1 percent (this in essence normalizes the initial step).
2   The value of α found above is then changed in a doubling fashion (1, 2, 4, 8, . . .) as long as the function is decreasing up to and including the first time the function increases.
3   One fits a quadratic to the last three values of the function; its minimum is then related to the optimum step size.

The flow chart of the quadratic interpolation algorithm is shown in Fig. 3.7.7. The top loop accomplishes the normalization so that the first step yields less than 1 percent change in the value of the function $h(\mathbf{y})$. The middle part performs the search and keeps doubling the step size until the function stops decreasing. Next, the quadratic fit to the last three points is made; the value of d is the minimum of the quadratic. Finally, a decision of whether this minimum (at d) is smaller than the already computed ones is made in order to deduce the optimum step size.

One way of improving the choice of the step size α once a set of a_k, b_k, c_k has been produced, according to the flow chart of Fig. 3.7.7, is to use a direct elimination search. As implied by the name, such a scheme proceeds by eliminating intervals where a minimum cannot occur once it is known that the curve is

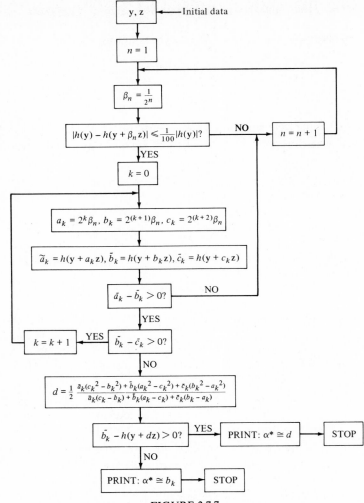

FIGURE 3.7.7
Algorithm for optimum step-size computation.

unimodal (nonincreasing on half the interval and nondecreasing on the other half). In our case, unimodality is plausible by the procedure of choosing a_k, b_k, c_k (Fig. 3.7.7); i.e., a valley is almost certain to be obtained.

While there are many direct elimination methods, with various degrees of sophistication, one possible scheme is the following version of the Fibonacci search. To implement the Fibonacci search one proceeds as follows.

STEP 1   Let a_k, b_k, c_k be defined as in Fig. 3.7.7. Define in addition

$$d_k \triangleq \tfrac{1}{2}(a_k + b_k) \tag{41}$$

$$e_k \triangleq \tfrac{1}{2}(b_k + c_k) \tag{42}$$

Note that

$$a_k \leq d_k \leq b_k \leq e_k \leq c_k \tag{43}$$

We also have (see Fig. 3.7.7)

$$\tilde{a}_k = h(\mathbf{y} + a_k\mathbf{z}) \tag{44}$$

$$\tilde{b}_k = h(\mathbf{y} + b_k\mathbf{z}) \tag{45}$$

$$\tilde{c}_k = h(\mathbf{y} + c_k\mathbf{z}) \tag{46}$$

We similarly define

$$\tilde{d}_k = h(\mathbf{y} + d_k\mathbf{z}) \tag{47}$$

$$\tilde{e}_k = h(\mathbf{y} + e_k\mathbf{z}) \tag{48}$$

STEP 2   If $\tilde{a}_k \geq \tilde{d}_k$ and $\tilde{d}_k \leq \tilde{b}_k$, make the following assignments:

$$c_k \leftarrow b_k \qquad b_k \leftarrow d_k \qquad a_k \text{ remains the same} \tag{49}$$

STEP 3   If $\tilde{b}_k \geq \tilde{e}_k$ and $\tilde{e}_k \leq \tilde{c}_k$, make the following assignments:

$$a_k \leftarrow b_k \qquad b_k \leftarrow e_k \qquad c_k \text{ remains the same} \tag{50}$$

STEP 4   Otherwise make the following assignments:

$$a_k \leftarrow d_k \qquad c_k \leftarrow e_k \qquad b_k \text{ remains the same} \tag{51}$$

STEP 5   Repeat steps 1 to 4, n times

STEP 6   Compute d as in Fig. 3.7.7.

The above procedure will reduce the region at which the minimum is located by a factor of $1/2^n$.

3.8 COMPARISON OF THE OPTIMIZATION ALGORITHMS

To illustrate the relative performance of

1 The steepest-descent algorithm
2 The conjugate-gradient algorithm
3 The Fletcher-Powell algorithm
4 Newton's method

it is instructive to compare their performance on the same function starting at the same initial guess.

A "standard" test function which has been often used in the literature to compare algorithms is the so-called *Rosenbrock banana function*, described analytically by

$$h(\mathbf{x}) = h(x_1, x_2) = 100(x_2 - x_1{}^2)^2 + (1 - x_1)^2 \tag{1}$$

Table 3.1   **PERFORMANCE OF STEEPEST-DESCENT ALGORITHM FOR THE ROSENBROCK BANANA FUNCTION**

Iteration	x_1	x_2	$h(x_1, x_2)$
0	−1.000	1.000	4.00
1	−0.995	1.000	3.99
2	−0.995	0.990	3.98
3	−0.990	0.990	3.97
4	−0.990	0.979	3.96
5	−0.984	0.979	3.95
6	−0.984	0.968	3.94
7	−0.979	0.968	3.93
8	−0.978	0.956	3.91
9	−0.973	0.956	3.90
10	−0.972	0.944	3.89
...	...	...	...
20	−0.935	0.872	3.75
...	...	...	...
30	−0.899	0.806	3.61
...	...	...	...
40	−0.862	0.741	3.47
...	...	...	...
50	−0.824	0.678	3.33
...	...	...	...
60	−0.784	0.613	3.18
...	...	...	...
70	−0.742	0.550	3.03
...	...	...	...
80	−0.698	0.487	2.88
...	...	...	...
90	−0.650	0.422	2.72
...	...	...	...
100†	−0.594	0.352	2.54

† Terminated at 100 iterations.

Table 3.2 **PERFORMANCE OF CONJUGATE-GRADIENT ALGORITHM FOR THE ROSENBROCK BANANA FUNCTION**

Iteration	x_1	x_2	$h(x_1, x_2)$
0	−1.000	1.000	4.00
1	−0.996	1.000	4.00
2	−0.983	0.988	3.99
3	−0.472	0.248	3.98
4	−0.451	0.204	2.23
5	−0.358	0.100	2.11
6	−0.300	0.049	1.92
7	−0.249	0.010	1.86
8	−0.213	−0.012	1.83
9	−0.171	−0.035	1.81
10	−0.146	−0.047	1.79
...	...	...	...
15	−0.011	−0.082	1.70
...	...	...	...
20	0.124	−0.077	1.63
...	...	...	...
30	0.365	0.029	1.52
...	...	...	...
40	0.646	0.307	1.34
...	...	...	...
45	0.829	0.579	1.20
46	0.885	0.677	1.18
47	0.914	0.730	1.14
48	0.972	0.842	1.11
49	1.002	0.903	1.06
50	1.061	1.027	1.04
51	1.118	1.156	0.96
52	1.174	1.288	0.90
53	1.227	1.420	0.84
54	1.323	1.675	0.77
55	1.402	1.901	0.65
56	1.465	2.094	0.58
57	1.544	2.364	0.48
58	1.556	2.415	0.34
59	1.556	2.420	0.31
60	1.555	2.420	0.31
61	1.555	2.420	0.31
62	1.555	2.420	0.31
63	1.554	2.418	0.31
64	1.551	2.412	0.31
65	1.541	2.387	0.31
66	1.237	1.537	0.31
67	1.227	1.505	0.06
68	1.219	1.490	0.05
69	1.130	1.269	0.05
70	1.087	1.173	0.02
71	1.007	1.012	0.01
72	1.004	1.009	0.00
73†	1.004	1.009	0.00

† Converged with accuracy of 10^{-3}.

This function looks like a twisting deep valley. Its minimum is unique and is at

$$x_1 = x_2 = 1 \qquad (2)$$

and the minimum value is clearly zero.

This function presents problems to certain algorithms because they may bounce around the steep walls of the valleys and be characterized by poor convergence rates.

We have coded the algorithms of Sec. 3.7, also using the Fibonacci search method for step control: Tables 3.1 to 3.4 show the performance for the same initial-guess vector

$$\mathbf{x}_0 = \begin{bmatrix} -1 \\ +1 \end{bmatrix}$$

As can be seen from Table 3.1, *the steepest-descent* algorithm is extremely slow. After 100 iterations, the iterates are still very far from the solution (at which time the algorithm was terminated). Such slow convergence, often observed in the steepest-descent algorithm, illustrates the fact that the simplest algorithms are not always the most practical or cost-effective.

Table 3.3 PERFORMANCE OF FLETCHER-POWELL ALGORITHM FOR THE ROSENBROCK BANANA FUNCTION

Iteration	x_1	x_2	$h(x_1, x_2)$
0	−1.000	1.000	4.00
1	−0.995	1.000	4.00
2	−0.775	0.562	3.99
3	−0.664	0.382	3.31
4	−0.254	0.029	3.11
5	−0.267	0.065	1.70
6	−0.082	−0.026	1.61
7	0.105	−0.029	1.28
8	0.128	0.021	0.96
9	0.269	0.050	0.76
10	0.371	0.106	0.58
11	0.457	0.221	0.50
12	0.559	0.300	0.31
13	0.621	0.366	0.21
14	0.836	0.690	0.18
15	0.828	0.685	0.04
16	0.912	0.825	0.03
17	0.961	0.927	0.01
18	0.986	0.972	0.00
19	1.000	1.000	0.00
20†	0.999	0.999	0.00

† Converged with accuracy of 10^{-3}.

Table 3.2 shows the performance of the *conjugate-gradient* algorithm. It only took about 4 iterations of the conjugate-gradient algorithm to get to the place that required 100 iterations for the steepest-descent algorithm. It should also be noted that at the forty-ninth iteration, it was very close to the solution ($x_1 = 1$, $x_2 = 1$) but the algorithm had to overshoot it (see iterations 60 to 65) before it came back. Convergence was achieved in 76 iterations.

Table 3.3 shows the performance of the *Fletcher-Powell method*. The convergence was rapid, and it did not exhibit the overshoot noted in the conjugate-gradient method.

Table 3.4 shows the extremely rapid convergence of *Newton's method*, even though at the first iteration it went sky high on the walls of the valley.

Table 3.4 PERFORMANCE OF NEWTON'S METHOD FOR THE ROSENBROCK BANANA FUNCTION

Iteration	x_1	x_2	$h(x_1, x_2)$
0	−1.000	1.000	4.00
1	1.000	−3.000	1,599.99
2	1.000	0.999	0.00
3†	1.000	1.000	0.00

† Converged with accuracy of 10^{-3}.

REFERENCES

1 LASDON, L. S.: "Optimization Theory for Large Systems," chap. 1, Macmillan, London, 1970.

2 WILDE, D. J.: "Optimum Seeking Methods," Prentice-Hall, Englewood Cliffs, N.J., 1964.

3 SOUTHWORTH, R. W., and S. L. DELEEUW: "Digital Computation and Numerical Analysis," chaps. 5–7, McGraw-Hill, New York, 1965.

4 MOURSUND, D. G., and C. S. DURIS: "Elementary Theory and Application of Numerical Analysis," chaps. 1 to 3, McGraw-Hill, New York, 1967.

5 RALSTON, A.: "A First Course in Numerical Analysis," chaps. 8 and 9, McGraw-Hill, New York, 1965.

6 TRAUB, J. F.: "Iterative Methods for the Solution of Equations," Prentice-Hall, Englewood Cliffs, N.J., 1964.

7 TODD, J. (ed.): "Survey of Numerical Analysis," chaps. 6 and 7, McGraw-Hill, New York, 1962.

8 VARGA, R. S.: "Matrix Iterative Analysis," chaps. 3 and 4, Prentice-Hall, Englewood Cliffs, N.J., 1962.

9 MCCALLA, T. R.: "Introduction to Numerical Methods and FORTRAN Programming," chaps. 2, 3, and 5, Wiley, New York, 1967.

10 KETTER, R. L., and S. P. PRAWEL, JR.: "Modern Methods of Engineering Computation," chaps. 5 and 8, McGraw-Hill, New York, 1969.

11 HAMMING, R. W.: "Numerical Methods for Scientists and Engineers," chaps. 28–31, McGraw-Hill, New York, 1962.

12 COOPER, L., and D. HEINBERG: "Introduction to Methods of Optimization," chaps. 4 and 5, Saunders, Philadelphia, 1970.

13 BEVERIDGE, G. S. G., and R. S. SCHECHTER: "Optimization: Theory and Practice," chaps. 1, 2, 4–6, and 8, McGraw-Hill, New York, 1970.

14 CURRY, H.: Methods of Steepest Descent for Nonlinear Minimization Problems, *Q. Appl. Math*, vol. 2, pp. 258–261, 1954.

15 FLETCHER, R. and M. J. D. POWELL: A Rapidly Convergent Descent Method for Minimization, *Br. Comput. J.*, vol. 6, pp. 163–168, 1963.

16 FLETCHER, R., and C. M. REEVES: Function Minimization by Conjugate Gradients, *Br. Comput. J.*, vol. 7, pp. 149–154, 1964.

17 POWELL, M. J. D.: An Efficient Method for Finding the Minimum of a Function of Several Variables without Using Derivatives, *Br. Comput. J.*, vol. 7, pp. 155–162, 1964.

18 ROSENBROCK, H. H.: An Automatic Method for Finding the Greatest or Least Value of a Function, *Br. Comput. J.*, vol. 3, p. 175, 1960.

EXERCISES

Section 3.2

3.2.1 Consider the LTI resistive network shown. Determine the values of the resistors R_1 and R_2 in ohms so that the power transferred to them

$$p = R_1 i_1^2 + R_2 i_2^2$$

is maximized.

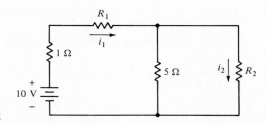

Fig. Ex. 3.2.1

3.2.2   Consider a scalar function $h(x)$. Suppose $x*$ is a minimum. Derive properties of

$$\left.\frac{\partial^3 h}{\partial x^3}\right|_{x=x^*} \quad \text{and} \quad \left.\frac{\partial^4 h}{\partial x^4}\right|_{x=x^*}$$

3.2.3   Write explicitly the necessary conditions for the minimum of the following function in terms of conditions on the first and second partial derivatives

$$h(x_1, x_2) = 1 - \frac{1}{x_1{}^2 + x_2{}^2 + 4}$$

3.2.4   (Hard!) This exercise lets you derive methods for optimization when certain of the variables to be optimized must satisfy certain constraints. You are being asked to develop a technique called the *Lagrange multiplier* method in a sequence of steps.

(*a*)   Let us first consider the following simple problem. We have a function $h(x_1, x_2)$ given by

$$h(x_1, x_2) = x_1{}^2 + x_2{}^2 \tag{1}$$

Obviously the minimum of the function is attained at $x_1 = 0$, $x_2 = 0$. This is an unconstrained minimization problem. Now suppose that we are not completely free to select the variables x_1 and x_2. Suppose that their possible values are constrained by the relation

$$x_1 + x_2 - 1 = 0 \tag{2}$$

We can now pose a *constrained minimization* problem. Find x_1^*, x_2^* such that

$$h(x_1^*, x_2^*) \leq h(x_1, x_2) \tag{3}$$

subject to the constraint

$$x_1^* + x_2^* - 1 = 0 \tag{4}$$

The problem can be solved in a brute-force way as follows. Solve for x_1 from Eq. (2)

$$x_1 = 1 - x_2 \tag{5}$$

and substitute in Eq. (1) to obtain the new function of a single variable

$$\phi(x_2) = (1 - x_2)^2 + x_2{}^2 \tag{6}$$

Now we can find the minimum of $\phi(x_2)$. Verify that it is given by

$$x_2^* = \tfrac{1}{2} \tag{7}$$

and so, in view of Eq. (5), we can deduce that

$$x_1^* = \tfrac{1}{2} \tag{8}$$

Draw a geometric interpretation of the above problem. The answer to this problem can also be obtained in the following manner. Let μ be an arbitrary scalar. In view of Eq. (2) the following equality holds:

$$\mu(x_1 + x_2 - 1) = 0 \tag{9}$$

Since the minimization of the function $h(x_1, x_2)$ is not affected by adding zero to it, let us define the function

$$m(x_1, x_2, \mu) = x_1{}^2 + x_2{}^2 + \mu(x_1 + x_2 - 1) \tag{10}$$

Let us now determine the condition that must hold at the minimum value of $m(x_1, x_2, \mu)$. We denote the optimal values by x_1^*, x_2^*, μ^*. These are

$$\left.\frac{\partial m}{\partial x_1}\right|_* = 2x_1^* + \mu^* = 0$$

$$\left.\frac{\partial m}{\partial x_2}\right|_* = 2x_2^* + \mu^* = 0 \tag{11}$$

$$\left.\frac{\partial m}{\partial \mu}\right|_* = x_1^* + x_2^* - 1 = 0$$

This set of three linear equations has the solution

$$x_1^* = \tfrac{1}{2} \qquad x_2^* = \tfrac{1}{2} \qquad \mu^* = 1 \tag{12}$$

We note that the values of x_1^* and x_2^* found with this method are identical to that found by the brute-force method.

(*b*) At this step you are asked to formalize the above method. Let $h(\mathbf{x})$ be a scalar-valued function of the vector $\mathbf{x} \in R_n$. Let $g(\mathbf{x})$ be another scalar-valued function of the vector $\mathbf{x}$. The constrained minimization problem is to minimize $h(\mathbf{x})$ subject to the constraint

$$g(\mathbf{x}) = 0$$

Let $\mathbf{x}^*$ denote the optimum. Then prove by imitating the development of Theorem 1 that if we define (μ scalar)

$$m(\mathbf{x}, \mu) = h(\mathbf{x}) + \mu g(\mathbf{x})$$

then

$$\left.\frac{\partial m}{\partial \mathbf{x}}\right|_* = \mathbf{0} \qquad \left.\frac{\partial m}{\partial \mu}\right|_* = 0$$

(*c*) Constrained optimization problems may involve a variety of equality constraints. A typical multiconstraint problem is to minimize $h(\mathbf{x})$ subject to

$$g_1(\mathbf{x}) = 0$$
$$g_2(\mathbf{x}) = 0$$
$$\vdots$$
$$g_m(\mathbf{x}) = 0$$

where in general $m < n$ (which is the dimension of $\mathbf{x}$). Let

$$\mathbf{g}(\mathbf{x}) \triangleq \begin{bmatrix} g_1(\mathbf{x}) \\ g_2(\mathbf{x}) \\ \vdots \\ g_m(\mathbf{x}) \end{bmatrix}$$

and let $\boldsymbol{\mu} \in R_m$. Then by imitating the development of step (b) prove that at the optimum point $\mathbf{x}*$ the following relations must hold:

$$\left. \frac{\partial m}{\partial \mathbf{x}} \right|_* = \mathbf{0} \in R_n \qquad \left. \frac{\partial m}{\partial \boldsymbol{\mu}} \right|_* = \mathbf{0} \in R_m$$

where $m(\mathbf{x}, \boldsymbol{\mu}) \triangleq h(\mathbf{x}) + \boldsymbol{\mu}' \mathbf{g}(\mathbf{x})$

3.2.5   Using the results of Exercise 3.2.4, determine the solution to the following constraint optimization problems:

(a)   Minimize $h(\mathbf{x}) = x_1^2 + x_2^2 + x_3^2$ subject to the constraint

$$x_1 + x_2 + x_3 - 1 = 0$$

(b)   Minimize $h(\mathbf{x}) = x_1^2 + x_2^2 + x_3^2$ subject to the constraints

$$x_1 + x_2 + x_3 - 1 = 0 \qquad x_1 + x_2 + 1 = 0$$

(c)   Minimize $h(\mathbf{x}) = x_1^2 + x_2^2 + x_3^2$ subject to the constraint

$$x_1^2 + x_2^2 - 9 = 0$$

(d)   Minimize $h(\mathbf{x}) = x_1^2 + x_2^2 + x_3^2$ subject to the constraints

$$x_1^2 + x_2^2 - 9 = 0 \qquad x_1 + x_2 + x_3 - 1 = 0$$

3.2.6   Let $\mathbf{A}$ be a positive definite symmetric matrix. Determine analytically the solution to the following constrained minimization problem. Minimize

$$h(\mathbf{x}) = \tfrac{1}{2}\mathbf{x}'\mathbf{A}\mathbf{x} + \mathbf{b}'\mathbf{x} + c \qquad \mathbf{x} \in R_n$$

subject to the constraints

$$\mathbf{d}_1' \mathbf{x} + e_1 = 0$$
$$\mathbf{d}_2' \mathbf{x} + e_2 = 0$$
$$\vdots$$
$$\mathbf{d}_m' \mathbf{x} + e_m = 0$$

where $m < n$.

Section 3.5

3.5.1   Consider the use of the Picard algorithm to solve the linear equation

$$\mathbf{x} = \mathbf{Ax}$$

What are sufficient conditions for convergence?

3.5.2   (Computer exercise)   Consider the use of the Picard algorithm to solve the simultaneous equations

$$x_1 = 0.2 x_1 \cos x_2$$
$$x_2 = 0.1 e^{-x_1} \sin 2x_2 + 0.2 \qquad 0 \le x_1 \le 1$$
$$0 \le x_2 \le 1$$

Verify that the sufficient conditions for convergence hold. Estimate the Lipschitz constant L and estimate the maximum number of iterations that you would expect for attaining an accuracy of 10^{-4}. Determine the solution for the following initial guesses:

(a) $x_1 = 0$ (b) $x_1 = 1$ (c) $x_1 = 0$ (d) $x_1 = 1$
    $x_2 = 0$ $x_2 = 1$ $x_2 = 1$ $x_2 = 0$

3.5.3   (Computer exercise) Consider the nonlinear resistive network shown. Each of the resistors is nonlinear. Their current-voltage characteristics are:

Resistor R_1: $v_1 = e^{i_1} - 1$ Resistor R_2: $v_2 = i_2|i_2|$ Resistor R_3: $v_3 = i_3^3$

Using Picard's method determine the currents i_1, i_2, and i_3. Construct the linearized network about this constant operating point (review Sec. 2.8).

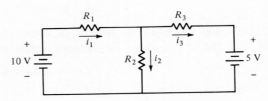

Fig. Ex. 3.5.3

3.5.4   We wish to minimize the scalar-valued function

$$h(\mathbf{x}) = \tfrac{1}{2}\mathbf{x}'\mathbf{Ax} + \mathbf{b}'\mathbf{x} + c \qquad \mathbf{x} \in R_n \tag{1}$$

where $\mathbf{A} = \mathbf{A}'$ is a positive definite matrix.

(*a*)   Find the global minimum **x*** such that

$$h(\mathbf{x}^*) \le h(\mathbf{x}) \qquad \text{for all } \mathbf{x} \in R_n$$

(*b*)   Consider the steepest-descent algorithm

$$\mathbf{x}(k+1) = \mathbf{x}(k) - \alpha \frac{\partial h}{\partial \mathbf{x}} \bigg|_{\mathbf{x}=\mathbf{x}(k)} \tag{2}$$

where α is a scalar size *independent* of the iteration index k. We wish to use (2) to find **x*** iteratively. Determine the range of α for which the algorithm (2) will converge to **x***, that is,

$$\lim_{k \to \infty} \mathbf{x}(k) = \mathbf{x}^*$$

HINTS

1   Consider the error $\mathbf{e}(k) = \mathbf{x}(k) - \mathbf{x}^*$.

2   Use the fact that the eigenvalues of the matrix $\mathbf{I} + \mathbf{B}$ are related to the eigenvalues of $\mathbf{B}$ by the formula

$$\lambda_i(\mathbf{I} + \mathbf{B}) = 1 + \lambda_i(\mathbf{B}) \qquad \text{for all } i$$

where
$$\lambda_i(\mathbf{I} + \mathbf{B}) = i\text{th eigenvalue of } \mathbf{I} + \mathbf{B}$$

$$\lambda_i(\mathbf{B}) = i\text{th eigenvalue of } \mathbf{B}$$

3   The answer will be in terms of **A**, **b**, and c.

3.5.5   (Computer exercise)   This twofold exercise is designed so that you model a nonlinear fluid network (read Sec. 2.9) and then determine its solution using Picard's method. An irrigation network can be modeled by the nonlinear flow network sketched, where *p* represents the pump, having pressure of 100 units. Elements 1, 2, and 3, representing the network in the fields to be irrigated, are fixed and modeled by

$$w_i = G_i p_i^{1/3} \qquad i = 1, 2, 3$$

where $G_1 = G_2 = G_3 = 1$. Elements 4, 5, and 6, representing the pipes connecting the pump and the three different fields, are modeled by

$$p_i = R_i |w_i| w_i \qquad i = 4, 5, 6$$

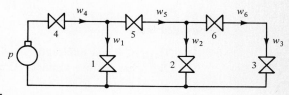

Fig. Ex. 3.5.5

Depending on scheduling considerations, i.e., how long the irrigation network is on, two possible pipe diameters for 4, 5, and 6 are acceptable corresponding to the following values of R_4, R_5, R_6:

Design 1: $\qquad\qquad\qquad\qquad R_4 = 0.5 \qquad R_5 = 1.0 \qquad R_6 = 1.5$

Design 2: $\qquad\qquad\qquad\qquad R_4 = 0.1 \qquad R_5 = 0.1 \qquad R_6 = 0.1$

Our concern is the flow w_3 delivered to the farthest field.

(a) Set up an equation of the form $\mathbf{p}_l = \mathbf{F}(\mathbf{p}_l)$ using branches 4, 5, and 6 as tree branches and 1, 2, and 3 as link branches.

(b) Using the R_i parameters of design 2 and with the starting values $p_1 = p_2 = p_3 = 10$ and $p_1 = p_2 = p_3 = 50$, go through 20 iterations of Picard's method in the equation you set up in part (a), also computing w_3 at each iteration. Sketch w_3 as a function of iteration number for both starting points. Comment on your results.

(c) Do the same for the R_i parameters of design 1.

(d) Determine the nonlinear characteristic of input pressure p vs. w_3 for the R_i parameters of design 2 by computing w_3 (using 20 iterations also) for $p = 5, 20, 40, 60, 80, 100$. Graph your results, and compute the linearized characteristic at $p = 10$ and 80, commenting on the effects of sensitivity to variations in p (or the accuracy of the linearizations).

3.5.6 In this exercise we consider an iterative method for computing the inverse of a nonsingular $n \times n$ matrix $\mathbf{A}$. Consider the matrix-valued function $\mathbf{f}(\mathbf{B})$ of an $n \times n$ matrix $\mathbf{B}$ given by

$$\mathbf{f}(\mathbf{B}) = \mathbf{B} + \alpha(\mathbf{AB} - \mathbf{I}) \qquad\qquad (1)$$

where α is a real scalar.

(a) Prove that when $\mathbf{B} = \mathbf{A}^{-1}$, then $\mathbf{f}(\mathbf{A}^{-1}) = \mathbf{A}^{-1}$.

(b) Apply the Picard algorithm to Eq. (1), that is,

$$\mathbf{B}_{k+1} = \mathbf{f}(\mathbf{B}_k) \qquad\qquad (2)$$

where $k = 0, 1, 2, \ldots$ is an iteration index and $\mathbf{B}_0$ is an arbitrary $n \times n$ matrix. Show that

$$\lim_{k \to \infty} \mathbf{B}_k = \mathbf{A}^{-1}$$

provided that:

(1) $\alpha < 0$ and sufficiently small in magnitude.

(2) $\mathbf{A}$ has positive distinct real eigenvalues.

HINTS

1 . Consider the evolution of the "error" matrix

$$\mathbf{E}_k \triangleq \mathbf{B}_k - \mathbf{A}^{-1}$$

2 The eigenvalues $\lambda_i(\mathbf{I} + \mathbf{C})$ of the matrix $\mathbf{I} + \mathbf{C}$ are given by

$$\lambda_i(\mathbf{I} + \mathbf{C}) = 1 + \lambda_i(\mathbf{C})$$

(you need not prove this fact).

3.5.7   In a network of pipes, valves, and pumps the flow of liquid is analogous to current in an electric network, and pressure is analogous to voltage. However, the relationship analogous to electric conductance is nonlinear:

$$F = C\sqrt{\Delta P} \tag{1}$$

where F = flow through branch
ΔP = pressure drop across branch
C = branch conductivity

This relationship is often used in industry as a basis for flow measurement. The pressure drop across an orifice is measured, and the flow is inferred from it. Pipes can be modeled as devices that satisfy Eq. (1). However, in many applications, some pipes are large enough for there to be negligible pressure; such pipes play the role of wires in an electric network. Pumps play the role of real voltage sources. Pumps produce a pressure differential ΔP, which for some pumps is modeled by

$$\Delta P = K_1 S^2 - K_2 F^2 \tag{2}$$

where S = pump speed
F = flow through pump
K_1, K_2 = positive constants

Show that this model is equivalent to an ideal source $K_1 S^2$ in series with a conductance. A valve can be thought of as being a pipe with adjustable area (or conductance). Hence, a valve satisfies Eq. (1), where C is given by

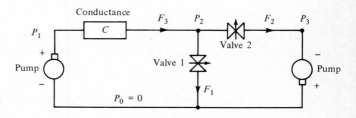

Fig. Ex. 3.5.7

$$C = C_{max} x \qquad 0 \le x \le 1 \tag{3}$$

and x is the valve position. Thus a valve is analogous to a variable conductance in an electric network. Consider now the network shown. Both pumps are identical with $K_1 S^2 = 80$ psi and $K_2 = 10^{-6}$.

For valve 1: $\qquad\qquad\qquad\qquad C_{1\,max} = 500 \qquad x_1 = 0.5$

For valve 2: $\qquad\qquad\qquad\qquad C_{2\,max} = 2{,}000 \qquad x_2 = 0.2$

The conductance is $C = 570$.

(a) (Analytical part) Determine the nonlinear equations that define P_1, P_2, P_3 and F_1, F_2, F_3 using the data given above. *Hint:* By combining conductances in series, you should be able to arrive at a scalar implicit equation on P_2 alone of the form $f(P_2) = 0$.

(b) (Computational part) Use the initial guess $P_2 = 20$ lb/in^2 and Newton's method on $f(P_2) = 0$ to determine P_1, P_2, P_3 and F_1, F_2, F_3. *Caution:* Put a stopping rule on number of iterations.

(c) (Design part, analytical) Suppose that we wish to have

$$P_2 = 20 \text{ lb/in}^2 \qquad P_3 = -50 \text{ lb/in}^2$$

where pressures in pounds per square inch are relative to the ground pressure. Determine the equations that will specify the valve positions x_1 and x_2 to attain this objective.

(d) (Computational part) Determine the values of x_1 and x_2 for part (c) using:

(1) Picard method
(2) Newton's method

In both cases use the initial guess

$$x_1 = 0.4 \qquad x_2 = 0.25$$

Caution: Put a stopping rule on number of iterations.

Section 3.6

3.6.1 (Computer exercise) Use Newton's method and the modified Newton method to determine the solution of the following minimization problem.

$$h(\mathbf{x}) = (\mathbf{x}'\mathbf{A}\mathbf{x})^2 \quad \text{where } \mathbf{x} \in R_3 \quad \text{and} \quad \mathbf{A} = \begin{bmatrix} 9 & 2 & 1 \\ 2 & 4 & 1 \\ 1 & 1 & 2 \end{bmatrix}$$

Limit the maximum number of iterations to 15. Use in each case the following initial guesses:

$$(a) \quad \mathbf{x}_0 = \begin{bmatrix} 100 \\ 100 \\ 100 \end{bmatrix} \qquad (b) \quad \mathbf{x}_0 = \begin{bmatrix} 10 \\ -10 \\ 10 \end{bmatrix} \qquad (c) \quad \mathbf{x}_0 = \begin{bmatrix} 0.1 \\ 0.1 \\ 0.1 \end{bmatrix}$$

3.6.2   It is desired to calculate $\sqrt{2}$ using an iterative procedure. Observe that $\hat{x} = \sqrt{2}$ is a solution to

$$x = x + \alpha(x^2 - 2) = f(x) \tag{1}$$

(a)   Write down Picard's algorithm to solve Eq. (1).
(b)   We know that $1 \leq \sqrt{2} \leq 2$; so pick $\Omega = \{x \in R: 1 \leq x \leq 2\}$. Let L be the Lipschitz constant of $f(.)$ in the set Ω. Determine L and deduce for what values of α Picard's algorithm will converge.
(c)   Pick a suitable value of α and verify that $f(x) \in \Omega$ for all $x \in \Omega$.
(d)   Pick an initial guess of $\sqrt{2}$ and calculate an upper bound on the error $|x_k - \sqrt{2}|$ at the kth step using α picked in part (c).
(e)   What is the best value of α for rapid convergence near the true solution?
(f)   Write down Newton's method to find $\sqrt{2}$, that is, solve $x^2 - 2 = 0$, and compare with part (e).

Section 3.7

3.7.1   (Hard)   Consider the use of the conjugate-gradient and Fletcher-Powell algorithms to determine the minimum of a quadratic function in two variables

$$h(\mathbf{x}) = \tfrac{1}{2}\mathbf{x}'\mathbf{A}\mathbf{x} + \mathbf{b}'\mathbf{x} + c$$

where $x \in R_2$ and $\mathbf{A}$ is 2×2 symmetric positive definite matrix. Present a graphical interpretation of what is going on at each step of the algorithm. Can you prove that convergence will be exact in two steps?

3.7.2   (Computer exercise)   Test:

(a)   The steepest-descent algorithm
(b)   The conjugate-gradient algorithm
(c)   The Fletcher-Powell algorithm
(d)   Newton's method

to find the minimum of the Rosenbrock banana function

$$h(x_1, x_2) = 100(x_2 - x_1^2)^2 + (1 - x_1)^2$$

Use a maximum number of 100 iterations and an accuracy level of 10^{-3} for the solution. Use the initial guesses

$$x_1 = -1 \qquad x_2 = 1$$
$$x_1 = -1 \qquad x_2 = 5$$
$$x_1 = -1 \qquad x_2 = -5$$
$$x_1 = 1 \qquad x_2 = 5$$

Obviously the minimum value is at

$$x_1 = x_2 = 1$$

Sketch the path of each algorithm towards the minimum in x_1, x_2 space.

4

DISCRETE-TIME
DYNAMICAL SYSTEMS

SUMMARY

In this chapter we begin our treatment of a special class of system with memory whose dynamical behavior is described by vector difference equations. This type of mathematical model is used to describe and simulate a wide variety of engineering, economic, and social systems. The importance of discrete-time systems in modern technology and system analysis makes the study of their fundamental properties, such as stability, and of our ability to control their time evolution of great interest. This chapter therefore provides the fundamental system-analysis tools associated with this class of problems.

4.1 INTRODUCTION

In the previous chapter we studied iterative techniques for solving vector equations and for finding the minima (or maxima) of functions of several variables. We have seen that all these methods had the following structure†

---

† We switch from the notation $\mathbf{x}_k$ to $\mathbf{x}(k)$ to allow us to write the components of a vector without resorting to double subscripts.

$$\mathbf{x}(k + 1) = \mathbf{f}(\mathbf{x}(k)) \qquad \mathbf{x}(k) \in R_n \tag{1}$$

where $\mathbf{x}(k)$ = guess to solution at iteration k
  $\mathbf{x}(k + 1)$ = guess to solution at iteration $k + 1$

We started with an initial guess vector $\mathbf{x}(0)$. Then the initial guess vector $\mathbf{x}(0)$ and the type of algorithm used, i.e., the function $\mathbf{f}(.)$, was sufficient information to generate all subsequent iterates $\mathbf{x}(1)$, $\mathbf{x}(2)$, ...

Now let us consider a different type of an example related to population growth. Let $x(k)$ denote the world population in year k, and let us assume that there is a constant birth rate b and a death rate d. Then the net increase (or decrease) in population can be described by the scalar relationship

$$x(k + 1) = x(k) + bx(k) - dx(k) = (1 + b - d)x(k) \tag{2}$$

Clearly, as long as the birth and death rates are known, if we know the population $x(k)$ in any year k, the simple recursive relation (difference equation) Eq. (2) can be used to compute the population for all other values of time.

Since Eq. (2) is a special case of Eq. (1), we can see that at least two different phenomena are described by the same type of mathematical relation.

Population-exchange problems are prime examples of discrete-time dynamical systems. By population, we do not necessarily mean human, animal, or plant population changes but anything that is capable of natural or artificial growth. Indeed several problems that arise in business and management decisions involve the interrelations, between themselves and in time, of several hundred variables that typically involve

1 Sales of different products
2 Inventory of these products
3 Transportation (from factories to warehouses) of these products
4 Production of these products
5 Type and number of machines required
6 Type and number of workers required
7 Type and quantity of natural resources needed
8 Capital-resource requirements

Such variables are related by recursive difference equations because of the time lag introduced by a sale order, changes in inventory, transportation delays, production delays, etc. The interested reader can find many illustrative examples in Refs. 1 to 5. In fact, the dynamics and time evolution of these types of variables, as a function of sales fluctuations and administrative decisions, are the bread-and-butter systems of many areas of economics, management science, and operations research.

Apart from population-exchange models in an ecological context, such systems arise in modeling interplay between population, quality of life, economics, pollution, etc. Reference 6 describes a complex model of assumed interrelations between several variables that contribute to urban decay. Reference 7 considers interrelations between population, natural resources, pollution, quality of life, etc., in a world environment. References 8 and 9 are excellent sources of prey-predator dynamic ecological systems.

Economic systems can also be described by difference equations. We postpone a general discussion of economic problems until Chap. 5, where we present the difference equations of a medium-complexity model of the United States economy derived on the basis of actual economic measurements from 1955 to 1968.

4.2 INPUT-STATE-OUTPUT DESCRIPTION OF DISCRETE-TIME SYSTEMS

Our conceptual and analytical description of discrete-time systems begins by going from an external input-output behavior to an internal state-variable behavior.†

4.2.1 The Notion of Time

In discrete-time systems we always have a scalar index, denoted by k, taking integer values

$$k = 0, 1, 2, \ldots \tag{1}$$

and we are concerned with the values of variables which are indexed by k, that is, the interrelationship of, say, the vectors

$$\mathbf{x}(0), \mathbf{x}(1), \mathbf{x}(2), \ldots$$

We refer to the index k as the time. This is not necessary and is purely for convenience. We saw in Chap. 3 that the index k is related to the iteration number. Similarly, k may represent some different physical quantity, e.g., a discrete distance measure on a highway, where $\mathbf{x}(k)$ may represent the number and average speed of automobiles between the kth and $(k + 1)$st marker.

4.2.2 Input-Output Description

In the external, black-box, input-output description of a dynamical system we distinguish between inputs to the system, which we shall denote by the vectors

---

† The conceptual development is analogous to that given in "Basic Concepts."

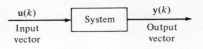

FIGURE 4.2.1
Block-diagram input-output description
of a discrete-time system.

$\mathbf{u}(0), \mathbf{u}(1), \mathbf{u}(2), \ldots, \mathbf{u}(k), \ldots,$ and outputs of the system, which we shall denote by the vectors $\mathbf{y}(0), \mathbf{y}(1), \ldots, \mathbf{y}(k), \ldots.$ It is useful to think of the input-vector sequence $\{\mathbf{u}(k)\}$ as being the *cause* or *external excitation* of the system and of the output sequence $\{\mathbf{y}(k)\}$ as being the *effect* or *response* of the system.

Loosely speaking, then, the input-vector $\mathbf{u}(k)$ summarizes all the things that are at our disposal to adjust or change at time k which will then influence the system response. On the other hand, the output $\mathbf{y}(k)$ represents all the quantities that we are interested in at time k and (usually) the ones we can measure. Figure 4.2.1 shows the usual abstract input-output representation of a discrete-time system.

Let us suppose that

$$\mathbf{u}(k) \in R_m \qquad \mathbf{y}(k) \in R_r \tag{2}$$

for all $k = 0, 1, 2, \ldots.$ Then a causal discrete-time system is characterized in general by the equation

$$\mathbf{y}(k) = \mathbf{f}(\mathbf{y}(k-1), \mathbf{y}(k-2), \ldots, \mathbf{y}(k-j), \mathbf{u}(k), \mathbf{u}(k-1), \ldots, \mathbf{u}(k-i)) \tag{3}$$

where $\mathbf{f}(.)$ is an "ordinary" vector-valued function of its arguments.

An equation of type (3) is called a *j*th-order vector difference equation. Its meaning is as follows: Let us think of the index k in (3) as the present time. Then, Eq. (3) implies that if we know the value of the output at the j previous times, i.e., if we know the vectors $\mathbf{y}(k-1), \mathbf{y}(k-2), \ldots, \mathbf{y}(k-j)$, and if we know the value of the input now and in the previous i times, i.e., if we know the vectors $\mathbf{u}(k), \mathbf{u}(k-1), \ldots, \mathbf{u}(k-i)$, then, through the function $\mathbf{f}(.)$, we can uniquely deduce the current values of the output, i.e., the vector $\mathbf{y}(k)$.

EXAMPLE 1   Let us consider a single input–single output system. Then Eq. (3) may take the form

$$y(k) = 3y(k-1) + u(k)y(k-2)y(k-3)^2 - \log[u(k-1)]\sin[y(k-1)]$$

A system characterized by Eq. (3) is said to have *memory* because the current value of the output $\mathbf{y}(k)$ depends on its own past values as well as the past values of the inputs.

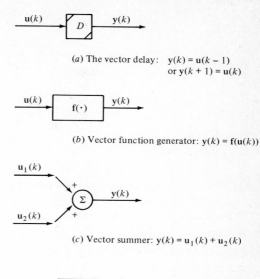

(a) The vector delay: $\mathbf{y}(k) = \mathbf{u}(k-1)$
or $\mathbf{y}(k+1) = \mathbf{u}(k)$

(b) Vector function generator: $\mathbf{y}(k) = \mathbf{f}(\mathbf{u}(k))$

(c) Vector summer: $\mathbf{y}(k) = \mathbf{u}_1(k) + \mathbf{u}_2(k)$

FIGURE 4.2.2
Information-processing elements for a
discrete-time system.

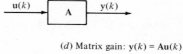

(d) Matrix gain: $\mathbf{y}(k) = \mathbf{A}\mathbf{u}(k)$

A familiar example of such a system with memory is the so-called *moving average*. For example, suppose that $\mathbf{u}(k)$ represents the average daily value of a particular stock. To remove some of the fluctuations one can use a 5-day moving average. This can be thought of as a discrete-time system whose input is the actual daily stock price $u(k)$ while its output $y(k)$ is

$$y(k) = \frac{1}{5} \sum_{j=0}^{4} u(k-j)$$

4.2.3 Basic Elements for a Discrete System

To enhance our basic understanding of a discrete-time system it is useful to think of certain mathematical operations that we shall visualize by information-processing elements. These are summarized in terms of their input-output behavior in Fig. 4.2.2.

Of these elements the *delay* represents the basic memory element. Its input $\mathbf{u}(k)$ and output $\mathbf{y}(k)$ are related by

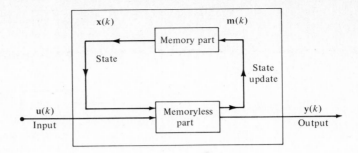

FIGURE 4.2.3
Conceptual subdivision of a system into its memory and memoryless subsystems.

$$\mathbf{y}(k) = \mathbf{u}(k - 1) \quad \text{or} \quad \mathbf{y}(k + 1) = \mathbf{u}(k) \tag{4}$$

The rest of the elements in Fig. 4.2.2 are used to illustrate familiar vector operations, which are *memoryless*.

4.2.4   The Internal State-Variable Representation of a Discrete-Time Dynamical System

Using the conceptual tools afforded by the memory and memoryless elements of Fig. 4.2.2, we can attempt to decompose the input-output description of a system into some conceptual internal subsystems that clearly divide the memory parts (or operations) from the memoryless ones.

The basic decomposition is shown in Fig. 4.2.3, where the input and output signals, $\mathbf{u}(k)$ and $\mathbf{y}(k)$, are the ones associated with the external input-output description. Note that in the internal decomposition:

1   The input $\mathbf{u}(k)$ acts on the memoryless part.
2   The output $\mathbf{y}(k)$ is generated by the memoryless part.

In the diagram of Fig. 4.2.3 another vector signal $\mathbf{m}(k)$, called the *state update*, is an output of the memoryless subsystem and acts as an input to the memory subsystem. Also, the output $\mathbf{x}(k)$ of the memory subsystem, called the *state*, acts as an additional input to the memoryless subsystem.

When we substitute for the memory and memoryless subsystems the general operations shown in Fig. 4.2.4, we can deduce the following mathematical definitions:

Variables:

State vector: $\mathbf{x}(k) \in R_n$ (*n*-dimensional column vector)
State-update vector: $\mathbf{m}(k) \in R_n$ (*n*-dimensional column vector)

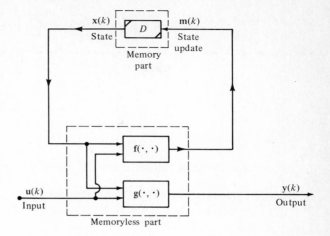

FIGURE 4.2.4
Detailed structure of memory and memoryless parts.

Input vector: $\mathbf{u}(k) \in R_m$ (m-dimensional column vector)
Output vector: $\mathbf{y}(k) \in R_r$ (r-dimensional column vector)

Memory subsystem:

$$\mathbf{x}(k + 1) = \mathbf{m}(k) \tag{5}$$

Memoryless subsystem:

$$\mathbf{m}(k) = \mathbf{f}(\mathbf{x}(k), \mathbf{u}(k)) \tag{6}$$

$$\mathbf{y}(k) = \mathbf{g}(\mathbf{x}(k), \mathbf{u}(k)) \tag{7}$$

We can eliminate the state-update vector $\mathbf{m}(k)$ between Eqs. (5) and (6) to obtain the standard mathematical description of a discrete-time system

$$
\begin{array}{ll}
\mathbf{x}(k + 1) = \mathbf{f}(\mathbf{x}(k), \mathbf{u}(k)) & \text{state equation} \\
\mathbf{y}(k) = \mathbf{g}(\mathbf{x}(k), \mathbf{u}(k)) & \text{output equation}
\end{array}
\tag{8}
$$

Thus, the state-vector description of a discrete-time system consists of a vector difference equation that relates the current state $\mathbf{x}(k)$ and the current input $\mathbf{u}(k)$ to obtain the state next time $\mathbf{x}(k + 1)$. Also, it involves a vector equality which relates the current state $\mathbf{x}(k)$ and the current input $\mathbf{u}(k)$ to obtain the current value of the output $\mathbf{y}(k)$.

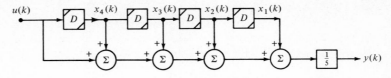

FIGURE 4.2.5
Simulation of a 5-day moving average.

From the above definition it is easy to see that given

1 The initial value $\mathbf{x}(0)$ of the state and
2 The values of the inputs $\mathbf{u}(0)$, $\mathbf{u}(1)$, $\mathbf{u}(2)$, . . . , $\mathbf{u}(k)$

we can uniquely deduce

1 The evolution of the state $\mathbf{x}(1)$, $\mathbf{x}(2)$, . . . , $\mathbf{x}(k + 1)$ and
2 The evolution of the output $\mathbf{y}(0)$, $\mathbf{y}(1)$, . . . , $\mathbf{y}(k)$

The mathematical description of a discrete-time system by Eq. (8) is much easier (for both analysis and digital-simulation purposes) than the input-output description (3), because in the latter one has to keep track of many past values of the outputs and the inputs. On the other hand, in the system representation (8) one need not store the past values $\mathbf{u}(k - 1)$, $\mathbf{u}(k - 2)$, . . . and $\mathbf{y}(k - 1)$, $\mathbf{y}(k - 2)$, . . . ; *the current state* $\mathbf{x}(\mathrm{k})$ *summarizes all this past information.*

To illustrate the state-variable description let us return to the generation of the 5-day moving average $y(k)$ of a particular stock whose daily value is $u(k)$. Figure 4.2.5 shows a simulation of this process using four scalar delay elements whose outputs are the state variables $x_1(k)$, $x_2(k)$, $x_3(k)$, $x_4(k)$; the input to the system is the actual daily price. In this case the state variables satisfy the difference equations

$$
\begin{aligned}
x_1(k + 1) &= x_2(k) \\
x_2(k + 1) &= x_3(k) \\
x_3(k + 1) &= x_4(k) \\
x_4(k + 1) &= u(k)
\end{aligned}
\qquad \text{state equation}
$$

and the output is given by

$$
y(k) = \frac{x_1(k) + x_2(k) + x_3(k) + x_4(k) + u(k)}{5}
\qquad \text{output equation}
$$

4.3   AN EXAMPLE

In this section we present yet another societal type of *dynamical* system having to do with general population interchange dynamics involving

1   The general population
2   The number of hard-core heroin addicts
3   The number of heroin addicts undergoing a methadone maintenance program
4   The number of heroin addicts who have been jailed

The basic time unit is 1 year. We define our basic state variables as follows:

$x_1(k)$ = general population at year k
$x_2(k)$ = number of heroin addicts at year k
$x_3(k)$ = number of heroin addicts undergoing methadone treatment in year k
$x_4(k)$ = number of heroin addicts arrested, convicted, and jailed in year k
$x_5(k)$ = number of heroin addicts released from jail in year k.

Before setting up the dynamical equations we make some remarks and state some assumptions.

1   In order to support his habit, an average heroin addict must steal a tremendous amount per year (to support a $50 per day habit, he has to steal about $200 worth of goods and fence them). The only way to reduce this dollar cost is for the addict to become a pusher and try to recruit new addicts from the general population. It is not unrealistic to imagine that an average heroin addict recruits a new addict every year. This potential doubling of the heroin population is close to recent statistics.
2   A methadone maintenance program will block the heroin craving; it also prevents a heroin high if an addict takes heroin on top of methadone. Hence, for all practical purposes a methadone patient can be a useful member of society. A methadone dropout returns to the heroin-addict population.
3   The existence of a methadone program naturally attracts a certain proportion of the heroin population (the rate depends on the degree of general advertisement); however, many heroin addicts join a methadone maintenance program only after consulting a current methadone patient to find out his reactions.
4   When a heroin addict is jailed, his average sentence is about 1 year; after that he returns to the heroin population.

Parameter definitions

b_1 = birth rate of normal population

d_1 = death rate of normal population

a = rate at which a heroin addict converts general public into heroin addiction.

d_2 = death rate of heroin addicts (in general higher than d_1)

c = percentage rate of jailed heroin addicts (depends on number of policemen, tips, judicial, etc.)

e = percentage rate of heroin addicts attracted to methadone program (depends on advertising budget)

f = rate at which a methadone patient converts a heroin addict to a methadone patient

g = rate of methadone dropouts

Under the above assumptions we can write the difference equations

$$x_1(k + 1) = x_1(k) + (b_1 - d_1)x_1(k) - ax_1(k)x_2(k)$$

$$x_2(k + 1) = x_2(k) + ax_1(k)x_2(k) + x_5(k) + gx_3(k)$$

$$- d_2 x_2(k) - cx_2(k) - ex_2(k) - fx_3(k)x_2(k)$$

$$x_3(k + 1) = x_3(k) + ex_2(k) + fx_3(k)x_2(k) - d_1 x_3(k) - gx_3(k) \qquad (1)$$

$$x_4(k + 1) = cx_2(k)$$

$$x_5(k + 1) = x_4(k)$$

This model of course can be made more complex by including the dynamics of heroin supply and price, education programs, halfway houses, rehabilitation centers, etc. Such mathematical models are becoming increasingly popular to simplify decision making. Besides being capable of simulating population changes, they are easily adaptable in evaluating the short- and long-term implications of a program. For example, let

m_2 = average yearly value of stolen goods per heroin addict

m_1 = average yearly cost of jailing a heroin addict

m_3 = average yearly cost of methadone program per patient

Then the dollar cost (considered as an output) to society in year k is given by

$$y(k) = m_2 x_2(k) + m_1[cx_2(k) - x_4(k)] + m_3 x_3(k) \qquad (2)$$

while the cumulative cost from year $k = 0$ to now is

$$z(k) = \sum_{j=0}^{k} y(j) \tag{3}$$

One can then examine the variation in economic losses to society for different values of parameters to obtain a rough idea of the sensitivity of the model to parameter variations.

For this type of a system there are several ways of exercising control. Some control can be exercised by changing the values of the parameters which appear as constants in the model of Eq. (1). For example,

1 Money allocated as a function of time to increase education should decrease parameter a. In this case $a = a(k)$ can be viewed as an input control variable.

2 Increased arrests, convictions, and elimination of suspended sentences will increase the parameter c; in this case, the parameter $c = c(k)$ will also act as an input control variable.

3 A change in the outlook of the police could mean that all those released from jail would have to undergo a methadone treatment. This would cause a change in the model structure. Under this policy, the second equation in (1) would be replaced by

$$x_2(k + 1) = x_2(k) + ax_1(k)x_2(k) + gx_3(k) - d_2 x_2(k)$$
$$- cx_2(k) - ex_2(k) - fx_3(k)x_2(k)$$

and the third equation of (1) by

$$x_3(k + 1) = x_3(k) + ex_2(k) + fx_3(k)x_2(k) + x_5(k)$$
$$- d_1 x_3(k) - gx_3(k)$$

The remaining three equations in (1) would be unchanged.

4 Additional money expenditures can be used to increase the attractiveness of the methadone treatment program to the heroin addict, by offering educational, job training, and even monetary incentives. Such incentives will increase the values of the parameters e and f and probably decrease the dropout parameter g. All these parameters now become time functions $e(k)$, $f(k)$, $g(k)$ and can be considered as additional input control variables.

We close this section with a warning to the reader about the proper use of such models in scientific investigation. To be sure, *if* (and this is a very big if) we could trust the structure of the model, i.e., the functional forms in Eq. (1), and *if* we could trust the numerical values of the parameters that we use in the model, then we could simulate this set of equations on a digital computer and thus obtain a true picture of the changes in the populations and the cost effectiveness of any contemplated decisions.

However, although such models contain all the terms one feels should be there, there is no natural physical law (such as Newton's or Ohm's law) that we can appeal to in order to verify the structure of the equations. Also, there are great difficulties associated with measuring all the variables of interest and computing the parameters. For example, how do we know the number $x_2(k)$ of heroin addicts in a city? If we do not know that number, how can we estimate, say, key parameter values a and f among others? One can say that someone will know, but expert opinions can differ drastically.

For this reason, apparently quantitative models of socioeconomic systems should be viewed more or less in a qualitative manner. One may argue that they are no better than the mental models a politician or social worker may have. This indeed may be true; however, there is still value in setting down equations, because at the very least, one is forced to state clearly all hypotheses and assumptions and subject them to scrutiny and constructive criticism. Even if the functional forms and parameter values are seriously in error, a large number of simulations using different assumptions can present a *qualitative* feel for the cause-and-effect relationship in both the short and long run. The serious engineer interested in socioeconomic problems should always be very careful in communicating the results of his simulations to the general public because it tends to think that numbers generated by a computer are somehow magically correct.

4.4 LINEAR DISCRETE-TIME SYSTEMS

We now begin examining the mathematical aspects of discrete-time systems. In the general nonlinear case the mathematical description is

$$\mathbf{x}(k+1) = \mathbf{f}(\mathbf{x}(k), \mathbf{u}(k)) \tag{1}$$

$$\mathbf{y}(k) = \mathbf{g}(\mathbf{x}(k), \mathbf{u}(k)) \tag{2}$$

and there is very little that we can say, except by explicit numerical simulation. However, there are subclasses of discrete-time systems that enjoy many interesting analytical properties thanks to their special structure.

We focus our attention on a special class of discrete-time systems which satisfy the *principle of superposition*. We first define them from a purely mathematical viewpoint and then illustrate their superposition and time-invariance properties.

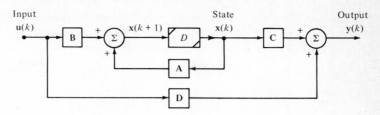

FIGURE 4.4.1
Block-diagram representation of an LTI discrete-time system.

Definition 1 *We say that a discrete-time system is linear and time invariant if the state* $\mathbf{x}(k)$, *input* $\mathbf{u}(k)$, *and output* $\mathbf{y}(k)$ *for all* $k = 0, 1, 2, \ldots$ *are related by the pair of equations (see Fig. 4.4.1):*

$$\mathbf{x}(k + 1) = \mathbf{A}\mathbf{x}(k) + \mathbf{B}\mathbf{u}(k) \qquad \text{state difference equation} \qquad (3)$$

$$\mathbf{y}(k) = \mathbf{C}\mathbf{x}(k) + \mathbf{D}\mathbf{u}(k) \qquad \text{output equation} \qquad (4)$$

where $\mathbf{x}(k)$ = *column n-vector, the* state
$\qquad \mathbf{u}(k)$ = *column m-vector, the* input *or* control
$\qquad \mathbf{y}(k)$ = *column r-vector, the* output
$\qquad \mathbf{A}$ = *constant n × n matrix (system matrix)*
$\qquad \mathbf{B}$ = *constant n × m matrix (input matrix)*
$\qquad \mathbf{C}$ = *constant r × n matrix*
$\qquad \mathbf{D}$ = *constant r × m matrix*

for all $k = 0, 1, 2, \ldots.$

Remark   By constant we mean that the elements of the matrices $\mathbf{A}$, $\mathbf{B}$, $\mathbf{C}$, $\mathbf{D}$ do not change as a function of time, i.e., as the index k changes.

To understand the specific properties of linear discrete systems it is easier to examine them first in the unforced case, i.e., when no input is applied, and then examine them when we apply a nonzero input sequence.

4.4.1   Unforced LTI Discrete-Time Systems

Suppose that no input is applied to the system (3); that is, we set

$$\mathbf{u}(0) = \mathbf{u}(1) = \mathbf{u}(2) = \cdots = \mathbf{u}(k) = \cdots = \mathbf{0} \qquad (5)$$

in Eq. (3). Then the state evolves according to the difference equation

$$\mathbf{x}(k + 1) = \mathbf{Ax}(k) \tag{6}$$

In order for the unforced system (6) to do something interesting, it must start from a nonzero initial state $\mathbf{x}(0)$.

The response of the system can then be readily deduced from the recursive set of linear equations[†]

$$\mathbf{x}(1) = \mathbf{Ax}(0)$$

$$\mathbf{x}(2) = \mathbf{Ax}(1) = \mathbf{A}^2\mathbf{x}(0)$$

$$\mathbf{x}(3) = \mathbf{Ax}(2) = \mathbf{A}^2\mathbf{x}(1) = \mathbf{A}^3\mathbf{x}(0) \tag{7}$$

$$\vdots$$

$$\mathbf{x}(k + 1) = \mathbf{Ax}(k) = \cdots = \mathbf{A}^{k+1}\mathbf{x}(0)$$

From this system of equations we can readily deduce the following facts.

PROPERTY 1   If $\mathbf{x}(0) = \mathbf{0}$, then $\mathbf{x}(k) = \mathbf{0}$ for all k.

PROPERTY 2

$$\mathbf{x}(j) = \mathbf{A}^j\mathbf{x}(0) \tag{8}$$

PROPERTY 3   Let k and j index arbitrary discrete values of time; then for $k \geq j$

$$\boxed{\mathbf{x}(k) = \mathbf{A}^{k-j}\mathbf{x}(j)} \tag{9}$$

Property 3 leads to the following useful definition.

Definition 2   *Let* $\mathbf{A}$ *be an* $n \times n$ *constant matrix and let* k *and* j *be integers,* $k \geq j$. *Then the matrix* $\mathbf{\Phi}(k,j)$, *an* $n \times n$ *matrix,*

$$\boxed{\mathbf{\Phi}(k,j) \triangleq \mathbf{A}^{k-j}} \tag{10}$$

is called the state transition matrix *for the system* $\mathbf{x}(k + 1) = \mathbf{Ax}(k)$.

[†] Recall that $\mathbf{A}^2 \triangleq \mathbf{A} \cdot \mathbf{A}$, $\mathbf{A}^3 \triangleq \mathbf{A} \cdot \mathbf{A} \cdot \mathbf{A}$, etc. The matrix $\mathbf{A}^0$ is by definition the identity matrix.

Although defined in a very simple manner, the state transition matrix has some very interesting properties.

IDENTITY PROPERTY

$$\mathbf{\Phi}(k,k) = \mathbf{I} \qquad \text{for all } k = 0, 1, 2, \ldots \tag{11}$$

SEMIGROUP PROPERTY

$$\mathbf{\Phi}(k,j)\mathbf{\Phi}(j,i) = \mathbf{\Phi}(k,i) \qquad \text{for all } k \geq j \geq i \tag{12}$$

INVERSE PROPERTY   If the inverse matrix $\mathbf{A}^{-1}$ exists, then $\mathbf{\Phi}^{-1}(k,j)$ exists and

$$\mathbf{\Phi}^{-1}(k,j) = \mathbf{\Phi}(j,k) \tag{13}$$

Under these conditions, we can see that the semigroup property (12) holds for all k, j, and i.

COMMUTATIVITY PROPERTY

$$\mathbf{A}\mathbf{\Phi}(k,j) = \mathbf{\Phi}(k,j)\mathbf{A} \tag{14}$$

The verification of all these properties is extremely easy and simply involves the use of Definition 2 for the transition matrix.

Let us now prove how the superposition property holds with respect to the initial state. We formally state the result as a theorem.

Theorem 1   (Superposition)   *Consider the LTI discrete-time system*

$$\mathbf{x}(k + 1) = \mathbf{A}\mathbf{x}(k) \tag{15}$$

Let $\{\hat{\mathbf{x}}(1), \hat{\mathbf{x}}(2), \ldots, \hat{\mathbf{x}}(k), \ldots\}$ denote the state sequence generated by the initial condition $\mathbf{x}(0) = \hat{\mathbf{x}}(0)$. Let $\{\tilde{\mathbf{x}}(1), \tilde{\mathbf{x}}(2), \ldots, \tilde{\mathbf{x}}(k), \ldots\}$ denote the state sequence generated by the initial condition $\mathbf{x}(0) = \tilde{\mathbf{x}}(0)$. Then if system (15) starts at the initial state

$$\mathbf{x}(0) = \alpha\hat{\mathbf{x}}(0) + \beta\tilde{\mathbf{x}}(0) \tag{16}$$

then the resultant state sequence $\{\mathbf{x}(1), \mathbf{x}(2), \ldots, \mathbf{x}(k), \ldots\}$ has the property

$$\mathbf{x}(k) = \alpha\hat{\mathbf{x}}(k) + \beta\tilde{\mathbf{x}}(k) \qquad \text{for all } k = 0, 1, 2, \ldots \tag{17}$$

PROOF   By definition

$$\hat{\mathbf{x}}(k) = \mathbf{\Phi}(k,0)\hat{\mathbf{x}}(0) \qquad \tilde{\mathbf{x}}(k) = \mathbf{\Phi}(k,0)\tilde{\mathbf{x}}(0)$$

But

$$\mathbf{x}(k) = \mathbf{\Phi}(k,0)\mathbf{x}(0) = \mathbf{\Phi}(k,0)[\alpha\hat{\mathbf{x}}(0) + \beta\tilde{\mathbf{x}}(0)]$$
$$= \alpha\mathbf{\Phi}(k,0)\hat{\mathbf{x}}(0) + \beta\mathbf{\Phi}(k,0)\tilde{\mathbf{x}}(0) = \alpha\hat{\mathbf{x}}(k) + \beta\tilde{\mathbf{x}}(k)$$

Next, we clarify the notion of time invariance. Once more we formally state this property as a theorem.

Theorem 2   (Time invariance)   *Consider the discrete-time system*

$$\mathbf{x}(k + 1) = \mathbf{A}\mathbf{x}(k) \tag{18}$$

Suppose that $\mathbf{x}(j) = \boldsymbol{\xi}$ *and let* $\{\mathbf{x}(j + 1), \mathbf{x}(j + 2), \ldots\}$ *denote the resulting state sequence. Suppose that* $\mathbf{x}(j + m) = \boldsymbol{\xi}$, *and let* $\{\mathbf{x}(j + m + 1), \mathbf{x}(j + m + 2), \ldots\}$ *denote the resulting state sequence. Then*

$$\mathbf{x}(j + k) = \mathbf{x}(j + m + k) \qquad \textit{for all } k = 0, 1, 2, \ldots \tag{19}$$

PROOF   When $\mathbf{x}(j) = \boldsymbol{\xi}$, then

$$\mathbf{x}(j + k) = \mathbf{\Phi}(j + k, j)\mathbf{x}(j) = \mathbf{\Phi}(j + k, j)\boldsymbol{\xi}$$

When $\mathbf{x}(j + m) = \boldsymbol{\xi}$, then

$$\mathbf{x}(j + m + k) = \mathbf{\Phi}(j + m + k, j + m)\mathbf{x}(j + m)$$
$$= \mathbf{\Phi}(j + m + k, j + m)\boldsymbol{\xi}$$

But by Eq. (10)

$$\mathbf{\Phi}(j + k, j) = \mathbf{A}^{j+k-j} = \mathbf{A}^k$$
$$\mathbf{\Phi}(j + m + k, j + m) = \mathbf{A}^{j+m+k-(j+m)} = \mathbf{A}^k$$

Hence,

$$\mathbf{\Phi}(j + m + k, j + m) = \mathbf{\Phi}(j + k, j)$$

which implies

$$\mathbf{x}(j + k) = \mathbf{x}(j + m + k)$$

4.4.2   Forced LTI Discrete-Time Systems

Let us now return to the case that the input sequence is nonzero. The state difference equation is

$$\mathbf{x}(k+1) = \mathbf{Ax}(k) + \mathbf{Bu}(k) \tag{20}$$

If we know the initial state $\mathbf{x}(0)$ and the input sequence $\{\mathbf{u}(0), \mathbf{u}(1), \mathbf{u}(2), \ldots\}$, we can readily compute the resultant state sequence $\{\mathbf{x}(1), \mathbf{x}(2), \ldots, \mathbf{x}(k), \ldots\}$. This can be seen immediately by writing the recursive equations

$$
\begin{aligned}
\mathbf{x}(1) &= \mathbf{Ax}(0) + \mathbf{Bu}(0) = \boldsymbol{\Phi}(1,0)\mathbf{x}(0) + \mathbf{Bu}(0) \\
\mathbf{x}(2) &= \mathbf{Ax}(1) + \mathbf{Bu}(1) = \mathbf{A}[\mathbf{Ax}(0) + \mathbf{Bu}(0)] + \mathbf{Bu}(1) \\
&= \mathbf{A}^2\mathbf{x}(0) + \mathbf{ABu}(0) + \mathbf{Bu}(1) \\
&= \boldsymbol{\Phi}(2,0)\mathbf{x}(0) + \boldsymbol{\Phi}(1,0)\mathbf{Bu}(0) + \mathbf{Bu}(1) \\
\mathbf{x}(3) &= \mathbf{Ax}(2) + \mathbf{Bu}(2) = \mathbf{A}[\mathbf{A}^2\mathbf{x}(0) + \mathbf{ABu}(0) + \mathbf{Bu}(1)] + \mathbf{Bu}(2) \\
&= \mathbf{A}^3\mathbf{x}(0) + \mathbf{A}^2\mathbf{Bu}(0) + \mathbf{ABu}(1) + \mathbf{Bu}(2) \\
&= \boldsymbol{\Phi}(3,0)\mathbf{x}(0) + \boldsymbol{\Phi}(2,0)\mathbf{Bu}(0) + \boldsymbol{\Phi}(1,0)\mathbf{Bu}(1) + \mathbf{Bu}(2)
\end{aligned}
\tag{21}
$$

$$\vdots$$

Let us examine these equations and see if there is a general pattern; our task will be easier if we recall that

$$\mathbf{A}^0 \triangleq \mathbf{I} \tag{22}$$

or

$$\boldsymbol{\Phi}(k,k) \triangleq \mathbf{A}^{k-k} = \mathbf{I} \tag{23}$$

One way of writing the jth equation of (21) is

$$\mathbf{x}(j) = \mathbf{A}^j\mathbf{x}(0) + \mathbf{A}^{j-1}\mathbf{Bu}(0) + \mathbf{A}^{j-2}\mathbf{Bu}(1) + \cdots + \mathbf{A}^0\mathbf{Bu}(j-1) \tag{24}$$

Equation (24) can be written, in particular, in the form

$$\mathbf{x}(j) = \mathbf{A}^j\mathbf{x}(0) + \sum_{i=0}^{j-1} \mathbf{A}^{j-(i+1)}\mathbf{Bu}(i) \tag{25}$$

In turn, Eq. (25) can be written in terms of transition matrices as

$$\boxed{\mathbf{x}(j) = \boldsymbol{\Phi}(j,0)\mathbf{x}(0) + \sum_{i=0}^{j-1} \boldsymbol{\Phi}(j,\,i+1)\mathbf{Bu}(i)} \tag{26}$$

The usefulness of this representation becomes apparent when we attempt to relate the state $\mathbf{x}_k$ to the state $\mathbf{x}_j$ for $k \geq j$. To see this, let us write the state $\mathbf{x}_k$ in terms of $\mathbf{x}_0$; using Eq. (26) and letting $j = k$, we obtain

$$\mathbf{x}(k) = \mathbf{\Phi}(k,0)\mathbf{x}(0) + \sum_{i=0}^{k-1} \mathbf{\Phi}(k,\ i+1)\mathbf{Bu}(i) \tag{27}$$

or, by splitting the sum

$$\mathbf{x}(k) = \mathbf{\Phi}(k,0)\mathbf{x}(0) + \sum_{i=0}^{j-1} \mathbf{\Phi}(k,\ i+1)\mathbf{Bu}(i) + \sum_{i=j}^{k-1} \mathbf{\Phi}(k,\ i+1)\mathbf{Bu}(i) \tag{28}$$

Recall that the semigroup property allows us to write

$$\mathbf{\Phi}(k,0) = \mathbf{\Phi}(k,j)\mathbf{\Phi}(j,0) \qquad k \geq j \tag{29}$$

$$\mathbf{\Phi}(k,\ i+1) = \mathbf{\Phi}(k,j)\mathbf{\Phi}(j,\ i+1) \tag{30}$$

Substituting Eqs. (29) and (30) into the right-hand side of Eq. (28), we obtain

$$\mathbf{x}(k) = \underbrace{\mathbf{\Phi}(k,j)\mathbf{\Phi}(j,0)}_{\mathbf{\Phi}(k,0)}\mathbf{x}(0) + \sum_{i=0}^{j-1} \underbrace{\mathbf{\Phi}(k,j)\mathbf{\Phi}(j,\ i+1)}_{\mathbf{\Phi}(k-1,i)}\mathbf{Bu}(i)$$

$$+ \sum_{i=j}^{k-1} \mathbf{\Phi}(k,i+1)\mathbf{Bu}(i) \tag{31}$$

Since $\mathbf{\Phi}(k-1,j-1)$ is independent of the summation index i, it can be factored out to obtain

$$\mathbf{x}(k) = \mathbf{\Phi}(k,j)\mathbf{\Phi}(j,0)\mathbf{x}(0) + \mathbf{\Phi}(k,j)\sum_{i=0}^{j-1} \mathbf{\Phi}(j,i+1)\mathbf{Bu}(i)$$

$$+ \sum_{i=j}^{k-1} \mathbf{\Phi}(k,i+1)\mathbf{Bu}(i) \tag{32}$$

The matrix $\mathbf{\Phi}(k,j)$ can now be factored out in Eq. (32) to yield

$$\mathbf{x}(k) = \mathbf{\Phi}(k,j)\left[\mathbf{\Phi}(j,0)\mathbf{x}(0) + \sum_{i=0}^{j-1} \mathbf{\Phi}(j,i+1)\mathbf{Bu}(i)\right] + \sum_{i=j}^{k-1} \mathbf{\Phi}(k-1,i)\mathbf{Bu}(i) \tag{33}$$

However, the vector in brackets is precisely $\mathbf{x}(j)$, given by Eq. (26). Hence, we have established the relation

$$\boxed{\mathbf{x}(k) = \mathbf{\Phi}(k,j)\mathbf{x}(j) + \sum_{i=j}^{k-1} \mathbf{\Phi}(k,i+1)\mathbf{Bu}(i)} \tag{34}$$

The linearity of the system allows us to develop its superposition properties.

Theorem 3 (Superposition property) *Consider the LTI system starting at rest, that is* $\mathbf{x}(0) = \mathbf{0}$,

$$\mathbf{x}(k + 1) = \mathbf{A}\mathbf{x}(k) + \mathbf{B}\mathbf{u}(k) \qquad \mathbf{x}(0) = \mathbf{0} \tag{35}$$

Let $\{\hat{\mathbf{x}}(1), \hat{\mathbf{x}}(2), \ldots\}$ *denote the state sequence resulting from the input sequence* $\{\hat{\mathbf{u}}(0), \hat{\mathbf{u}}(1), \ldots\}$. *Let* $\{\tilde{\mathbf{x}}(1), \tilde{\mathbf{x}}(2), \ldots\}$ *denote the state sequence resulting from the input sequence* $\{\tilde{\mathbf{u}}(0), \tilde{\mathbf{u}}(1), \ldots\}$. *Let* $\{\mathbf{u}(0), \mathbf{u}(1), \ldots\}$ *denote an input sequence such that*

$$\mathbf{u}(k) = \alpha\hat{\mathbf{u}}(k) + \beta\tilde{\mathbf{u}}(k) \qquad k = 0, 1, 2, \ldots \tag{36}$$

Then the resultant state sequence $\{\mathbf{x}(1), \mathbf{x}(2), \ldots\}$ *due to the input* (36) *has the property that*

$$\mathbf{x}(k) = \alpha\hat{\mathbf{x}}(k) + \beta\tilde{\mathbf{x}}(k) \qquad k = 1, 2, \ldots \tag{37}$$

PROOF The state sequences are given [in view of (34) and $\mathbf{x}(0) = \mathbf{0}$] by

$$\tilde{\mathbf{x}}(k) = \sum_{i=0}^{k-1} \mathbf{\Phi}(k, i + 1)\mathbf{B}\tilde{\mathbf{u}}(i)$$

$$\hat{\mathbf{x}}(k) = \sum_{i=0}^{k-1} \mathbf{\Phi}(k, i + 1)\mathbf{B}\hat{\mathbf{u}}(i)$$

$$\mathbf{x}(k) = \sum_{i=0}^{k-1} \mathbf{\Phi}(k, i + 1)\mathbf{B}\mathbf{u}(i)$$

Hence, from Eq. (36),

$$\mathbf{x}(k) = \sum_{i=0}^{k-1} \mathbf{\Phi}(k, i + 1)\mathbf{B}\{\alpha\hat{\mathbf{u}}(i) + \beta\tilde{\mathbf{u}}(i)\}$$

$$= \alpha \sum_{i=0}^{k-1} \mathbf{\Phi}(k, i + 1)\mathbf{B}\hat{\mathbf{u}}(i) + \beta \sum_{i=0}^{k-1} \mathbf{\Phi}(k, i + 1)\mathbf{B}\tilde{\mathbf{u}}(i)$$

$$= \alpha\hat{\mathbf{x}}(k) + \beta\tilde{\mathbf{x}}(k)$$

The superposition property with respect to the inputs is valid only if the initial state $\mathbf{x}(0) = \mathbf{0}$; that is, the system starts at rest.

4.5 SOLUTION METHOD USING EIGENVALUES

In this section, we approach the solution of LTI discrete-time systems by another avenue, namely, one that brings into focus the role of the eigenvalues of the **A** matrix in the time evolution of the solution. This in turn will enable us later to investigate the notion of the stability of an LTI discrete-time system.

The eigenvalue analysis is not crucial in our ability to obtain numerical solutions to difference equations, since this is an easy task with modern computers, but it enables us to understand certain key properties of linear systems that are important in both system analysis and system design.

Let us return to the unforced LTI system

$$\boxed{\mathbf{x}(k + 1) = \mathbf{A}\mathbf{x}(k)} \tag{1}$$

We have seen that we can express its solution by

$$\boxed{\mathbf{x}(k) = \mathbf{\Phi}(k,j)\mathbf{x}(j) = \mathbf{A}^{k-j}\mathbf{x}(j) \qquad k \geq j} \tag{2}$$

This solution formula is very nice from the viewpoint of obtaining numbers. However, it does not bring into focus certain properties of the dynamical system; such properties can be found by examining the dependence of the solution upon the eigenvalues and eigenvectors of the system matrix $\mathbf{A}$.

Let us denote the eigenvalues of the matrix $\mathbf{A}$ by $\lambda_1, \lambda_2, \ldots, \lambda_n$ and its eigenvectors by $\mathbf{v}_1, \mathbf{v}_2, \ldots, \mathbf{v}_n$. We recall (see Appendix A, Secs. A.9 and A.10) that

$$\mathbf{A}\mathbf{v}_i = \lambda_i \mathbf{v}_i \qquad i = 1, 2, \ldots, n \tag{3}$$

For simplicity we shall assume in the remainder of this section that the eigenvalues are distinct.

First, let us see what happens if the initial state $\mathbf{x}(0)$ of system (1) happens to be equal to one of the eigenvectors of the $\mathbf{A}$ matrix. Suppose that the initial state equals the eigenvector $\mathbf{v}_i$, that is,

$$\mathbf{x}(0) = \mathbf{v}_i \tag{4}$$

Then, in view of Eq. (3), we can see that

$$\begin{aligned}
\mathbf{x}(1) &= \mathbf{A}\mathbf{x}(0) = \mathbf{A}\mathbf{v}_i = \lambda_i \mathbf{v}_i \\
\mathbf{x}(2) &= \mathbf{A}\mathbf{x}(1) = \mathbf{A}\lambda_i \mathbf{v}_i = \lambda_i \mathbf{A}\mathbf{v}_i = \lambda_i^2 \mathbf{v}_i \\
&\vdots
\end{aligned} \tag{5}$$

and, in general, $\qquad \mathbf{x}(k) = \lambda_i^k \mathbf{v}_i$

Hence we make the important observation that if the initial state $\mathbf{x}(0)$ coincides with one of the eigenvectors of the $\mathbf{A}$ matrix, then the state vector $\mathbf{x}(k)$ simply scales the eigenvector by the corresponding eigenvalue raised to the kth power. In other words, the state always moves along the same direction as the eigenvector.

EXAMPLE 1 To illustrate this idea let us consider the second-order system

$$\underbrace{\begin{bmatrix} x_1(k+1) \\ x_2(k+1) \end{bmatrix}}_{\mathbf{x}(k+1)} = \underbrace{\begin{bmatrix} 0 & 1 \\ -2 & 3 \end{bmatrix}}_{\mathbf{A}} \underbrace{\begin{bmatrix} x_1(k) \\ x_2(k) \end{bmatrix}}_{\mathbf{x}(k)}$$

The characteristic polynomial of $\mathbf{A}$ is

$$p(\lambda) = \det(\lambda\mathbf{I} - \mathbf{A}) = \lambda(\lambda - 3) + 2 = \lambda^2 - 3\lambda + 2$$
$$= (\lambda - 1)(\lambda - 2) \tag{7}$$

Hence the eigenvalues of the matrix $\mathbf{A}$ are at

$$\lambda_1 = 1 \qquad \lambda_2 = 2 \tag{8}$$

Let $\mathbf{v}_1$ and $\mathbf{v}_2$ denote the corresponding eigenvectors of $\mathbf{A}$ defined by

$$\mathbf{A}\mathbf{v}_1 = \mathbf{v}_1 \qquad \mathbf{A}\mathbf{v}_2 = 2\mathbf{v}_2 \tag{9}$$

It is easy to find out that (recall that the magnitude of the eigenvectors is not unique)

$$\mathbf{v}_1 = \begin{bmatrix} 1 \\ 1 \end{bmatrix} \qquad \mathbf{v}_2 = \begin{bmatrix} 1 \\ 2 \end{bmatrix} \tag{10}$$

Suppose $\mathbf{x}(0) = \mathbf{v}(1)$; then, according to Eq. (5), we have

$$\mathbf{x}(k) = (1)^k \mathbf{v}_1 = \mathbf{v}_1 \qquad \text{for all } k \tag{11}$$

To double-check this we see that

$$\mathbf{x}(1) = \begin{bmatrix} 0 & 1 \\ -2 & 3 \end{bmatrix} \begin{bmatrix} 1 \\ 1 \end{bmatrix} = \begin{bmatrix} 1 \\ 1 \end{bmatrix} \tag{12}$$

$$\mathbf{x}(2) = \mathbf{A}\mathbf{x}(1) = \begin{bmatrix} 0 & 1 \\ -2 & 3 \end{bmatrix} \begin{bmatrix} 1 \\ 1 \end{bmatrix} = \begin{bmatrix} 1 \\ 1 \end{bmatrix} \tag{13}$$

and so on.

Now suppose that $\mathbf{x}(0) = \mathbf{v}_2$; then, according to Eq. (5),

$$\mathbf{x}(k) = (2)^k \begin{bmatrix} 1 \\ 2 \end{bmatrix} \tag{14}$$

To double-check we compute

$$\mathbf{x}(1) = \mathbf{A}\mathbf{x}(0) = \begin{bmatrix} 0 & 1 \\ -2 & 3 \end{bmatrix}\begin{bmatrix} 1 \\ 2 \end{bmatrix} = \begin{bmatrix} 2 \\ 4 \end{bmatrix} = (2)\begin{bmatrix} 1 \\ 2 \end{bmatrix} \tag{15}$$

$$\mathbf{x}(2) = \mathbf{A}\mathbf{x}(1) = \begin{bmatrix} 0 & 1 \\ -2 & 3 \end{bmatrix}(2)\begin{bmatrix} 1 \\ 2 \end{bmatrix} = (2)^2\begin{bmatrix} 1 \\ 2 \end{bmatrix} \tag{16}$$

and so on.

Now recall that the eigenvectors of $\mathbf{A}$ span the euclidean space R_n; that is, they form a linearly independent set.[†] Hence, they form a basis. Therefore, any vector in R_n can be written as a linear combination of the eigenvectors of $\mathbf{A}$.

In particular, let us express the initial-state vector $\mathbf{x}(0)$ as a linear combination of the eigenvectors $\mathbf{v}_1, \mathbf{v}_2, \ldots, \mathbf{v}_n$ of the system matrix $\mathbf{A}$. Hence, suppose that

$$\mathbf{x}(0) = \alpha_1 \mathbf{v}_1 + \alpha_2 \mathbf{v}_2 + \cdots + \alpha_n \mathbf{v}_n = \sum_{i=1}^{n} \alpha_i \mathbf{v}_i \tag{17}$$

where $\alpha_1, \alpha_2, \ldots, \alpha_n$ are the coordinates of $\mathbf{x}(0)$ in the eigenvector coordinate system. In general, we know that

$$\mathbf{x}(k) = \mathbf{A}^k \mathbf{x}(0) \tag{18}$$

Substituting Eq. (17) into Eq. (18), we obtain

$$\begin{aligned} \mathbf{x}(k) &= \mathbf{A}^k \sum_{i=1}^{n} \alpha_i \mathbf{v}_i = \mathbf{A}^{k-1} \sum_{i=1}^{n} \alpha_i \mathbf{A}\mathbf{v}_i \\ &= \mathbf{A}^{k-1} \sum_{i=1}^{n} \alpha_i \lambda_i \mathbf{v}_i \\ &= \mathbf{A}^{k-2} \sum_{i=1}^{n} \alpha_i \lambda_i \mathbf{A}\mathbf{v}_i \\ &= \mathbf{A}^{k-2} \sum_{i=1}^{n} \alpha_i \lambda_i^2 \mathbf{v}_i = \cdots \end{aligned} \tag{19}$$

Proceeding in this manner, we arrive at the following theorem.

Theorem 1 *Consider the LTI discrete-time system*

$$\boxed{\mathbf{x}(k + 1) = \mathbf{A}\mathbf{x}(k)} \tag{20}$$

Let $\lambda_1, \lambda_2, \ldots, \lambda_n$ denote the (assumed distinct) eigenvalues of $\mathbf{A}$. Let $\mathbf{v}_1, \mathbf{v}_2, \ldots, \mathbf{v}_n$ be the corresponding eigenvectors of $\mathbf{A}$. If the initial state is

---

† This is always true when the eigenvalues of $\mathbf{A}$ are distinct, as we have assumed.

$$\mathbf{x}(0) = \sum_{i=1}^{n} \alpha_i \mathbf{v}_i \tag{21}$$

then the state at time k is

$$\mathbf{x}(k) = \sum_{i=1}^{n} \alpha_i (\lambda_i)^k \mathbf{v}_i \tag{22}$$

This theorem establishes a fundamental result. It implies that the state $\mathbf{x}(k)$ at time k involves the addition of vectors which are multiplied by the eigenvalues of the $\mathbf{A}$ matrix raised to the kth power. This property will be exploited in the sequel to establish some easy tests for stability.

EXAMPLE 1 (*continued*) Suppose that the initial-state vector $\mathbf{x}(0)$ was the vector

$$\mathbf{x}(0) = \begin{bmatrix} 1 \\ -2 \end{bmatrix} \tag{23}$$

This vector $\mathbf{x}(0)$ can be written in terms of the eigenvectors $\mathbf{v}_1$ and $\mathbf{v}_2$ [see Eq. (10)] as follows:

$$\underbrace{\begin{bmatrix} 1 \\ -2 \end{bmatrix}}_{\mathbf{x}(0)} = \underbrace{4 \begin{bmatrix} 1 \\ 1 \end{bmatrix}}_{\alpha_1 \mathbf{v}_1} - \underbrace{3 \begin{bmatrix} 1 \\ 2 \end{bmatrix}}_{\alpha_2 \mathbf{v}_2} \tag{24}$$

Suppose that $k = 3$. Then according to the theorem,

$$\mathbf{x}(3) = \sum_{i=1}^{2} \alpha_i (\lambda_i)^3 \mathbf{v}_i = 4(1)^3 \begin{bmatrix} 1 \\ 1 \end{bmatrix} + (-3)(2)^3 \begin{bmatrix} 1 \\ 2 \end{bmatrix}$$

$$= \begin{bmatrix} 4 \\ 4 \end{bmatrix} - \begin{bmatrix} 24 \\ 48 \end{bmatrix} = \begin{bmatrix} -20 \\ -44 \end{bmatrix} \tag{25}$$

Let us double-check; we know that $\mathbf{x}(3) = \mathbf{A}^3 \mathbf{x}(0)$. We find that

$$\mathbf{A}^3 = \begin{bmatrix} -6 & 7 \\ -14 & 15 \end{bmatrix} \tag{26}$$

Hence,

$$\mathbf{x}(3) = \begin{bmatrix} -6 & 7 \\ -14 & 15 \end{bmatrix} \begin{bmatrix} 1 \\ -2 \end{bmatrix} = \begin{bmatrix} -20 \\ -44 \end{bmatrix} \tag{27}$$

which checks.

4.6 STABILITY OF DISCRETE-TIME SYSTEMS

In this section we examine the notion of stability for both linear and nonlinear discrete-time dynamical systems. The notion of stability is of paramount importance in a great variety of engineering and nonengineering systems because it deals with the following type of question: as time goes on, will the actual system response stay near its desired response, or will it diverge? If a system is unstable, external means of stabilizing it are necessary. This often requires the use of *feedback*.

The type of stability questions we shall examine should be viewed with some care. If a linear system is unstable, then *mathematically* the magnitude $\|\mathbf{x}(k)\|$ of its state vector tends to infinity as the time unit $k \to \infty$. If we think of the difference equation as a mathematical model for a real process, it should be obvious that no matter what type of physical system we are dealing with, it cannot possibly have the property that its variables become infinite. In such problems it is important to realize that if we find that a physical system is unstable, then as time goes on, its variables become so large that the physical system may break, burn, saturate, and so on. Often the mathematical model for a physical system is valid only for a certain range of its variables; if that mathematical model is unstable, it may mean that eventually the growth of its variables will bring it into a mode of operation which requires a different mathematical model for an adequate mathematical description.

The instability of a mathematical model points out that eventually the physical process may get in trouble or we may require a different mathematical model for its representation.

To study this intuitive notion of stability in a somewhat precise framework it becomes necessary to offer certain definitions. Since these definitions are common for both linear and nonlinear systems, we shall state them in the context of nonlinear time-invariant systems described by the vector difference equation

$$\mathbf{x}(k + 1) = \mathbf{f}(\mathbf{x}(k), \mathbf{u}(k)) \qquad k = 0, 1, 2, \ldots \tag{1}$$

where $\mathbf{x}(k)$ is the state vector at time k and $\mathbf{u}(k)$ is the input vector at time k.

First, let us consider the case of an unforced nonlinear system, i.e., a system with $\mathbf{u}(k) = \mathbf{0}$ for all k. In this case, the time evolution of the state $\mathbf{x}(k)$ is primarily governed by its initial state $\mathbf{x}(0)$ and its dynamics.

Definition 1 **(Equilibrium state for unforced systems)** *Consider a discrete-time system described by the difference equation*

$$\mathbf{x}(k + 1) = \mathbf{f}(\mathbf{x}(k)) \qquad k = 0, 1, 2, \ldots \tag{2}$$

with $\mathbf{x}(k) \in R_n$ for all k. We say that a vector $\mathbf{x}_e \in R_n$ is an equilibrium state for the system (2) if it satisfies the vector equality

$$\mathbf{x}_e = \mathbf{f}(\mathbf{x}_e) \tag{3}$$

Remark A nonlinear system may have more than one equilibrium state. The existence and uniqueness of an equilibrium state are related to the existence and uniqueness of solutions to the nonlinear vector equation $\mathbf{x} = \mathbf{f}(\mathbf{x})$.

The reader should notice how nonlinear vector algebraic equations pop up once more. Algebraic equations of the form $\mathbf{x} = \mathbf{f}(\mathbf{x})$ were examined in Sec. 3.5, and sufficient conditions for the existence and uniqueness of solutions were given when we discussed the Picard algorithm.

Theorem 1 **(Equilibrium state for unforced linear systems)** *Consider the LTI system*

$$\mathbf{x}(k + 1) = \mathbf{A}\mathbf{x}(k) \tag{4}$$

Then

$$\mathbf{x}_e = \mathbf{0} \tag{5}$$

is always an equilibrium state. If $\mathbf{I} - \mathbf{A}$ is nonsingular, then $\mathbf{0}$ is the unique equilibrium state. If $\mathbf{I} - \mathbf{A}$ is singular, then all elements of the null space of $\mathbf{I} - \mathbf{A}$, that is, the set of vectors such that $(\mathbf{I} - \mathbf{A})\mathbf{x} = \mathbf{0}$, are equilibrium states.

The proof is left as an exercise for the reader.

In general, the notion of stability is intimately related to the behavior of the system with respect to its equilibrium state. The questions that typically arise in stability theory are of the following type.

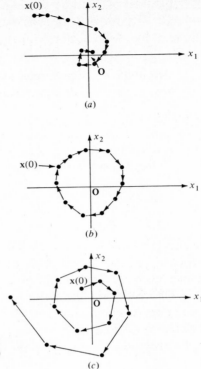

FIGURE 4.6.1
Visualization of stability and instability:
(*a*) asymptotic stability; (*b*) stability; (*c*)
instability.

Suppose that the initial state $\mathbf{x}(0)$ of the system is "near" an equilibrium state $\mathbf{x}_e$. Then, as the time index k increases, does the state $\mathbf{x}(k)$ approach the equilibrium state? Does it stay close to it? Does it diverge?

Since nonlinear systems may in general have many equilibrium states, the majority of the stability questions one can reasonably ask are *local*. By local we mean properties of the system in the vicinity of the equilibrium state. However, as the reader may suspect, in the case of linear systems one can usually say much more. Thus, in linear systems one may ask *global* stability questions, especially when the origin $\mathbf{0}$ of the state space is the only equilibrium state. Thus, for linear systems one can ask whether for any initial state the system state would tend to the origin (strict or asymptotic stability), would remain bounded (stability), or would diverge toward infinity (instability), as shown in Fig. 4.6.1.

These intuitive notions can be formalized by the following definitions.

Definition 2   *Consider the system*

$$\mathbf{x}(k + 1) = \mathbf{f}(\mathbf{x}(k)) \tag{6}$$

and let $\mathbf{x}_e$ *denote an equilibrium state.*

(a)   If for $\|\mathbf{x}(0) - \mathbf{x}_e\| \leq \epsilon$,

$$\lim_{k \to \infty} \mathbf{x}(k) = \mathbf{x}_e$$

then we say that the system is locally strictly or asymptotically stable with respect to $\mathbf{x}_e$.
(b)   If for all $\mathbf{x}(0) \in R_n$,

$$\lim_{k \to \infty} \mathbf{x}(k) = \mathbf{x}_e$$

then we say that the system is globally strictly stable with respect to $\mathbf{x}_e$.

Let us now examine the class of LTI discrete-time systems and present some results pertaining to their stability with respect to the origin. We state the main result in the following theorem.

Theorem 2   (Stability of LTI systems)   *Consider the system*

$$\mathbf{x}(k + 1) = \mathbf{A}\mathbf{x}(k) \tag{7}$$

The system (7) *is globally strictly or asymptotically stable with respect to the origin, i.e.,*

$$\lim_{k \to \infty} \mathbf{x}(k) = \mathbf{0} \tag{8}$$

if and only if the magnitude of all of the eigenvalues of $\mathbf{A}$ *is strictly less than unity.*

PROOF   The theorem is true even if some of the eigenvalues of $\mathbf{A}$ are repeated. We shall present the proof only in the case where the eigenvalues are assumed distinct.

We have seen in the previous section that the solution of Eq. (7) can be written in the form

$$\mathbf{x}(k) = \sum_{i=1}^{n} \alpha_i (\lambda_i)^k \mathbf{v}_i \tag{9}$$

where $\lambda_1, \lambda_2, \ldots, \lambda_n$ are the eigenvalues of $\mathbf{A}$ and $\mathbf{v}_1, \mathbf{v}_2, \ldots, \mathbf{v}_n$ are the corresponding eigenvectors. To obtain Eq. (9) we wrote the initial state $\mathbf{x}(0)$ as a linear combination of the eigenvectors

$$\mathbf{x}(0) = \sum_{i=1}^{n} \alpha_i \mathbf{v}_i \tag{10}$$

Using the properties of norms (see Appendix A, Sec. A.15), we know that

$$\|\mathbf{x}(k)\| \leq \sum_{i=1}^{n} |\alpha_i| |\lambda_i|^k \|\mathbf{v}_i\| \tag{11}$$

In general, if $\|\mathbf{x}(0)\| < \infty$, then $|\alpha_i|$ and $\|\mathbf{v}_i\|$ are finite; we can now see that

$$\lim_{k \to \infty} \sum_{i=1}^{n} |\alpha_i| |\lambda_i|^k \|\mathbf{v}_i\| = 0 \tag{12}$$

if and only if

$$|\lambda_i| < 1 \qquad \text{for all } i = 1, 2, \ldots, n \tag{13}$$

Hence, Eq. (13) is necessary and sufficient to have

$$\lim_{k \to \infty} \|\mathbf{x}(k)\| = 0 \tag{14}$$

which implies in turn that

$$\lim_{k \to \infty} \mathbf{x}(k) = \mathbf{0} \tag{15}$$

The proof of this theorem allows us to make certain observations which we also state for the sake of completeness as corollaries. Their formal proof hinges on examination of Eq. (9) or (11).

Corollary 1 *Consider the system $\mathbf{x}(k + 1) = \mathbf{A}\mathbf{x}(k)$ and let $\lambda_1, \lambda_2, \ldots, \lambda_n$ be the eigenvalues of $\mathbf{A}$, which are assumed to be distinct. The system is stable, i.e.,*

$$\lim_{k \to \infty} \|\mathbf{x}_k\| < \infty \tag{16}$$

if and only if

$$|\lambda_i| \leq 1 \qquad \text{for all } i = 1, 2, \ldots, n \tag{17}$$

In this case, the system is called stable. If any one of the eigenvalues has magnitude greater than unity, i.e., if

$$|\lambda_i| > 1 \qquad \text{for some } i \tag{18}$$

then

$$\lim_{k \to \infty} \|\mathbf{x}(k)\| \to \infty \tag{19}$$

and the system is called unstable.

The fact that the determination of the eigenvalues of a matrix is a relatively simple task with current digital computers means that tests of stability for LTI systems are very easy to make even for high-dimensional systems.

The basic stability result expressed in terms of whether the state of an unforced system does tend to the origin can be used to resolve stability questions that arise when one examines forced linear systems. To illustrate this point, let us suppose that we deal with a forced linear discrete-time system described by

$$\mathbf{x}(k+1) = \mathbf{A}\mathbf{x}(k) + \mathbf{B}\mathbf{u}(k) \tag{20}$$

Suppose that we fix the input sequence $\{\mathbf{u}(0), \mathbf{u}(1), \dots\}$. If the system starts at the initial state

$$\mathbf{x}(0) = \hat{\mathbf{x}}(0) \tag{21}$$

we denote its resultant response by $\{\hat{\mathbf{x}}(1), \hat{\mathbf{x}}(2), \dots\}$. If the system starts at the initial state

$$\mathbf{x}(0) = \tilde{\mathbf{x}}(0) \tag{22}$$

we denote its resultant response by $\{\tilde{\mathbf{x}}(1), \tilde{\mathbf{x}}(2), \dots\}$. In both cases the *same* input sequence is applied. We can now ask: Suppose that $\tilde{\mathbf{x}}(0) \neq \hat{\mathbf{x}}(0)$; under what conditions will the two sets of responses get close to each other?

In this type of problem we are interested in properties of the difference of the solution vectors

$$\hat{\mathbf{x}}(k) - \tilde{\mathbf{x}}(k) \tag{23}$$

as k changes. For example, if

$$\lim_{k \to \infty} [\hat{\mathbf{x}}(k) - \tilde{\mathbf{x}}(k)] = \mathbf{0} \tag{24}$$

we can conclude that the long-term system response is relatively insensitive to different initial conditions.

To see how such questions can be resolved let us write the difference equations satisfied by the two different responses

$$\hat{\mathbf{x}}(k+1) = \mathbf{A}\hat{\mathbf{x}}(k) + \mathbf{B}\mathbf{u}(k) \tag{25}$$

$$\tilde{\mathbf{x}}(k+1) = \mathbf{A}\tilde{\mathbf{x}}(k) + \mathbf{B}\mathbf{u}(k) \tag{26}$$

Subtracting, we obtain

$$\hat{\mathbf{x}}(k+1) - \tilde{\mathbf{x}}(k+1) = \mathbf{A}[\hat{\mathbf{x}}(k) - \tilde{\mathbf{x}}(k)] \tag{27}$$

which is now an *unforced* system with respect to the difference vector $\hat{\mathbf{x}}(k) - \tilde{\mathbf{x}}(k)$. Hence, if all eigenvalues λ_i of $\mathbf{A}$ have the property $|\lambda_i| < 1$, we can conclude that

$$\lim_{k \to \infty} [\hat{\mathbf{x}}(k) - \tilde{\mathbf{x}}(k)] = \mathbf{0} \qquad (28)$$

i.e., the two solutions will approach each other, while if any one of the eigenvalues has magnitude greater than unity, the two solutions will be diverging from each other.

EXAMPLE 1 The second-order system

$$\begin{bmatrix} x_1(k + 1) \\ x_2(k + 1) \end{bmatrix} = \begin{bmatrix} 1 & 1 \\ 0 & 2 \end{bmatrix} \begin{bmatrix} x_1(k) \\ x_2(k) \end{bmatrix}$$

is unstable because the eigenvalues of the system matrix, which are the roots of the characteristic polynomial

$$p(\lambda) = (\lambda - 1)(\lambda - 2)$$

are at $\lambda = 1$, $\lambda = 2$. Since one of the eigenvalues is greater than unity, the system is unstable.

On the other hand, the second-order system

$$\begin{bmatrix} x_1(k + 1) \\ x_2(k + 1) \end{bmatrix} = \begin{bmatrix} 0.8 & 1 \\ 0 & 0.9 \end{bmatrix} \begin{bmatrix} x_1(k) \\ x_2(k) \end{bmatrix}$$

is strictly stable because its eigenvalues are at $\lambda = 0.8$ and $\lambda = 0.9$ and both are less than 1 in magnitude.

4.7 STATIC LINEARIZATION AND EQUILIBRIUM STABILITY OF NONLINEAR SYSTEMS

In Sec. 4.6 we analyzed the notions of stability for nonlinear and linear systems. For linear systems we deduced a very simple test for stability involving the eigenvalues of the system matrix. In this section we present techniques which are useful in the stability analysis of nonlinear forced discrete-time systems about an input-state equilibrium pair. In general, it is impossible to establish *global* stability results for nonlinear systems. Hence, we shall settle for *local* stability results.

The method of attack is called *static linearization*. In essence, we linearize a nonlinear system about a possible equilibrium condition. The method of linearization essentially involves using Taylor series expansions and retaining only the linear terms. This technique will yield a *linear system which approximates the behavior of the nonlinear system in the immediate vicinity of the equilibrium*. Then, by studying the stability of the linear system we can deduce *local* stability properties of the nonlinear system.†

4.7.1 Equilibrium Pairs

We start our treatment by considering a nonlinear, time-invariant, forced, discrete-time system described by the vector difference equation

$$\mathbf{x}(k+1) = \mathbf{f}(\mathbf{x}(k), \mathbf{u}(k)) \tag{1}$$

Our first task is to make precise what we mean by an equilibrium input-state pair.

Let us suppose that we apply a *constant* input sequence to our system, i.e., that

$$\mathbf{u}(0) = \mathbf{u}(1) = \cdots = \mathbf{u}(k) = \cdots = \mathbf{u}^0 \qquad \text{for all } k = 0, 1, 2, \ldots \tag{2}$$

Let us suppose that the system (1) reaches a *steady state* under the influence of this constant input sequence; by this we mean that

$$\lim_{k \to \infty} \mathbf{x}(k) = \mathbf{x}^0 \qquad \text{when } \mathbf{u}(k) = \mathbf{u}^0 \text{ for all } k \tag{3}$$

Let us assume that $\mathbf{f}(.,.)$ is a continuous function. Then our assumptions imply that

$$
\begin{aligned}
\mathbf{x}^0 = \lim_{k \to \infty} \mathbf{x}(k+1) &= \lim_{k \to \infty} \mathbf{f}(\mathbf{x}(k), \mathbf{u}^0) && \text{from Eq. (1)} \\
&= \mathbf{f}(\lim_{k \to \infty} \mathbf{x}(k), \mathbf{u}^0) && \text{by continuity} \\
&= \mathbf{f}(\mathbf{x}^0, \mathbf{u}^0) && \text{from Eq. (3)}
\end{aligned}
\tag{4}
$$

This establishes in a precise manner what we mean by an equilibrium input-state pair. We formalize this concept by the following definition.

Definition 1 (Input-state equilibrium pair) *Consider the system* $\mathbf{x}(k+1)$ $= \mathbf{f}(\mathbf{x}(k), \mathbf{u}(k))$. *Assume that*

$$\mathbf{x}(k) \in R_n \qquad \mathbf{u}(k) \in R_m \tag{5}$$

We say that the constant vectors $(\mathbf{x}^0, \mathbf{u}^0)$ *constitute an* equilibrium pair *if the following equality holds*:

$$\mathbf{x}^0 = \mathbf{f}(\mathbf{x}^0, \mathbf{u}^0) \tag{6}$$

---

† Precisely the same idea will be used later (Chap. 7) for continuous-time nonlinear systems which are described by vector differential equations.

Remark   In general, the equilibrium state $\mathbf{x}^0$ depends on $\mathbf{u}^0$; that is, the system may reach different steady-state conditions for different values of the constant input sequence $\mathbf{u}^0$. Furthermore, if we apply the same input sequence $\mathbf{u}^0$ to our system, whether the system reaches the same equilibrium state $\mathbf{x}^0$ will in general depend on the initial system state $\mathbf{x}(0)$. For some initial states the system may reach $\mathbf{x}^0$, while for others it may reach different equilibrium states or diverge altogether. This means that the vector equation

$$\mathbf{x} = \mathbf{f}(\mathbf{x}, \mathbf{u}^0) \qquad \mathbf{u}^0 = \text{fixed} \tag{7}$$

may have more than one solution. Let us suppose that $\mathbf{x}_1^0, \mathbf{x}_2^0, \ldots, \mathbf{x}_N^0$ denote all solutions of Eq. (7), that is,

$$
\begin{aligned}
\mathbf{x}_1^0 &= \mathbf{f}(\mathbf{x}_1^0, \mathbf{u}^0) \\
\mathbf{x}_2^0 &= \mathbf{f}(\mathbf{x}_2^0, \mathbf{u}^0) \\
&\vdots \\
\mathbf{x}_N^0 &= \mathbf{f}(\mathbf{x}_N^0, \mathbf{u}^0)
\end{aligned}
\tag{8}
$$

Then, according to our above definitions, the pairs $(\mathbf{x}_1^0, \mathbf{u}^0), (\mathbf{x}_2^0, \mathbf{u}^0), \ldots, (\mathbf{x}_N^0, \mathbf{u}^0)$ all constitute input-state equilibrium pairs. Which one will actually be attained strongly depends on the initial state of the system.

To carry this train of thought somewhat further let us assume that $(\mathbf{x}^0, \mathbf{u}^0)$ does constitute an equilibrium pair. Let us consider the system

$$\mathbf{x}(k + 1) = \mathbf{f}(\mathbf{x}(k), \mathbf{u}^0) \qquad \mathbf{x}(0) = \mathbf{x}^0 \tag{9}$$

In other words, the initial state $\mathbf{x}(0)$ is precisely the equilibrium state $\mathbf{x}^0$. Then

$$
\begin{aligned}
\mathbf{x}(1) &= \mathbf{f}(\mathbf{x}(0), \mathbf{u}^0) = \mathbf{f}(\mathbf{x}^0, \mathbf{u}^0) \Rightarrow \mathbf{x}_1 = \mathbf{x}^0 \\
\mathbf{x}(2) &= \mathbf{f}(\mathbf{x}(1), \mathbf{u}^0) = \mathbf{f}(\mathbf{x}^0, \mathbf{u}^0) \Rightarrow \mathbf{x}_2 = \mathbf{x}^0 \\
&\vdots
\end{aligned}
\tag{10}
$$

which means that $\mathbf{x}(k) = \mathbf{x}^0$ for all k. This means that if we start at an equilibrium state, we shall remain at an equilibrium state (theoretically).

4.7.2   Stability of Equilibrium States

The notion of the *stability* of an equilibrium state arises when we consider the behavior of the system

$$\mathbf{x}(k + 1) = \mathbf{f}(\mathbf{x}(k), \mathbf{u}^0) \qquad \mathbf{x}^0 + \boldsymbol{\delta}\mathbf{x}(0) = \mathbf{x}(0) \tag{11}$$

where $\boldsymbol{\delta}\mathbf{x}(0)$ is a very small but arbitrary vector. In this case, the initial state is *very close* to the equilibrium state. What will be the behavior of $\mathbf{x}(k)$? It will either tend to the equilibrium state $\mathbf{x}^0$, or it will not. In the former case, we can say that $\mathbf{x}^0$ is a locally strictly stable equilibrium state, and in the latter it is not. We formalize these notions by the following definition.

Definition 2   *Consider the system* $\mathbf{x}(k + 1) = \mathbf{f}(\mathbf{x}(k), \mathbf{u}(k))$. *Let* $(\mathbf{x}^0, \mathbf{u}^0)$ *denote an equilibrium pair, that is,* $\mathbf{x}^0 = \mathbf{f}(\mathbf{x}^0, \mathbf{u}^0)$. *Then we say that* $(\mathbf{x}^0, \mathbf{u}^0)$ *is a* locally strictly stable equilibrium pair *if the system*

$$\mathbf{x}(k + 1) = \mathbf{f}(\mathbf{x}(k), \mathbf{u}^0) \qquad \mathbf{x}(0) = \mathbf{x}^0 + \boldsymbol{\delta}\mathbf{x}(0) \qquad (12)$$

has the property that

$$\lim_{k \to \infty} \mathbf{x}(k) = \mathbf{x}^0 \qquad (13)$$

for all sufficiently small $\boldsymbol{\delta}\mathbf{x}(0)$, *that is,* $\|\boldsymbol{\delta}\mathbf{x}(0)\| < \epsilon$.

In discussing the intuitive notions of locally stable equilibrium pairs in terms of the properties of the nonlinear system response in the vicinity of such equilibrium pairs we have seen that for linear systems the *qualitative* notion of stability can be analyzed in a *quantitative* manner by checking eigenvalues. Hence, it is natural to ask: Is there a quantitative method that can be used to check the local stability properties of an equilibrium pair for nonlinear systems? The answer is yes, and the remainder of this section is devoted to the development of the necessary tools.

4.7.3   Linearization

The basic idea in dealing with the stability properties of a nonlinear-system equilibrium pair is to obtain a *linear system* whose behavior *approximates* that of the nonlinear system in the *immediate vicinity* of the equilibrium pair. Since we know how to check for stability of linear systems, we can utilize these tools to answer stability properties for the nonlinear system. However, we must pay a price; namely, we must assume that the nonlinear system has some additional properties; these will be clarified later.

The rule of thumb in going from a nonlinear system to a linear system is to *use Taylor series* (Appendix B) *and retain only the linear terms*. Let us see how this is accomplished.

We start with our nonlinear system with $\mathbf{u}(k) = \mathbf{u}^0$ for all k

$$\mathbf{x}(k + 1) = \mathbf{f}(\mathbf{x}(k), \mathbf{u}^0) \qquad (14)$$

We assume that $(\mathbf{x}^0, \mathbf{u}^0)$ constitute an equilibrium pair, i.e.,

$$\mathbf{x}^0 = \mathbf{f}(\mathbf{x}^0, \mathbf{u}^0) \tag{15}$$

Now, let us arbitrarily write

$$\mathbf{x}(k) \triangleq \mathbf{x}^0 + \boldsymbol{\delta}\mathbf{x}(k) \tag{16}$$

$$\mathbf{x}(k+1) \triangleq \mathbf{x}^0 + \boldsymbol{\delta}\mathbf{x}(k+1) \tag{17}$$

Hence, the *state perturbation vector* $\boldsymbol{\delta}\mathbf{x}(k)$ measures the difference between the actual state $\mathbf{x}(k)$ and the equilibrium state $\mathbf{x}^0$. The initial-state perturbation vector

$$\boldsymbol{\delta}\mathbf{x}(0) = \mathbf{x}(0) - \mathbf{x}^0 \tag{18}$$

measures how far the system (14) has started from the equilibrium state $\mathbf{x}^0$.

Clearly if for sufficiently small $\boldsymbol{\delta}\mathbf{x}(0)$

$$\lim_{k \to \infty} \boldsymbol{\delta}\mathbf{x}(k) = \mathbf{0} \tag{19}$$

we can conclude that $(\mathbf{x}^0, \mathbf{u}^0)$ is a locally strictly stable equilibrium pair.

Let us now subtract Eq. (15) from Eq. (14) to obtain

$$\boldsymbol{\delta}\mathbf{x}(k+1) = \mathbf{x}(k+1) - \mathbf{x}^0 = \mathbf{f}(\mathbf{x}(k), \mathbf{u}^0) - \mathbf{f}(\mathbf{x}^0, \mathbf{u}^0) \tag{20}$$

We can see that the right-hand side of Eq. (20) involves the difference of two nonlinear functions which are evaluated at two different arguments, $(\mathbf{x}(k), \mathbf{u}^0)$ and $(\mathbf{x}^0, \mathbf{u}^0)$. If we *assume* that $\mathbf{f}(.,.)$ has the requisite differentiability properties, we can expand $\mathbf{f}(\mathbf{x}(k), \mathbf{u}^0)$ about $(\mathbf{x}^0, \mathbf{u}^0)$ using a Taylor series to obtain

$$\mathbf{f}(\mathbf{x}(k), \mathbf{u}^0) = \mathbf{f}(\mathbf{x}^0, \mathbf{u}^0) + \left. \frac{\partial \mathbf{f}(\mathbf{x}(k), \mathbf{u}^0)}{\partial \mathbf{x}(k)} \right|_{\mathbf{x}(k) = \mathbf{x}^0} (\mathbf{x}(k) - \mathbf{x}^0) + \text{higher-order terms} \tag{21}$$

In Eq. (21)

1 $\partial \mathbf{f}(\mathbf{x}(k), \mathbf{u}(0)) / \partial \mathbf{x}(k)$ is the Jacobian matrix of $\mathbf{f}(\mathbf{x}(k), \mathbf{u}^0)$ with respect to $\mathbf{x}(k)$. Note that the elements of the Jacobian matrix depend on $\mathbf{x}(k)$ and $\mathbf{u}^0$.

2 $\partial \mathbf{f}(\mathbf{x}(k), \mathbf{u}^0) / \partial \mathbf{x}(k)|_{\mathbf{x}(k) = \mathbf{x}^0}$ means that all the elements of the Jacobian matrix are evaluated at $\mathbf{x}(k) = \mathbf{x}^0$. If we assume that we know $\mathbf{x}^0$ and $\mathbf{u}^0$, then we know this matrix. To indicate this fact and to stress the dependence of this matrix upon the values of $\mathbf{x}^0$ and $\mathbf{u}^0$ we denote it by $\mathbf{A}_0$, that is,

$$\mathbf{A}_0 \triangleq \left. \frac{\partial \mathbf{f}(\mathbf{x}(k), \mathbf{u}^0)}{\partial \mathbf{x}(k)} \right|_{\mathbf{x}(k) = \mathbf{x}^0} \tag{22}$$

where the subscript 0 is used to stress the fact that the value of $\mathbf{A}_0$ depends on the values of both $\mathbf{x}^0$ and $\mathbf{u}^0$.

3 The higher-order terms involve second and higher partial derivatives and involve quadratic and higher dependence on the vector

$$\delta\mathbf{x}(k) = \mathbf{x}(k) - \mathbf{x}^0 \tag{23}$$

Let us now substitute Eq. (21) into Eq. (20) to obtain

$$\delta\mathbf{x}(k+1) = \mathbf{A}_0\,\delta\mathbf{x}(k) + \text{higher-order terms} \tag{24}$$

This is almost a linear difference equation except that the higher-order terms involve quadratic, cubic, etc., terms on $\delta\mathbf{x}(k)$.

Now let us suppose that the equilibrium pair $(\mathbf{x}^0, \mathbf{u}^0)$ is locally strictly stable. This means that

$$\lim_{k\to\infty} \delta\mathbf{x}(k) = \mathbf{0} \Rightarrow \lim_{k\to\infty} \|\delta\mathbf{x}(k)\| = 0$$

In this case,

$$\lim_{k\to\infty} \|\text{higher-order terms}\| = 0$$

because they involve functions of $\delta\mathbf{x}(k)$. However, the higher-order terms go to zero much faster than $\delta\mathbf{x}(k)$ itself; a precise statement of this is

$$\lim_{k\to\infty} \frac{\|\text{higher-order terms}\|}{\|\delta\mathbf{x}(k)\|} = 0$$

Intuitively, this is due to the fact that the cube of a small number is much smaller than its square, the square is much smaller than the number itself, and so on.

Hence, for arbitrary small (but nonzero) $\delta\mathbf{x}(k)$ one can *approximate* Eq. (24) by the *linear time-invariant* vector difference equation

$$\delta\mathbf{x}(k+1) = \mathbf{A}_0\,\delta\mathbf{x}(k) \tag{25}$$

If the linear system (25) is strictly stable, the higher-order terms become smaller and smaller and (25) becomes a better and better approximation to (24), which (in the higher-order terms) hides the full nonlinearity of our original system. Therefore, *the strict stability of the linear system (25) is sufficient to ensure the locally strict stability of the equilibrium pair* $(\mathbf{x}^0, \mathbf{u}^0)$.

If, on the other hand, the system (25) is not strictly stable, then the $\|\boldsymbol{\delta}\mathbf{x}(k)\|$ do not converge to zero; hence, one is not justified in neglecting the higher-order terms in (24), and hence the instability of (25) is *sufficient* to ensure that $(\mathbf{x}^0, \mathbf{u}^0)$ is *not* a locally strictly stable equilibrium pair.

We summarize these results in the following theorem.

Theorem 1 *Consider the nonlinear system*

$$\mathbf{x}(k + 1) = \mathbf{f}(\mathbf{x}(k), \mathbf{u}(k)) \tag{26}$$

Let $(\mathbf{x}^0, \mathbf{u}^0)$ constitute an equilibrium pair, i.e.,

$$\mathbf{x}^0 = \mathbf{f}(\mathbf{x}^0, \mathbf{u}^0) \tag{27}$$

Suppose that $\mathbf{f}(.,.)$ is at least continuously twice differentiable with respect to $\mathbf{x}(k)$. Consider the matrix

$$\mathbf{A}_0 \triangleq \left. \frac{\partial \mathbf{f}(\mathbf{x}(k), \mathbf{u}^0)}{\partial \mathbf{x}(k)} \right|_{\mathbf{x}(k) = \mathbf{x}^0} \tag{28}$$

and let $\lambda_1^0, \lambda_2^0, \ldots, \lambda_n^0$ denote its eigenvalues. If

$$|\lambda_i^0| < 1 \qquad \text{for all } i = 1, 2, \ldots, n \tag{29}$$

then the equilibrium pair $(\mathbf{x}^0, \mathbf{u}^0)$ is locally strictly stable.

The above discussion points out the value of linear systems in understanding certain properties of nonlinear systems. Although the stability of an equilibrium pair of a nonlinear system can always be deduced by exhaustive simulation of the actual system and looking at its actual response, this is a time-consuming and perhaps expensive task. Theorem 1 allows us to investigate the stability properties of the system simply by carrying out some analysis and by computing the eigenvalues of a constant matrix.

4.7.4 An Ecological Example

In analyzing a very simple ecological example our aim is twofold: first, to illustrate the ideas of this section with a concrete numerical example and, second, to introduce a different class of problems that arise in some ecological systems. We also deal with ecological problems in later chapters, for example, Sec. 7.4.

Let us suppose that we deal with two species that have to coexist in the same environment and compete for the same resources for survival. We define

$$x_1(k) = \text{number of individuals in species 1 at time } k$$
$$x_2(k) = \text{number of individuals in species 2 at time } k$$

We shall assume that man hunts only species 1. A reasonable mathematical model for these two species is

$$x_1(k + 1) = x_1(k) + x_1(k)[a_1 - u(k) - b_{11}x_1(k) - b_{12}x_2(k)]$$
$$x_2(k + 1) = x_2(k) + x_2(k)[a_2 - b_{21}x_1(k) - b_{22}x_2(k)]$$
(30)

Thus $x_1(k)$, $x_2(k)$ are the state variables, and $u(k)$ is the control variable.

If we examine Eq. (30), we can see that

$$a_1 - u(k) - b_{11}x_1(k) - b_{12}x_2(k) = \text{rate of increase or decrease}$$

$$\text{in time } k \text{ of species 1}$$

$$a_2 - b_{21}x_1(k) - b_{22}x_2(k) = \text{rate of increase or decrease}$$

$$\text{in time } k \text{ of species 2}$$

The control variable $u(k)$ is the rate at which species 1 is hunted by man. The constants a_1 and a_2 are the reproduction rates of each species if infinite resources are available. The terms $b_{11}x_1{}^2(k)$ and $b_{22}x_2{}^2(k)$ are the effects of overcrowding of each individual species upon itself, while the terms $b_{12}x_1(k)x_2(k)$ and $b_{21}x_1(k)x_2(k)$ are the influences of overcrowding of each species upon the other.

First, let us assign some numerical values to the parameters. Suppose that

$$a_1 = 3 \qquad\qquad a_2 = 2 \tag{31}$$

$$b_{11} = 10^{-3} \qquad\qquad b_{22} = 10^{-3} \tag{32}$$

$$b_{12} = \tfrac{1}{4} \times 10^{-3} \qquad b_{21} = 10^{-3} \tag{33}$$

Then, Eq. (30) becomes

$$x_1(k + 1) = x_1(k) + x_1(k)[3 - u(k) - 10^{-3}x_1(k) - \tfrac{1}{4} \times 10^{-3}x_2(k)]$$
$$x_2(k + 1) = x_2(k) + x_2(k)[2 - 10^{-3}x_1(k) - 10^{-3}x_2(k)]$$
(34)

Equilibrium analysis   To study the possible equilibria we must specify an equilibrium value for the hunting-rate control $u(k)$. Suppose that we fix it to be

$$u(k) = u^o = 2 \tag{35}$$

Under this assumption Eq. (34) becomes

$$x_1(k + 1) = 2x_1(k) - 10^{-3}x_1{}^2(k) - \tfrac{1}{4} \times 10^{-3}x_1(k)x_2(k) = f_1(\mathbf{x}(k))$$
$$x_2(k + 1) = 3x_2(k) - 10^{-3}x_1(k)x_2(k) - 10^{-3}x_2{}^2(k) = f_2(\mathbf{x}(k))$$
(36)

Equation (36) is of the form

$$\mathbf{x}(k + 1) = \mathbf{f}(\mathbf{x}(k), \mathbf{u}^o) \tag{37}$$

We now seek the equilibrium states x_1^o, x_2^o of Eq. (36). These must be related by

$$
\begin{aligned}
x_1^o &= 2x_1^o - 10^{-3}x_1^{o2} - \tfrac{1}{4} \times 10^{-3}x_1^o x_2^o \\
x_2^o &= 3x_2^o - 10^{-3}x_1^o x_2^o - 10^{-3}x_2^{o2}
\end{aligned}
\tag{38}
$$

The system of equations (38) has the equilibria

Condition A:	$x_1^o = x_2^o = 0$		both species die
Condition B:	$x_1^o = 0$	$x_2^o = 2{,}000$	species 2 wins out, 1 dies
Condition C:	$x_1^o = 1{,}000$	$x_2^o = 0$	species 1 wins out, 2 dies
Condition D:	$x_1^o = 666$	$x_2^o = 1{,}333$	both species coexist

illustrating that for the same equilibrium input we may have many possible equilibrium states.

Perturbation analysis To examine the stability of equilibria, we must do a perturbation analysis. For each equilibrium we let

$$
\begin{aligned}
\delta x_1(k) &= x_1(k) - x_1^o \\
\delta x_2(k) &= x_2(k) - x_2^o
\end{aligned}
\tag{39}
$$

Our previous analysis states that the perturbation variable dynamics must have the form

$$\delta \mathbf{x}(k + 1) = \mathbf{A}_o \, \delta \mathbf{x}(k) \tag{40}$$

where

$$\mathbf{A}_o = \left. \frac{\partial \mathbf{f}}{\partial \mathbf{x}(k)} \right|_o \tag{41}$$

Hence, we must evaluate the Jacobian matrix $\partial \mathbf{f}/\partial \mathbf{x}(k)$. For our problem

$$\frac{\partial \mathbf{f}}{\partial \mathbf{x}(k)} = \begin{bmatrix} \dfrac{\partial f_1}{\partial x_1} & \dfrac{\partial f_1}{\partial x_2} \\[2mm] \dfrac{\partial f_2}{\partial x_1} & \dfrac{\partial f_2}{\partial x_2} \end{bmatrix} \tag{42}$$

where $f_1(.)$ and $f_2(.)$ are defined in Eq. (36). Hence

$$\frac{\partial \mathbf{f}}{\partial \mathbf{x}(k)} = \begin{bmatrix} 2 - 2 \times 10^{-3} x_1(k) - \dfrac{10^{-3} x_2(k)}{4} & -\dfrac{10^{-3} x_1(k)}{4} \\ -10^{-3} x_2(k) & 3 - 10^{-3} x_1(k) - 2 \times 10^{-3} x_2(k) \end{bmatrix} \quad (43)$$

The analysis of each of the four possible equilibria requires the evaluation of the matrix $\mathbf{A}_o$, defined by Eq. (41), and of its eigenvalues.

ANALYSIS OF EQUILIBRIUM CONDITION A The equilibrium values are

$$x_1^o = x_2^o = 0 \quad (44)$$

From Eqs. (41), (43), and (44) we find that for this equilibrium

$$\mathbf{A}_o = \begin{bmatrix} 2 & 0 \\ 0 & 3 \end{bmatrix} \quad (45)$$

The eigenvalues of $\mathbf{A}_o$ are at $\lambda = 2$, $\lambda = 3$. Hence, this is an unstable equilibrium; this means that a small population of either species 1 or 2 or both will not die out.

ANALYSIS OF EQUILIBRIUM CONDITION B The equilibrium values are

$$x_1^o = 0 \qquad x_2^o = 2{,}000 \quad (46)$$

From Eqs. (41), (43), and (46) we find that

$$\mathbf{A}_o = \begin{bmatrix} 1.5 & 0 \\ -2 & -1 \end{bmatrix} \quad (47)$$

The eigenvalues of $\mathbf{A}_o$ are at $\lambda = -1$ and $\lambda = 1.5$. Once more this is an unstable equilibrium. Physically this means that a small number of species 1 will grow in a predominant population of species 2.

ANALYSIS OF EQUILIBRIUM CONDITION C The equilibrium values are

$$x_1^o = 1{,}000 \qquad x_2^o = 0 \quad (48)$$

From Eqs. (41), (43), and (48) we find that

$$\mathbf{A}_o = \begin{bmatrix} 0 & -0.25 \\ 0 & 2 \end{bmatrix} \quad (49)$$

The eigenvalues of $\mathbf{A}_o$ are at $\lambda = 0$ and $\lambda = 2$. Once more, this is an unstable equilibrium. Physically it means that a small number of species 2 will grow in a predominant population of species 1.

ANALYSIS OF EQUILIBRIUM CONDITION D The equilibrium values are

$$x_1^o = 666 \qquad x_2^o = 1{,}333 \tag{50}$$

From Eqs. (41), (43), and (50) we find that

$$\mathbf{A}_o = \begin{bmatrix} \frac{1}{3} & -\frac{1}{6} \\ -\frac{4}{3} & -\frac{1}{3} \end{bmatrix} \tag{51}$$

The eigenvalues of $\mathbf{A}_o$ are at $\lambda = \pm 1/\sqrt{3}$. Since both are less than unity, this represents a stable equilibrium.

⋆ 4.8   APPLICATION OF STABILITY RESULTS TO ITERATION ALGORITHMS

In this section we examine in a general way the interrelation between the convergence properties of iterative algorithms for solving vector algebraic equations and the stability results established in the previous section.

Let us consider the problem of finding the solution vector $\mathbf{x}^*$ of a vector algebraic equation. In other words, we are faced with the following problem: given a vector-valued function $\mathbf{g}(.)$, find $\mathbf{x}^* \in R_n$ (if it exists) such that

$$\mathbf{g}(\mathbf{x}^*) = \mathbf{0} \tag{1}$$

We shall adopt the same type of trick used to deduce Newton's method from Picard's method. We seek a Picard type algorithm

$$\mathbf{x}(k + 1) = \mathbf{f}(\mathbf{x}(k)) \tag{2}$$

such that

$$\lim_{k \to \infty} \mathbf{x}(k) = \mathbf{x}^* \tag{3}$$

Let $\mathbf{A}(\mathbf{x})$ be an arbitrary bounded $n \times n$ matrix whose elements $a_{ij}(\mathbf{x})$ are arbitrary, continuous, and sufficiently differentiable functions of the vector $\mathbf{x}$. Let us construct the function $\mathbf{f}(.)$ appearing in Eq. (2) as follows:

$$\mathbf{f}(\mathbf{x}) = \mathbf{x} + \mathbf{A}(\mathbf{x})\mathbf{g}(\mathbf{x}) \tag{4}$$

Then the Picard algorithm (2) has the more explicit form

$$\mathbf{x}(k + 1) = \mathbf{x}(k) + \mathbf{A}(\mathbf{x}(k))\mathbf{g}(\mathbf{x}(k)) \tag{5}$$

Clearly, we can view (5) on a nonlinear discrete-time system. We now claim that the solution vector $\mathbf{x}^*$ which satisfies (1) is an equilibrium state of the system (5). To see this, we recall that if $\mathbf{x}^*$ is an equilibrium state of (5), then the equality

$$\mathbf{x}^* = \mathbf{x}^* + \mathbf{A}(\mathbf{x}^*)\mathbf{g}(\mathbf{x}^*) \tag{6}$$

must be true; clearly, Eq. (6) holds in view of Eq. (1) as long as $\mathbf{A}(\mathbf{x}^*)$ is defined.

As we have remarked in Chap. 3, a good iterative algorithm must converge to the solution if the initial guess happens to be near the solution (recall that Newton's method had that property). This now can be interpreted in terms of local stability; we want the equilibrium state $\mathbf{x}^*$ of (5) to be locally strictly stable.

Relying on the linearization techniques developed in the previous section, we can see that the local stability of $\mathbf{x}^*$, and hence the local convergence of the algorithm (5), is assured if all the eigenvalues of the $n \times n$ matrix

$$\left.\frac{\partial \mathbf{f}}{\partial \mathbf{x}}\right|_{\mathbf{x}=\mathbf{x}^*} = \frac{\partial}{\partial \mathbf{x}}[\mathbf{x} + \mathbf{A}(\mathbf{x})\mathbf{g}(\mathbf{x})]_{\mathbf{x}=\mathbf{x}^*} \tag{7}$$

have magnitude less than unity. It is easy to see that

$$\left.\frac{\partial \mathbf{f}}{\partial \mathbf{x}}\right|_{\mathbf{x}=\mathbf{x}^*} = \mathbf{I} + \mathbf{A}(\mathbf{x}^*)\left.\frac{\partial \mathbf{g}}{\partial \mathbf{x}}\right|_{\mathbf{x}=\mathbf{x}^*} \tag{8}$$

Hence, the algorithm can be designed, and the design involves selecting the arbitrary matrix $\mathbf{A}(\mathbf{x})$ by attempting to have all the eigenvalues of matrix (8) be within the unit circle in the complex plane.

4.9 INTRODUCTION TO FEEDBACK

This section introduces one of the most powerful techniques that can be used to make systems behave in a way specified by the designer.

4.9.1 General Concepts

To illustrate the basic idea of feedback let us consider a general discrete-time dynamical system whose state vector $\mathbf{x}(k)$, input vector $\mathbf{u}(k)$, and output vector $\mathbf{y}(k)$ are related by

$$\mathbf{x}(k + 1) = \mathbf{f}(\mathbf{x}(k), \mathbf{u}(k)) \qquad \mathbf{x}(0) \text{ known} \tag{1}$$

$$\mathbf{y}(k) = \mathbf{g}(\mathbf{x}(k)) \tag{2}$$

Let us think of the output vector $\mathbf{y}(k)$ as representing the response variables of the system that are of direct interest with respect to a specific task the system has to accomplish.

Often, the system's specific goal is specified by defining a desired output sequence denoted by

$$\mathbf{y}^*(0), \mathbf{y}^*(1), \ldots, \mathbf{y}^*(k), \ldots \tag{3}$$

Then, roughly speaking, we wish the actual vector $\mathbf{y}(k)$ to be close to the desired value $\mathbf{y}^*(k)$, for $k = 0, 1, 2, \ldots, N$, where N is the final time unit of interest (it may be infinity).

If no external input is applied to the system (1), (2), i.e., if

$$\mathbf{u}(k) = \mathbf{0} \qquad \text{for all } k = 0, 1, 2, \ldots \tag{4}$$

then its *natural or unforced response*, denoted by $\mathbf{x}_o(k)$ for the state and $\mathbf{y}_o(k)$ for the output, can be found from the equations

$$\begin{aligned} \mathbf{x}_o(k + 1) &= \mathbf{f}(\mathbf{x}_o(k), \mathbf{0}) \qquad \mathbf{x}(0) \text{ known} \\ \mathbf{y}_o(k) &= \mathbf{g}(\mathbf{x}_o(k)) \end{aligned} \tag{5}$$

It often happens, as illustrated in Fig. 4.9.1, that the response sequence $\{\mathbf{y}_o(k)\}$ has certain undesirable characteristics compared with the desired response sequence $\{\mathbf{y}^*(k)\}$. For example, in Fig. 4.9.1a the unforced response has the same general character as the desired response, but the system is too sluggish. In Fig. 4.9.1b the unforced response oscillates about the desired response. In Fig. 4.9.1c the unforced system response diverges from the desired one.

Ideally, of course, we would like to have

$$\mathbf{y}_o(k) = \mathbf{y}^*(k) \qquad \text{for all } k \tag{6}$$

but in general this cannot be accomplished. The only way that we can make the system output sequence $\{\mathbf{y}(k)\}$ approximate the desired sequence $\{\mathbf{y}^*(k)\}$ is to influence the system by applying an input sequence $\{\mathbf{u}(k)\}$ to the system. It is clear that by selecting different input sequences $\{\mathbf{u}(k)\}$ we can generate different output sequences $\{\mathbf{y}(k)\}$.

The control problem in general is: *given the system described by Eqs. (1) and (2), and given the desired output sequence $\{\mathbf{y}^*(k)\}$, determine the input sequence $\{\mathbf{u}(k)\}$ such that the output sequence $\{\mathbf{y}(k)\}$ is close, in some sense, to the desired one, $\{\mathbf{y}^*(k)\}$.*

The issue of closeness between the actual and desired responses can be qualitative or made more precise. In modern optimal-control theory one measures

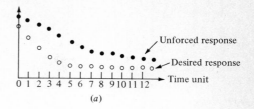

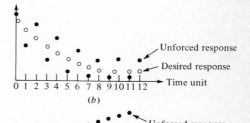

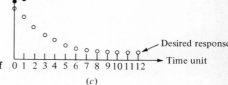

FIGURE 4.9.1
Three types of undesirable response of
the unforced system.

the goodness of fit by forming a single scalar of the form

$$J = \sum_{k=0}^{N} \|\mathbf{y}(k) - \mathbf{y}^*(k)\| \tag{7}$$

or

$$J = \sum_{k=0}^{N} \|\mathbf{y}(k) - \mathbf{y}^*(k)\|^2 \tag{8}$$

Such scalars are called *cost functionals*; then in general one seeks a control sequence

$$\mathbf{u}(k) \qquad 0 \le k \le N - 1 \tag{9}$$

so that the resultant value of the cost functional J is as small as possible.

If one picks the input sequence a priori, the scheme is called *open-loop control*. On the other hand, it is often possible to select the current value of the input $\mathbf{u}(k)$ as a function of current and past measurements. For example, if we can measure the output vector $\mathbf{y}(k)$, then we can form the control $\mathbf{u}(k)$ at time k as a function of past and current measurements of the output

$$\mathbf{y}(0), \mathbf{y}(1), \ldots, \mathbf{y}(k) \tag{10}$$

and we write

$$\mathbf{u}(k) = \mathbf{u}(\mathbf{y}(0), \mathbf{y}(1), \ldots, \mathbf{y}(k)) \tag{11}$$

Similarly, if we can measure the state vector, we can express the current control $\mathbf{u}(k)$ as a function of current and past state measurememts, i.e.,

$$\mathbf{u}(k) = \mathbf{u}(\mathbf{x}(0), \mathbf{x}(1), \ldots, \mathbf{x}(k)) \tag{12}$$

In either case, because the control depends on the actual observed measurements, we call this mode of operation *feedback control*. Feedback control is preferable to open-loop control for reasons to be discussed.

4.9.2 Feedback Strategies for Linear Systems

To illustrate the use of feedback control in a specific context let us suppose that we have an LTI discrete-time system characterized by the equation

$$\mathbf{x}(k + 1) = \mathbf{A}\mathbf{x}(k) + \mathbf{B}\mathbf{u}(k) \qquad \mathbf{x}(0) \text{ known} \tag{13}$$

where $\mathbf{x}(k)$ = n-dimensional vector
  $\mathbf{u}(k)$ = m-dimensional vector
  $\mathbf{A}$ = $n \times n$ constant matrix
  $\mathbf{B}$ = $n \times m$ constant matrix

Let us now suppose that if the system starts in a specific initial state, denoted by $\mathbf{x}*(0)$, we have found a specific input sequence denoted by

$$\mathbf{u}^*(0), \mathbf{u}^*(1), \mathbf{u}^*(2), \ldots$$

generating a state sequence

$$\mathbf{x}^*(0), \mathbf{x}^*(1), \mathbf{x}^*(2), \ldots$$

which is in some sense *desirable*.

On the other hand, if the initial state of the system $\mathbf{x}(0)$ happens to be different from the one thought of, $\mathbf{x}*(0)$, or if during the past the actual inputs $\mathbf{u}(j)$, $0 \leq j \leq k$, were different from $\mathbf{u}*(j)$, at time k one may find that the actual state $\mathbf{x}(k)$ at the present time k is different from the desired one, $\mathbf{x}*(k)$.

Such deviations between the actual state $\mathbf{x}(k)$ and the desired state $\mathbf{x}*(k)$, at the present time k, may then be an unavoidable event about which nothing can be done. What can be done, however, is to demand that from now on such deviations shall eventually be nulled out. Roughly speaking, we would like to see the difference vector $\mathbf{x}(k) - \mathbf{x}^*(k)$ go to zero as $k \to \infty$.

To understand in somewhat more precise terms what is going on and what we can do about it let us examine the equations in some detail. By definition, the desirable state sequence $\{x^*(k)\}$ and the control sequence $\{u^*(k)\}$ that generated it are related by the difference equation

$$x^*(k + 1) = Ax^*(k) + Bu^*(k) \tag{14}$$

Now let $\{x(k)\}$ denote the actual system state sequence and $\{u(k)\}$ the actual input sequence; once more these must be related by the difference equation

$$x(k + 1) = Ax(k) + Bu(k) \tag{15}$$

Let us define the *state perturbation vector* $\delta x(k)$ by

$$\delta x(k) \triangleq x(k) - x^*(k) \qquad k = 0, 1, 2, \ldots \tag{16}$$

and let us define the *control correction vector* $\delta u(k)$ by

$$\delta u(k) = u(k) - u^*(k) \tag{17}$$

Thus, at each instant of time k, $\delta x(k)$ measures how far the actual system state is from its desired value; similarly, $\delta u(k)$ measures how much the actual input differs from the precomputed one $u^*(k)$.

It is possible to obtain the dynamics that relate $\delta x(k)$ to $\delta u(k)$. By subtracting Eq. (14) from Eq. (15) and using (16) and (17) we obtain the vector difference equation

$$\delta x(k + 1) = A\delta x(k) + B\delta u(k) \tag{18}$$

Equation (18) brings into focus how $\delta u(k)$ can be used to control the $\delta x(k)$'s. Note that if

$$\delta u(k) = 0 \qquad \text{that is } u(k) = u^*(k) \tag{19}$$

then

$$\delta x(k + 1) = A\delta x(k) \tag{20}$$

and the future evolution of the $\delta x(k)$'s hinges on the stability properties of A (compare with the discussion in Sec. 4.6).

Depending on the eigenvalues of A, the system (20) will be stable or unstable. Even if it is stable, its response, i.e., the way the error vector propagates, may be too sluggish or oscillatory and we should apply some control to make it behave better.

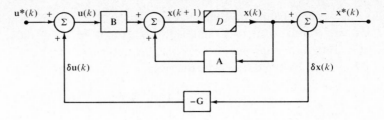

FIGURE 4.9.2
The structure of the feedback control system.

A simple way of generating the contol $\delta\mathbf{u}(k)$ is by *linear negative feedback*. This means that the control is generated by

$$\delta\mathbf{u}(k) = -\mathbf{G}\delta\mathbf{x}(k) \tag{21}$$

where $\mathbf{G}$ is a constant $m \times n$ gain matrix. In other words, we measure the error vector $\delta\mathbf{x}(k)$ at time k and multiply it by a constant gain matrix $\mathbf{G}$. The block diagram of the resultant system is shown in Fig. 4.9.2.

What is the effect? By substituting Eq. (21) into Eq. (18) we obtain

$$\delta\mathbf{x}(k + 1) = \mathbf{A}\delta\mathbf{x}(k) - \mathbf{BG}\delta\mathbf{x}(k) \tag{22}$$

or

$$\delta\mathbf{x}(k + 1) = (\mathbf{A} - \mathbf{BG})\delta\mathbf{x}(k) \tag{23}$$

By introducing feedback we can obtain the error-vector propagation from the unforced equation (23). How the error vector $\delta\mathbf{x}(k)$ will propagate depends on the *closed-loop matrix* $\overline{\mathbf{A}}$

$$\overline{\mathbf{A}} \overset{\triangle}{=} \mathbf{A} - \mathbf{BG} \tag{24}$$

One can then interpret the use of feedback as modifying the open-loop system error dynamics, as dictated by the $\mathbf{A}$ matrix in Eq. (20), to the closed-loop dynamics, as dictated by the closed-loop matrix $\mathbf{A} - \mathbf{BG}$ in Eq. (23).

In particular, we can use feedback to stabilize the system. Suppose that one or more of the eigenvalues of $\mathbf{A}$ has magnitude greater than unity so that the unforced original system (13) and the error system (20) are unstable. Some thought should convince the reader that, except for certain pathological cases, one should be able to find at least one matrix $\mathbf{G}$ such that the eigenvalues of the closed-loop matrix $\mathbf{A} - \mathbf{BG}$ all have magnitudes less than unity, so that the system (23) is strictly stable.

In fact it turns out, except for certain pathological cases again, that for any given A and B matrices we can find a matrix G such that all the n eigenvalues of $A - BG$ can be placed at prespecified values (subject to the constraint that for real A, B, G complex eigenvalues come in complex-conjugate pairs). Since we know (see Sec. 4.6) that the eigenvalues of the system matrix govern the time evolution of all the states, we can see that by using feedback we can take the dynamics of a given system and change them at will.

We now state in a precise form the conditions that guarantee that we can always find at least one matrix G such that the eigenvalues of $A - BG$ have prespecified values. Before we state the theorem, we shall need a definition.

Definition 1 *Let A be an arbitrary real $n \times n$ matrix and let B be an arbitrary real $n \times m$ matrix. We say that the matrix pair $[A, B]$ is a controllable pair if the $n \times nm$ matrix M defined by*

$$M \triangleq [B \vdots AB \vdots A^2B \vdots \cdots \vdots A^{n-1}B] \qquad (25)$$

has rank n, that is, out of the total nm columns of M we can find n that are linearly independent.

Remark The notation of Eq. (25) means that the first m columns of M are defined by the columns of the $n \times m$ matrix B; the next m columns of M are defined by the columns of the $n \times m$ matrix AB and so on.

We now state the theorem.

Theorem 1 *Let A be an arbitrary real $n \times n$ matrix and let B be an arbitrary real $n \times m$ matrix. Suppose that $[A,B]$ is a controllable pair. Then there exists at least one $m \times n$ real matrix G such that the eigenvalues $\mu_1, \mu_2, \ldots, \mu_n$ of the (closed-loop) matrix $A - BG$ are equal to prespecified values. Since $A - BG$ is real, the eigenvalues μ_i must either be real or come in complex-conjugate pairs.*

The proof of the theorem is rather involved and is omitted.

★ **4.9.3 Output Feedback**

In the development above we assumed that we could measure all the state variables of our system. However, there are many problems in which we can measure only the output variables. In such problems one can again use feedback from the output vector to construct the control vector and thus modify the open-loop dynamics. However, one has more limited freedom in placing the eigenvalues of the closed-loop system at will.

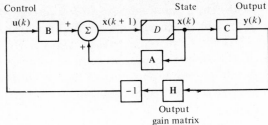

FIGURE 4.9.3
Structure of feedback output regulator.

To illustrate, consider the system

$$x(k + 1) = Ax(k) + Bu(k)$$
$$y(k) = Cx(k)$$

(26)

where $x(k) = n$-vector
$u(k) = m$-vector
$y(k) = r$-vector
$A = n \times n$ constant matrix
$B = n \times m$ constant matrix
$C = r \times n$ constant matrix

Suppose that the objective is to regulate or steer the vector $x(k)$ near zero from an arbitrary initial condition; i.e., the desired value of $x(k)$ is the **0** vector. The open-loop system, i.e., the one with $u(k) = 0$ in (26) may be unstable, sluggish, or oscillatory. Hence, we seek methods for improving its performance.

If we assume that we can measure the output vector $y(k)$ [but not the state vector $x(k)$] then we can generate the control vector $u(k)$ by *linear ouput feedback* i.e., we compute $u(k)$ by

$$u(k) = -Hy(k)$$

(27)

where **H** is a *constant, output-gain* $m \times r$ matrix. The structure of the resultant system is illustrated in Fig. 4.9.3.

To deduce the characteristics of the closed-loop system, we can see from Eqs. (20) and (21) that

$$u(k) = -Hy(k) = -HCx(k)$$

(28)

and that the closed-loop state evolves according to

$$x(k + 1) = (A - BHC)x(k)$$

(29)

The dynamical behavior of the feedback system hinges on the character and eigenvalues of the *closed-loop matrix* $\mathbf{A} - \mathbf{BHC}$. For stability, for any given matrices $\mathbf{A}, \mathbf{B}, \mathbf{C}$ we would like to select the matrix $\mathbf{H}$ so that all the eigenvalues of $\mathbf{A} - \mathbf{BHC}$ have magnitude less than unity. *This may or may not be possible*, as illustrated by the following two examples.

EXAMPLE 1   Consider a second-order discrete-time single-input–single-output system.

$$\begin{aligned} x_1(k + 1) &= x_2(k) \\ x_2(k + 1) &= 2x_1(k) + x_2(k) + u(k) \end{aligned} \tag{30}$$

$$y(k) = x_1(k) \tag{31}$$

In vector form the system takes the form

$$\underbrace{\begin{bmatrix} x_1(k + 1) \\ x_2(k + 1) \end{bmatrix}}_{\mathbf{x}(k+1)} = \underbrace{\begin{bmatrix} 0 & 1 \\ 2 & 1 \end{bmatrix}}_{\mathbf{A}} \underbrace{\begin{bmatrix} x_1(k) \\ x_2(k) \end{bmatrix}}_{\mathbf{x}(k)} + \underbrace{\begin{bmatrix} 0 \\ 1 \end{bmatrix}}_{\mathbf{B}} \underbrace{u(k)}_{\mathbf{u}(k)} \tag{32}$$

$$y(k) = \underbrace{\begin{bmatrix} 1 & 0 \end{bmatrix}}_{\mathbf{C}} \underbrace{\begin{bmatrix} x_1(k) \\ x_2(k) \end{bmatrix}}_{\mathbf{x}(k)} \tag{33}$$

Note that the eigenvalues of the open-loop system matrix $\mathbf{A}$ are at

$$\lambda_1 = 2 \qquad \lambda_2 = -1 \tag{34}$$

so that the open-loop system is unstable. To stabilize the system by output feedback we seek a value of the scalar H such that

$$u(k) = -Hy(k) \tag{35}$$

and the eigenvalues of the closed-loop matrix

$$\begin{aligned} \overline{\mathbf{A}} \triangleq \mathbf{A} - \mathbf{BHC} &= \begin{bmatrix} 0 & 1 \\ 2 & 1 \end{bmatrix} - \begin{bmatrix} 0 \\ 1 \end{bmatrix} H \begin{bmatrix} 1 & 0 \end{bmatrix} \\ &= \begin{bmatrix} 0 & 1 \\ 2 - H & 1 \end{bmatrix} \end{aligned} \tag{36}$$

have magnitudes less than unity. The eigenvalues of $\overline{\mathbf{A}}$ are the roots of

$$
\det(\lambda \mathbf{I} - \overline{\mathbf{A}}) = \det \begin{bmatrix} \lambda & -1 \\ H - 2 & \lambda - 1 \end{bmatrix}
$$

$$
= \lambda^2 - \lambda + H - 2 \tag{37}
$$

Hence,

$$
\lambda = \tfrac{1}{2} \pm \tfrac{1}{2}\sqrt{1 - 4(H - 2)} \tag{38}
$$

It is easy to verify that there is a whole range of H that yields magnitudes of λ less than 1; in particular the value

$$
H = \tfrac{9}{4} \tag{39}
$$

yields a double eigenvalue at $\tfrac{1}{2}$, so that $\overline{\mathbf{A}}$ is stable.

EXAMPLE 2 Consider another second-order system

$$
\begin{bmatrix} x_1(k + 1) \\ x_2(k + 1) \end{bmatrix} = \underbrace{\begin{bmatrix} 0 & 1 \\ -2 & -3 \end{bmatrix}}_{\mathbf{A}} \begin{bmatrix} x_1(k) \\ x_2(k) \end{bmatrix} + \underbrace{\begin{bmatrix} 0 \\ 1 \end{bmatrix}}_{\mathbf{B}} u(k) \tag{40}
$$

$$
y(k) = \underbrace{\begin{bmatrix} 1 & 0 \end{bmatrix}}_{\mathbf{C}} \begin{bmatrix} x_1(k) \\ x_2(k) \end{bmatrix} \tag{41}
$$

The eigenvalues of the **A** matrix are at

$$
\lambda_1 = -2 \qquad \lambda_2 = -1 \tag{42}
$$

so that the open-loop system is unstable. Also note that *in magnitude* these eigenvalues are the same as in Example 1 [see Eq. (34)].

To stabilize the system we seek a (scalar) H such that using the output feedback

$$
u(k) = -Hy(k) \tag{43}
$$

the eigenvalues of the closed-loop matrix

$$\overline{\mathbf{A}} = \mathbf{A} - \mathbf{B}HC = \begin{bmatrix} 0 & 1 \\ -2 & -3 \end{bmatrix} - \begin{bmatrix} 0 \\ 1 \end{bmatrix} H[1 \quad 0]$$

$$= \begin{bmatrix} 0 & 1 \\ -2 - H & -3 \end{bmatrix} \tag{44}$$

are less than unity in magnitude.

The characteristic polynomial of $\overline{\mathbf{A}}$ is

$$\det(\lambda\mathbf{I} - \overline{\mathbf{A}}) = \det \begin{bmatrix} \lambda & -1 \\ 2 + H & \lambda + 3 \end{bmatrix}$$

$$= \lambda^2 + 3\lambda + 2 + H \tag{45}$$

with roots at

$$\lambda = -\tfrac{3}{2} \pm \tfrac{1}{2}\sqrt{9 - 4(2 + H)} \tag{46}$$

Clearly, there is no value of H that can make both roots (46) less than unity in magnitude, and hence the system (40), (41) cannot be stabilized using output feedback.

4.9.4   Feedback Control for Nonlinear Systems

We now extend the techniques used for feedback control of linear systems to nonlinear systems, and in particular the systematic procedure one must follow to maintain a nonlinear system about a desired equilibrium condition.

In essence, the same linearization technique that enabled us to check the stability of a nonlinear system about an equilibrium by looking at the properties of a linear system, as explained in Sec. 4.7, allows us to use linear feedback to ensure that a possibly unstable equilibrium condition is maintained, provided that the initial deviation from the equilibrium is not large enough to destroy the validity of the linearized model.

In this section we shall assume that we can measure all the state variables for the purposes of feedback control. The ideas can be easily modified to handle the case where only output variables can be measured.

Throughout the remainder of this section we shall be concerned with the control of the nonlinear system

$$\mathbf{x}(k + 1) = \mathbf{f}(\mathbf{x}(k), \mathbf{u}(k)) \tag{47}$$

about the *constant* equilibrium state $\mathbf{x}*$ and *constant* control $\mathbf{u}*$, which, by definition, satisfy the relation

$$\mathbf{x^*} = \mathbf{f(x^*, u^*)} \tag{48}$$

where $\mathbf{x}(k), \mathbf{x^*} = n$-vectors
$\mathbf{u}(k), \mathbf{u^*} = m$-vectors

We assume that $\mathbf{x^*}$ and $\mathbf{u^*}$ are known. We also assume that $\mathbf{f}(.,.)$ is at least twice differentiable with respect to both its arguments.

We first clarify what we mean by *open-loop control* in this case. For open-loop control, the control sequence $\{\mathbf{u}(k)\}$ in (47) is constant and equal to the equilibrium value $\mathbf{u^*}$. Thus the open-loop system is characterized by the difference equation

$$\mathbf{x}(k+1) = \mathbf{f(x}(k), \mathbf{u^*}) \tag{49}$$

If

$$\mathbf{x}(0) = \mathbf{x^*} \tag{50}$$

then the open-loop system has the property that (theoretically)†

$$\mathbf{x}(k) = \mathbf{x^*} \qquad \text{for all } k = 0, 1, 2, \ldots \tag{51}$$

We are interested in the system behavior when

$$\mathbf{x}(0) = \mathbf{x^*} + \boldsymbol{\delta}\mathbf{x}(0) \tag{52}$$

i.e., the initial state is slightly different from the equilibrium state. In this case, the open-loop system (49) may exhibit undesirable characteristics, e.g., instability, oscillation, sluggishness. In order to provide a cure for such undesirable behavior, we have no choice but to change the actual control input $\mathbf{u}(k)$ to our system to something different from the equilibrium value.

Before we proceed with the mathematical analysis, it is useful to set forth certain notation and definitions, anticipating that since we are going to tackle nonlinear problems with linear tools, we shall use the Taylor series at one stage or another.

Definition 2   *We define the* state perturbation vector $\boldsymbol{\delta}\mathbf{x}(k)$, *at time k, in terms of the equilibrium state* $\mathbf{x^*}$ *and the actual system state* $\mathbf{x}(k)$ *as*

$$\mathbf{x}(k) \stackrel{\triangle}{=} \mathbf{x^*} + \boldsymbol{\delta}\mathbf{x}(k) \tag{53}$$

---

† This is like saying that an inverted broom will stay inverted.

Definition 3 *We define the* control perturbation vector $\boldsymbol{\delta}\mathbf{u}(k)$, *at time* k, *in terms of the equilibrium control* $\mathbf{u}^*$ *and the actual control* $\mathbf{u}(k)$ *applied to the system as*

$$\mathbf{u}(k) \triangleq \mathbf{u}^* + \boldsymbol{\delta}\mathbf{u}(k) \tag{54}$$

Definition 4 *We define the constant* $n \times n$ *matrix* $\mathbf{A}_*$ *to be the Jacobian matrix of* $\mathbf{f}(\mathbf{x}, \mathbf{u})$ *with respect to* $\mathbf{x}$, *evaluated at the equilibrium pair* $(\mathbf{x}^*, \mathbf{u}^*)$, *that is,*

$$\mathbf{A}_* \triangleq \left. \frac{\partial \mathbf{f}(\mathbf{x}, \mathbf{u})}{\partial \mathbf{x}} \right|_{\mathbf{x}=\mathbf{x}^*, \mathbf{u}=\mathbf{u}^*} \tag{55}$$

Similarly, we define the constant $n \times m$ *matrix* $\mathbf{B}_*$ *to be the Jacobian matrix of* $\mathbf{f}(\mathbf{x}, \mathbf{u})$ *with respect to* $\mathbf{u}$ *evaluated at the equilibrium pair* $(\mathbf{x}^*, \mathbf{u}^*)$ *that is,*

$$\mathbf{B}_* \triangleq \left. \frac{\partial \mathbf{f}(\mathbf{x}, \mathbf{u})}{\partial \mathbf{u}} \right|_{\mathbf{x}=\mathbf{x}^*, \mathbf{u}=\mathbf{u}^*} \tag{56}$$

We are interested in an approximation to the system dynamics in the vicinity of the equilibrium defined by $\mathbf{x}^*$ and $\mathbf{u}^*$. We can obtain such a description by deriving the interrelationship between the state perturbation vector $\boldsymbol{\delta}\mathbf{x}(k)$ and the control perturbation vector $\boldsymbol{\delta}\mathbf{u}(k)$. If we can obtain such a relationship, we can get a handle on how wiggles, as expressed by $\boldsymbol{\delta}\mathbf{u}(k)$ about $\mathbf{u}^*$, can be generated and influence wiggles, as expressed by $\boldsymbol{\delta}\mathbf{x}(k)$ about $\mathbf{x}^*$.

To derive the desired relationship between $\boldsymbol{\delta}\mathbf{x}(k)$ and $\boldsymbol{\delta}\mathbf{u}(k)$ we start by expanding $\mathbf{f}(\mathbf{x}(k), \mathbf{u}(k))$ about $\mathbf{x}^*$ and $\mathbf{u}^*$ in a Taylor series (see Appendix B). Using the definitions (53) to (56), we obtain

$$\mathbf{f}(\mathbf{x}(k), \mathbf{u}(k)) = \mathbf{f}(\mathbf{x}^*, \mathbf{u}^*) + \mathbf{A}_* \boldsymbol{\delta}\mathbf{x}(k) + \mathbf{B}_* \boldsymbol{\delta}\mathbf{u}(k) + \text{higher-order terms} \tag{57}$$

where higher-order terms are the quadratic, cubic, quartic, etc., terms in $\boldsymbol{\delta}\mathbf{x}(k)$ and $\boldsymbol{\delta}\mathbf{u}(k)$.

The next step is to get a difference equation. Subtracting Eq. (47) from Eq. (48) and using the notation of Eq. (53), we obtain

$$\boldsymbol{\delta}\mathbf{x}(k + 1) = \mathbf{f}(\mathbf{x}(k), \mathbf{u}(k)) - \mathbf{f}(\mathbf{x}^*, \mathbf{u}^*) \tag{58}$$

We now substitute Eq. (57) into Eq. (58) to get the exact relationship.

$$\boldsymbol{\delta}\mathbf{x}(k + 1) = \mathbf{A}_* \boldsymbol{\delta}\mathbf{x}(k) + \mathbf{B}_* \boldsymbol{\delta}\mathbf{u}(k) + \text{higher-order terms} \tag{59}$$

Equation (59) is almost a linear difference equation except for the higher-order terms. We would be justified in neglecting the higher-order terms in (59) if *both*

$\delta x(k)$ were small *and* $\delta u(k)$ were small. We stress that *both the state perturbation vector $\delta x(k)$ and the control perturbation vector $\delta u(k)$ must be small*. If the $\delta x(k)$ were small *but we applied large control perturbations*, we could not neglect the higher-order terms in (59).

At any rate, if we can guarantee that the higher-order terms terms can be neglected, we can approximate Eq. (59) by the LTI discrete-time system

$$\delta x(k + 1) = A_* x(k) + B_* \delta u(k) \tag{60}$$

These remarks show that in order to generate the control perturbation $\delta u(k)$ we should use feedback and monitor the actual state perturbations $\delta x(k)$ because if we applied $\delta u(k)$ without regard to the measurements of $\delta x(k)$, we could not guarantee that small $\delta u(k)$ would *always* produce small $\delta x(k)$. We should then attempt to generate $\delta u(k)$ by a relationship of the form (feedback control)

$$\delta u(k) = g(\delta x(k)) \tag{61}$$

where $g(.)$ is some function.

The next stage of development deals with the specification of the feedback function $g(.)$ in Eq. (61). It should have the property that[†] small $\delta x(k)$ yield small $\delta u(k)$ so that consistency is maintained.

One way to accomplish this is to use a *linear* relationship of the form

$$\delta u(k) = -G\delta x(k) \tag{62}$$

where G is a constant $m \times n$ gain matrix. Apart from the obvious advantages of easy implementation, the functional relationship (62) has all the desired small-to-small properties.

Figure 4.9.4 shows the structure of the feedback control system. Its mathematical properties can be analyzed by substituting Eq. (62) into Eq. (60) to obtain the approximate linearized model

$$\delta x(k + 1) = (A_* - B_* G)\delta x(k) \tag{63}$$

or by substituting Eq. (62) into Eq. (59) to obtain the exact model

$$\delta x(k + 1) = (A_* - B_* G)\delta x(k) + \text{higher-order terms} \tag{64}$$

Note that in Eq. (64) higher-order terms now involve only the state perturbation vector $\delta x(k)$.

[†] Thus, in the scalar case, functions of the form $\delta u(k) = 1/\delta x(k)$ should not be considered.

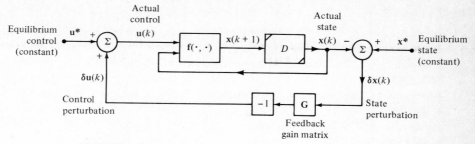

FIGURE 4.9.4
Structure of feedback system for control about equilibrium state and control pair $(\mathbf{x}^*, \mathbf{u}^*)$.

If we select the feedback gain matrix $\mathbf{G}$ so that the matrix

$$\overline{\mathbf{A}}_* \triangleq \mathbf{A}_* - \mathbf{B}_* \mathbf{G} \qquad (65)$$

has eigenvalues with magnitude less than unity, we can appeal to the discussion presented in Sec. 4.7 to prove that the system of Fig. 4.9.4 is locally stable about the equilibrium condition. This can be done provided that $[\mathbf{A}_*, \mathbf{B}_*]$ is a controllable pair.

4.10 CONCLUDING REMARKS

We have touched upon some of the fundamental ideas associated with linear and nonlinear discrete-time systems. These concepts, in particular those associated with stability and feedback, are extremely useful not only in the analysis and design of systems naturally described by difference equations but also in understanding iterative algorithms (Chap. 3) and numerical integration algorithms (Chap. 9).

We would like to emphasize that we have only scratched the surface of the theory of discrete-time systems. The interested reader should consult Refs. 10 to 26 for more details. Nevertheless, these simple ideas are sufficient to allow the reader to develop certain additional systems concepts, e.g., controllability, observability, observers, on his own; for this reason we urge the reader to try the appropriate exercises.

From a conceptual viewpoint, we shall make yet another pass at the concepts presented in this chapter (linearity, superposition, role of eigenvalues and eigenvectors, stability, linearization, and feedback) in Chap. 7, where we examine systems that are described by vector differential equations. Although the mathematics of difference and differential equations look different, the concepts remain the same.

REFERENCES

1 HOLT, C. C., et al.: "Planning Production, Inventories, and Work Force," Prentice-Hall, Englewood Cliffs, N.J., 1960.

2 WAGNER, H. M.: "Principles of Management Science," Prentice-Hall, Englewood Cliffs, N.J., 1970.

3 HOROWITZ, I.: "An Introduction to Quantitative Business Analysis," McGraw-Hill, New York, 1965.

4 HADLEY, G., and T. M. WHITIN: "Analysis of Inventory Systems," Prentice-Hall, Englewood Cliffs, N.J., 1963.

5 FORRESTER, J. W.: "Industrial Dynamics," The M.I.T. Press, Cambridge, Mass., 1961.

6 FORRESTER, J.W.: "Urban Dynamics," The M.I.T. Press, Cambridge, Mass., 1969.

7 FORRESTER, J.W.: "World Dynamics," Wright-Allen Press, Cambridge, Mass., 1971.

8 PIELOU, E. C.: "An Introduction to Mathematical Ecology," Wiley-Interscience, New York, 1969.

9 GOEL, N. S., et al.: On the Volterra and Other Nonlinear Models of Interacting Populations, *Rev. Mod. Phys.*, vol. 43, no. 2, pt. I, pp. 231–276, April 1971.

10 ZADEH, L. A., and C. A. DESOER: "Linear System Theory: The State Space Approach," McGraw-Hill, New York, 1963.

11 TIMOTHY, L., and B. BONA: "State Space Analysis: An Introduction," McGraw-Hill, New York, 1968.

12 DERUSSO, P., R. ROY, and C. CLOSE: "State Variables for Engineers," Wiley, New York, 1965.

13 GUPTA, S.: "Transform and State Variable Methods in Linear Systems," Wiley, New York, 1966.

14 OGATA, K.: "State Space Analysis of Control Systems," Prentice-Hall, Englewood Cliffs, N.J., 1967.

15 SCHULTZ, D., and J. MELSA: "State Functions and Linear Control Systems," McGraw-Hill, New York, 1967.

16 BROCKETT, R. W.: "Finite Dimensional Linear Systems," Wiley, New York, 1970.

17 KUO, B. C.: "Discrete-Data Control Systems," Prentice-Hall, Englewood Cliffs, N.J., 1970.

18 KUO, F. F , and J. F. KAISER (eds.): "System Analysis by Digital Computer," Wiley, New York, 1966.

19 HENRICI, P.: "Discrete Variable Methods in Ordinary Differential Equations," Wiley, New York, 1962.

20 ROSENBROCK, H. H.: "State Space and Multivariable Theory," Wiley, New York, 1970.

21 ROSENBROCK, H. H., and C. STOREY: "Mathematics of Dynamical Systems," Wiley, New York, 1970.

22 MACFARLANE, A. G. J.: "Engineering Systems Analysis," Addison-Wesley, Reading, Mass., 1964.

23 CHEN, C. T.: "Introduction to Linear System Theory," Holt, New York, 1970.

24 CADZOW, J. A., and H. R. MARTENS: "Discrete-Time and Computer Control Systems," Prentice-Hall, Englewood Cliffs, N.J., 1970.

25 BOX, G. E. P., and G. M. JENKINS: "Time Series Analysis, Forecasting, and Control," Holden-Day, San Francisco, 1970.

26 DIRECTOR, S. W., and R. A. ROHRER: "Introduction to System Theory," McGraw-Hill, New York, 1972.

EXERCISES

Section 4.4

4.4.1   This exercise lets you extend some of the results presented for LTI discrete-time systems to the class of linear *time-varying* discrete-time systems. The system

$$\mathbf{x}(k + 1) = \mathbf{A}(k)\mathbf{x}(k) + \mathbf{B}(k)\mathbf{u}(k) \qquad k = 0, 1, 2, \ldots \tag{1}$$

is called a *linear time-varying discrete-time system*. The index k is used to denote the fact that the matrix $\mathbf{A}(k)$ is different for different values of k (hence, time-varying). Thus in general

$$\mathbf{A}(0) \neq \mathbf{A}(1) \neq \mathbf{A}(2) \tag{2}$$

Let us define the state transition matrix by

$$\mathbf{\Phi}(k,j) \triangleq \mathbf{A}(k)\mathbf{A}(k - 1)\cdots\mathbf{A}(j) \qquad \text{with } \mathbf{\Phi}(k,k) \triangleq \mathbf{I} \tag{3}$$

For example,

$$\mathbf{\Phi}(5,3) \triangleq \mathbf{A}(5)\mathbf{A}(4)\mathbf{A}(3) \tag{4}$$

Prove the following properties:

(*a*)   The transition matrix satisfies the semigroup property

$$\mathbf{\Phi}(k,j)\mathbf{\Phi}(j,i) = \mathbf{\Phi}(k,i) \qquad \text{for all } k \geq j \geq i \tag{5}$$

(*b*)   The transition matrix satisfies the matrix difference equation

$$\mathbf{\Phi}(j + 1, k) = \mathbf{A}(j)\mathbf{\Phi}(j,k) \qquad j \geq k \qquad \mathbf{\Phi}(k,k) = \mathbf{I} \tag{6}$$

(*c*)   The solution of Eq. (1) can be written

$$\mathbf{x}(k) = \mathbf{\Phi}(k,j)\mathbf{x}(j) + \sum_{i=j}^{k-1} \mathbf{\Phi}(k, i + 1)\mathbf{B}(i)\mathbf{u}(i) \tag{7}$$

(*d*)   Prove that when $\mathbf{x}(0) = \mathbf{0}$, the system (1) satisfies the superposition principle with respect to the inputs.

(*e*)   Suppose that the matrix $\mathbf{A}(k)$ is given by

$$A(k) = \begin{bmatrix} 1 & 2 - k \\ k^2 & 3 + k \end{bmatrix} \qquad k = 0, 1, 2, \ldots \qquad (8)$$

Determine $\Phi(5,3)$.

Section 4.5

4.5.1   Consider a third-order LTI system described by the difference equation

$$\mathbf{x}(k + 1) = \mathbf{A}\mathbf{x}(k) \qquad \text{where } \mathbf{x}(k) \in R_3$$

and **A** has the form

$$\mathbf{A} = \begin{bmatrix} 0 & 1 & 0 \\ 0 & 0 & 1 \\ 0 & 1 & 0 \end{bmatrix}$$

Given the initial state

$$\mathbf{x}(0) = \begin{bmatrix} 1 \\ 0 \\ 0 \end{bmatrix}$$

(*a*)   Compute $\mathbf{x}(k)$ for $k = 1, 2,$ and 3 using the method described in Sec. 4.4.1.

(*b*)   Determine the eigenvalues and eigenvectors of the **A**-matrix.

(*c*)   Recompute $\mathbf{x}(k)$ for $k = 1, 2,$ and 3 using the eigenvalue and eigenvector method described in Sec. 4.5.

4.5.2   Consider an LTI second-order discrete system

$$\mathbf{x}(k + 1) = \mathbf{A}\mathbf{x}(k)$$

where $\mathbf{x}(k) \in R_2$ and **A** has the form

$$\mathbf{A} = \begin{bmatrix} 0 & \omega \\ -\omega & 0 \end{bmatrix} \qquad \omega = \text{real and positive}$$

Suppose that the initial state is

$$\mathbf{x}(0) = \begin{bmatrix} 1 \\ 0 \end{bmatrix}$$

Describe the nature of the solution vector $\mathbf{x}(k)$ for:

(*a*) $\omega = 0$

(*b*) $0 < \omega < 1$

(*c*) $\omega = 1$

(*d*) $\omega > 1$

Draw qualitative plots and state your conclusions clearly.

4.5.3   Consider the LTI system

$$\mathbf{x}(k + 1) = \mathbf{A}\mathbf{x}(k) \qquad (1)$$

where $\mathbf{x}(k)$ is a 3-dimensional vector and $\mathbf{A}$ is the matrix

$$\mathbf{A} = \begin{bmatrix} 0 & 1 & 0 \\ 0 & 0 & 1 \\ 2 & 1 & -2 \end{bmatrix}$$

It is given that the eigenvalues of $\mathbf{A}$ are

$$\lambda_1 = -1 \qquad \lambda_2 = +1 \qquad \lambda_3 = -2$$

and that the corresponding eigenvectors are

$$\mathbf{v}_1 = \begin{bmatrix} 1 \\ -1 \\ 1 \end{bmatrix} \qquad \mathbf{v}_2 = \begin{bmatrix} 1 \\ 1 \\ 1 \end{bmatrix} \qquad \mathbf{v}_3 = \begin{bmatrix} 1 \\ -2 \\ 4 \end{bmatrix}$$

Find the initial state vector $\mathbf{x}(0)$ such that at $k = 4$, the solution of Eq. (1) is

$$\mathbf{x}(k + 1) = -16 \begin{bmatrix} 2 \\ -4 \\ 8 \end{bmatrix}$$

4.5.4   Consider the unforced third-order LTI system

$$\mathbf{x}(k + 1) = \mathbf{A}\mathbf{x}(k)$$

Suppose that $\mathbf{A}$ is the 3×3 matrix

$$\mathbf{A} = \begin{bmatrix} -1 & 2 & 2 \\ 2 & 2 & 2 \\ -3 & -6 & -6 \end{bmatrix}$$

(a)   Show that the eigenvalues of $\mathbf{A}$ are

$$\lambda_1 = -2 \qquad \lambda_2 = -3 \qquad \lambda_3 = 0$$

(b)   Show that the corresponding eigenvectors are

$$\mathbf{v}_1 = \begin{bmatrix} 2 \\ -1 \\ 0 \end{bmatrix} \qquad \mathbf{v}_2 = \begin{bmatrix} 1 \\ 0 \\ -1 \end{bmatrix} \qquad \mathbf{v}_3 = \begin{bmatrix} 0 \\ 1 \\ -1 \end{bmatrix}$$

(c)   Consider the initial state vector $\mathbf{x}(0)$

$$\mathbf{x}(0) = \begin{bmatrix} 3 \\ 0 \\ -2 \end{bmatrix}$$

Find $\mathbf{x}(3)$ without using the eigenvalue method.

(*d*)   Now express $\mathbf{x}(0)$ as a linear combination of the eigenvectors, i.e., find α, β, γ such that

$$\mathbf{x}(0) = \alpha \mathbf{v}_1 + \beta \mathbf{v}_2 + \gamma \mathbf{v}_3$$

Determine $\mathbf{x}(3)$ once more using the eigenvalue method. Convince yourself that you get the same answer as in part (*c*).

4.5.5   Consider the discrete-time system

$$\mathbf{x}(k + 1) = \mathbf{A}\mathbf{x}(k) \qquad \mathbf{x}(k) \in R_3 \tag{1}$$

The matrix $\mathbf{A}$ is given by

$$\mathbf{A} = \begin{bmatrix} 0 & 1 & 0 \\ 0 & 0 & 1 \\ 2 & 1 & -2 \end{bmatrix}$$

(*a*)   Write the characteristic polynomial of $\mathbf{A}$. Verify that the eigenvalues of $\mathbf{A}$ are given by

$$\lambda_1 = -1 \qquad \lambda_2 = +1 \qquad \lambda_3 = -2$$

(*b*)   Verify that the associated eigenvectors of $\mathbf{A}$ are given by

$$\mathbf{v}_1 = \begin{bmatrix} 1 \\ -1 \\ 1 \end{bmatrix} \qquad \mathbf{v}_2 = \begin{bmatrix} 1 \\ 1 \\ 1 \end{bmatrix} \qquad \mathbf{v}_3 = \begin{bmatrix} 1 \\ -2 \\ 4 \end{bmatrix}$$

(*c*)   Suppose that the initial state $\mathbf{x}(0)$ of the system (1) is given by

$$\mathbf{x}(0) = \begin{bmatrix} 13 \\ 13 \\ 13 \end{bmatrix}$$

Determine $\mathbf{x}(5)$.

(*d*)   Suppose that the initial state $\mathbf{x}(0)$ of the system (1) is given by

$$\mathbf{x}(0) = \begin{bmatrix} 3 \\ 6 \\ 12 \end{bmatrix}$$

Determine $\mathbf{x}(5)$. *Hint*: Think before you plunge into millions of multiplications.

4.5.6   Consider a LTI system

$$\mathbf{x}(k + 1) = \mathbf{A}\mathbf{x}(k) \tag{1}$$

where $\mathbf{x}(k) \in R_n$ and $\mathbf{A}$ is a constant $n \times n$ matrix in *companion* form. Suppose that $\mathbf{A}$ has n real *identical* eigenvalues all equal to λ, that is,

$$\lambda_1 = \lambda_2 = \cdots = \lambda_n = \lambda \tag{2}$$

Write the solution of (1) in terms of the eigenvalue λ. Read Sec. A.10 of Appendix A before doing this exercise.

4.5.7 Using the ideas of Appendix Sec. A.10 and of Exercise 4.5.6, determine the solution using the eigenvalue method of the system

$$\mathbf{x}(k+1) = \begin{bmatrix} 0 & 1 & 0 \\ 0 & 0 & 1 \\ 4 & 0 & -3 \end{bmatrix} \mathbf{x}(k)$$

for $k = 1, 2, 3$ given that

$$\mathbf{x}(0) = \begin{bmatrix} 1 \\ 1 \\ 1 \end{bmatrix}$$

Hint: The eigenvalues of the **A** matrix are at 1 and a repeated one at -2.

Section 4.7

4.7.1 Consider the ecological example presented in Sec. 4.7.4.

(*a*) Suppose conservationists that love both species convinced the legislature to abolish hunting. Thus set

$$u(k) = 0$$

in Eq. (34) and

(*1*) Find all equilibrium conditions.
(*2*) Analyze the stability of each.
(*3*) Do you think the conservationists will be pleased with the outcome?

(*b*) Consider the situation with hunting but suppose that the effect of a chemical in the environment cuts the birth rate of species 1 so that the constant a_1 becomes

$$a_1 = 2$$

(*1*) Examine in full detail once more all equilibrium conditions.
(*2*) Analyze the stability of each.

(*c*) Consider the general model of Eq. (30) with no hunting allowed $[u(k) = 0]$; all parameters are positive.

(*1*) Derive general expressions for all possible equilibria.
(*2*) Derive general expressions for the stability of all equilibria in terms of a_1, a_2, b_{11}, b_{12}, b_{21}, and b_{22}.

Section 4.9

4.9.1   Consider a discrete-time linear dynamic system described by

$$\begin{bmatrix} x_1(k+1) \\ x_2(k+1) \end{bmatrix} = \begin{bmatrix} 0 & 1 \\ 0 & 2 \end{bmatrix} \begin{bmatrix} x_1(k) \\ x_2(k) \end{bmatrix} + \begin{bmatrix} 0 \\ 1 \end{bmatrix} u(k)$$

(a)   Prove that if $u(k) = 0$, the resulting system is unstable.

(b)   To stabilize the system an engineer proposes to use a feedback control arrangement of the form

$$u(k) = k_1 x_1(k)$$

where k_1 is some constant. Is it possible to find a value of k_1 such that the resultant system is strictly stable?

(c)   Another engineer proposes to use a feedback control of the form

$$u(k) = k_1 x_1(k) + k_2 x_2(k)$$

where k_1 and k_2 are also constants. Determine the range of values of k_1 and k_2 and their interrelation such that the resulting closed-loop system is stable.

(d)   In part (c) is it possible to find values of k_1 and k_2 such that the eigenvalues of the closed-loop systems are at

$$\bar{\lambda} = \tfrac{1}{2} \pm j\tfrac{1}{2}$$

If so, determine the values of k_1 and k_2.

4.9.2   Consider the following problem faced by a cattle rancher. He has a particular breed of cattle:

$$x_1(k) = \text{average number of 1-year-old cattle (young)}$$

$$x_2(k) = \text{average number of 2-year-old cattle (mature)}$$

$$x_3(k) = \text{average number of 3-year old and older cattle (old)}$$

In the absence of slaughtering, the cattle propagate and die according to the equations

$$x_1(k+1) = 0.8x_2(k) + 0.4x_3(k)$$

$$x_2(k+1) = x_1(k)$$

$$x_3(k+1) = x_2(k) + 0.7x_3(k)$$

which means that one-year-olds do not multiply, each two-year-old produces on the average 0.8 young cattle per year, and old cattle produce on the average 0.4 young cattle per year. Also note that only old cattle die from natural causes, at the rate of 30 percent per year.

The cattle rancher can slaughter only mature and old cattle. Let $S_2 x_2(k)$,

$0 \leq S_2 \leq 1$, denote the average number of mature cattle he slaughters and sells at the average price of \$400 per head. Let $S_3 x_3(k)$, $0 \leq S_3 \leq 0.7$, denote the average number of old cattle he slaughters and sells at the average price of \$300 per head. At time $k = 0$ he has inherited the following cattle

$$x_1(0) = 100 \qquad x_2(0) = 100 \qquad x_3(0) = 100$$

He would like to establish the constant slaughter rates S_2 and S_3 such that after a sufficient number of years *he maximizes his yearly profit and maintains a steady supply of cattle*, given that the dynamics are now

$$x_1(k + 1) = 0.8 x_2(k) + 0.4 x_3(k)$$
$$x_2(k + 1) = x_1(k)$$
$$x_3(k + 1) = x_2(k) + 0.7 x_3(k) - S_2 x_2(k) - S_3 x_3(k)$$

4.9.3   Consider the following discrete-time nonlinear system

$$x_1(k + 1) = \tfrac{1}{4} x_2^2(k)$$
$$x_2(k + 1) = x_3(k)$$
$$x_3(k + 1) = 2 x_2(k) - \tfrac{1}{4} x_3^2(k) + u(k) - 1$$

(*a*)   Set $u(k) = 0$ for all k. Determine an equilibrium state for the system. Show that the equilibrium is unstable.

(*b*)   Let $\mathbf{x}^*$ denote the equilibrium state found in part (*a*). Let

$$\delta \mathbf{x}(k) \triangleq \mathbf{x}(k) - \mathbf{x}^*$$

denote the state perturbation. It is desired to use feedback so that the closed-loop system is stable about the equilibrium state $\mathbf{x}^*$. The form of the feedback is

$$u(k) = -[g_1\, \delta x_1(k) + g_2\, \delta x_2(k) + g_3\, \delta x_3(k)]$$

where g_1, g_2, g_3 are constants. Determine g_1, g_2, g_3 such that the linearized closed-loop system has eigenvalues of $\tfrac{1}{2}$, $-\tfrac{1}{2}$, and $\tfrac{1}{4}$.

4.9.4   (Very long)   The purpose of this exercise is to analyze the problems associated with the growth characteristics and logging operations of redwood trees. The problem is based in part upon an article[†] you are encouraged to read. In the following model the redwoods are divided into three classes at each unit of time:

Class 1:   trees 0 to 200 years old (young)

Class 2:   200 to 800 years old (mature)

Class 3:   More than 800 years old (old)

† C. A. Bosch, Redwoods: A Population Model, *Science*, vol. 172, pp. 345–349, Apr. 23, 1971.

We shall adopt as *50 years* the basic time unit. Thus $k = 1$ means 50 years, $k = 2$ means 100 years, etc. In the absence of logging operations we can assume the following characteristics. Natural resistance to disease and environmental conditions means that the redwoods die only of old age. Thus in every 50-year period, we assume that one-third of the redwood trees in class 3 die of old age. We assume that no trees in class 1 or 2 ever die (they just grow old). Due to genetic factors we assume that *in each 50-year period*:

1   Each tree in class 1 can produce on the average 10.25 new trees.
2   Each tree in class 2 can produce on the average 25 new trees.
3   Each tree in class 3 can produce on the average 5 new trees.

(The variability in reproduction rates is related to the ability of the mature trees to produce more resistant and viable seed.)

(*a*)   (Mathematical Modeling)   Each unit of k represents 50 years. Use the following state variables:

$$x_1(k) = \text{number of redwoods in class 1 in period } k$$
$$x_2(k) = \text{number of redwoods in class 2 in period } k$$
$$x_3(k) = \text{number of redwoods in class 3 in period } k$$

Let

$$\mathbf{x}(k) = \begin{bmatrix} x_1(k) \\ x_2(k) \\ x_3(k) \end{bmatrix}$$

denote the state vector for the problem. To simplify the model still further, we shall not make any distinction between the specific ages of the redwoods in each class. Thus, we specify the following transitions. Each 50 years

1   One-fourth of the trees in class 1 graduate to class 2.
2   One-twelfth of the trees in class 2 graduate to class 3.

Develop the discrete equations of the redwood-population model in the absence of any external factors in the form

$$\mathbf{x}(k + 1) = \mathbf{A}\mathbf{x}(k)$$

Specify completely all the elements of the matrix $\mathbf{A}$.

(*b*)   Discuss the stability properties of the model derived in part (*a*). Draw conclusions about the unchecked redwood growth. *Hint*: To find roots of third-order polynomials use Newton's method.

(*c*)   Next we discuss the effects of logging. We assume that at time $k = 0$, the *R* logging company has purchased land which contains the following population:

$$x_1(0) = 1{,}000 \qquad x_2(0) = 500 \qquad x_3(0) = 50$$

The company is interested in developing a policy for cutting down redwoods, but suppose that the laws of California forbid felling redwoods over 800 years old. The company then decides to choose the following logging policy:

$$u_1(k) = c_1 x_1(k) = \text{trees cut down from class 1 in a 50-year period}$$
$$u_2(k) = c_2 x_2(k) = \text{trees cut down from class 2 in a 50-year period}$$

Note that c_1 and c_2 are constants (independent of k). Let

$$\bar{x}_1 \triangleq \lim_{k \to \infty} x_1(k)$$

$$\bar{x}_2 \triangleq \lim_{k \to \infty} x_2(k) \quad \text{and} \quad \bar{\mathbf{x}} = \begin{bmatrix} \bar{x}_1 \\ \bar{x}_2 \\ \bar{x}_3 \end{bmatrix}$$

$$\bar{x}_3 \triangleq \lim_{k \to \infty} x_3(k)$$

Then the model for the population change is

$$x_1(k+1) = 11[x_1(k) - u_1(k)] + 25[x_2(k) - u_2(k)] + 5x_3(k)$$
$$x_2(k+1) = \tfrac{1}{4}[x_1(k) - u_1(k)] + \tfrac{11}{12}[x_2(k) - u_2(k)]$$
$$x_3(k+1) = \tfrac{1}{12}[x_2(k) - u_2(k)] + \tfrac{2}{3}x_3(k)$$

Determine the constants c_1 and c_2 subject to the constraints

$$0 \le c_1 \le 1 \qquad 0 \le c_2 \le 1$$

such that the company objectives are met and as $k \to \infty$, the redwood population stabilizes.

(d) Since the values of c_1 and c_2 may not be unique, select them on the basis of the following argument. Suppose each class 1 redwood sells for $100 and that each class 2 redwood sells for $5,000. The company then is interested in maximizing the profit $\bar{p}$ when steady state is reached:

$$\bar{p} = 100c_1 \bar{x}_1 + 5{,}000c_2 \bar{x}_2$$

Find the values of c_1 and c_2 which will maximize the profit and meet all other objectives.

(e) Double-check that the model that you have arrived at is stable in the sense of maintaining a constant steady-state population.

(f) Plot the resultant changes starting from the given initial population. How many years will it take until steady state is achieved for all practical purposes?

4.9.5 This exercise introduces an extremely important concept in modern system theory, the notion of *controllability*. Consider an LTI discrete-time system described by the vector difference equation

$$\mathbf{x}(k + 1) = \mathbf{A}\mathbf{x}(k) + \mathbf{b}u(k) \qquad k = 0, 1, 2, \ldots \tag{1}$$

where $\mathbf{x}(k) \in R_n$ for all k, $u(k)$ is a scalar, $\mathbf{A}$ is a constant $n \times n$ matrix, and $\mathbf{b}$ is a constant n-column vector. Suppose that the initial state $\mathbf{x}(0)$ is an arbitrary vector in R_n. The question arises:

> Is it possible to find an input sequence $u(0), u(1), \ldots, u(N)$, with N finite, such that $\mathbf{x}(N + 1) = \mathbf{0}$? If so, the system (1) is called *controllable*.

Basically the issue is one of *control*. We can interpret the origin of the n-dimensional euclidean space as being the desired state of the system (1). The initial state $\mathbf{x}(0)$ is different from the origin. Then we wish to find the input sequence $\{u(k)\}$ so that the system is transferred to the desired state $(\mathbf{0})$ in a finite number of steps.

(*a*)   Prove that system (1) is controllable if and only if the $n \times n$ matrix

$$[\mathbf{b} \mid \mathbf{A}\mathbf{b} \mid \mathbf{A}^2\mathbf{b} \mid \cdots \mid \mathbf{A}^{n-1}\mathbf{b}] \tag{2}$$

is nonsingular, where $\mathbf{b}$ is the first column vector, $\mathbf{A}\mathbf{b}$ is the second column vector, $\mathbf{A}^2\mathbf{b}$ is the third column vector, etc. *Hint*: Try to get n equations in n unknowns.

(*b*)   Suppose that system (1) is controllable. Show that the origin can be reached (by some control sequence) in *at most n* steps, that is, $N - 1 \leq n$.

(*c*)   Suppose that in Eq. (1)

$$\mathbf{A} = \begin{bmatrix} 0 & 4 & 3 \\ 0 & 20 & 16 \\ 0 & -25 & -20 \end{bmatrix} \qquad \mathbf{b} = \begin{bmatrix} -1 \\ 3 \\ 0 \end{bmatrix}$$

Is the system controllable?

4.9.6   In Exercise 4.9.5 the notion of controllability was established for a specific target state, namely, the origin. Prove that if such a system is controllable, one can find a control sequence that will transfer any initial state $\boldsymbol{\xi}$ to any final state $\boldsymbol{\theta}$ (not necessarily the origin) in *at most n* steps (time units.)

4.9.7   This exercise extends the results of Exercises 4.9.5 and 4.9.6 to systems with an arbitrary number of inputs. Consider a system $[\mathbf{x}(k) \in R_n, \mathbf{u}(k) \in R_m]$

$$\mathbf{x}(k + 1) = \mathbf{A}\mathbf{x}(k) + \mathbf{B}\mathbf{u}(k) \tag{1}$$

Let $\boldsymbol{\xi}$ be any initial state, and let $\boldsymbol{\theta}$ be any final state. The system is controllable if there exists a sequence of input vectors $\{\mathbf{u}(k)\}$ such that the system state of (1) is transferred from $\boldsymbol{\xi}$ to $\boldsymbol{\theta}$ in a finite number of steps.

(*a*)   Prove that a necessary and sufficient condition for controllability is that the rank of the following $n \times nm$ matrix is n

$$[\mathbf{B} \mid \mathbf{A}\mathbf{B} \mid \mathbf{A}^2\mathbf{B} \mid \cdots \mid \mathbf{A}^{n-1}\mathbf{B}]$$

(*b*) Prove that if the system is controllable, the transfer from $\boldsymbol{\xi}$ to $\boldsymbol{\theta}$ can be accomplished in at most *n* time units. *Hint*: The Cayley-Hamilton theorem, Theorem 16, Sec. A.10, Appendix A, may be helpful.

4.9.8 Consider a linear *n*th-order time-invariant system whose state vector $\mathbf{x}(k)$ satisfies the difference equation

$$\mathbf{x}(k+1) = \mathbf{A}\mathbf{x}(k) \qquad \begin{array}{l} k = 0, 1, 2, \dots \\ \mathbf{x}(k) \in R_n \end{array} \tag{1}$$

Let us suppose that we cannot directly measure the state $\mathbf{x}(k)$ of the system. Furthermore, suppose that the only quantity we can measure is a *scalar* output, $y(k)$, which is related to the state vector $\mathbf{x}(k)$ by the relation

$$y(k) = c_1 x_1(k) + c_2 x_2(k) + c_3 x_3(k) + \cdots + c_n x_n(k) = \mathbf{c}'\mathbf{x}(k) \tag{2}$$

Clearly, if we could determine the initial state vector $\mathbf{x}(0)$ of Eq. (1), we could compute $\mathbf{x}(k)$ for all $k = 1, 2, \dots$ simply by solving the difference equation (1).

(*a*) An interesting question then is: Can we compute $\mathbf{x}(0)$ simply by knowing the value of $y(k)$ at several instants of time (also assuming that we know the $n \times n$ matrix $\mathbf{A}$ and the *n*-vector $\mathbf{c}$)? If so, how many measurements of $y(k)$ are enough to determine $\mathbf{x}(0)$ uniquely?

(*b*) Prove that condition (3) below is both a *necessary and sufficient* condition for uniquely determining an arbitrary initial state $\mathbf{x}(0)$ given a *finite* number of measurements.

$$\text{Rank} \begin{bmatrix} \mathbf{c}' \\ \mathbf{c}' & \mathbf{A} \\ \vdots \\ \mathbf{c}' & \mathbf{A}^{n-1} \end{bmatrix} = n \tag{3}$$

HINTS

1 Since $\mathbf{x}(0)$ is *n*-dimensional, look for *n* linearly independent equations in *n* unknowns.

2 The Cayley-Hamilton theorem (Theorem 16, Sec. A.10, Appendix A) may be helpful.

A system that has this property is called *observable*.

4.9.9 This exercise extends the notion of observability to systems with more than scalar measurements. Consider the LTI system

$$\begin{array}{l} \mathbf{x}(k+1) = \mathbf{A}\mathbf{x}(k) + \mathbf{B}\mathbf{u}(k) \\ \mathbf{y}(k) = \mathbf{C}\mathbf{x}(k) \end{array} \qquad k = 0, 1, 2, \dots \tag{1}$$

Suppose that
$$\mathbf{x}(k) \in R_n \qquad \mathbf{u}(k) \in R_m \qquad \mathbf{y}(k) \in R_r$$
Suppose that we know **A**, **B**, **C** and that we can measure *exactly*
$$\mathbf{u}(0), \mathbf{u}(1), \mathbf{u}(2), \ldots \qquad \mathbf{y}(0), \mathbf{y}(1), \mathbf{y}(2), \ldots \tag{2}$$

The system (1) is called *observable* if on the basis of a *finite* number of measurements (2) one can *uniquely* calculate the initial state $\mathbf{x}(0)$ of (1) and hence all subsequent $\mathbf{x}(k)$'s.

(*a*) Prove that the system (1) is observable if and only if the rank of the following $n \times nr$ matrix is equal to n:
$$[\mathbf{C}' \quad \mathbf{A}'\mathbf{C}' \quad \mathbf{A}'^2\mathbf{C}' \quad \cdots \quad \mathbf{A}'^{n-1}\mathbf{C}']$$

(*b*) Determine the smallest number of measurements, in the sense of time units, required to calculate $\mathbf{x}(0)$ for an observable system.

4.9.10 A linear, constant, discrete-time, asymptotically stable, dynamical system is described by
$$\mathbf{x}(k+1) = \mathbf{A}\mathbf{x}(k) + \mathbf{B}\mathbf{u}(x) \qquad \mathbf{x}(0) = \mathbf{x}_0$$
$$\mathbf{y}(k) = \mathbf{C}\mathbf{x}(k)$$

From measurements on the output $\mathbf{y}(k)$ and the input $\mathbf{u}(k)$ it is desired to estimate the state $\mathbf{x}(k)$. A scheme for achieving this goal is to construct another linear dynamical system driven by $\mathbf{u}(k)$ and $\mathbf{y}(k)$ and use the output of this second system to estimate $\mathbf{x}(k)$. The second system is called an *observer*. Let $\mathbf{z}(k)$ be the state vector of the observer. Then the evolution of $\mathbf{z}(k)$ is governed by
$$\mathbf{z}(k+1) = \mathbf{F}\mathbf{z}(k) + \mathbf{G}\mathbf{y}(k) + \mathbf{H}\mathbf{u}(k)$$

It is desired to develop design equations for the matrices **F**, **G**, and **H** which determine the observer behavior. The design criteria are as follows:

(*a*) Let $\mathbf{e}(k) \triangleq \mathbf{z}(k) - \mathbf{x}(k)$ represent the observer *error*. Then the difference equation governing $\mathbf{e}(k)$ must be independent of $\mathbf{u}(k)$, that is, $\mathbf{e}(k)$ must satisfy an equation of the form
$$\mathbf{e}(k+1) = \mathbf{E}\mathbf{e}(k)$$

(*b*) As $k \to \infty$, the observer output $\mathbf{z}(k)$ must asymptotically approach $\mathbf{x}(k)$, that is, $\mathbf{z}(k) \to \mathbf{x}(k)$, regardless of the initial condition, $\mathbf{z}(0) = \mathbf{z}_0$.

(*c*) The rate of approach of $\mathbf{z}(k)$ to $\mathbf{x}(k)$ must be fast compared with the natural settling time of $\mathbf{x}(k)$ [the natural settling time is roughly the time required for $\mathbf{x}(k)$ to fall to a small fraction, say 0.10, of its initial value when $\mathbf{u}(k) \equiv \mathbf{0}$].

Develop design equations which can be used to construct the observer. In particular, how can **F**, **G**, and **H** be determined? Comment on which of these matrices are uniquely determined and which are not. Comment on how you would select the nonunique matrices in view of the design specifications stated.

5

AN EXAMPLE OF A DISCRETE-TIME LINEAR SYSTEM: A QUARTERLY ECONOMETRIC MODEL OF THE UNITED STATES ECONOMY

SUMMARY[†]

This chapter provides a nontrivial example of a discrete-time dynamical system. The system we shall consider is the United States economy. A fundamental understanding of how the United States economy evolves is important not only to the welfare of the citizens of the United States but also for its worldwide effects.

The model we shall consider is an LTI discrete-time model of the United States economy from 1955 to 1968. It is a quarterly model, in the sense that the values of the economic variables are described, or defined, every 3 months. It is characterized by 28 state variables and 3 control variables, representing an econometric model of medium complexity.

[†] This chapter may be omitted without loss of continuity. The material has been adapted from R. S. Pindyck, Optimal Economic Stabilization Policy, Ph.D. dissertation, Massachusetts Institute of Technology, June 1971. We are grateful to Prof. Pindyck for his permission to use his results and for his constructive comments on the material of this chapter.

Our purpose in examining this example is to acquaint the reader with the types of large-scale dynamical models we must be concerned with. We shall use the issues that arise from this model to motivate a host of questions relevant to modern system and control science.

5.1 INTRODUCTION

Models of economic systems deal with the interrelations of variables of physical interest. Typical variables are gross national product (GNP), government expenditures, unemployment rates, interest rates, wage levels, money supply (as controlled by the Federal Reserve Board), etc.

Mathematical models of the economy can be used for several purposes and by different agencies. They can be used for forecasting or prediction purposes to deduce the probable outcome of contemplated actions, e.g., raising taxes, freezing wages, by the national government. They can also be used for control purposes, to adjust frequently a typical policy variable, e.g., money supply, which can have a short-term effect upon other economic variables, e.g., interest rates.

Mathematical economic models come in different shapes and sizes. In general, *macroeconomic models*† are models that utilize aggregate measures, e.g., national income, employment, price level, to describe the interaction of economic units, e.g., the consumer, labor, business, and government. They attempt to answer such questions as stability of prices or oscillatory economic behavior and perhaps to deduce monetary and fiscal policies for achieving certain goals. On the other hand, *microeconomic models* attempt to model the behavior of individual atomic economic units, e.g., a particular type of business or the buying habits of a particular profession.

Since millions of individual economic units make up an economic system, questions on monetary and fiscal policy at the national government level cannot be decided upon the basis of a huge microeconomic model. Instead the individual economic units are all represented together in economic groups with relatively well-defined objectives; in this manner one obtains a macroeconomic model with far fewer variables but of course subject to more uncertainty and error.

This method of aggregation is not restricted to economic systems. Every

---

† See, for example, K. C. Kogiku, "An Introduction to Macroeconomic Models," McGraw-Hill, New York, 1968; J. M. Culbertson, "Macroeconomic Theory and Stabilization Policy," McGraw-Hill, New York, 1968; R. G. D. Allen, "Macroeconomic Theory," St. Martin's Press, New York, 1967.

complex engineering system can be represented by either *macroeconomic* or *microeconomic* equations. For example, consider the Saturn booster vehicle;[†] to describe its equations of motion exactly one must consider not only the point-mass equations of motion but also the dynamic response characteristics of electronic amplifiers, hydraulic actuators, bending modes, torsional modes, fuel sloshing, gyro dynamics, accelerometer dynamics, etc. To be able to make *on-line* predictions upon the rocket trajectory and to use such information to make *on-line* decisions about guidance and control, e.g., time and magnitude of midcourse corrections, one must use a much simpler mathematical model for computer simulation. This is analogous to the use of macroeconomic models.

Macroeconomics and economic theory in general attempt to establish some structure in the interrelationship of economic variables. These are based upon empirical observations and logical reasoning, and they provide the counterpart of physics laws, e.g., Kirchhoff's or Newton's laws. Typical economic laws are those of *supply and demand* and the *Phillips curve*, which roughly relates unemployment and inflation;[‡] they are general functional guidelines used to interrelate economic variables in a macroeconomic model.

In general, macroeconomic models are described by difference equations; it is widely recognized that the exact equations that describe a dynamical economy must be *nonlinear* (also time-varying and stochastic). This is also the case with almost all engineering systems. The key difference between many complex nonlinear engineering systems and complex nonlinear macroeconomic models is the ease with which the *coefficients* in difference equations can be obtained in a relatively accurate form.

In physical systems extensive knowledge of many fundamental constants is extensive, e.g., the acceleration of gravity, the mass of the earth, the speed of light. Standard nominal tables of the atmospheric density as a function of altitude are available. Such information is coupled with manufacturing data, e.g., value in ohms of a resistor, weight of a component, gain of an amplifier, specific impulse of a particular propellant; the availability of such quantitative information, coupled with extensive knowledge of the laws of nature, makes engineering systems relatively easy to describe by a quantitative dynamical model.

This is not true of economic systems. There is no universal agreement on economic laws. There are few, if any, economic constants. Thus, to deduce the structure of an economic model and the coefficients associated with it is by no means a trivial task.

[†] W. Haeussermann, Saturn Launch Vehicle's Navigation, Guidance, and Control System, *Automatica*, vol. 7, no. 5, pp. 537–556, September 1971.

[‡] A. W. Philips, A Simple Model of Employment, Money, and Prices in a Growing Economy, *Economica*, November 1961; P. A. Samuelson and R. M. Solow, Analytical Aspects of Anti-inflation Policy, *Am. Econ. Rev.*, vol. 50, pp. 177–194, May 1960.

The field of economics that attempts to produce quantitative mathematical models is that of *econometrics*.† Loosely speaking, econometrics is the science that transforms past measurements of economic variables into a set of mathematical equations with numerical values for the coefficients. It is based upon statistical theory and time-series analysis; thus it provides not only a deterministic mathematical model but also some statistical information which can be used to deduce the level of uncertainty associated with each coefficient in the equations and the equation itself.

At present the state of the art in econometrics is roughly as follows:

1 One can obtain linear or nonlinear time-invariant, discrete-time difference equations relating the economic variables.
2 These models may be valid only for short periods.
3 In view of the aggregation techniques, complexity assumptions, and the structure assumptions, the mathematical models represent only an approximation to reality.

5.2 AN EXAMPLE OF A SIMPLE MACROECONOMIC QUALITATIVE MODEL

In this section we examine an ad hoc but typical macroeconomic model. There is no claim whatsoever that this model has any realism; as we shall see in the remainder of this chapter, it neglects many variables which are important.

The reason for introducing this simple qualitative model now is to set the stage for the more complex quantitative model which we consider later and to show how one can translate standard econometric difference equations into state-variable form.

We remark, parenthetically, that the use of state-variable representations for econometric models is still at its infancy. Many economists utilize state variables simply as an alternate mathematical representation. *The advantages of state-variable representation in conjunction with the powerful system-theoretic techniques, e.g., dynamical linearization, stability, static and dynamical optimization, parameter estimation, and adaptive control, are not yet fully appreciated by the economic community.*

† See, for example, J. Johnston, "Econometric Methods," McGraw-Hill, New York, 1963; A. S. Goldberg, "Econometric Theory," Wiley, New York, 1964; J. S. Duesenberry, G. Fromm, L. R. Klein, and E. Kuh (eds.), "The Brookings Quarterly Econometric Model of the United States," Rand McNally, Chicago, 1965.

5.2.1 A Standard Macroeconomic Model

We shall consider the interrelationship of the following four economic variables (all units are in dollars):

C = consumption expenditures for goods and services
P = price level of goods and services (related to inflation)
W = wage level
M = money supply (regulated by the Federal Reserve Board)

We are interested in the interrelation of these four variables at different instants of time. We shall use the time index

$$k = 0, 1, 2, \ldots \tag{1}$$

to denote discrete time instants. Each k represents the passage of a quarter (3 months).† Thus $k = 0$ may index the start of the macroeconomic model, say, third quarter 1971. Then $k = 1$ indexes the fourth quarter of 1971, $k = 2$ the first quarter of 1972, and so on.

To stress the dependence of the economic variables upon time we use the notation

$$
\begin{aligned}
C(k) &\triangleq \text{consumption at time } k \\[4pt]
P(k) &\triangleq \text{price level at time } k \\[4pt]
W(k) &\triangleq \text{wage level at time } k \\[4pt]
M(k) &\triangleq \text{money supply at time } k
\end{aligned}
\tag{2}
$$

A typical set of equations that can be used to describe the interdependence of these four variables is

$$C(k) = \alpha_1 C(k-1) + \alpha_2 P(k-1) + \alpha_3 W(k-1) + \alpha_4 W(k-2) \tag{3}$$

$$P(k) = \beta_1 P(k-1) + \beta_2 W(k-1) + \beta_3 W(k-2) + \beta_4 M(k-1) \tag{4}$$

$$W(k) = \gamma_1 P(k-3) + \gamma_2 C(k-1) \tag{5}$$

Where α_1, α_2, α_3, α_4, β_1, β_2, β_3, β_4, γ_1, and γ_2 are constant parameters (they do not change with k). Although we have pulled the model of Eqs. (3) to (5) out of a hat, its structure makes some sense.

---

† The reason for this is that many important variables, for example GNP, are measured only every 3 months.

For example, let us examine Eq. (5). It states that the present wage level $W(k)$ depends on the price level of 9 months ago (to allow for negotiation time between unions and management) as well as upon the consumption level of 3 months ago $C(k-1)$ (because the most immediate expenditures, and hence the standard of living, are freshest in the minds of the consumer). Equation (4) governs the price level (inflationary trend). The current price level $P(k)$ depends upon the previous quarter price level $P(k-1)$, past wage levels $W(k-1)$ and $W(k-2)$, and the total money supply in the previous quarter $M(k-1)$, based on the argument that the more money around the higher the prices. One can also make similar comments about Eq. (3).

In such macroeconomic models the variables C, P, W are called *endogenous* variables; M is called an *exogenous* or *policy* variable.

5.2.2 State-Variable Formulation

Let us now attempt to reduce the macroeconomic model (3) to (5) to a state-variable representation. First let us recall that in a discrete-time state-variable representation we in general seek a vector difference equation of the form (see Chap. 4)

$$\mathbf{x}(k+1) = \mathbf{f}(\mathbf{x}(k), \mathbf{u}(k)) \tag{6}$$

where $\mathbf{x}(k)$ = value of state vector at time k
$\mathbf{u}(k)$ = value of input vector at time k

If the system is linear and time-invariant, we seek a vector difference equation of the form

$$\mathbf{x}(k+1) = \mathbf{A}\mathbf{x}(k) + \mathbf{B}\mathbf{u}(k) \qquad k = 0, 1, 2, \ldots \tag{7}$$

where $\mathbf{A}$ and $\mathbf{B}$ are constant matrices. An alternate form of Eq. (7) is

$$\mathbf{x}(k) = \mathbf{A}\mathbf{x}(k-1) + \mathbf{B}\mathbf{u}(k-1) \qquad k = 1, 2, \ldots \tag{8}$$

Intuitively we expect that the variables $C(k)$, $P(k)$, and $W(k)$ will be state variables. Hence, the first question to be considered is whether Eqs. (3) to (5) are already in a state-variable form (8). The answer is no, because $W(k-2)$ appears in the right-hand side of (3) and (4) and because $P(k-3)$ appears in the right-hand side of (5). Note that our state-variable equation (8) allows only state variables $x_i(k-1)$, with a single time lag, to appear in the right-hand side of the equation.

It is a relatively straightforward task to construct a correct state-variable model for the macroeconomic system (3) to (5). To do this we proceed as follows.

First let us define

$$x_1(k) \triangleq C(k) \tag{9}$$

$$x_2(k) \triangleq P(k) \tag{10}$$

$$x_3(k) \triangleq W(k) \tag{11}$$

From Eq. (11) we deduce that

$$x_3(k-1) = W(k-1) \tag{12}$$

and

$$x_3(k-2) = W(k-2) \tag{13}$$

Let us define

$$x_4(k-1) \triangleq x_3(k-2) = W(k-2) \tag{14}$$

which implies

$$x_4(k) = x_3(k-1) = W(k-1) \tag{15}$$

In a similar manner we have

$$x_2(k-1) = P(k-1) \tag{16}$$

$$x_2(k-2) = P(k-2) \tag{17}$$

$$x_2(k-3) = P(k-3) \tag{18}$$

Let us define

$$x_5(k-1) \triangleq x_2(k-2) = P(k-2) \tag{19}$$

which implies

$$x_5(k) = x_2(k-1) = P(k-1) \tag{20}$$

Also let us define

$$x_6(k-1) \triangleq x_5(k-2) = P(k-3) \tag{21}$$

which implies

$$x_6(k) = x_5(k - 1) = P(k - 2) \tag{22}$$

Now we substitute Eqs. (9), (10), (11), (14), and (21) into the macroeconomic model (3) to (5) using

$$u(k) \triangleq M(k) \tag{23}$$

to obtain

$$x_1(k) = \alpha_1 x_1(k - 1) + \alpha_2 x_2(k - 1) + \alpha_3 x_3(k - 1) + \alpha_4 x_4(k - 1) \tag{24}$$

$$x_2(k) = \beta_1 x_2(k - 1) + \beta_2 x_3(k - 1) + \beta_3 x_4(k - 1) + \beta_4 u(k - 1) \tag{25}$$

$$x_3(k) = \gamma_2 x_1(k - 1) + \gamma_1 x_6(k - 1) \tag{26}$$

Note that Eqs. (24) to (26) have the correct structure of a state-variable representation because the right-hand sides of the equations contain values of variables only at time $k - 1$. However, they are not complete because they do not tell us how to find $x_4(k - 1)$ and $x_6(k - 1)$. These additional relations are obtained from our defining relations (15), (20), and (22), which yield

$$x_4(k) = x_3(k - 1) \tag{27}$$

$$x_5(k) = x_2(k - 1) \tag{28}$$

$$x_6(k) = x_5(k - 1) \tag{29}$$

The six equations (24) to (29) provide us with a complete representation in state-variable form of the macroeconomic model (3) to (5). To see this more clearly we rewrite Eqs. (24) to (29) in vector form

$$
\underbrace{\begin{bmatrix} x_1(k) \\ x_2(k) \\ x_3(k) \\ x_4(k) \\ x_5(k) \\ x_6(k) \end{bmatrix}}_{\mathbf{x}(k)} = \underbrace{\begin{bmatrix} \alpha_1 & \alpha_2 & \alpha_3 & \alpha_4 & 0 & 0 \\ 0 & \beta_1 & \beta_2 & \beta_3 & 0 & 0 \\ \gamma_2 & 0 & 0 & 0 & 0 & \gamma_1 \\ 0 & 0 & 1 & 0 & 0 & 0 \\ 0 & 1 & 0 & 0 & 0 & 0 \\ 0 & 0 & 0 & 0 & 1 & 0 \end{bmatrix}}_{\mathbf{A}} \underbrace{\begin{bmatrix} x_1(k-1) \\ x_2(k-1) \\ x_3(k-1) \\ x_4(k-1) \\ x_5(k-1) \\ x_6(k-1) \end{bmatrix}}_{\mathbf{x}(k-1)} + \underbrace{\begin{bmatrix} 0 \\ \beta_4 \\ 0 \\ 0 \\ 0 \\ 0 \end{bmatrix}}_{\mathbf{B}} \underbrace{u(k-1)}_{\mathbf{u}(k-1)}
\tag{30}
$$

which is clearly of the desired form

$$
\boxed{\mathbf{x}(k) = \mathbf{A}\mathbf{x}(k-1) + \mathbf{B}u(k-1)}
\tag{31}
$$

Thus, from a system-theoretic point of view the macroeconomic system (3) to (5) is a six-state-variable, one-input-variable, LTI discrete-time system. *The physical significance of the state variables is that they are related to the current and lagged values of the endogenous variables.* Thus the state vector $\mathbf{x}(k)$ is

$$
\mathbf{x}(k) = \begin{bmatrix} x_1(k) \\ x_2(k) \\ x_3(k) \\ x_4(k) \\ x_5(k) \\ x_6(k) \end{bmatrix} = \begin{bmatrix} C(k) \\ P(k) \\ W(k) \\ W(k-1) \\ P(k-1) \\ P(k-2) \end{bmatrix}
\tag{32}
$$

and the input variable $u(k) = M(k) =$ money supply.

5.2.3 Block-Diagram Visualization

The relative complexity of this econometric model, the multiple time lags, and the feedback loops can be best visualized by constructing a block diagram of the system. The basic memory element is the *delay element*, shown in Fig. 5.2.1. The

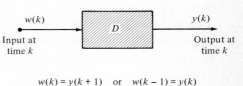

FIGURE 5.2.1
Visual representation of the basic memory element, the delay, for discrete-time systems. The input $w(k)$ at time k is the output at the next instant $k+1$.

$w(k)$
Input at time k

$y(k)$
Output at time k

$w(k) = y(k+1)$ or $w(k-1) = y(k)$

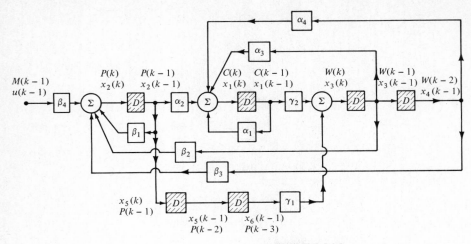

FIGURE 5.2.2
Block diagram of macroeconomic model.

combination of six delay elements with other memoryless elements can be used to construct the block-diagram representation of the macroeconomic model. This is illustrated in Fig. 5.2.2, which shows the state variables as well as the corresponding economic variables. The reader should examine this figure carefully to see how the delay elements are used to generate the lagged versions of the economic variables and how they are fed back to reintroduce their effect into the system.

Block diagrams like Fig. 5.2.2 are often useful in tracing the cause-and-effect relationships and helping intuitive understanding of system behavior.

5.3 A QUANTITATIVE MACROECONOMIC MODEL OF THE UNITED STATES ECONOMY†

This section outlines the techniques Pindyck used to derive a specific econometric model of the United States economy on a quarterly basis for 1955 to 1968. The macroeconomic model was restricted to 10 basic economic variables (endogenous variables). As we shall see, all these endogenous variables are state variables (but there are 18 additional state variables). Different economic variables could have been used as state variables; as always, there is a certain freedom of choice (even in economic systems) in what one selects as a state variable and what one considers to be an output variable.

---

† This is the model derived by R. S. Pindyck in his thesis.

5.3.1   The Ten Basic Economic Variables

The 10 endogenous economic variables considered are†

1   C, total personal *consumption* expenditures, adjusted for inflation, in billions of dollars

2   INR, *nonresidential investment*, i.e., gross private domestic investment in nonresidential structures and producers of durable equipment, adjusted for inflation, in billions of dollars

3   IR, *residential investment*, i.e., value of new additions and alterations to private nonfarm residential buildings, adjusted for inflation, in billions of dollars

4   IIN, *change in business inventories*, adjusted for inflation, in billions of dollars, from quarter to quarter

5   R, *short-term interest rate*, i.e., average market yield (bank discount rate) on United States government 3-month bills, in percent per annum, average during quarter

6   RL, *long-term interest rate*, i.e., yield on long-term United States government bonds, maturing or callable in 10 or more years, in percent per annum, average during quarter

7   P, *price level*, i.e., implicit price deflator on GNP, normalized so that during the first quarter of 1958 its value was set at 100

8   UR, *unemployment rate* (UR = 0.03 means 3 percent unemployment rate)

9   W, *hourly wage rate*, in dollars per hour

10   YD, *disposable income* after taxes, adjusted for inflation, in billions of dollars

5.3.2   The Three Basic Control Variables

The 10 endogenous variables and their time evolution are taken as fundamental consequences of setting other control or policy variables. The three basic variables assumed to affect the economy are

1   G, *government spending*, i.e., purchases of goods and services by the government, adjusted for inflation, in billions of dollars per year.

2   DM, *change in money supply* per quarter. Money supply equals the sum of (*a*) currency outside the treasury, the Federal Reserve system, and the vaults of all commercial banks and (*b*) demand deposits at all commercial banks

---

† These variables were selected as being the most appropriate for formulating economic policies.

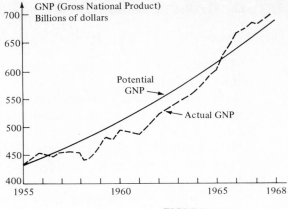

FIGURE 5.3.1
Actual and potential GNP.

other than those due to domestic commercial banks and the United States government. All units are in billions of dollars, adjusted for inflation.

3 TO, *tax surcharge*, in billions of dollars, adjusted for inflation.

The model assumed a constant 15 percent tax rate on the GNP, given by the equality

$$GNP = C + INR + IR + IIN + G$$

(Note that GNP is *not* one of the fundamental state variables; however, it is an output variable which can be computed from the four state variables, C, INR, IR, and IIN and the control or policy variable G.) The tax surcharge was either positive or negative, thus adding to or subtracting from the taxes collected.

Some additional assumptions were made in the model. As in real life, government spending is not constrained to equal collected taxes.

In addition, it was assumed that there was an exogenous variable that affected the dynamical evolution of the economy. This exogenous variable is related to the difference between the actual GNP aud the potential GNP[†] (if the difference is negative, one has the so-called *Okun gap*); loosely speaking, this variable describes the potential capacity of the economy. Figure 5.3.1 shows for 1955 to 1968 the actual GNP and the potential GNP.

---

† A. M. Okun, Potential GNP: Its Measurement and Significance, *1962 Proc. Bus. Econ. Statist. Sec. Am. Statist. Ass.*, pp. 98–104.

5.3.3 State-Variable Formulation

The state vector of the economy was described by a 28-dimensional column vector, $\mathbf{x}(k)$. The time index k, $k = 0, 1, 2, \ldots$, has the following interpretation:

$$k = 0 \text{ corresponds to first quarter of 1955}$$

$$k = 1 \text{ corresponds to second quarter of 1955}$$

$$\vdots$$

$$k = 4 \text{ corresponds to first quarter of 1956}$$

$$\vdots$$

$$k = 7 \text{ corresponds to fourth quarter of 1956}$$

and so on.

The notation $x_i(k)$ denotes the value of the ith state variable, $i = 1, 2, \ldots, 28$, at time k.

Table 5.1 shows the state-variable definition of the basic 10 economic variables already discussed. Table 5.2 defines the additional 18 state variables; the left column of Table 5.2 shows the definition of the state variables $x_i(k)$, $i = 11, 12, \ldots, 28$ in terms of the economic variables in previous quarters. The right column of Table 5.2 shows the equivalent mathematical definition of the state variables $x_i(k)$, $i = 11, 12, \ldots, 28$, in terms of the state variable $x_j(k - 1)$, $j = 1, 2, \ldots, 28$.

EXAMPLE 1 Suppose that $k = 7$, that is, fourth quarter of 1956. Then, the value of the state variable $x_{25}(7)$ is identical to UR(4), that is, the unemployment rate in the first quarter of 1956.

Table 5.1 THE BASIC 10 ENDOGENEOUS VARIABLES, THEIR STATE-VARIABLE NAME, AND THEIR ECONOMIC INTERPRETATION[†]

Symbol	Definition	Unit
$x_1(k) = \mathrm{C}(k)$	Personal consumption	Billions of dollars
$x_2(k) = \mathrm{INR}(k)$	Nonresidential investment	Billions of dollars
$x_3(k) = \mathrm{IR}(k)$	Residential investment	Billions of dollars
$x_4(k) = \mathrm{IIN}(k)$	Change in business inventories	Billions of dollars
$x_5(k) = \mathrm{R}(k)$	Short-term interest rate	Percent
$x_6(k) = \mathrm{RL}(k)$	Long-term interest rate	Percent
$x_7(k) = \mathrm{P}(k)$	Implicit price deflator	1958 level $= 100$
$x_8(k) = \mathrm{UR}(k)$	Unemployment rate	Percent
$x_9(k) = \mathrm{W}(k)$	Hourly wage rate	Dollars
$x_{10}(k) = \mathrm{YD}(k)$	After-taxes disposable income	Billions of dollars

[†] Variables are adjusted to take deflation or inflation into account.

Table 5.2 ADDITIONAL 18 STATE VARIABLES IN TERMS OF THE BASIC 10 ENDOGENOUS VARIABLES AND THEIR DYNAMICAL EQUATIONS

Definition	Equivalent equation
$x_{11}(k) = C(k-1)$	$x_{11}(k) = x_1(k-1)$
$x_{12}(k) = INR(k-1)$	$x_{12}(k) = x_2(k-1)$
$x_{13}(k) = IIN(k-1)$	$x_{13}(k) = x_4(k-1)$
$x_{14}(k) = R(k-1)$	$x_{14}(k) = x_5(k-1)$
$x_{15}(k) = R(k-2)$	$x_{15}(k) = x_{14}(k-1)$
$x_{16}(k) = RL(k-1)$	$x_{16}(k) = x_6(k-1)$
$x_{17}(k) = RL(k-2)$	$x_{17}(k) = x_{16}(k-1)$
$x_{18}(k) = RL(k-3)$	$x_{18}(k) = x_{17}(k-1)$
$x_{19}(k) = RL(k-4)$	$x_{19}(k) = x_{18}(k-1)$
$x_{20}(k) = RL(k-5)$	$x_{20}(k) = x_{19}(k-1)$
$x_{21}(k) = P(k-1)$	$x_{21}(k) = x_7(k-1)$
$x_{22}(k) = P(k-2)$	$x_{22}(k) = x_{21}(k-1)$
$x_{23}(k) = UR(k-1)$	$x_{23}(k) = x_8(k-1)$
$x_{24}(k) = UR(k-2)$	$x_{24}(k) = x_{23}(k-1)$
$x_{25}(k) = UR(k-3)$	$x_{25}(k) = x_{24}(k-1)$
$x_{26}(k) = YD(k-1)$	$x_{26}(k) = x_{10}(k-1)$
$x_{27}(k) = YD(k-2)$	$x_{27}(k) = x_{26}(k-1)$
$x_{28}(k) = YD(k-3)$	$x_{28}(k) = x_{27}(k-1)$

Table 5.3 summarizes the three components of the control or policy vector $\mathbf{u}(k)$, whose three components are denoted by $u_1(k)$, $u_2(k)$, $u_3(k)$.

Besides the three variables $u_1(k)$, $u_2(k)$, $u_3(k)$, the model contains two exogenous variables $z_1(k)$ and $z_2(k)$, which form the components of a two-dimensional vector $\mathbf{z}(k)$. The first component $z_1(k)$ of $\mathbf{z}(k)$ is purely an auxiliary variable whose value is unity, i.e.

$$z_1(k) = 1.0 \quad \text{for all } k \tag{1}$$

The second exogenous variable $z_2(k)$ is related to the potential GNP (see Fig. 5.3.1) and is computed as follows:

$$z_2(k) = 0.85 GNP(k)_{\text{potential}} \tag{2}$$

Table 5.3 THE BASIC THREE POLICY-CONTROL (EXOGENOUS) VARIABLES†

Symbol	Definition	Unit
$u_1(k) = TO(k)$	Tax surcharge	Billions of dollars
$u_2(k) = G(k)$	Government spending	Billions of dollars
$u_3(k) = DM(k)$	Changes in money supply during quarter	Billions of dollars

† All variables adjusted.

5.3.4   The Structure of the State-Variable Model

From a state-variable viewpoint, we seek the following type of relationship between the state vector $\mathbf{x}(k)$, the control or policy vector $\mathbf{u}(k)$, and the exogenous vector $\mathbf{z}(k)$:

$$\mathbf{x}(k + 1) = \mathbf{A}\mathbf{x}(k) + \mathbf{B}\mathbf{u}(k) + \mathbf{C}\mathbf{z}(k) \tag{3}$$

or, equivalently,

$$\mathbf{x}(k) = \mathbf{A}\mathbf{x}(k - 1) + \mathbf{B}\mathbf{u}(k - 1) + \mathbf{C}\mathbf{z}(k - 1) \tag{4}$$

where $\mathbf{x}(k)$, $\mathbf{x}(k - 1)$ = 28-dimensional vectors
$\mathbf{u}(k)$ = 3-dimensional vector
$\mathbf{z}(k)$ = 2-dimensional vector
$\mathbf{A}$ = 28 × 28 matrix
$\mathbf{B}$ = 28 × 3 matrix
$\mathbf{C}$ = 28 × 2 matrix

Hence, the complete specification of the economic model reduces to the numerical specification of the elements of the matrices $\mathbf{A}$, $\mathbf{B}$, and $\mathbf{C}$.

A large portion of the $\mathbf{A}$, $\mathbf{B}$, and $\mathbf{C}$ matrices is implicitly defined by the interrelationship between the different state variables explicitly given in the second column of Table 5.2.

The basic quantitative equations that relate the 10 basic state variables to themselves, to the remaining 18 state variables, to the 3 policy variables, and to the 2 exogenous variables are summarized in Table 5.4; these equations, together with those of Table 5.2, relate $\mathbf{x}(k)$, $\mathbf{x}(k - 1)$, $\mathbf{u}(k - 1)$, and $\mathbf{z}(k - 1)$ by an expression of the form

$$\mathbf{x}(k) = \mathbf{H}\mathbf{x}(k) + \mathbf{F}\mathbf{x}(k - 1) + \mathbf{G}\mathbf{u}(k - 1) + \mathbf{K}\mathbf{z}(k - 1) \tag{5}$$

We remark that Eq. (5) is *not* in the standard state-variable form (4); however, it can be transformed into (4) in view of the manipulations

$$(\mathbf{I} - \mathbf{H})\mathbf{x}(k) = \mathbf{F}\mathbf{x}(k - 1) + \mathbf{G}\mathbf{u}(k - 1) + \mathbf{K}\mathbf{z}(k - 1) \tag{6}$$

$$\mathbf{x}(k) = (\mathbf{I} - \mathbf{H})^{-1}\mathbf{F}\mathbf{x}(k - 1) + (\mathbf{I} - \mathbf{H})^{-1}\mathbf{G}\mathbf{u}(k - 1) + (\mathbf{I} - \mathbf{H})^{-1}\mathbf{K}\mathbf{z}(k - 1) \tag{7}$$

By comparing Eqs. (4) and (7) we readily deduce the correspondence

$$\mathbf{A} = (\mathbf{I} - \mathbf{H})^{-1}\mathbf{F} \qquad \mathbf{B} = (\mathbf{I} - \mathbf{H})^{-1}\mathbf{G} \qquad \mathbf{C} = (\mathbf{I} - \mathbf{H})^{-1}\mathbf{K} \tag{8}$$

5.3.5　Comparison of the Econometric Model with Actual Data

The quantitative model was obtained by assuming (based on past experience, economic laws, and common sense) first a type of dependence of each endogenous variable upon others (present and lagged) using models of the type discussed in Sec. 5.2, that is, linear difference equation with undetermined constant coefficients.

Next the undetermined coefficients were determined by a technique known as time-series regression analysis by the MIT TROLL system. The details of regression analysis are beyond our scope; it suffices to state that from actual economic data and from the assumed structure of the economic difference equation one can obtain a "best" (in the sense of least-squares) fit to the actual data and obtain numerical values for the undetermined coefficients.

We stress that this technique yields mathematical models that are only *approximate*. To illustrate, we present a sequence of graphs showing the actual values of some of the 10 endogenous variables and those generated by the econometric model specified in state-variable form in Tables 5.2 and 5.4.

To generate the time evolution of the state variables one must, of course, specify the time evolution of the control or policy variables. For the data to be presented

$$u_1(k) = 0 \qquad \text{for all } k$$

Table 5.4　BASIC ECONOMETRIC EQUATIONS (PINDYCK)

$x_1(k) = -2.368 x_7(k) + 0.415 x_{10}(k) + 0.7596 x_1(k-1) + 2.368 x_7(k-1) + 8.174 x_9(k-1)$
$\qquad - 0.282 x_{10}(k-1) + 5.299 z_1(k-1)$

$x_2(k) = 0.157 x_{10}(k) + 1.336 x_2(k-1) - 0.157 x_{10}(k-1) - 0.344 x_{12}(k-1) - 1.356 x_{19}(k-1)$
$\qquad + 1.356 x_{20}(k-1) + 0.044 x_{27}(k-1) - 0.044 x_{28}(k-1)$

$x_3(k) = 0.0127 x_{10}(k) + 0.603 x_3(k-1) - 0.55 x_{14}(k-1) - 0.55 x_{15}(k-1) - 0.465 x_{26}(k-1)$
$\qquad + 6.65 - 2.462 z_1(k-1)$

$x_4(k) = -0.60 x_1(k) + 0.4763 x_{10}(k) + 0.422 x_4(k-1) + 0.6 x_{11}(k-1) - 0.465 x_{26}(k-1)$
$\qquad - 2.462 z_1(k-1)$

$x_5(k) = 0.479 x_7(k) + 0.0415 x_{10}(k) + 0.375 x_5(k-1) - 0.479 x_7(k-1) - 0.0344 x_{10}(k-1)$
$\qquad - 0.165 u_3(k-1) - 1.473 z_1(k-1)$

$x_6(k) = 0.06 x_5(k) + 0.0055 x_{10}(k) + 0.871 x_6(k-1) - 0.0055 x_{26}(k) + 0.313 z_1(k-1)$

$x_7(k) = -0.0156 x_{10}(k) + 0.804 x_7(k-1) + 6.28 x_9(k-1) + 0.0195 x_{10}(k-1)$
$\qquad - 0.033 x_{13}(k-1) + 14.55 z_1(k-1) - 0.0195 z_2(k-1)$

$x_8(k) = -0.00043 x_{10}(k) + 0.805 x_8(k-1) + 0.0024 x_9(k-1) - 0.00003 x_{10}(k-1)$
$\qquad + 0.00032 x_{26}(k-1) + 0.0065 z_1(k-1) + 0.00014 z_2(k-1)$

$x_9(k) = 0.0012 x_{10}(k) + 0.627 x_9(k-1) - 0.0001 x_{10}(k-1) + 0.011 x_{22}(k-1) - 0.828 x_{25}(k-1)$
$\qquad - 0.685 z_1(k-1)$

$x_{10}(k) = 0.85 x_1(k) + 0.85 x_2(k) + 0.85 x_3(k) + 0.85 x_4(k) - u_1(k-1) + 0.85 u_2(k-1)$

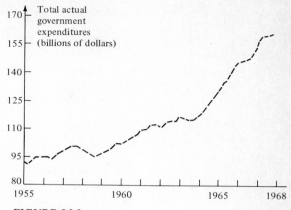

FIGURE 5.3.2
Time plot of $u_2(k) = G(k)$, total government expenditures.

i.e., no tax surcharge was used. The second policy variable, $u_2(k)$, is the government spending; its time evolution is shown in Fig. 5.3.2. The third policy variable, $u_3(k)$, the quarterly change in money supply, can be deduced from the actual money-supply time evolution shown in Fig. 5.3.3.

Figures 5.3.4 to 5.3.7 show a comparison between the actual historical time evolution of some of the endogenous economic variables and the time evolution obtained from digital-computer simulation of the linear discrete-time econometric model.

The minor and major deviations between the actual variables and the ones predicted by the econometric model are due to several sources. First, they are due to aggregation; i.e., many economic interrelationships of variables not taken into account are combined in the behavior of other variables. Presumably an economet-

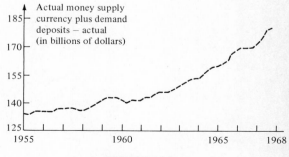

FIGURE 5.3.3
Actual-time evolution of money supply.

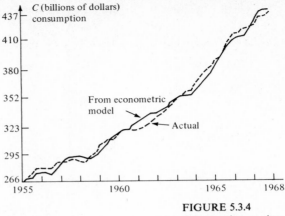

FIGURE 5.3.4
Consumption vs. time.

ric model with more endogenous and state variables would have produced better accuracy. Second, it should be noted that an LTI structure was imposed for a prolonged time. The United States economy was characterized by a nonuniform set of political, fiscal, and economic policies during that time. There was a recession at about 1958 to 1959; naturally, even in an aggregate mode the United States economy behaves differently in a recession era and in an expansionist era. A large proportion of the errors of the mathematical model occur during that time, especially as they pertain to residential investment, interest rates, and unemployment. The performance of the model as a whole is perhaps best for 1961 to 1966. After that time the economy took a much more inflationary turn, perhaps triggered by the Vietnam war, and once more the economy was behaving in a different mode.

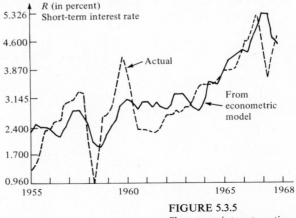

FIGURE 5.3.5
Short-term interest vs. time.

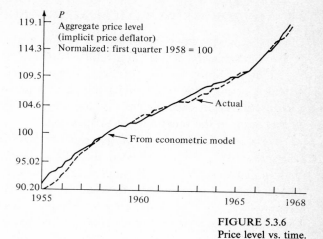

FIGURE 5.3.6
Price level vs. time.

All in all, despite its shortcomings, the model seems to be a relatively realistic one, given the wide span of time involved and its small dimensionality. We hope that the reader has obtained the impression that

1 Mathematical models of nonengineering systems can indeed be constructed.

2 Even simple models can be quite complicated.

3 The accuracy of the mathematical model suffers from the lack of precise laws and cause-and-effect relationships and, of course, our inability to conduct several experiments with the United States economy.

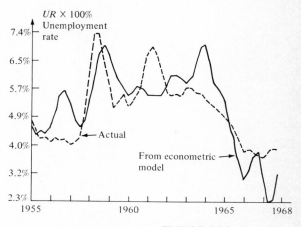

FIGURE 5.3.7
Unemployment rate vs. time.

5.4 WHY ARE MATHEMATICAL ECONOMIC MODELS USEFUL?

The reader may wonder: So what? The period 1955–1968 has gone, and apart from intellectual curiosity, is there any value in fitting a bunch of equations to past data?

It is certainly true that understanding the past dynamical behavior of any system is of no value in changing the past. However, such information can be of great usefulness in selecting policies and decisions in the future. We shall elaborate on this point in the remainder of this section.

A fundamental assumption that must be made is that if we have a quantitative description of a past phenomenon, with a given level of accuracy, the same quantitative description will remain approximately valid in the near future. If indeed this is the case, then by simulating the state-variable description of the economy we can change the control or policy variables from their contemplated levels in order to study how the economy will react.

We call this mode of experimentation *open-loop control*. To illustrate the type of information that can be obtained we give some quantitative data, again as done by Pindyck.

5.4.1 Open-Loop Control

Consider the following experiment. Suppose that in the period from the first quarter of 1960 until 1968 government expenditures were increased by 10 percent above their historical value (see Fig. 5.3.2). We assume that all other policy variables, i.e., surtax and money supply, followed exactly their historical values; similarly for the exogenous variables.

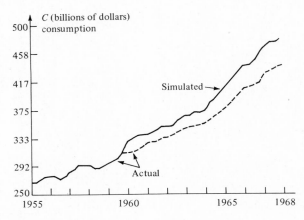

FIGURE 5.4.1
Effects on consumption of a 10 percent increase in government spending.

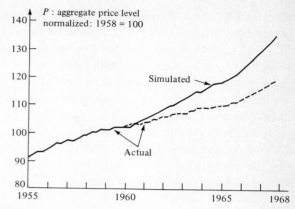

FIGURE 5.4.2
Effect on price level of a 10 percent increase in government spending.

Even a noneconomist can anticipate *in a qualitative way* what would have happened in this case. By pumping more money in the economy one would stimulate the economy, increase consumer spending, reduce unemployment, and probably increase inflation.

However, now we have a fairly accurate tool with which to study more quantitatively the consequences of such a fiscal decision. Figures 5.4.1 to 5.4.3 show the simulation results for selected variables, namely, consumption, price level, and unemployment rate. The econometric model was quite accurate in predicting these variables, and so we should expect that if this change in government spending had indeed taken place, the simulation results would probably have been relatively

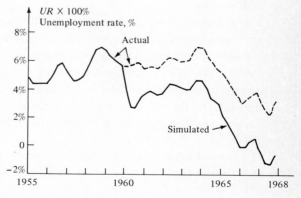

FIGURE 5.4.3
Effect on unemployment rate of a 10 percent increase in government spending.

realistic. Of course, the simulation results have the same qualitative aspects as those discussed above. However, the simulation points out the probable inability of the model to predict drastic changes in unemployment rate since the model predicts a negative unemployment rate from 1966 on.

Naturally, one can construct a host of such *open-loop control* experiments with the simulation model by inventing changes (individual or simultaneous) in the policy variables and examining their implications. Although admittedly inaccurate, such simulation results can be useful in economic planning and fiscal-policy studies. Of course, one must be very careful not to trust the simulation predictions too much, especially for the indefinite future, because of the inherent inaccuracy of the dimensionality, structure, and parameters of the economic mathematical model.

5.5 CONCLUDING REMARKS

We have provided a concrete illustration of a linear, time-invariant, dynamical system by presenting the detailed equations of a 28-state-variable–3-control-variable model of the United States economy, our main purposes being to illustrate that for nonphysical systems, it is possible to combine available "laws," common sense, and parameter-estimation methods to arrive at systems that are relatively good mathematical models of reality.

It is important to realize that the modeling attempted by an engineer or physicist is far easier than that faced by an economist or a sociologist. In a physical system, the laws of nature, prior experimentation, laboratory breadboard models, and prototype models provide far more flexibility and accuracy in modeling a physical system by a mathematical model. The task is far more complex when one deals with socioeconomic systems.

The authors believe that mathematical system-analysis techniques will find more and more applicability in methods for understanding and modeling nonphysical processes. Economic systems are but one type possible to consider. There has been recent work in modeling social systems. Perhaps one of the most interesting and controversial efforts pertains to modeling urban systems.[†] Such systems are characterized by far more complex interrelationships than economic systems; furthermore, the quantitative data (statistics) available are meager and often nonexistent.

The type of mathematical modeling for urban systems, as exemplified by the work of Forrester and his associates, is based on the development of a mathematical model that attempts to duplicate intuitive understanding of the cause-and-effect

---

[†] J. W. Forrester, "Urban Dynamics," The M.I.T. Press, Cambridge, Mass., 1969.

relationships between socioeconomic variables. The resultant urban-dynamics models are essentially discrete-time state-variable models.†

There is no doubt that nonlinear socioeconomic models are more realistic. However, there is little reason to believe that just because one selects a nonlinear model, with subjective structure and parameter values, it is automatically capable of better prediction. This danger can be illustrated by the simulation experiment described in Sec. 5.4. Recall that a relatively accurate linear model, with a small perturbation in one of its policy variables (10 percent increase in government expenditures) produced a negative unemployment rate 7 years after the inception of the experiment. Since the time scale of urban-dynamics simulations runs to several decades, one must be extremely careful in interpreting the quantitative simulation results, especially when *open-loop control* simulation experiments are conducted.

What is needed, of course, is much more fundamental research in the issue of both modeling and parameter identification of complex socioeconomic systems. The current trend toward state-variable formulations is healthy, because modern control theory provides‡ the system analyst with many new parameter-estimation techniques that appear superior to traditional econometric time-series analysis methods. Such techniques are ideally suited for parameter identification for systems in state-variable form even under very unreliable (noisy) measurements. Thus, by providing a common system description, using a state-variable approach, one would be able to see whether the great variety of parameter-estimation methods used in engineering systems is applicable to socioeconomic systems.

---

† In urban-dynamics nomenclature *level variables* are state variables, and *rate variables* are control, policy, or other exogenous variables.

‡ See, for example, A. H. Jazwinski, "Stochastic Processes and Filtering Theory," Academic, New York, 1970; K. Åström, "Introduction to Stochastic Control Theory," Academic, New York, 1970.

6
CONTINUOUS-TIME DYNAMIC NETWORKS

SUMMARY

So far we have considered dynamic systems characterized by difference equations. However, the majority of the physical systems we are familiar with are most naturally modeled continuously in time, and the most natural mathematical description of such systems is by differential equations.

In this chapter we focus our attention on one such class of systems, namely, electric networks containing resistors, capacitors, and inductors. From a utilitarian viewpoint, we shall demonstrate systematic ways of obtaining the network differential equations, using the cut-set method, and in state-variable form. From a motivational point of view, we shall use linear and nonlinear dynamic networks to illustrate the structure and form of the resultant vector differential equations. This material, together with the description of other physical systems described by vector differential equations, will form the basis for the theoretical treatment of multivariable continuous-time systems in the next chapter.

6.1 INTRODUCTION

In a vast variety of physical processes and systems the time rate of change of a particular variable at any time t depends on the instantaneous values of other variables at the same instant of time. For example, the current $i(t)$ through a 1-H linear inductor is related to the voltage across it $v(t)$ by the equation

$$\frac{d}{dt}i(t) = v(t) \qquad (1)$$

As another simple example, the position $x(t)$, velocity $v(t)$, and force $f(t)$ acting on a mass M undergoing rectilinear motion is described by (Newton's law)

$$\frac{d}{dt}x(t) = v(t) \qquad M\frac{d}{dt}v(t) = f(t) \qquad (2)$$

If we consider interconnection of such elements in systems of arbitrary complexity, it is clear that time derivatives of several variables will depend on the instantaneous values of other variables. Such systems are then described by sets of simultaneous ordinary differential equations.

Our main purpose is to convince the reader that a common mathematical framework, that of *vector differential equations*, is useful for analyzing systems of diverse physical properties.

Our plan of attack will be to start by analyzing dynamic (linear and nonlinear) electric networks that contain resistors, capacitors, and inductors. We shall follow a systematic approach, utilizing the cut-set method presented in Chap. 2, to obtain the differential equations of such dynamic networks, using the state-variable approach.

Following this approach we shall see that a network with linear R, L, and C's is completely described by two equations of the form

$$\dot{\mathbf{x}}(t) = \mathbf{A}\mathbf{x}(t) + \mathbf{B}\mathbf{u}(t) \qquad (3)$$

$$\mathbf{y}(t) = \mathbf{C}\mathbf{x}(t) + \mathbf{D}\mathbf{u}(t) \qquad (4)$$

Equation (3) is called an LTI vector differential equation. In Eq. (3) the components of the *state vector* $\mathbf{x}(t)$ are the

1. Capacitor voltages
2. Inductor currents

The components of the *input vector* $\mathbf{u}(t)$ are the instantaneous values of

1. Independent source voltages
2. Independent source currents

The components of the *output vector* $\mathbf{y}(t)$ can consist of all other currents and voltages, namely, resistor currents and voltages, capacitor currents, and inductor voltages.

Next, we shall analyze nonlinear networks, e.g., networks whose resistors and/ or capacitors and/or inductors are nonlinear; these will be characterized by equations of the form

$$\dot{\mathbf{x}}(t) = \mathbf{f}(\mathbf{x}(t), \mathbf{u}(t)) \tag{5}$$

$$\mathbf{y}(t) = \mathbf{g}(\mathbf{x}(t), \mathbf{u}(t)) \tag{6}$$

Equation (5) is a nonlinear vector differential equation. In nonlinear electric networks the most convenient set of state variables [the components of the *state vector* $\mathbf{x}(t)$] are

1 The capacitor charges
2 The inductor flux linkages

The components of the *input vector* $\mathbf{u}(t)$ are once more related to the currents and voltages provided by the independent sources. The components of the *output vector* $\mathbf{y}(t)$ can then be related to all the branch currents and branch voltages of the network.

In Chap. 7, we present a collection of other physical systems described by mathematical models of the form (3) and (4) or (5) and (6).

6.2 DEFINITIONS

Before we start to interconnect our basic network elements into circuits of arbitrary complexity, we summarize below the common terminology used in the chapter.

6.2.1 The Physical Variables

Throughout this chapter $v(t)$ will denote the voltage at time t across an element and $i(t)$ the current through it. The passive polarity (sign) convention will be assumed (see Sec. 2.3).

The flux linkages associated with an inductor will be denoted by $\phi(t)$; the units are weber-turns (Wb). The charge associated with a capacitor will be denoted by $q(t)$; the units are coulombs (C).

6.2.2 Linear and Nonlinear Resistors

A resistor is a device whose mathematical model constrains the voltage $v(t)$ and the current $i(t)$ by means of an algebraic (memoryless) relation

$$\alpha(v(t), i(t)) = 0 \tag{1}$$

A *linear time-invariant resistor* satisfies the linear equation

$$v(t) = Ri(t) \qquad R = \text{const} \tag{2}$$

or

$$i(t) = Gv(t) \qquad G = \text{const} \tag{3}$$

R is called the *resistance*, and $G = 1/R$ is called the *conductance*.

A *nonlinear time-invariant current-controlled resistor* satisfies the equation

$$v(t) = r(i(t)) \tag{4}$$

where $r(.)$ is a scalar-valued (memoryless) algebraic function. The quantity

$$\left. \frac{\partial r(i(t))}{\partial i(t)} \right|_{i(t)=i_0} \tag{5}$$

where i_0 is some constant current value, is called the *incremental resistance* at i_0.

A *nonlinear time-invariant voltage-controlled resistor* satisfies the equation

$$i(t) = g(v(t)) \tag{6}$$

where $g(.)$ is a scalar-valued (memoryless) algebraic function. The quantity

$$\left. \frac{\partial g(v(t))}{\partial v(t)} \right|_{v(t)=v_0} \tag{7}$$

where v_0 is some constant voltage value, is called the *incremental conductance*.

If a resistor described by (4) has the property that $r^{-1}(.)$ exists, it can also be described by

$$i(t) = r^{-1}(v(t)) \tag{8}$$

and hence it is both current- and voltage-controlled; and similarly if $g^{-1}(.)$ exists in Eq. (6).

Figure 6.2.1 summarizes this information in graphical terms.

6.2.3 Linear and Nonlinear Capacitors

A capacitor is a device whose mathematical model constrains the voltage $v(t)$, current $i(t)$, and charge $q(t)$ by the relations

$$\frac{d}{dt}q(t) = i(t) \qquad \text{or} \qquad q(t) = q(t_0) + \int_{t_0}^{t} i(\tau)\, d\tau \tag{9}$$

and

$$f(q(t), v(t)) = 0 \tag{10}$$

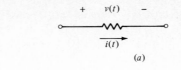

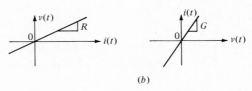

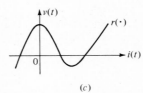

FIGURE 6.2.1
Characteristics of resistors: (*a*) symbol for resistor; (*b*) characteristics of linear resistors; (*c*) current-controlled nonlinear resistor; (*d*) voltage-controlled nonlinear resistor.

A *linear time-invariant capacitor* satisfies the linear equations

$$\frac{d}{dt}q(t) = i(t) \qquad q(t) = Cv(t) \tag{11}$$

or

$$C\frac{dv(t)}{dt} = i(t) \tag{12}$$

or

$$v(t) = v(t_0) + \frac{1}{C}\int_{t_0}^{t} i(\tau)\,d\tau \tag{13}$$

The constant C is called the *capacitance*.

A *nonlinear time-invariant voltage-controlled capacitor* satisfies the equations

$$\frac{d}{dt}q(t) = i(t) \qquad q(t) = c(v(t)) \tag{14}$$

where $c(\,.\,)$ is a scalar-valued algebraic function.

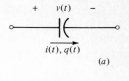

(a)

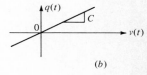

(b)

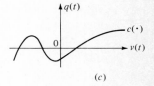

(c)

FIGURE 6.2.2
Characteristics of capacitors: (a) symbol
for a capacitor; (b) linear capacitor; (c)
voltage-controlled capacitor; (d) charge-
controlled capacitor.

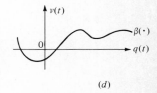

(d)

A *nonlinear time-invariant charge-controlled capacitor* satisfies the equations

$$\frac{d}{dt}q(t) = i(t) \qquad v(t) = \beta(q(t)) \tag{15}$$

where $\beta(.)$ is a scalar-valued algebraic function. If $c^{-1}(.)$ or $\beta^{-1}(.)$ exists, the
capacitor is both voltage- and charge-controlled. The quantity

$$\left.\frac{\partial c(v(t))}{\partial v(t)}\right|_{v(t)=v_0} \tag{16}$$

is called the *incremental capacitance* at the constant voltage v_0. The quantity

$$\left.\frac{\partial \beta(q(t))}{\partial q(t)}\right|_{q(t)=q_0} \tag{17}$$

is called the *incremental elastance* at the constant charge q_0.
Figure 6.2.2 shows the characteristics of capacitors in graphical form.

6.2.4   Linear and Nonlinear Inductors

An inductor is a device whose mathematical model constrains the voltage $v(t)$, current $i(t)$, and flux linkages $\phi(t)$ by the relations

$$\frac{d\phi(t)}{dt} = v(t) \qquad \text{or} \qquad \phi(t) = \phi(t_0) + \int_{t_0}^{t} \phi(\tau)\, d\tau \tag{18}$$

and

$$g(\phi(t), i(t)) = 0 \tag{19}$$

A *linear time-invariant inductor* satisfies the linear equations

$$\frac{d}{dt}\phi(t) = v(t) \qquad \phi(t) = Li(t) \tag{20}$$

or

$$L\frac{di(t)}{dt} = v(t) \tag{21}$$

or

$$i(t) = i(t_0) + \frac{1}{L}\int_{t_0}^{t} v(\tau)\, d\tau \tag{22}$$

The constant L is called the inductance.

A *nonlinear time-invariant current-controlled inductor* satisfies the equations

$$\frac{d\phi(t)}{dt} = v(t) \qquad \phi(t) = l(i(t)) \tag{23}$$

where $l(\,.\,)$ is a scalar-valued algebraic function.

A *nonlinear time-invariant flux-controlled inductor* satisfies the equations

$$\frac{d}{dt}\phi(t) = v(t) \qquad i(t) = \gamma(\phi(t)) \tag{24}$$

where $\gamma(\,.\,)$ is a scalar-valued function.

If $l^{-1}(\,.\,)$ or $\gamma^{-1}(\,.\,)$ exists, the nonlinear inductor is both current- and flux-controlled.

The quantity

$$\left.\frac{\partial l(i(t))}{\partial i(t)}\right|_{i(t)=i_0} \tag{25}$$

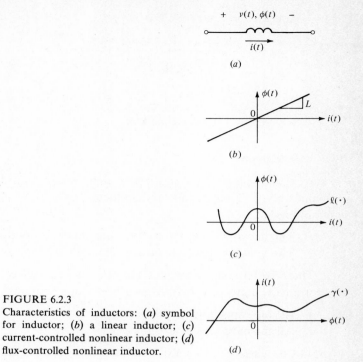

FIGURE 6.2.3
Characteristics of inductors: (*a*) symbol for inductor; (*b*) a linear inductor; (*c*) current-controlled nonlinear inductor; (*d*) flux-controlled nonlinear inductor.

is called the *incremental inductance* at the constant current value of i_0. Similarly, the quantity

$$\left.\frac{\partial \gamma(\phi(t))}{\partial \phi(t)}\right|_{\phi(t)=\phi_0} \tag{26}$$

is called the *incremental inverse inductance* at the constant flux value ϕ_0.

Figure 6.2.3 illustrates the graphical implications of the above relations.

6.3 STATE-VARIABLE EQUATIONS, VIA CUT-SET ANALYSIS, OF LTI *LC* NETWORKS

In Chap. 2 we discussed the cut-set analysis of networks and the results obtained when the network contained resistors together with voltage and current sources. In this and the following sections we shall illustrate one of the major advantages of cut-set analysis as we determine network equations using the state-variable approach.

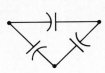

FIGURE 6.3.1
A capacitor loop.

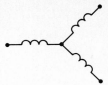

FIGURE 6.3.2
An inductor cut set.

6.3.1 Assumptions

To start our discussion, let us suppose that we are considering a network which contains only LTI capacitors and inductors. We shall make the following two assumptions:

1   The network does not contain any capacitor loops (see Fig. 6.3.1).
2   The network does not contain any inductor cut sets (see Fig. 6.3.2).

These two assumptions guarantee that we can select a *proper tree*. A proper tree for an *LC* network is defined as a tree which contains

1   All the capacitors in the tree branches
2   All the inductors in the links

It should be clear from an examination of Figs. 6.3.1 and 6.3.2 that if we have capacitor loops and/or inductor cut sets, we cannot select a proper tree. For example, in Fig. 6.3.2 one of the inductance branches must be a tree branch to connect the middle node to the remainder of the network.

Let us suppose that we have selected a proper tree. As in Chap. 2 we let

$$\mathbf{i}_l = \text{link current vector}$$
$$\mathbf{v}_l = \text{link voltage vector}$$
$$\mathbf{i}_t = \text{tree-branch current vector}$$
$$\mathbf{v}_t = \text{tree-branch voltage vector}$$

6.3.2 Structure of Branches

Since all capacitors are in the tree branches of the proper tree, a typical tree branch is shown in Fig. 6.3.3. A typical link is shown in Fig. 6.3.4 since by the selection of the proper tree all inductors are in the links. Thus, the branch currents and voltages will be constrained by:

For proper tree branches:
$$i_j = C_j \frac{dv_j}{dt} \tag{1}$$

For links:
$$v_k = L_k \frac{di_k}{dt} \tag{2}$$

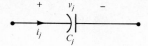

FIGURE 6.3.3
A typical (proper) tree branch for an
LC network.

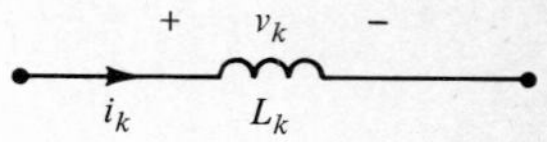

FIGURE 6.3.4
A typical link for an *LC* network.

We now define a diagonal *capacitance matrix* $\mathbf{C}$ and a diagonal *inductance matrix* $\mathbf{L}$ such that the following relations hold:

Links:
$$\mathbf{v}_l = \mathbf{L}\frac{d\mathbf{i}_l}{dt} \qquad \mathbf{L} \triangleq \begin{bmatrix} L_1 & 0 & 0 & \cdots \\ 0 & L_2 & 0 & \cdots \\ 0 & 0 & L_3 & \cdots \\ \vdots & \vdots & \vdots & \vdots \end{bmatrix} \tag{3}$$

Tree branches:
$$\mathbf{i}_t = \mathbf{C}\frac{d\mathbf{v}_t}{dt} \qquad \mathbf{C} \triangleq \begin{bmatrix} C_1 & 0 & 0 & \cdots \\ 0 & C_2 & 0 & \cdots \\ 0 & 0 & C_3 & \cdots \\ \vdots & \vdots & \vdots & \vdots \end{bmatrix} \tag{4}$$

Note that $\mathbf{C}^{-1}$ and $\mathbf{L}^{-1}$ exist.

If we know, at some initial time t_0, the voltages across the capacitors and the currents through the inductors, this implies that we know the vectors

$$\mathbf{v}_t(t_0) \qquad \text{and} \qquad \mathbf{i}_l(t_0) \tag{5}$$

and that we can find the link currents from the link voltages, because the vector differential equation (3)

$$\boxed{\frac{d}{dt}\mathbf{i}_l = \mathbf{L}^{-1}\mathbf{v}_l} \tag{6}$$

can be integrated (component by component) to yield

$$\mathbf{i}_l(t) = \mathbf{i}_l(t_0) + \mathbf{L}^{-1}\int_{t_0}^{t} \mathbf{v}_l(\tau)\,d\tau \tag{7}$$

Similarly, the vector differential equation (4)

$$\boxed{\frac{d}{dt}\mathbf{v}_t = \mathbf{C}^{-1}\mathbf{i}_t} \tag{8}$$

integrates to

$$v_t(t) = v_t(t_0) + C^{-1} \int_{t_0}^{t} i_t(\tau) \, d\tau \tag{9}$$

so that the tree-branch voltages can be found from the tree-branch currents.

6.3.3 Complete Set of Equations

In order to determine all our variables we have the relations

$$\frac{d}{dt} i_l = L^{-1} v_l \qquad i_l(t_0) \text{ known} \tag{10}$$

$$\frac{d}{dt} v_t = C^{-1} i_t \qquad v_t(t_0) \text{ known} \tag{11}$$

from the topology of the network since we have selected a (proper) tree, we can find the fundamental cut sets. From a cut-set analysis, we can then find the implications of KVL and KCL (see Sec. 2.6). Recall that the structure of the KVL and KCL equations is

$$\underbrace{\begin{bmatrix} F & \vdots & I \end{bmatrix}}_{Q} \underbrace{\begin{bmatrix} i_l \\ - - - \\ i_t \end{bmatrix}}_{i} = 0 \qquad \text{KCL} \tag{12}$$

$$\underbrace{\begin{bmatrix} v_l \\ - - - \\ v_t \end{bmatrix}}_{v} = \underbrace{\begin{bmatrix} F' \\ - - - \\ I \end{bmatrix}}_{Q'} v_t \qquad \text{KVL} \tag{13}$$

We want to eliminate the vectors v_l and i_t between Eqs. (10) to (13). From Eq. (13) we have

$$v_l = F' v_t \tag{14}$$

Substituting Eq. (14) into Eq. (10), we have

$$\frac{d}{dt} i_l = L^{-1} F' v_t \tag{15}$$

Now the KCL equation (12) can be written

$$F i_l + I i_t = F i_l + i_t = 0 \tag{16}$$

Thus, the tree-branch currents can be found from the link currents by

$$\mathbf{i}_t = -\mathbf{F}\mathbf{i}_l \tag{17}$$

Substituting Eq. (17) into (11), we obtain

$$\frac{d}{dt}\mathbf{v}_t = -\mathbf{C}^{-1}\mathbf{F}\mathbf{i}_l \tag{18}$$

The two vector differential equations (15) and (18) can be combined:

$$\frac{d}{dt}\begin{bmatrix} \mathbf{i}_l \\ ---- \\ \mathbf{v}_t \end{bmatrix} = \begin{bmatrix} \mathbf{0} & \vdots & \mathbf{L}^{-1}\mathbf{F}' \\ ---- & --- & ---- \\ -\mathbf{C}^{-1}\mathbf{F} & \vdots & \mathbf{0} \end{bmatrix}\begin{bmatrix} \mathbf{i}_l \\ --- \\ \mathbf{v}_t \end{bmatrix} \tag{19}$$

6.3.4 Discussion

Let us now review the development of this section. Our objective was to analyze an arbitrary LC network with LTI elements. We omitted the effect of voltage and current sources to avoid cluttering the equations. In seeking the form of the network equations, we decided to focus our attention on the inductor currents and the capacitor voltages (the common state variables of LTI networks). We remind the reader that knowledge of these variables at any instant of time is both necessary and sufficient for the evaluation of the total energy (electric and magnetic) stored in the network. Since capacitors and inductors are memory elements, with respect to current and voltage relations, we expect that the network equations will be differential equations.

The state-variable differential equations we obtain must relate the time rate of change (time derivative) of each state variable to the other state variables and the inputs (which we assumed to be zero in this section). This is precisely what Eq. (19) means. To see this let us define the network state vector $\mathbf{x}(t)$ as the vector whose components are the inductor currents and capacitor voltages; thus, $\mathbf{x}(t)$ is given by

$$\mathbf{x}(t) \triangleq \begin{bmatrix} \mathbf{i}_l(t) \\ --- \\ \mathbf{v}_t(t) \end{bmatrix} \tag{20}$$

If we also define the time-invariant matrix $\mathbf{A}$ by

$$\mathbf{A} \triangleq \begin{bmatrix} \mathbf{0} & \vdots & \mathbf{L}^{-1}\mathbf{F}' \\ ---- & --- & ---- \\ -\mathbf{C}^{-1}\mathbf{F} & \vdots & \mathbf{0} \end{bmatrix} \tag{21}$$

then Eq. (19) takes the form

$$\dot{\mathbf{x}}(t) = \mathbf{A}\mathbf{x}(t) \tag{22}$$

where $\dot{\mathbf{x}}(t)$ is the time derivative of the state vector $\mathbf{x}(t)$. Equation (22) then means that each state variable $x_i(t)$ satisfies the first-order differential equation

$$\dot{x}_i(t) = \sum_{j=1}^{b} a_{ij} x_j(t) \qquad i = 1, 2, \ldots, b \tag{23}$$

where b is the number of the network branches and the a_{ij}'s are the elements of the matrix $\mathbf{A}$. If we know the value of the state vector $\mathbf{x}(t_0)$ at any time t_0, then we *assert* (and shall prove in Chap. 7) that this set of differential equations admits a unique solution. Therefore, if we know the inductor currents and capacitor currents at any time t_0, Eq. (19) can be solved to yield these variables at any other time t. Since in this case the inductor currents are the link currents $\mathbf{i}_l$ and the capacitor voltages are the tree-branch voltages $\mathbf{v}_t$, it follows that all other network variables, namely, the tree-branch currents $\mathbf{i}_t$ and the link voltages $\mathbf{v}_l$, can be evaluated from Eqs. (17) and (14), respectively.

The choice of a proper tree and the use of cut-set analysis is useful for this development because KVL leads to Eq. (14) and KCL leads to Eq. (17); thus, knowledge of the state variables $\mathbf{v}_t$ and $\mathbf{i}_l$ determines in a trivial way the rest of the network variables, $\mathbf{i}_t$ and $\mathbf{v}_l$.

In order to clarify these concepts we shall present an example at the end of this section. In the meantime, we shall demonstrate an important property of *LC* networks.

6.3.5 Energy Considerations in Pure *LC* Networks

If we consider any pure *LC* network, the absence of resistances and sources means that there is no way of introducing energy into the network or of absorbing energy from it. Thus, intuition would lead us to believe that the total energy $E(t)$ stored in the network is a constant. We shall now prove this fact mathematically.

Let us recall that the (magnetic) energy stored in an inductor is given by $\frac{1}{2}Li^2(t)$ and the (electric) energy stored in a capacitor is given by $\frac{1}{2}Cv^2(t)$. Using the definitions of this section, we can see that[†]

$$\text{Total magnetic energy} = \tfrac{1}{2}\langle \mathbf{i}_l(t), \mathbf{L}\mathbf{i}_l(t)\rangle \tag{24}$$

$$\text{Total electric energy} = \tfrac{1}{2}\langle \mathbf{v}_t(t), \mathbf{C}\mathbf{v}_t(t)\rangle \tag{25}$$

[†] $\langle \mathbf{x}, \mathbf{y}\rangle$ is the scalar product of the vectors $\mathbf{x}$ and $\mathbf{y}$; see Appendix A, Sec. A.11.

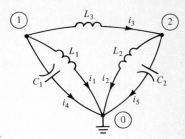

FIGURE 6.3.5
LC network for Example 1.

where $\mathbf{L}$ and $\mathbf{C}$ are the diagonal inductance and capacitance matrices of Eqs. (3) and (4).

The total stored energy $E(t)$ is then given by the sum of the magnetic and electric energies, i.e.

$$E(t) = \tfrac{1}{2}\langle \mathbf{i}_l(t), \mathbf{L}\mathbf{i}_l(t)\rangle + \tfrac{1}{2}\langle \mathbf{v}_t(t), \mathbf{C}\mathbf{v}_t(t)\rangle \tag{26}$$

To prove that $E(t) = $ const, it suffices to show that $dE(t)/dt = 0$ for all t. From Eq. (26) we compute

$$\frac{dE(t)}{dt} = \left\langle \frac{d\mathbf{i}_l(t)}{dt}, \mathbf{L}\mathbf{i}_l(t)\right\rangle + \left\langle \frac{d\mathbf{v}_t(t)}{dt}, \mathbf{C}\mathbf{v}_t(t)\right\rangle \tag{27}$$

But from the state-vector differential equation (19) we have

$$\frac{d}{dt}\mathbf{i}_l(t) = \mathbf{L}^{-1}\mathbf{F}'\mathbf{v}_t(t) \qquad \frac{d}{dt}\mathbf{v}_t(t) = -\mathbf{C}^{-1}\mathbf{F}\mathbf{i}_l(t) \tag{28}$$

Substituting Eq. (28) into Eq. (27) we obtain, using the fact that $\mathbf{L}$ and $\mathbf{C}$ are symmetric matrices,

$$\begin{aligned}
\frac{dE(t)}{dt} &= \langle \mathbf{L}^{-1}\mathbf{F}'\mathbf{v}_t(t), \mathbf{L}\mathbf{i}_l(t)\rangle - \langle \mathbf{C}^{-1}\mathbf{F}\mathbf{i}_l(t), \mathbf{C}\mathbf{v}_t(t)\rangle \\
&= \langle \mathbf{L}\mathbf{L}^{-1}\mathbf{F}'\mathbf{v}_t(t), \mathbf{i}_l(t)\rangle - \langle \mathbf{C}\mathbf{C}^{-1}\mathbf{F}\mathbf{i}_l(t), \mathbf{v}_t(t)\rangle \\
&= \langle \mathbf{F}'\mathbf{v}_t(t), \mathbf{i}_l(t)\rangle - \langle \mathbf{F}\mathbf{i}_l(t), \mathbf{v}_t(t)\rangle \\
&= \langle \mathbf{v}_t(t), \mathbf{F}\mathbf{i}_l(t)\rangle - \langle \mathbf{F}\mathbf{i}_l(t), \mathbf{v}_t(t)\rangle = 0
\end{aligned} \tag{29}$$

which establishes the fact that $E(t) = $ const.

EXAMPLE 1 To illustrate these ideas we consider the network of Fig. 6.3.5. This network is characterized by 3 nodes, 5 branches, 2 LTI capacitors, and 3 LTI inductors. It does not contain any capacitor loops or inductor cut sets (it has an inductor loop, but this is all right).

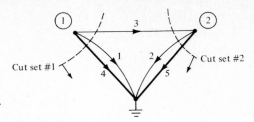

FIGURE 6.3.6
The proper tree for the network of Fig.
6.3.5.

The proper tree for this network is shown in Fig. 6.3.6. We have also identified the two fundamental cut sets and their reference directions. Our link and tree-branch current and voltage vectors are

$$\mathbf{i}_l = \begin{bmatrix} i_1 \\ i_2 \\ i_3 \end{bmatrix} \qquad \mathbf{v}_l = \begin{bmatrix} v_1 \\ v_2 \\ v_3 \end{bmatrix} \qquad \text{links}$$

$$\mathbf{i}_t = \begin{bmatrix} i_4 \\ i_5 \end{bmatrix} \qquad \mathbf{v}_t = \begin{bmatrix} v_4 \\ v_5 \end{bmatrix} \qquad \text{tree branches}$$

The two differential equations are [see Eqs. (3) and (4)]

$$\underbrace{\begin{bmatrix} L_1 & 0 & 0 \\ 0 & L_2 & 0 \\ 0 & 0 & L_3 \end{bmatrix}}_{\mathbf{L}} \frac{d}{dt} \underbrace{\begin{bmatrix} i_1 \\ i_2 \\ i_3 \end{bmatrix}}_{\mathbf{i}_l} = \underbrace{\begin{bmatrix} v_1 \\ v_2 \\ v_3 \end{bmatrix}}_{\mathbf{v}_l} \qquad \text{or} \qquad \mathbf{L} \frac{d}{dt} \mathbf{i}_l = \mathbf{v}_l$$

$$\underbrace{\begin{bmatrix} C_1 & 0 \\ 0 & C_2 \end{bmatrix}}_{\mathbf{C}} \frac{d}{dt} \underbrace{\begin{bmatrix} v_4 \\ v_5 \end{bmatrix}}_{\mathbf{v}_t} = \underbrace{\begin{bmatrix} i_4 \\ i_5 \end{bmatrix}}_{\mathbf{i}_t} \qquad \text{or} \qquad \mathbf{C} \frac{d}{dt} \mathbf{v}_t = \mathbf{i}_t$$

We note that since $\mathbf{L}$ and $\mathbf{C}$ are both diagonal matrices, their inverses are simply

$$\mathbf{L}^{-1} = \begin{bmatrix} \dfrac{1}{L_1} & 0 & 0 \\ 0 & \dfrac{1}{L_2} & 0 \\ 0 & 0 & \dfrac{1}{L_3} \end{bmatrix} \qquad \mathbf{C}^{-1} = \begin{bmatrix} \dfrac{1}{C_1} & 0 \\ 0 & \dfrac{1}{C_2} \end{bmatrix}$$

Now we perform a cut-set analysis to find the matrix $\mathbf{Q} = [\mathbf{F} \quad \mathbf{I}]$. Using KCL across the two fundamental cut sets, we have

Cut set 1: $\qquad\qquad\qquad i_1 + i_3 + i_4 = 0$

Cut set 2: $\qquad\qquad\qquad i_2 + i_3 + i_5 = 0$

Thus, the fundamental cut-set matrix $\mathbf{Q}$ is

$$\mathbf{Q} = \begin{bmatrix} 1 & 0 & 1 & \vdots & 1 & 0 \\ 0 & 1 & 1 & \vdots & 0 & 1 \end{bmatrix}$$

$$\underbrace{\qquad\qquad}_{\mathbf{F}} \quad \underbrace{\qquad}_{\mathbf{I}}$$

and therefore

$$\mathbf{F} = \begin{bmatrix} 1 & 0 & 1 \\ 0 & 1 & 1 \end{bmatrix}$$

In order to find the implications of Eq. (19) we compute the matrices

$$\mathbf{L}^{-1}\mathbf{F}' = \begin{bmatrix} \dfrac{1}{L_1} & 0 & 0 \\ 0 & \dfrac{1}{L_2} & 0 \\ 0 & 0 & \dfrac{1}{L_3} \end{bmatrix} \begin{bmatrix} 1 & 0 \\ 0 & 1 \\ 1 & 1 \end{bmatrix} = \begin{bmatrix} \dfrac{1}{L_1} & 0 \\ 0 & \dfrac{1}{L_2} \\ \dfrac{1}{L_3} & \dfrac{1}{L_3} \end{bmatrix}$$

$$-\mathbf{C}^{-1}\mathbf{F} = -\begin{bmatrix} \dfrac{1}{C_1} & 0 \\ 0 & \dfrac{1}{C_2} \end{bmatrix} \begin{bmatrix} 1 & 0 & 1 \\ 0 & 1 & 1 \end{bmatrix} = \begin{bmatrix} -\dfrac{1}{C_1} & 0 & -\dfrac{1}{C_1} \\ 0 & -\dfrac{1}{C_2} & -\dfrac{1}{C_2} \end{bmatrix}$$

Thus, Eq. (19) yields

$$\frac{d}{dt}\begin{bmatrix} i_1 \\ i_2 \\ i_3 \\ \text{---} \\ v_4 \\ v_5 \end{bmatrix} = \begin{bmatrix} 0 & 0 & 0 & \vdots & \dfrac{1}{L_1} & 0 \\ 0 & 0 & 0 & \vdots & 0 & \dfrac{1}{L_2} \\ 0 & 0 & 0 & \vdots & \dfrac{1}{L_3} & \dfrac{1}{L_3} \\ \cdots & \cdots & \cdots & & \cdots & \cdots \\ -\dfrac{1}{C_1} & 0 & -\dfrac{1}{C_1}' & \vdots & 0 & 0 \\ 0 & -\dfrac{1}{C_2} & -\dfrac{1}{C_2} & \vdots & 0 & 0 \end{bmatrix} \begin{bmatrix} i_1 \\ i_2 \\ i_3 \\ \text{---} \\ v_4 \\ v_5 \end{bmatrix}$$

which is simply the vector form of the set of five coupled first-order, differential equations:

$$\frac{d}{dt}i_1 = \frac{1}{L_1}v_4$$

$$\frac{d}{dt}i_2 = \frac{1}{L_2}v_5$$

$$\frac{d}{dt}i_3 = \frac{1}{L_3}v_4 + \frac{1}{L_3}v_5$$

$$\frac{d}{dt}v_4 = -\frac{1}{C_1}i_1 - \frac{1}{C_1}i_3$$

$$\frac{d}{dt}v_5 = -\frac{1}{C_2}i_2 - \frac{1}{C_2}i_3$$

Thus, the state variables i_1, i_2, i_3, v_4, and v_5 for this network are coupled by five first-order differential equations. The remaining network variables, namely, v_1, v_2, v_3, i_4, and i_5 can easily be determined once the state variables have been found because

$$\mathbf{v}_l = \begin{bmatrix} v_1 \\ v_2 \\ v_3 \end{bmatrix} = \mathbf{F}'\mathbf{v}_t = \begin{bmatrix} 1 & 0 \\ 0 & 1 \\ 1 & 1 \end{bmatrix}\begin{bmatrix} v_4 \\ v_5 \end{bmatrix} = \begin{bmatrix} v_4 \\ v_5 \\ v_4 + v_5 \end{bmatrix}$$

so that

$$v_1 = v_4 \qquad v_2 = v_5 \qquad v_3 = v_4 + v_5$$

From Eq. (17) we have

$$\mathbf{i}_t = \begin{bmatrix} i_4 \\ i_5 \end{bmatrix} = -\mathbf{F}\mathbf{i}_l = -\begin{bmatrix} 1 & 0 & 1 \\ 0 & 1 & 1 \end{bmatrix}\begin{bmatrix} i_1 \\ i_2 \\ i_3 \end{bmatrix} = \begin{bmatrix} -i_1 & -i_3 \\ -i_2 & -i_3 \end{bmatrix}$$

so that

$$i_4 = -i_1 - i_3 \qquad i_5 = -i_2 - i_3$$

6.4 STATE-VARIABLE DESCRIPTION OF LTI *RLC* NETWORKS

In Sec. 6.3 we developed a systematic procedure for deriving the differential equations satisfied by the state variables (inductor currents and capacitor voltages) of an arbitrary network containing only LTI inductors and capacitors. In this section, we extend these basic ideas to develop a systematic approach to the development of network equations. We shall consider independent current and voltage sources as well as LTI resistors, capacitors, and inductors.

Any systematic development of the equations that characterize such a general network involves a set of precise definitions for the various vectors and matrices one must deal with. As the reader will soon observe, the terminology and equations are somewhat complicated in spite of the shorthand advantages offered by the vector notation. To avoid loosing track of the essential ideas we start with an outline of the procedure we shall follow.

6.4.1 The General Objective

Our basic objective is to determine the vector differential equation satisfied by the state vector of the network. As before, the components of the state vector will be the inductor currents and the capacitor voltages. We stress that once the state variables are determined, all other remaining variables, i.e., the inductor voltages, the capacitor currents, and the resistor currents and voltages, can be found as algebraic (memoryless) linear functions of the state variables and of the source voltages and currents generated by the independent sources.

6.4.2 Assumptions

We make several assumptions, outlined below.

1 There are no capacitor loops, and there are no inductor cut sets. This assumption was also made in the previous section; we emphasize that the implication of this assumption is that we can select a tree so that the capacitive branches form a subset of the tree branches while the inductive branches form a subset of the links.

2 It is assumed that all the network branches have the general structure shown in Fig. 6.4.1. Thus, we allow only independent current sources across the LTI capacitors and independent voltage sources in series with the LTI inductors.

3 It is assumed that a tree can be selected such that

    a All capacitive and conductive[†] branches form the tree branches.

    b All inductive and resistive[‡] branches form the links.

Such a tree is called a proper tree for this class of network.

These assumptions are not essential for carrying out state-variable network analysis but are made to keep the formulas as simple as possible.

[†] If this cannot be done, one must transform some resistive branches into equivalent conductive branches by Norton's theorem (see the exercises).

[‡] If this cannot be done, one must transform some conductive branches into equivalent resistive branches by Thevenin's theorem (see the exercises).

Capacitive branches : $i_{Ck} = C_k \frac{d}{dt} v_{Ck} + j_{Ck}$

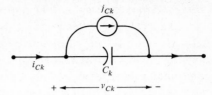

Conductive branches : $i_{Gk} = G_k v_{Gk} + j_{Gk}$

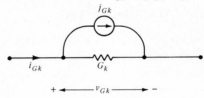

Inductive branches : $v_{Lk} = L_k \frac{d}{dt} i_{Lk} + e_{Lk}$

Resistive branches : $v_{Rk} = R_k i_{Rk} + e_{Rk}$

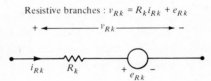

FIGURE 6.4.1
The assumed structure of the network branches. Source currents are denoted by j, source voltages by e, branch currents by i, and branch voltages by v.

6.4.3   Notation and Terminology

We start by picking a *proper tree* for the network. As indicated above, the proper tree is selected so that all the capacitive and conductive branches define the proper tree branches; all resistive and inductive branches then are in the links.

As before, we let **i** and **v** denote the branch-current and -voltage vectors, respectively. We shall label the branches as follows: first we index the link resistive branches, then the link inductive branches, then the capacitive tree branches, and finally the conductive tree branches. We define the following current and voltage vectors, respectively, according to the name of the branch.

Resistive link branches: $\mathbf{i}_R, \mathbf{v}_R$

Inductive link branches: $\mathbf{i}_L, \mathbf{v}_L$

Capacitive tree branches: $\mathbf{i}_C, \mathbf{v}_C$

Conductive tree branches: $\mathbf{i}_G, \mathbf{v}_G$

As before, we let $\mathbf{i}_l$ and $\mathbf{v}_l$ denote the link-current and link-voltage vectors, respectively. Similarly, we let $\mathbf{i}_t$ and $\mathbf{v}_t$ denote the tree-branch current and voltage vectors. According to our method of indexing branches, we have the vector partitions

$$
\mathbf{i} = \begin{bmatrix} \mathbf{i}_l \\ --- \\ \mathbf{i}_t \end{bmatrix} = \begin{bmatrix} \mathbf{i}_R \\ --- \\ \mathbf{i}_L \\ --- \\ \mathbf{i}_C \\ --- \\ \mathbf{i}_G \end{bmatrix}
\qquad
\mathbf{v} = \begin{bmatrix} \mathbf{v}_l \\ --- \\ \mathbf{v}_t \end{bmatrix} = \begin{bmatrix} \mathbf{v}_R \\ --- \\ \mathbf{v}_L \\ --- \\ \mathbf{v}_C \\ --- \\ \mathbf{v}_G \end{bmatrix}
\qquad (1)
$$

Thus, both the branch-current and -voltage vectors are subdivided into four subvectors according to the nature of the four types of branches illustrated in Fig. 6.4.1.

As indicated in Fig. 6.4.1, each branch is characterized by its capacitance, conductance, inductance, or resistance. These values, together with the number of like branches, can be used to define the four *diagonal* matrices:

Resistance matrix:
$$
\mathbf{R} \triangleq \begin{bmatrix} R_1 & 0 & 0 & \cdots \\ 0 & R_2 & 0 & \cdots \\ 0 & 0 & R_3 & \cdots \\ \vdots & \vdots & \vdots & \vdots \end{bmatrix}
\qquad (2)
$$

Inductance matrix:
$$
\mathbf{L} \triangleq \begin{bmatrix} L_1 & 0 & 0 & \cdots \\ 0 & L_2 & 0 & \cdots \\ 0 & 0 & L_3 & \cdots \\ \vdots & \vdots & \vdots & \vdots \end{bmatrix}
\qquad (3)
$$

Capacitance matrix:
$$
\mathbf{C} \triangleq \begin{bmatrix} C_1 & 0 & 0 & \cdots \\ 0 & C_2 & 0 & \cdots \\ 0 & 0 & C_3 & \cdots \\ \vdots & \vdots & \vdots & \vdots \end{bmatrix}
\qquad (4)
$$

Conductance matrix:
$$\mathbf{G} = \begin{bmatrix} G_1 & 0 & 0 & \cdots \\ 0 & G_2 & 0 & \cdots \\ 0 & 0 & G_3 & \cdots \\ \vdots & \vdots & \vdots & \vdots \end{bmatrix} \tag{5}$$

Finally, the presence of the independent current and voltage sources (see Fig. 6.4.1) leads to the definition of

Current-source vector:
$$\mathbf{j} = \begin{bmatrix} \mathbf{j}_C \\ \text{- - -} \\ \mathbf{j}_G \end{bmatrix} \tag{6}$$

Voltage-source vector:
$$\mathbf{e} = \begin{bmatrix} \mathbf{e}_R \\ \text{- - -} \\ \mathbf{e}_L \end{bmatrix} \tag{7}$$

6.4.4 Implications of Cut-Set Analysis

We have indicated that a proper tree has been chosen by having the *capacitive and conductive branches be the tree branches* and by having the *resistive and inductive branches be the links*. It follows that the choice of the tree defines the fundamental cut sets. By performing a standard cut-set analysis we deduce that KCL yields

$$\boxed{\mathbf{Qi} = \mathbf{0} \quad \text{or} \quad [\mathbf{F} \mid \mathbf{I}]\mathbf{i} = \mathbf{0} \quad \text{or} \quad \mathbf{Fi}_l = -\mathbf{i}_t} \tag{8}$$

KVL yields

$$\boxed{\mathbf{v} = \mathbf{Q}'\mathbf{v}_t \quad \text{or} \quad \mathbf{v}_l = \mathbf{F}'\mathbf{v}_t} \tag{9}$$

where $\mathbf{Q}$ is the fundamental cut-set matrix (see Chap. 2). Recall that the number of rows of $\mathbf{F}$ equals the number of tree branches and the number of columns of $\mathbf{F}$ equals the number of links.

Since the link- and tree-branch vectors are composed of two subvectors, as evidenced by Eq. (1), Eqs. (8) and (9) can be also expressed in the form

$$\underbrace{\begin{bmatrix} \mathbf{F}_1 & \vdots & \mathbf{F}_2 \\ \text{- - - -} & \vdots & \text{- - - -} \\ \mathbf{F}_3 & \vdots & \mathbf{F}_4 \end{bmatrix}}_{\mathbf{F}} \underbrace{\begin{bmatrix} \mathbf{i}_R \\ \text{- - -} \\ \mathbf{i}_L \end{bmatrix}}_{\mathbf{i}_l} = - \underbrace{\begin{bmatrix} \mathbf{i}_C \\ \text{- - -} \\ \mathbf{i}_G \end{bmatrix}}_{\mathbf{i}_t} \tag{10}$$

$$
\begin{bmatrix} \mathbf{v}_R \\ \text{- - -} \\ \mathbf{v}_L \end{bmatrix} = \begin{bmatrix} \mathbf{F}'_1 & \vdots & \mathbf{F}'_3 \\ \text{- - - -} & \vdots & \text{- - - -} \\ \mathbf{F}'_2 & \vdots & \mathbf{F}'_4 \end{bmatrix} \begin{bmatrix} \mathbf{v}_C \\ \text{- - -} \\ \mathbf{v}_G \end{bmatrix} \tag{11}
$$

$$\underbrace{}_{\mathbf{v}_t} \qquad \underbrace{}_{\mathbf{F}'} \quad \underbrace{}_{\mathbf{v}_t}$$

where $\mathbf{F}_1$, $\mathbf{F}_2$, $\mathbf{F}_3$, and $\mathbf{F}_4$ are submatrices defined by $\mathbf{F}$. By carrying out the block-matrix–vector multiplications, we obtain

$$
\mathbf{F}_1 \mathbf{i}_R + \mathbf{F}_2 \mathbf{i}_L = -\mathbf{i}_C \tag{12}
$$

$$
\mathbf{F}_3 \mathbf{i}_R + \mathbf{F}_4 \mathbf{i}_L = -\mathbf{i}_G \tag{13}
$$

$$
\mathbf{F}'_1 \mathbf{v}_C + \mathbf{F}'_3 \mathbf{v}_G = \mathbf{v}_R \tag{14}
$$

$$
\mathbf{F}'_2 \mathbf{v}_C + \mathbf{F}'_4 \mathbf{v}_G = \mathbf{v}_L \tag{15}
$$

6.4.5 Branch Constraints

An examination of the branches in Fig. 6.4.1 and use of the definitions of the vectors and matrices provided by Eqs. (1) to (7) yield the following set of constraint equations expressed in vector form:

Resistive branches: $\qquad \mathbf{v}_R = \mathbf{R}\mathbf{i}_R + \mathbf{e}_R$ (16)

Inductive branches: $\qquad \mathbf{v}_L = \mathbf{L}\dfrac{d\mathbf{i}_L}{dt} + \mathbf{e}_L$ (17)

Capacitive branches: $\qquad \mathbf{i}_C = \mathbf{C}\dfrac{d\mathbf{v}_C}{dt} + \mathbf{j}_C$ (18)

Conductive branches: $\qquad \mathbf{i}_G = \mathbf{G}\mathbf{v}_G + \mathbf{j}_G$ (19)

6.4.6 The State-Vector Equations

As we have indicated many times, the network state variables are the inductor currents and the capacitor voltages. The state variables form the components of the state vector $\mathbf{x}(t)$, which is given by

$$
\mathbf{x}(t) = \begin{bmatrix} \mathbf{i}_L \\ \text{- - -} \\ \mathbf{v}_C \end{bmatrix} \tag{20}
$$

Our objective is to determine the differential equation of the state vector $\mathbf{x}(t)$. To do this we must eliminate the non-state-variable vectors v_R, i_R, v_L, i_C, v_G, and i_G from the vector equations (12) to (15). The required algebraic manipulations are straightforward but somewhat messy.

If one substitutes (19) into (13) and (16) into (14), one can solve for i_R to find

$$i_R = (R + F_3' G^{-1} F_3)^{-1} (F_1' v_C - F_3' G^{-1} F_4 i_L - F_3' G^{-1} j_G - e_R) \qquad (21)$$

and for v_G to find

$$v_G = (G + F_3 R^{-1} F_3')^{-1} (-F_3 R^{-1} F_1' v_C - F_4 i_L + F_3 R^{-1} e_R - j_G) \qquad (22)$$

Note that both i_R and v_G are expressed as linear combinations of the state variables and the (source) input variables.

From Eqs. (17) and (15) we have

$$L \frac{d i_L}{dt} = v_L - e_L = F_2' v_C + F_4' v_G - e_L \qquad (23)$$

Similarly from Eqs. (18) and (12) we deduce that

$$C \frac{d v_C}{dt} = i_C - j_C = -F_1 i_R - F_2 i_L - j_C \qquad (24)$$

If, for the sake of simplicity, we define the matrices Φ and Ψ by

$$\Phi \triangleq (R + F_3' G^{-1} F_3)^{-1} \qquad (25)$$

$$\Psi \triangleq (G + F_3 R^{-1} F_3')^{-1} \qquad (26)$$

then substitution of Eqs. (21), (22), (25), and (26) into Eqs. (23) and (24) yields

$$L \frac{d i_L}{dt} = -F_4' \Psi F_4 i_L + (F_2' - F_4' \Psi F_3 R^{-1} F_1') v_C + F_4' \Psi F_3 R^{-1} e_R - F_4' \Psi j_G - e_L \qquad (27)$$

$$C \frac{d v_C}{dt} = -(F_2 - F_1 \Phi F_3' G^{-1} F_4) i_L - F_1 \Phi F_1' v_C + F_1 \Phi F_3' G^{-1} j_G + F_1 \Phi e_R - j_C \qquad (28)$$

Digression Equations (27) and (28) can be written in a suggestive, somewhat standard, form by proving the (not so obvious) equality

$$G^{-1} F_3 \Phi = \Psi F_3 R^{-1} \qquad (29)$$

To prove this we shall prove the equivalent equality

$$F_3 = G \Psi F_3 R^{-1} \Phi^{-1} \qquad (30)$$

The proof proceeds as follows:

$$\mathbf{G\Psi F_3 R^{-1} \Phi^{-1}} = \mathbf{G\Psi F_3 R^{-1}(R + F_3' G^{-1} F_3)} = \mathbf{G\Psi(F_3 R^{-1} R + F_3 R^{-1} F_3' G^{-1} F_3)}$$

$$= \mathbf{G\Psi(F_3 + F_3 R^{-1} F_3' G^{-1} F_3)} = \mathbf{G\Psi(I + F_3 R^{-1} F_3' G^{-1})F_3}$$

$$= \mathbf{G\Psi(GG^{-1} + F_3 R^{-1} F_3' G^{-1})F_3}$$

$$= \mathbf{G\Psi(G + F_3 R^{-1} F_3')G^{-1} F_3} = \mathbf{G\Psi\Psi^{-1} G^{-1} F_3} = \mathbf{GG^{-1} F_3} = \mathbf{F_3}$$

By using Eq. (29) one can easily see that the matrix coefficient of $\mathbf{v}_C$ in Eq. (27) is the negative transpose of the matrix coefficient of $\mathbf{i}_L$ in Eq. (28). Thus if we define the matrices

$$\mathbf{P_1} \triangleq \mathbf{F_4' \Psi F_4} \tag{31}$$

$$\mathbf{P_2} \triangleq \mathbf{F_2' - F_4' \Psi F_3 R^{-1} F_1'} \tag{32}$$

$$\mathbf{P_4} \triangleq \mathbf{F_1 \Phi F_1'} \tag{33}$$

and

$$\mathbf{W} = \left[\begin{array}{c|c|c|c} \mathbf{F_4' \Psi F_3 R^{-1}} & -\mathbf{I} & \mathbf{0} & -\mathbf{F_4' \Psi} \\ \hline \mathbf{F_1 \Phi} & \mathbf{0} & -\mathbf{I} & \mathbf{F_1 \Phi F_3' G^{-1}} \end{array} \right] \tag{34}$$

then Eqs. (27) and (28) take the form

$$\left[\begin{array}{c|c} \mathbf{L} & \mathbf{0} \\ \hline \mathbf{0} & \mathbf{C} \end{array} \right] \frac{d}{dt} \left[\begin{array}{c} \mathbf{i}_L \\ \hline \mathbf{v}_C \end{array} \right] = \left[\begin{array}{c|c} -\mathbf{P_1} & \mathbf{P_2} \\ \hline -\mathbf{P_2'} & -\mathbf{P_4} \end{array} \right] \left[\begin{array}{c} \mathbf{i}_L \\ \hline \mathbf{v}_C \end{array} \right] + \mathbf{W} \left[\begin{array}{c} \mathbf{e}_R \\ \hline \mathbf{e}_L \\ \hline \mathbf{j}_C \\ \hline \mathbf{j}_G \end{array} \right] \tag{35}$$

It therefore follows that the state vector $\mathbf{x}(t)$ of the network

$$\mathbf{x}(t) = \left[\begin{array}{c} \mathbf{i}_L \\ \hline \mathbf{v}_C \end{array} \right] \tag{36}$$

satisfies a vector differential equation of the form

$$\boxed{\dot{\mathbf{x}}(t) = \mathbf{Ax}(t) + \mathbf{Bu}(t)} \tag{37}$$

where **A** is the matrix

$$
\mathbf{A} \triangleq
\begin{bmatrix}
\mathbf{L}^{-1} & \vdots & \mathbf{0} \\
\cdots & \vdots & \cdots \\
\mathbf{0} & \vdots & \mathbf{C}^{-1}
\end{bmatrix}
\begin{bmatrix}
-\mathbf{P}_1 & \vdots & \mathbf{P}_2 \\
\cdots & \vdots & \cdots \\
-\mathbf{P}_2' & \vdots & -\mathbf{P}_4
\end{bmatrix}
\tag{38}
$$

B is the matrix

$$
\mathbf{B} \triangleq
\begin{bmatrix}
\mathbf{L}^{-1} & \vdots & \mathbf{0} \\
\cdots & \vdots & \cdots \\
\mathbf{0} & \vdots & \mathbf{C}^{-1}
\end{bmatrix}
\mathbf{W}
\tag{39}
$$

and **u**(t) is the (input) vector

$$
\mathbf{u}(t) \triangleq
\begin{bmatrix}
\mathbf{e}_R \\
\cdots \\
\mathbf{e}_L \\
\cdots \\
\mathbf{j}_C \\
\cdots \\
\mathbf{j}_G
\end{bmatrix}
\tag{40}
$$

We interpret the state-vector differential equation (35) or (37) in the following way: the time derivative of each state variable, i.e., inductor current or capacitor voltage, is equal to a linear (constant-coefficient) combination of all the other state variables and inputs, i.e., the source voltages and source currents of the independent current sources. This is why a vector differential equation of this type is often called an LTI vector differential equation.

Techniques for the analytical and algorithmic solution of this type of differential equation will be given in subsequent chapters. For the time being, we assert that we can obtain a unique solution to this equation if we are given the value of the state variables at some time t_0 and the time-varying inputs provided by the voltage and current sources. We also wish to stress that *no time derivatives of the input variables appear in the state-variable differential equations, so that there is never any problem arising from the use of impulses, doublets, and other such nasty singularity functions.*

To recapitulate: solution of the state-vector differential equation specifies the inductor current vector $\mathbf{i}_L$ and the capacitor voltage vector $\mathbf{v}_L$ as functions of time. From knowledge of these vectors and of the voltage- and current-source vectors, one can determine the remaining network variables by linear algebraic (memoryless) operations. Thus, $\mathbf{i}_R$ is found from Eq. (21) and $\mathbf{v}_G$ from Eq. (22); next, $\mathbf{v}_R$ can be evaluated from Eq. (16) and $\mathbf{i}_G$ from Eq. (19); finally, one can determine $\mathbf{v}_L$ from

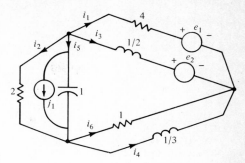

FIGURE 6.4.2
An LTI *RLC* network.

(15) and $\mathbf{i}_C$ from (12). We shall not write the detailed expressions here; but we stress that all the remaining voltages and currents are LTI combinations of the state variables and of the input variables.

EXAMPLE 1   We shall now present a detailed numerical example to illustrate the step-by-step procedure that we have followed in order to determine the state differential equations. The LTI network we shall analyze is shown in Fig. 6.4.2; it contains two inductors, one capacitor, three resistors, two voltage sources, and a current source. This network satisfies the assumption outlined in Sec. 6.4.2. The capacitive branch and the branch with the 1-Ω resistor define the proper tree, as shown in the graph of Fig. 6.4.3. The branches are indexed following the sequence described in Sec. 6.4.3. Thus [see Eq. (1)]

$$\mathbf{i}_R = \begin{bmatrix} i_1 \\ i_2 \end{bmatrix} \qquad \mathbf{i}_L = \begin{bmatrix} i_3 \\ i_4 \end{bmatrix} \qquad \mathbf{i}_C = i_5 \qquad \mathbf{i}_G = i_6 \tag{41}$$

$$\mathbf{v}_R = \begin{bmatrix} v_1 \\ v_2 \end{bmatrix} \qquad \mathbf{v}_L = \begin{bmatrix} v_3 \\ v_4 \end{bmatrix} \qquad \mathbf{v}_C = i_5 \qquad \mathbf{v}_G = v_6 \tag{42}$$

For this network the resistance, inductance, capacitance, and conductance matrices are given by [see Eqs. (2) to (5)]

$$\mathbf{R} = \begin{bmatrix} 4 & 0 \\ 0 & 2 \end{bmatrix} \qquad \mathbf{L} = \begin{bmatrix} \frac{1}{2} & 0 \\ 0 & \frac{1}{3} \end{bmatrix} \qquad \mathbf{C} = 1 \qquad \mathbf{G} = 1 \tag{43}$$

The current-source and voltage-source vectors [see Eqs. (6) and (7)] are given by

$$\mathbf{j} = \begin{bmatrix} j_1 \\ \text{- - -} \\ 0 \end{bmatrix} \begin{matrix} \mathbf{j}_c \\ \\ \mathbf{j}_G \end{matrix} \qquad \mathbf{e} = \begin{bmatrix} e_1 \\ 0 \\ \text{- - -} \\ e_2 \\ 0 \end{bmatrix} \begin{matrix} \mathbf{e}_R \\ \\ \mathbf{e}_L \end{matrix} \tag{44}$$

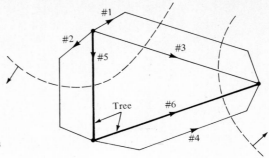

FIGURE 6.4.3
The proper tree and fundamental cut sets
for the network of Fig. 6.4.2.

Figure 6.4.3 shows the two fundamental cut sets defined by the proper tree. Use of KCL across the cut sets yields the equations

$$i_1 + i_2 + i_3 + i_5 = 0$$
$$i_1 + i_3 + i_4 + i_6 = 0 \tag{45}$$

which can be written [see Eq. (8)] in the form

$$
\underbrace{
\begin{bmatrix}
\overbrace{1 \quad 1}^{\mathbf{F_1}} & \vdots & \overbrace{1 \quad 0}^{\mathbf{F_2}} \\
\hdashline
\underbrace{1 \quad 0}_{\mathbf{F_3}} & \vdots & \underbrace{1 \quad 1}_{\mathbf{F_4}}
\end{bmatrix}
}_{\mathbf{F}}
\underbrace{
\begin{bmatrix}
i_1 \\ i_2 \\ \hdashline i_3 \\ i_4
\end{bmatrix}
\begin{matrix} {\scriptstyle\mathbf{i}_R} \\[6pt] {\scriptstyle\mathbf{i}_L} \end{matrix}
}_{\mathbf{i}_t}
= -
\underbrace{
\begin{bmatrix}
i_5 \\ \hdashline i_6
\end{bmatrix}
\begin{matrix} {\scriptstyle\mathbf{i}_C} \\[6pt] {\scriptstyle\mathbf{i}_G} \end{matrix}
}_{\mathbf{i}_l}
\tag{46}
$$

The matrix $\mathbf{F}$ can be thus decomposed to yield the four submatrices $\mathbf{F_1}$, $\mathbf{F_2}$, $\mathbf{F_3}$, $\mathbf{F_4}$ and their transposes [see Eqs. (10) and (11)]

$$\mathbf{F_1} = [1 \quad 1] \qquad \mathbf{F_2} = [1 \quad 0] \qquad \mathbf{F_3} = [1 \quad 0] \qquad \mathbf{F_4} = [1 \quad 1] \tag{47}$$

$$\mathbf{F'_1} = \begin{bmatrix} 1 \\ 1 \end{bmatrix} \qquad \mathbf{F'_2} = \begin{bmatrix} 1 \\ 0 \end{bmatrix} \qquad \mathbf{F'_3} = \begin{bmatrix} 1 \\ 0 \end{bmatrix} \qquad \mathbf{F'_4} = \begin{bmatrix} 1 \\ 1 \end{bmatrix} \tag{48}$$

The state vector $\mathbf{x}(t)$ for this network has three components; its first two components are the inductor currents i_3 and i_4, and its third component is the capacitor voltage v_5. Thus [see Eq. (20)]

$$\mathbf{x}(t) = \begin{bmatrix} i_3 \\ i_4 \\ v_5 \end{bmatrix} \tag{49}$$

Toward the determination of the state differential equations we defined the matrices $\mathbf{\Phi}$ and $\mathbf{\Psi}$ [see Eqs. (25) and (26)]. In our example, in view of Eq. (43), we have

$$\mathbf{R}^{-1} = \begin{bmatrix} \frac{1}{4} & 0 \\ 0 & \frac{1}{2} \end{bmatrix} \qquad \mathbf{G}^{-1} = 1 \tag{50}$$

Thus

$$\begin{aligned} \mathbf{\Phi} = (\mathbf{R} + \mathbf{F}_3' \mathbf{G}^{-1} \mathbf{F}_3)^{-1} &= \left[\begin{bmatrix} 4 & 0 \\ 0 & 2 \end{bmatrix} + \begin{bmatrix} 1 \\ 0 \end{bmatrix} (1)[1 \quad 0] \right]^{-1} \\ &= \begin{bmatrix} 5 & 0 \\ 0 & 2 \end{bmatrix}^{-1} = \begin{bmatrix} \frac{1}{5} & 0 \\ 0 & \frac{1}{2} \end{bmatrix} \end{aligned} \tag{51}$$

$$\begin{aligned} \mathbf{\Psi} = (\mathbf{G} + \mathbf{F}_3 \mathbf{R}^{-1} \mathbf{F}_3')^{-1} &= \left[1 + [1 \quad 0] \begin{bmatrix} \frac{1}{4} & 0 \\ 0 & \frac{1}{2} \end{bmatrix} \begin{bmatrix} 1 \\ 0 \end{bmatrix} \right]^{-1} \\ &= \left(1 + \frac{1}{4} \right)^{-1} = \frac{4}{5} \qquad \text{a scalar} \end{aligned} \tag{52}$$

We can now evaluate the matrices $\mathbf{P}_1$, $\mathbf{P}_2$, and $\mathbf{P}_4$ [see Eqs. (31) to (33)].

$$\mathbf{P}_1 = \mathbf{F}_4' \mathbf{\Psi} \mathbf{F}_4 = \begin{bmatrix} 1 \\ 1 \end{bmatrix} \left(\frac{4}{5} \right) [1 \quad 1] = \frac{4}{5} \begin{bmatrix} 1 & 1 \\ 1 & 1 \end{bmatrix} \tag{53}$$

$$\mathbf{P}_2 = \mathbf{F}_2' - \mathbf{F}_4' \mathbf{\Psi} \mathbf{F}_3 \mathbf{R}^{-1} \mathbf{F}_1' = \begin{bmatrix} 1 \\ 0 \end{bmatrix} - \begin{bmatrix} 1 \\ 1 \end{bmatrix} \left(\frac{4}{5} \right) [1 \quad 0] \begin{bmatrix} \frac{1}{4} & 0 \\ 0 & \frac{1}{2} \end{bmatrix} \begin{bmatrix} 1 \\ 1 \end{bmatrix} = \begin{bmatrix} \frac{4}{5} \\ -\frac{1}{5} \end{bmatrix} \tag{54}$$

$$\mathbf{P}_4 = \mathbf{F}_1 \mathbf{\Phi} \mathbf{F}_1' = [1 \quad 1] \begin{bmatrix} \frac{1}{5} & 0 \\ 0 & \frac{1}{2} \end{bmatrix} \begin{bmatrix} 1 \\ 1 \end{bmatrix} = \frac{7}{10} \tag{55}$$

Next we evaluate the $\mathbf{W}$ matrix in Eq. (34). We first note that

$$\mathbf{e}_R = \begin{bmatrix} e_1 \\ 0 \end{bmatrix} \qquad \mathbf{e}_L = \begin{bmatrix} e_2 \\ 0 \end{bmatrix} \qquad \mathbf{j}_C = j_1 \qquad \mathbf{j}_G = 0 \tag{56}$$

Thus, the input vector $\mathbf{u}(t)$ [see Eq. (40)] takes the form

$$\mathbf{u}(t) = \begin{bmatrix} e_1 \\ 0 \\ e_2 \\ 0 \\ j_1 \\ 0 \end{bmatrix} \tag{57}$$

The $\mathbf{W}$ matrix is easily evaluated to be

$$\mathbf{W} = \left[\begin{array}{cc:cc:c:c} \frac{1}{5} & 0 & -1 & 0 & 0 & -\frac{4}{5} \\ \frac{1}{5} & 0 & 0 & -1 & 0 & -\frac{4}{5} \\ \hdashline \frac{1}{5} & \frac{1}{2} & 0 & 0 & -1 & \frac{1}{5} \end{array}\right] \tag{58}$$

Thus, Eq. (35) is

$$\left[\begin{array}{cc:c} \frac{1}{2} & 0 & 0 \\ 0 & \frac{1}{3} & 0 \\ \hdashline 0 & 0 & 1 \end{array}\right] \frac{d}{dt} \left[\begin{array}{c} i_3 \\ i_4 \\ \hdashline v_5 \end{array}\right] = \left[\begin{array}{cc:c} -\frac{4}{5} & -\frac{4}{5} & \frac{4}{5} \\ -\frac{4}{5} & -\frac{4}{5} & -\frac{1}{5} \\ \hdashline -\frac{4}{5} & \frac{1}{5} & \frac{7}{10} \end{array}\right] \left[\begin{array}{c} i_3 \\ i_4 \\ \hdashline v_5 \end{array}\right]$$

$$+ \left[\begin{array}{ccc} \frac{1}{5} & -1 & 0 \\ \frac{1}{5} & 0 & 0 \\ \frac{1}{5} & 0 & -1 \end{array}\right] \left[\begin{array}{c} e_1 \\ e_2 \\ j_1 \end{array}\right] \tag{59}$$

which reduces to

$$\underbrace{\frac{d}{dt}\begin{bmatrix} i_3 \\ i_4 \\ v_5 \end{bmatrix}}_{\dot{\mathbf{x}}(t)} = \frac{1}{10}\underbrace{\begin{bmatrix} -16 & -16 & 16 \\ -24 & -24 & -6 \\ -8 & 2 & 7 \end{bmatrix}}_{\mathbf{A}}\underbrace{\begin{bmatrix} i_3 \\ i_4 \\ v_5 \end{bmatrix}}_{\mathbf{x}(t)} + \underbrace{\frac{1}{5}\begin{bmatrix} 2 & -10 & 0 \\ 3 & 0 & 0 \\ 1 & 0 & -5 \end{bmatrix}\begin{bmatrix} e_1 \\ e_2 \\ j_1 \end{bmatrix}}_{\mathbf{B}\mathbf{u}(t)} \tag{60}$$

Clearly this vector differential equation can be decomposed into the set of the three simultaneous first-order differential equations

$$\frac{d}{dt}i_3 = -\tfrac{16}{10}i_3 - \tfrac{16}{10}i_4 + \tfrac{16}{10}v_5 + \tfrac{2}{5}e_1 - 2e_2$$

$$\frac{d}{dt}i_4 = -\tfrac{24}{10}i_3 - \tfrac{24}{10}i_4 - \tfrac{6}{10}v_5 + \tfrac{3}{5}e_1 \tag{61}$$

$$\frac{d}{dt}v_5 = -\tfrac{8}{10}i_3 + \tfrac{2}{10}i_4 + \tfrac{7}{10}v_5 + \tfrac{1}{5}e_1 - j_1$$

6.4.7 Concluding Remarks

The analysis of LTI networks presented in Secs. 6.3 and 6.4 has motivated the study of linear vector differential equations of the form $\dot{\mathbf{x}}(t) = \mathbf{A}\mathbf{x}(t) + \mathbf{B}\mathbf{u}(t)$, where $\mathbf{x}(t)$ is the state vector and $\mathbf{u}(t)$ is the input vector. We shall examine analytical and algorithmic techniques for the solution of this class of differential equations in Chaps. 7 and 9. In the remainder of this chapter we focus our attention on the systematic development of the equations characterizing networks which contain nonlinear circuit elements.

The analysis of *RLC* nonlinear networks leads to nonlinear vector differential equations of the form $\dot{\mathbf{x}}(t) = \mathbf{f}(\mathbf{x}(t), \mathbf{u}(t))$, where $\mathbf{x}(t)$ is the state vector, $\mathbf{u}(t)$ is the input vector, and $\mathbf{f}(.\,,.)$ is a nonlinear vector-valued function of its two arguments $\mathbf{x}(t)$ and $\mathbf{u}(t)$. This motivates the study of nonlinear vector differential equations as well as the numerical integration algorithms, which we shall study later.

Thus, by examining a single class of physical systems, namely, electric networks, we provide the motivation for examining the general, system-oriented mathematical tools of later chapters. We reiterate that these tools are useful for the analysis of a multitude of physical systems (aerospace systems, chemical process systems, etc.) as well as for the analytical and/or algorithmic solution of the equations that describe such systems.

6.5 STATE-VARIABLE EQUATIONS FOR NONLINEAR *RLC* NETWORKS

We conclude this chapter with the development of the state differential equations for a class of nonlinear networks consisting of nonlinear capacitors, inductors, and resistors. Our objective is threefold:

1 To show that the choice of the inductor flux linkages and of the capacitor charges as state variables is suitable

2 To demonstrate that the *state vector* $\mathbf{x}(t)$ of the network satisfies a nonlinear vector differential equation of the form

$$\dot{\mathbf{x}}(t) = \mathbf{f}(\mathbf{x}(t), \mathbf{u}(t)) \tag{1}$$

where $\mathbf{u}(t)$ is the *input vector*, whose components are defined by the network independent voltage and current sources

3 To show that all other network variables of interest which form the components of an *output vector* $\mathbf{y}(t)$ are related to the state and input vectors

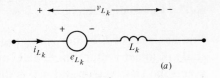

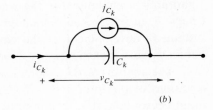

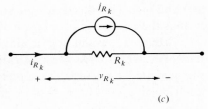

FIGURE 6.5.1
The three types of branches containing nonlinear elements that form the network: (a) inductive branch; (b) capacitive branch; (c) resistive branch.

$\mathbf{x}(t)$ and $\mathbf{u}(t)$, respectively, by a nonlinear memoryless algebraic equation of the form

$$\mathbf{y}(t) = \mathbf{h}(\mathbf{x}(t), \mathbf{u}(t)) \qquad (2)$$

The reader should recall that when we studied LTI networks in Secs. 6.3 and 6.4, we chose as state variables the inductor currents and the capacitor voltages. For reasons that will become obvious, a more natural choice for state variables in nonlinear networks is

1 The inductor flux linkages
2 The capacitor charges†

We shall assume that the branches of the nonlinear networks under consideration fall into the three categories shown in Fig. 6.5.1. Thus, we assume that

1 All inductors are flux-controlled.
2 All capacitors are charge-controlled.
3 All resistors are current-controlled.

Furthermore, we shall assume that the network is such that one can select a (proper) tree with the property that

---

† These can also be used as state variables in linear networks, but they are more awkward.

1   The capacitive and resistive branches define the tree branches.
2   The inductive branches define the links.

A typical inductive branch (see Fig. 6.5.1*a*) is described by the equations

$$i_{Lk} = \gamma_k(\phi_k) \qquad \dot{\phi}_k = v_{Lk} - e_{Lk} \tag{3}$$

where $\gamma_k(.)$ is a nonlinear scalar-valued function associated with this inductor. Letting ϕ denote the flux-linkage vector, $\mathbf{i}_L$ the inductor-current vector, $\mathbf{v}_L$ the inductor-voltage vector, and $\mathbf{e}_L$ the voltage-source vector, we obtain the following vector equations, which completely characterize the inductive branches:

$$\boxed{\dot{\phi} = \mathbf{v}_L - \mathbf{e}_L} \tag{4}$$

$$\boxed{\mathbf{i}_L = \gamma(\phi)} \tag{5}$$

where $\gamma(.)$ is now a vector-valued function whose components are defined by the current-flux characteristics of each inductor.

A typical capacitive branch (see Fig. 6.5.1*b*) is described by the equations

$$v_{Ck} = \beta_k(q_k) \qquad \dot{q}_k = i_{Ck} - j_{Ck} \tag{6}$$

where $\beta_k(.)$ is a nonlinear scalar-valued function associated with this capacitor. We shall define $\mathbf{q}$ to be the charge vector, $\mathbf{i}_C$ the capacitor-current vector, $\mathbf{j}_C$ the current-source vector, and $\mathbf{v}_C$ the capacitor-voltage vector. Then, the capacitor branches are described by the equations

$$\boxed{\dot{\mathbf{q}} = \mathbf{i}_C - \mathbf{j}_C} \tag{7}$$

$$\boxed{\mathbf{v}_C = \beta(\mathbf{q})} \tag{8}$$

Finally, the resistive branches (see Fig. 6.5.1*c*) are described by

$$v_{Rk} = r_k(i_{Rk} - j_{Rk}) \tag{9}$$

The vector $\mathbf{v}_R$ will describe the resistor-voltage vector, $\mathbf{i}_R$ the resistor-current vector, and $\mathbf{j}_R$ the current-source vector. Hence, the current-controlled resistive branches are described by

$$\boxed{\mathbf{v}_R = \mathbf{r}(\mathbf{i}_R - \mathbf{j}_R)} \tag{10}$$

According to our assumptions the tree-branch current and voltage vectors $\mathbf{i}_t$ and $\mathbf{v}_t$, respectively, are

$$
\mathbf{i}_t = \begin{bmatrix} \mathbf{i}_C \\ \text{---} \\ \mathbf{i}_R \end{bmatrix} \qquad \mathbf{v}_t = \begin{bmatrix} \mathbf{v}_C \\ \text{---} \\ \mathbf{v}_R \end{bmatrix} \tag{11}
$$

while the link-current and -voltage vectors, $\mathbf{i}_l$ and $\mathbf{v}_l$, respectively, are simply

$$
\mathbf{i}_l = \mathbf{i}_L \qquad \mathbf{v}_l = \mathbf{v}_L \tag{12}
$$

The implications of a cut-set analysis are the equations

$$
\underbrace{\begin{bmatrix} \mathbf{F}_a \\ \mathbf{F}_b \end{bmatrix}}_{\mathbf{F}} \underbrace{\mathbf{i}_L}_{\mathbf{i}_l} = -\underbrace{\begin{bmatrix} \mathbf{i}_C \\ \mathbf{i}_R \end{bmatrix}}_{\mathbf{i}_t} \qquad \text{KCL} \tag{13}
$$

$$
\underbrace{\mathbf{v}_L}_{\mathbf{v}_l} = \underbrace{[\mathbf{F}_a' \mid \mathbf{F}_b']}_{\mathbf{F}'} \underbrace{\begin{bmatrix} \mathbf{v}_C \\ \mathbf{v}_R \end{bmatrix}}_{\mathbf{v}_t} \qquad \text{KVL} \tag{14}
$$

The KCL and KVL equations above yield

$$
\mathbf{F}_a \mathbf{i}_L = -\mathbf{i}_C \tag{15}
$$

$$
\mathbf{F}_b \mathbf{i}_L = -\mathbf{i}_R \tag{16}
$$

$$
\mathbf{v}_L = \mathbf{F}_a' \mathbf{v}_C + \mathbf{F}_b' \mathbf{v}_R \tag{17}
$$

We are now ready to deduce the differential equations satisfied by the flux vector $\boldsymbol{\phi}$ and the charge vector $\mathbf{q}$. To do this we note that

$$
\begin{aligned}
\dot{\boldsymbol{\phi}} &= \mathbf{v}_L - \mathbf{e}_L && \text{by Eq. (4)} \\
&= \mathbf{F}_a' \mathbf{v}_C + \mathbf{F}_b' \mathbf{v}_R - \mathbf{e}_L && \text{by Eq. (17)} \\
&= \mathbf{F}_a' \beta(\mathbf{q}) + \mathbf{F}_b' \mathbf{r}(\mathbf{i}_R - \mathbf{j}_R) - \mathbf{e}_L && \text{by Eqs. (8) and (10)} \\
&= \mathbf{F}_a' \beta(\mathbf{q}) + \mathbf{F}_b' \mathbf{r}(-\mathbf{F}_b \mathbf{i}_L - \mathbf{j}_R) - \mathbf{e}_L && \text{by Eq. (16)} \\
&= \mathbf{F}_a' \beta(\mathbf{q}) + \mathbf{F}_b' \mathbf{r}(-\mathbf{F}_b \gamma(\boldsymbol{\phi}) - \mathbf{j}_R) - \mathbf{e}_L && \text{by Eq. (5)}
\end{aligned} \tag{18}
$$

which evidently states that $\dot{\boldsymbol{\phi}}$ is a nonlinear function of $\boldsymbol{\phi}$ and $\mathbf{q}$ and of the input (source) variables $\mathbf{j}_R$ and $\mathbf{e}_L$.

In an analogous manner we can compute $\dot{\mathbf{q}}$:

$$
\begin{aligned}
\dot{\mathbf{q}} &= \mathbf{i}_C - \mathbf{j}_C && \text{by Eq. (7)} \\
&= -\mathbf{F}_a\mathbf{i}_L - \mathbf{j}_C && \text{by Eq. (15)} \\
&= -\mathbf{F}_a\gamma(\boldsymbol{\phi}) - \mathbf{j}_C && \text{by Eq. (5)}
\end{aligned}
\tag{19}
$$

which implies that $\dot{\mathbf{q}}$ is a nonlinear function of the flux vector $\boldsymbol{\phi}$ and is linearly dependent upon the input (source) vector $\mathbf{j}_C$.

If we let $\mathbf{x}(t)$ be the state vector and $\mathbf{u}(t)$ the input vector, i.e.,

$$
\mathbf{x}(t) \triangleq \begin{bmatrix} \boldsymbol{\phi} \\ \mathbf{q} \end{bmatrix} \qquad \mathbf{u}(t) \triangleq \begin{bmatrix} \mathbf{e}_L \\ \mathbf{j}_C \\ \mathbf{j}_R \end{bmatrix}
\tag{20}
$$

then the pair of differential equations (18) and (19), i.e.,

$$
\boxed{
\begin{aligned}
\dot{\boldsymbol{\phi}} &= \mathbf{F}'_b\mathbf{r}(-\mathbf{F}_b\gamma(\boldsymbol{\phi}) - \mathbf{j}_R) + \mathbf{F}'_a\beta(\mathbf{q}) - \mathbf{e}_L \\
\dot{\mathbf{q}} &= -\mathbf{F}_a\gamma(\boldsymbol{\phi}) - \mathbf{j}_C
\end{aligned}
}
\tag{21}
$$

are of the form

$$
\dot{\mathbf{x}}(t) = \mathbf{f}(\mathbf{x}(t), \mathbf{u}(t))
\tag{22}
$$

which is a nonlinear vector differential equation. In general, such differential equations cannot be solved analytically, and one must employ numerical integration procedures. (We shall consider the topic of numerical integration in Chap. 9.) However, as we shall see in Chap. 7, given an initial value of the state vector, say at $t = t_0$, $\mathbf{x}(t_0)$, and given the input vector $\mathbf{u}(\tau)$ for all $t_0 \leq \tau \leq t$, one can evaluate the state $\mathbf{x}(t)$ at time t.

If we accept that $\boldsymbol{\phi}$ and $\mathbf{q}$ can be determined for any time t by solving Eq. (21), it is obvious that all other currents and branches can be found by means of memoryless algebraic (albeit nonlinear) equations from knowledge at time t of the state variables $\boldsymbol{\phi}(t)$ and $\mathbf{q}(t)$ and of the input variables $\mathbf{e}_L(t)$, $\mathbf{j}_C(t)$, and $\mathbf{j}_R(t)$. To see this, we note that $\mathbf{i}_L$ is found from (5) and $\mathbf{v}_C$ from (8); then, one can determine $\mathbf{i}_C$ from (15) and $\mathbf{i}_R$ from (16); next, one can find $\mathbf{v}_R$ from (10) and $\mathbf{v}_L$ from (17). Therefore, all branch voltages and currents (the outputs) are immediately determined from the state variables and input variables.

REFERENCES

1 CRUZ, J. B., and M. E. VAN VALKENBURG: "Introductory Signals and Circuits," Blaisdell, Waltham, Mass., 1967.

2 DESOER, C. A., and E. S. KUH: "Basic Circuit Theory," McGraw-Hill, New York, 1969.

3 STERN, T. E.: "Theory of Nonlinear Networks and Systems," Addison-Wesley, Reading, Mass., 1965.

4 WEINBERG, L.: "Network Analysis and Synthesis," McGraw-Hill, New York, 1962.

5 CHUA, L. O.: "Introduction to Nonlinear Network Theory," McGraw-Hill, New York, 1969.

6 KUO, F. F., and W. G. MAGNUSON, JR.: "Computer Oriented Circuit Design," Prentice-Hall, Englewood Cliffs, N. J., 1969.

7 SESHU, S., and M. B. REED: "Linear Graphs and Electric Networks," Addison-Wesley, Reading, Mass., 1961.

EXERCISES

Section 6.3

6.3.1   Consider the LTI network shown. Let the source voltage $u(t)$ be the input to this system; let the voltage $y(t)$ denote the output of this system. Let $\mathbf{x}(t)$ denote the state vector of this network. [Use as the first three components of $\mathbf{x}(t)$ the currents in the inductors 1, 2, and 3 H, respectively, and use as the last two components of $\mathbf{x}(t)$ the voltages in the capacitors 2 and 3 F, respectively.]

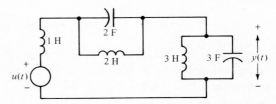

Fig. Ex. 6.3.1

(a)   Show that the state vector $\mathbf{x}(t)$ satisfies a vector differential equation of the form

$$\dot{\mathbf{x}}(t) = \mathbf{A}\mathbf{x}(t) + \mathbf{b}u(t)$$

(b)   Evaluate $\mathbf{A}$ and $\mathbf{b}$ in part (a).

(c)   Show that the output $y(t)$ is of the form

$$y(t) = \mathbf{c}'\mathbf{x}(t) + du(t)$$

Evaluate the column vector $\mathbf{c}$ and the scalar d.

(*d*) Suppose that $u(t)$ is a complex exponential of the form

$$u(t) = \hat{u}e^{st}$$

Suppose that the network is initially at rest. Assume that s is dominant with respect to the natural frequencies. Then show that at steady state

$$\mathbf{x}(t) = \hat{\mathbf{x}}e^{st} \qquad y(t) = \hat{y}e^{st}$$

Determine the vector $\hat{\mathbf{x}}$ and the scalar $\hat{y}$. Evaluate in this manner the system function $H(s)$, that is,

$$\hat{y} = H(s)\hat{u}$$

in terms of matrices and vectors involving $\mathbf{A}$, $\mathbf{b}$, $\mathbf{c}$, and d.

(*e*) Replace the L's and C's by their impedance and directly derive the system function $H(s)$. Convince yourself that it is identical to that found in step (*d*). Determine the poles and zeros of $H(s)$.

(*f*) Prove that the poles of $H(s)$ found above are identical to the eigenvalues of the matrix $\mathbf{A}$ [see parts (*a*) and (*b*)].

6.3.2 (Tricky) Consider a pure LTI LC network whose state vector $\mathbf{x}(t)$ satisfies the vector differential equation $\dot{\mathbf{x}}(t) = \mathbf{A}\mathbf{x}(t)$. Prove that the eigenvalues λ_k of the matrix $\mathbf{A}$ have the property

$$\mathrm{Re}(\lambda_k) = 0$$

for all k.

6.3.3. One of the assumptions we have made in setting up equations is that of no capacitor loops or inductor cut sets. Often a network does contain inductor cut sets and/or capacitor loops. In such problems, before one applies the procedures of this section it is worthwhile to eliminate these violations of assumptions beforehand. Suppose that one is given a network that contains a set of inductors connected in series as in Fig. Ex. 6.3.3a. Then clearly we cannot put all the inductors in the links. However, by KCL we know that all the inductive branches have the same current i. Prove that

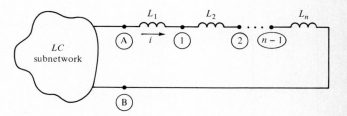

Fig. Ex. 6.3.3 (*a*)

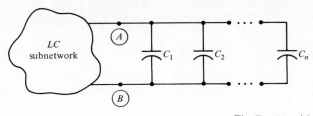

Fig. Ex. 6.3.3 (*b*)

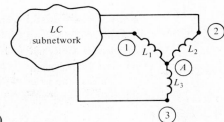

Fig. Ex. 6.3.3 (*c*)

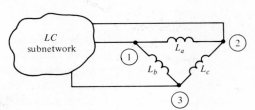

Fig. Ex. 6.3.3 (*d*)

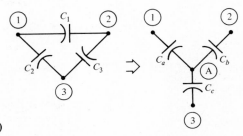

Fig. Ex. 6.3.3 (*e*)

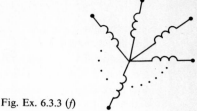

Fig. Ex. 6.3.3 (*f*)

for the purposes of analyzing the network we can replace the *n* inductors by an equivalent inductor

$$L = L_1 + L_2 + \cdots + L_n$$

between nodes *A* and *B* and still get the right answer. Develop the analogous concept when we have *n* capacitors in parallel, as shown in Fig. Ex. 6.3.3*b*. A more complex situation occurs when inductors form a Y (Fig. Ex. 6.3.3*c*). Once more we cannot place all the inductors in the links because the node labeled *A* is connected to the rest of the network only by inductors. In this case, the simplest procedure is to replace the Y inductance network by an equivalent Δ, to obtain Fig. Ex. 6.3.3*d*. Determine the values of the inductors L_a, L_b, and L_c, in terms of L_1, L_2, L_3, so that the overall network equations remain the same. Develop the analogous situation for capacitor loops. Extend the ideas to the case where inductor cut set is of the form in Fig. Ex. 6.3.3*f* (this is hard).

6.3.4   Write the state-variable equations for an *RL* ladder network of the form illustrated.

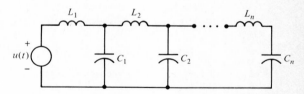

Fig. Ex. 6.3.4

6.3.5   Write the state-variable equations for the network illustrated.

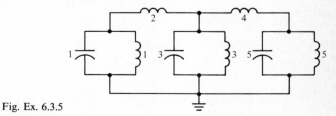

Fig. Ex. 6.3.5

Section 6.4

6.4.1   Suppose that all resistors, inductors, and capacitors in the LTI *RLC* network considered are positive. Then prove that the matrices Φ and Ψ [see Eqs. (25) and (26)] are positive definite matrices.

6.4.2   Consider the state equation of the LTI *RLC* network [see Eq. (35)]. Set all independent voltage and current sources to zero. Letting $\mathbf{x}(t)$ denote the state vector at time t, show that the total (electric plus magnetic) energy $E(t)$ stored in the network at time t is given by the quadratic form

$$E(t) = \tfrac{1}{2}\langle \mathbf{x}(t), \mathbf{M}\mathbf{x}(t)\rangle \quad \text{where } \mathbf{M} = \begin{bmatrix} \mathbf{L} & 0 \\ \hline 0 & \mathbf{C} \end{bmatrix}$$

Compute the time derivative $\dot{E}(t)$ of $E(t)$. Use the results of Exercise 6.4.1 to show that

$$\dot{E}(t) < 0 \quad \text{for all } t \geq 0$$

and so conclude that in this unforced network any initial stored energy at $t = 0$ is dissipated (in this case we deal with a stable network). This method of deducing the stability of a system (or network) is a special case of a very general stability-theoretic approach called *Lyapunov's second method.*

6.4.3   Consider the LTI network illustrated and use the reference directions shown.

    (*a*)   Find a proper tree.
    (*b*)   Find the fundamental cut sets for your tree.
    (*c*)   Find the fundamental cut-set matrix $\mathbf{Q}$.
    (*d*)   Let $\mathbf{x}$ be a state vector for this network. Find $\mathbf{A}$ and $\mathbf{b}$ such that

$$\dot{\mathbf{x}}(t) = \mathbf{A}\mathbf{x}(t) + \mathbf{b}j(t)$$

Identify your state vector with network variables.

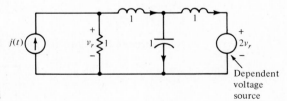

Fig. Ex. 6.4.3

6.4.4   Consider the LTI network shown. Note that it contains a *dependent* current source in parallel with the 2-F capacitor. Write the state-variable differential equations for this

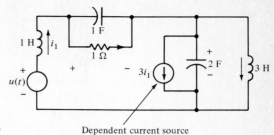

Fig. Ex. 6.4.4   Dependent current source

network. Use the current directions and polarities indicated. Use the following state variables

$$x_1(t) = \text{current through 1-H inductor}$$
$$x_2(t) = \text{current through 3-H inductor}$$
$$x_3(t) = \text{voltage across 1-F capacitor}$$
$$x_4(t) = \text{voltage across 2-F capacitor}$$

which form the components of the state vector $\mathbf{x}(t)$. Write the state equations in the form

$$\dot{\mathbf{x}}(t) = \mathbf{A}\mathbf{x}(t) + \mathbf{b}u(t)$$

Evaluate the matrix $\mathbf{A}$ and the vector $\mathbf{b}$. Let $y(t)$ denote the output of the network. Suppose that $y(t)$ is the current through the 1-Ω resistor. Show that

$$y(t) = \mathbf{c}'\mathbf{x}(t) + du(t)$$

Determine the column vector $\mathbf{c}$ and the scalar d.

6.4.5   Write the state-variable equations for the network sketched.

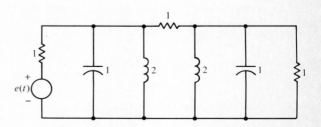

Fig. Ex. 6.4.5

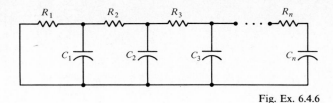

Fig. Ex. 6.4.6

6.4.6 Write the state-variable equations for the RC ladder network sketched, in the form $\dot{\mathbf{x}}(t) = \mathbf{A}\mathbf{x}(t)$. Can you prove that all eigenvalues of $\mathbf{A}$ are real and negative (hard part)?

Section 6.5

6.5.1 Using the inductor flux linkages and the capacitor charges as state variables, deduce the state differential equations for nonlinear networks such that:

(*a*) All capacitors are charge-controlled and belong to the tree branches of a proper tree.
(*b*) All inductors are flux-controlled and belong to the links of a proper tree.
(*c*) All resistors are voltage-controlled and belong to the links of a proper tree.

Include independent current sources across the capacitors and independent voltage sources in series with the inductors and the resistors.

6.5.2 Suppose that the assumptions of Exercise 6.5.1 still hold but in addition one has current-controlled resistors in the tree branches of the proper tree. Make the *further assumption* that the network is such that each fundamental loop defined by a (voltage-controlled) resistive link includes no (current-controlled) resistive tree branches. Once more derive the vector differential equations satisfied by the flux vector $\boldsymbol{\phi}$ and the charge vector $\mathbf{q}$.

6.5.3 Consider the statement of Exercise 6.5.2 without the further assumption. Convince yourself that you cannot obtain the state-vector differential equations in an *explicit* form. Explain clearly where the source of the difficulty lies.

6.5.4 Consider the nonlinear network shown. Using the notation of this section, the elements are characterized by

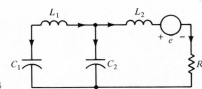

Fig. Ex. 6.5.4

$$L_1: i_{L_1} = 2\phi_1 \qquad C_1: v_{C_1} = q_1 \qquad R: v_R = i_R^3$$
$$L_2: i_{L_2} = \phi_2^2 \qquad C_2: v_{C_2} = e^{q_2}$$

Derive the state differential equations, i.e., find $\dot\phi_1$, $\dot\phi_2$, $\dot q_1$, $\dot q_2$. Write them in the form suggested by Eq. (22) and identify each nonlinearity in Eq. (22).

6.5.5   Consider the general LTI *RLC* network analyzed in Sec. 6.4. Instead of using the inductor currents and capacitor voltages as state variables, use the flux linkages and charges. Derive the state differential equations and relate them to the previously obtained ones.

7
CONTINUOUS-TIME DYNAMIC SYSTEMS

SUMMARY

The analysis of linear and nonlinear networks in Chap. 6 showed that the complete description of such systems is based upon vector differential equations. In this chapter we shall discuss examples of other types of physical systems described by vector differential equations.

From a theoretical point of view our main objective is to expose the reader to certain system-analysis concepts and techniques that can be used to predict the response of continuous-time dynamic systems qualitatively and quantitatively. In addition, we shall provide a minimal introduction to system design using feedback, to show the reader that it is possible to alter the dynamic behavior of an existing system and in so doing make the system behave in a more desirable manner.

From a logical and a conceptual point of view, the material in this chapter is the analog (in a continuous-time sense) of the point of view, concepts, and techniques of Chap. 4. Thus, we shall meet familiar ideas once again—superposition, stability, equilibrium states, linearization, and feedback.

7.1 INTRODUCTION

This chapter contains an introduction to vector differential equations, both linear and nonlinear. More precisely, we shall deal with *ordinary* (rather than partial) vector differential equations. Since differential equations represent a well-defined mathematical discipline, we must often employ a somewhat rigorous treatment. On the other hand, we shall attempt to motivate some of the more abstract aspects of our study and try to present heuristic and intuitive explanations of the important results, hoping that the reader will appreciate the need for mathematical precision while realizing that formal and rigorous developments are not devoid of physical meaning.

In Chap. 6 we studied networks containing both linear and nonlinear circuit elements. We have shown that the choice of a proper tree, the application of cut-set analysis, and the subsequent manipulation of the resultant equations lead to the vector differential equation satisfied by the state vector $\mathbf{x}(t)$ of the network. We recall that the state variables, the components of the state vector, are the inductor currents and capacitor voltages in the case of LTI networks; on the other hand, when we studied nonlinear time-invariant networks, we found it more convenient to choose the inductor flux linkages and the capacitor charges as the state variables. We also demonstrated that all independent source variables can be used to define the components of an input vector $\mathbf{u}(t)$. Thus, as long as the network is *lumped* and contains a finite number of elements, its state vector satisfies a vector differential equation of the form

$$\dot{\mathbf{x}}(t) = \mathbf{f}(\mathbf{x}(t), \mathbf{u}(t)) \tag{1}$$

where $\mathbf{x}(t)$ = state vector
  $\mathbf{u}(t)$ = input vector
  $\dot{\mathbf{x}}(t)$ = time derivative of state vector

Here $\mathbf{f}(.,.)$ is a vector-valued nonlinear function (or mapping) defined by the network topology and the specific characteristics of the network elements. Equation (1) is a nonlinear vector differential equation.

When the network elements were LTI resistors, capacitors, and inductors, we demonstrated that the network state vector $\mathbf{x}(t)$ satisfies an equation of the form

$$\dot{\mathbf{x}}(t) = \mathbf{A}\mathbf{x}(t) + \mathbf{B}\mathbf{u}(t) \tag{2}$$

where the matrices $\mathbf{A}$ and $\mathbf{B}$ have constant elements. Equation (2) is also a vector differential equation; it is a special case of Eq. (1) obtained when the mapping $\mathbf{f}(.,.)$ has the specific form

$$\mathbf{f}(\mathbf{x}(t), \mathbf{u}(t)) = \mathbf{A}\mathbf{x}(t) + \mathbf{B}\mathbf{u}(t) \tag{3}$$

Equation (2) is called an LTI vector differential equation. It is called time-invariant because the matrices **A** and **B** are constant, i.e., not time-varying, matrices.† It is called a linear differential equation because the mapping represented by the right-hand side of Eq. (2) satisfies the superposition principle;‡ we shall elaborate further on this topic in Secs. 7.4 and 7.7.

Our main objective in this chapter is to determine methods of solution of vector differential equations. Explicit solutions will be derived for the LTI differential equation (2). It is impossible to obtain analytical solutions for the nonlinear differential equation (1). For this reason, we shall investigate the conditions which guarantee the existence and uniqueness of solutions of such differential equations so that we can confidently apply the numerical integration algorithms discussed in Chap. 9.

The reader has been exposed to one class of physical system, namely, electric networks, whose behavior is described by differential equations of the form suggested by Eq. (1) or (2). For this reason, we shall often employ networks to enhance the physical intuition of the reader by translating many of the mathematical properties of the equations into the corresponding network concepts. We shall also employ continuous-time information-processing systems of the analog-computer variety; for this purpose such systems are constructed by ideal integrators, summers, and memoryless function generators.

It is important, however, that the reader be aware of additional classes of physical systems whose behavior is also described by vector differential equations. Although some of the physical properties of such systems may not be familiar to the reader, it is worth presenting a very brief explanation of the physical variables which form the components of the state vector $\mathbf{x}(t)$ and the components of the input vector $\mathbf{u}(t)$.

Mechanical systems Such systems are often characterized by interconnecting various masses, springs, and dashpots (which represent friction forces). In such systems, the state variables often used correspond to the position and velocities of the various masses. The inputs represent the action of several external forces. The precise nature of the differential equations is deduced primarily through Newton's laws of motion. In the same category fall several types of torsional springs and dashpots. In this case, the angular displacements and angular velocities are used as state variables; the applied torques represent the inputs to such systems.

Aerospace-vehicle orientation systems In such systems one is interested in analyz-

---

† This is because all network elements (excluding the sources) are linear and time-invariant.
‡ This is because KCL and KVL yield linear relations and because the branch constraints defined by the linear network elements are also linear equations.

ing the orientation of a rigid body† in some reference coordinate system, e.g., an inertial coordinate system, as a function of the applied torques. In such systems, the state variables often correspond to roll, pitch, and yaw angles and roll rate, pitch rate, and yaw rate. In the absence of any atmosphere, e.g., for a satellite, the inputs correspond to torques generated by gas jets or small rockets, reaction wheels, and gyros. When the vehicle operates in the atmosphere, e.g., an airplane, the inputs are generated by torques produced by the several aerodynamic control surfaces, such as elevators, ailerons, etc.

Aerospace-vehicle trajectory systems   Typical systems in this category are aircraft and missiles. Their motion in a reference coordinate system is again described by vector differential equations. Typical state variables are the three components of their position vector, the three components of their velocity (or momentum) vector, and often the vehicle's mass. Typical input variables are the generated thrust, the angles of the thrust vector with respect to the centerline of the vehicle, the angle of attack, and the lift forces generated by the various aerodynamic control surfaces.

Chemical process systems   In such systems several reagents are combined to yield a new product after a chemical reaction. Such reactions may take place in a single vessel or in a set of vessels. Typical state variables for such problems are concentration of the various products in the different vessels. Typical inputs in such problems are the temperature and concentration of the incoming reagents and the pressure in the various vessels. Such problems are usually described by extremely nonlinear differential equations.

7.2   ABSTRACT VISUALIZATION OF CONTINUOUS-TIME SYSTEMS

Since we are already familiar with at least one class of physical system that gives rise to vector differential equations, i.e., electric networks, it is worthwhile to examine at this point an alternate viewpoint that stresses the mathematical representation. This approach is the exact analog of the discrete-time case presented in Sec. 4.2.

The first step is to visualize a continuous-time dynamical system as a "black box," as illustrated in Fig. 7.2.1. We assume that the inputs to the system, denoted by $u_1(t), u_2(t), \ldots, u_m(t)$, are piecewise continuous time functions, and they form the components of the *input vector* $\mathbf{u}(t)$. The system outputs (or responses) are denoted by $y_1(t), y_2(t), \ldots, y_r(t)$, and they are also piecewise continuous time functions; they define the components of the *output vector* $\mathbf{y}(t)$.

---

† The *rigid-body assumption* in mechanics plays the same role as the *lumped-network assumption* in circuits.

FIGURE 7.2.1
Input-output representation of a continuous-time dynamical system.

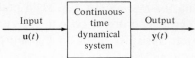

As with discrete-time systems, in a dynamic continuous-time system the instantaneous value of the output vector $\mathbf{y}(t)$ in general depends on past values of the input $\mathbf{u}(\tau)$, where τ represents values of past times, $t_0 \leq \tau \leq t$; t_0 is thought of as an initial time, and t is thought of as the present time. For example, consider the motion of a system, where the velocity $v(t)$ is viewed as the input and the position $z(t)$ is viewed as the output. Clearly the present position depends on the integral of the velocity and hence on the past values of the input to this system.

Next, let us examine in an abstract way the internal composition of the dynamical system depicted in Fig. 7.2.1. Basically, we imagine† that the internal structure of the dynamical system is made up of two subsystems, a memoryless subsystem and a memory system, as illustrated in Fig. 7.2.2.

7.2.1 Description of the Memoryless Subsystem

The *memoryless system* has two sets of inputs:

1 The actual input $\mathbf{u}(t)$ to the dynamical system (shown in Fig. 7.2.1).
2 Another vector, denoted by $\mathbf{x}(t)$, called the *state vector*; $\mathbf{x}(t)$ is internal to the system and is not reflected in its input-output representation.

† The same concept was expounded in "Basic Concepts," Chap. 2, for simple systems.

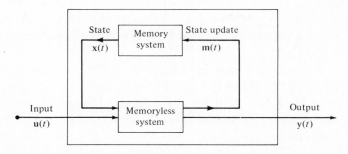

FIGURE 7.2.2
Internal conceptual subdivision of a continuous-time dynamical system.

The memoryless system has also two distinct sets of outputs:

1 The actual output vector $\mathbf{y}(t)$ of the dynamical system (shown in Fig. 7.2.1).

2 Another vector, denoted by $\mathbf{m}(t)$, called the *state-update vector*; $\mathbf{m}(t)$ is also internal to the system and is not reflected in its input-output representation.

From a mathematical point of view we can specify the memoryless subsystem by relating its outputs, $\mathbf{y}(t)$ and $\mathbf{m}(t)$, to its inputs, $\mathbf{u}(t)$ and $\mathbf{x}(t)$.

To be precise let us assume that at each instant of time t

$$\mathbf{u}(t) \in R_m \qquad \text{an } m\text{-dimensional vector}$$

$$\mathbf{x}(t) \in R_n \qquad \text{an } n\text{-dimensional vector}$$

$$\mathbf{m}(t) \in R_n \qquad \text{an } n\text{-dimensional vector}$$

$$\mathbf{y}(t) \in R_r \qquad \text{an } r\text{-dimensional vector}$$

By memoryless we mean that given, at any time t, the instantaneous values of $\mathbf{u}(t)$ and $\mathbf{x}(t)$, we can calculate the instantaneous values of $\mathbf{m}(t)$ and $\mathbf{y}(t)$. This property can be modeled by defining two vector-valued functions $\mathbf{f}(.,.)$ and $\mathbf{g}(.,.)$ such that

$$\mathbf{m}(t) = \mathbf{f}(\mathbf{x}(t), \mathbf{u}(t)) \tag{1}$$

$$\mathbf{y}(t) = \mathbf{g}(\mathbf{x}(t), \mathbf{u}(t)) \tag{2}$$

7.2.2 Description of the Memory Subsystem

As illustrated in Fig. 7.2.2, the memory subsystem provides the interrelationship between the two internal sets of variables:

1 The state vector $\mathbf{x}(t)$

2 The state-update vector $\mathbf{m}(t)$

Note that the state-update vector $\mathbf{m}(t)$ acts as the input to the memory subsystem and that the state vector $\mathbf{x}(t)$ acts as the output of the memory subsystem.

The term memory subsystem means that the value of its output, i.e., the value of the state vector $\mathbf{x}(t)$, at any time t, depends in general upon past values of its input, i.e., values of $\mathbf{m}(\tau)$, $\tau \leq t$.

Many types of memory relations can be used to define the memory subsystem in continuous-time systems.† We shall use a specific one, so that the overall dynamical system is described by a set of ordinary differential equations. To be specific, then, we assume that the *memory subsystem* is defined by

---

† For example, we may have combinations of delays, that is, $\mathbf{x}(t) = \mathbf{m}(t - 1) + \mathbf{m}(t - 3.9)$.

$$\mathbf{x}(t) = \mathbf{x}(t_0) + \int_{t_0}^{t} \mathbf{m}(\tau)\, d\tau \tag{3}$$

where $t \geq t_0$, t_0 is some initial time, and $\mathbf{x}(t_0)$ is called the *initial-state vector*. Clearly, this specific memory relationship arises from the fact that the current value $\mathbf{x}(t)$ of the state vector is related to the definite integral of the state-update vector.

We can readily obtain an alternate form of Eq. (3) by taking time derivatives

$$\frac{d}{dt}\mathbf{x}(t) = \frac{d}{dt}\int_{t_0}^{t} \mathbf{m}(\tau)\, d\tau = \mathbf{m}(t) \qquad \mathbf{x}(t_0) = \text{known} \tag{4}$$

and so an alternate way of describing the memory subsystem is to state that it constrains the time rate of change $\dot{\mathbf{x}}(t)$ of the state vector to equal the state-update vector $\mathbf{m}(t)$.

7.2.3 Complete Description of the Dynamical System

From the interconnections of Fig. 7.2.2, the description of the memoryless subsystem provided by Eqs. (1) and (2), and the description of the memory subsystem provided by Eq. (4) we can readily arrive [by substituting Eq. (1) into Eq. (4)] at the so-called state-variable description of a lumped dynamical system

$$\boxed{\dot{\mathbf{x}}(t) = \mathbf{f}(\mathbf{x}(t), \mathbf{u}(t)) \qquad \mathbf{x}(t_0)\ \text{known}} \tag{5}$$

$$\boxed{\mathbf{y}(t) = \mathbf{g}(\mathbf{x}(t), \mathbf{u}(t))} \tag{6}$$

It is implicit from the above representation that given

1 The value of the state vector at any time t_0
2 The time history of the input $\mathbf{u}(\tau)$ over an interval of time $t_0 \leq \tau \leq t$

then

1 Solution of the vector differential equation (5) will yield the state $\mathbf{x}(t)$.
2 Substitution of $\mathbf{x}(t)$ found above and of $\mathbf{u}(t)$ into the algebraic equation (6) will yield the value of the output vector $\mathbf{y}(t)$.

7.3 SOME ADDITIONAL EXAMPLES OF CONTINUOUS-TIME DYNAMIC SYSTEMS

In this section we consider several physical systems and derive the differential equations satisfied by their state variables. Our purpose is to motivate the need for a general treatment of vector differential equations since many different physical

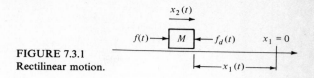

FIGURE 7.3.1
Rectilinear motion.

systems are described by the same type of differential equation. These examples, together with those of Chap. 6, should illustrate the structure of linear and nonlinear differential equations.

7.3.1 Linear Motion

Consider the motion of a very simple mechanical system. Let us imagine a mass M moving under the action of a force $f(t)$, as shown in Fig. 7.3.1. We let

$$x_1(t) = \text{position of mass at time } t \text{ as measured from reference point}$$
$$x_1 = 0$$
$$x_2(t) = \text{velocity of mass at time } t$$
$$f(t) = \text{force applied at time } t$$
$$f_d(t) = -ax_2(t) = \text{drag force due to environment}$$

Here we assume that the drag force is proportional to the velocity $x_2(t)$. The relationship between the position $x_1(t)$ and the velocity $x_2(t)$ is, by definition,

$$\frac{d}{dt}x_1(t) = x_2(t) \tag{1}$$

From Newton's law we also have

$$M\frac{d}{dt}x_2(t) = -ax_2(t) + f(t) \tag{2}$$

We shall now show that Eqs. (1) and (2) can be written in the form $\dot{\mathbf{x}}(t) = \mathbf{A}\mathbf{x}(t) + \mathbf{B}\mathbf{u}(t)$.

We observe that Eqs. (1) and (2) can be written

$$\underbrace{\frac{d}{dt}\begin{bmatrix} x_1(t) \\ x_2(t) \end{bmatrix}}_{\dot{\mathbf{x}}(t)} = \underbrace{\begin{bmatrix} 0 & 1 \\ 0 & -\dfrac{a}{M} \end{bmatrix}}_{\mathbf{A}}\underbrace{\begin{bmatrix} x_1(t) \\ x_2(t) \end{bmatrix}}_{\mathbf{x}(t)} + \underbrace{\begin{bmatrix} 0 \\ \dfrac{1}{M} \end{bmatrix}}_{\mathbf{B}}\underbrace{f(t)}_{\mathbf{u}(t)} \tag{3}$$

which is evidently a special case of the linear vector differential equation $\dot{\mathbf{x}}(t) = \mathbf{A}\mathbf{x}(t) + \mathbf{B}\mathbf{u}(t)$.

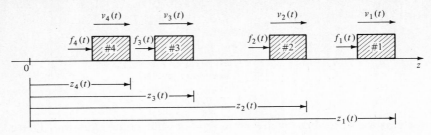

FIGURE 7.3.2
Single-lane motion of four vehicles.

The state variables for this system are the position $x_1(t)$ and the velocity $x_2(t)$, the two components of the state vector $\mathbf{x}(t)$. The input is the applied force $f(t)$.

7.3.2 A Transportation System

To illustrate the occurrence of multivariable systems in some transportation problems, let us consider the following situation. Suppose that we have four vehicles 1, 2, 3, and 4 moving from left to right in a single-lane level guideway, as illustrated in Fig. 7.3.2. Let 0 denote some reference position on the guideway. Define, for $i = 1, 2, 3, 4$,

$z_i(t) = $ distance of ith vehicle from the reference point 0
$v_i(t) = $ velocity of ith vehicle
$\quad m_i = $ mass of ith vehicle
$f_i(t) = $ force applied to ith vehicle

Let us assume that each vehicle obeys the equations of motion described in Sec. 7.3.1, that is,

$$\dot{z}_i(t) = v_i(t) \tag{4}$$

$$\dot{v}_i(t) = -\frac{a_i}{m_i}v_i(t) + \frac{1}{m_i}f_i(t) \tag{5}$$

where a_i denotes the friction coefficient associated with the ith vehicle.

Let us suppose that the objective of a transportation engineer would be to

1 Keep adjacent vehicles separated by a predetermined headway (distance) h^*

2 Keep the velocity of each vehicle close to a desired constant velocity v^*

Thus, he is interested in adjusting the applied forces $f_i(t)$ so that these objectives are met.

Since we are interested in controlling the headway and velocity errors, let us define the following error variables (for $i = 1, 2, 3, 4$)

$$y_{i,i+1}(t) \triangleq z_i(t) - z_{i+1}(t) - h^* \tag{6}$$

$$w_i(t) \triangleq v_i(t) - v^* \tag{7}$$

Thus, $y_{12}(t)$ denotes the headway error between vehicles 1 and 2 at time t [note that if $y_{12}(t) > 0$, the vehicles are too far apart; if $y_{12}(t) < 0$, the vehicles are too close]. Similarly $w_1(t)$ denotes the velocity error of the first vehicle.

Next we derive the dynamical equations of the error variables. By differentiating Eq. (7) and noting that the desired velocity v^* is constant we obtain, in view of Eqs. (5) and (7).

$$
\begin{aligned}
\dot{w}_i(t) = \dot{v}_i(t) &= -\frac{a_i}{m_i} v_i(t) + \frac{1}{m_i} f_i(t) \\
&= -\frac{a_i}{m_i} w_i(t) + \frac{a_i}{m_i} v^* + \frac{1}{m_i} f_i(t)
\end{aligned}
\tag{8}
$$

Now let

$$f_i(t) \triangleq f_i^* + g_i(t) \tag{9}$$

where

$$f_i^* = a_i v^* \tag{10}$$

The physical significance of f_i^* is that of the constant force required to just balance the drag force $a_i v^*$ experienced by the ith vehicle at the constant desired velocity v^*. The quantity $g_i(t)$ then denotes the incremental thrust, positive or negative, which can be used to accelerate or decelerate the ith vehicle. If we now substitute Eqs. (9) and (10) into Eq. (8), we obtain the differential equation

$$\dot{w}_i(t) = -\frac{a_i}{m_i} w_i(t) + \frac{1}{m_i} g_i(t) \tag{11}$$

which provides us with the dynamics of the velocity error $w_i(t)$.

To deduce the dynamics of the headway errors we differentiate Eq. (6), noting that the desired headway h^* is constant, and use Eq. (4) to obtain

$$\dot{y}_{i,i+1}(t) = \dot{z}_i(t) - \dot{z}_{i+1}(t) = v_i(t) - v_{i+1}(t) \tag{12}$$

Substitution of Eq. (7) into (12) yields

$$\dot{y}_{i,i+1}(t) = w_i(t) - w_{i+1}(t) \tag{13}$$

Let us now write the full implications of Eqs. (11) and (13) for all four vehicles:

$$\dot{w}_1(t) = -\frac{a_1}{m_1}w_1(t) + \frac{1}{m_1}g_1(t) \qquad \dot{y}_{12}(t) = w_1(t) - w_2(t)$$

$$\dot{w}_2(t) = -\frac{a_2}{m_2}w_2(t) + \frac{1}{m_2}g_2(t) \qquad \dot{y}_{23}(t) = w_2(t) - w_3(t)$$

$$\dot{w}_3(t) = -\frac{a_3}{m_3}w_3(t) + \frac{1}{m_3}g_3(t) \qquad \dot{y}_{34}(t) = w_3(t) - w_4(t) \qquad (14)$$

$$\dot{w}_4(t) = -\frac{a_4}{m_4}w_4(t) + \frac{1}{m_4}g_4(t)$$

We can now write Eq. (14) in vector form as

$$\frac{d}{dt}\underbrace{\begin{bmatrix} w_1(t) \\ y_{12}(t) \\ w_2(t) \\ y_{23}(t) \\ w_3(t) \\ y_{34}(t) \\ w_4(t) \end{bmatrix}}_{\dot{\mathbf{x}}(t)} = \underbrace{\begin{bmatrix} -\dfrac{a_1}{m_1} & 0 & 0 & 0 & 0 & 0 & 0 \\ 1 & 0 & -1 & 0 & 0 & 0 & 0 \\ 0 & 0 & -\dfrac{a_2}{m_2} & 0 & 0 & 0 & 0 \\ 0 & 0 & 1 & 0 & -1 & 0 & 0 \\ 0 & 0 & 0 & 0 & -\dfrac{a_3}{m_3} & 0 & 0 \\ 0 & 0 & 0 & 0 & 1 & 0 & -1 \\ 0 & 0 & 0 & 0 & 0 & 0 & -\dfrac{a_4}{m_4} \end{bmatrix}}_{\mathbf{A}} \underbrace{\begin{bmatrix} w_1(t) \\ y_{12}(t) \\ w_2(t) \\ y_{23}(t) \\ w_3(t) \\ y_{34}(t) \\ w_4(t) \end{bmatrix}}_{\mathbf{x}(t)}$$

$$+ \underbrace{\begin{bmatrix} \dfrac{1}{m_1} & 0 & 0 & 0 \\ 0 & 0 & 0 & 0 \\ 0 & \dfrac{1}{m_2} & 0 & 0 \\ 0 & 0 & 0 & 0 \\ 0 & 0 & \dfrac{1}{m_3} & 0 \\ 0 & 0 & 0 & 0 \\ 0 & 0 & 0 & \dfrac{1}{m_4} \end{bmatrix}}_{\mathbf{B}(t)} \cdot \underbrace{\begin{bmatrix} g_1(t) \\ g_2(t) \\ g_3(t) \\ g_4(t) \end{bmatrix}}_{\mathbf{u}(t)} \qquad (15)$$

which is clearly of the form $\dot{\mathbf{x}} = \mathbf{A}\mathbf{x} + \mathbf{B}\mathbf{u}$. The components of the state vector $\mathbf{x}$ are the interlaced velocity and headway errors, and the components of the input vector $\mathbf{u}$ are the incremental accelerating and decelerating forces.

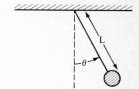

FIGURE 7.3.3
A simple pendulum.

This example illustrates that the choice of the state variables may well be dictated by the problem; in this case, we chose the velocity errors and headway errors.

It should be obvious that we can increase the number of state and input variables simply by considering more vehicles. In general, for a string of M vehicles the dimension of the state vector $\mathbf{x}(t)$ is $2M - 1$, the dimension of the input vector $\mathbf{u}(t)$ is M, $\mathbf{A}$ is a $(2M - 1) \times (2M - 1)$ constant matrix, and $\mathbf{B}$ is a $(2M - 1) \times M$ matrix.

7.3.3 Mechanical Systems

We now consider two different types of mechanical systems which are modeled by nonlinear differential equations.

The pendulum As a first example, consider the pendulum shown in Fig. 7.3.3. We assume for the sake of simplicity that the pendulum moves in a frictionless environment (vacuum), where θ is the displacement angle, ω is the rate of change of θ, m is the mass, and g is the constant of gravity. By the definition of angular velocity we have

$$\frac{d\theta(t)}{dt} = \omega(t) \tag{16}$$

From Newton's law we obtain

$$m\frac{d\omega(t)}{dt} = \frac{mg}{L} \sin \theta(t) \tag{17}$$

Thus, the pendulum is described by the system of the two first-order differential equations

$$\frac{d\theta(t)}{dt} = \omega(t) \qquad \frac{d\omega(t)}{dt} = \frac{g}{L} \sin \theta(t) \tag{18}$$

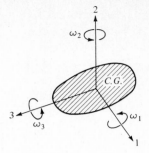

FIGURE 7.3.4
Definition of variables for tumbling satellite.

This system is nonlinear because of the $\sin \theta$ dependence. Note that if θ is small, then $\sin \theta \approx \theta$; and in this case (18) reduces to a set of linear differential equations.

For this example, the state vector $\mathbf{x} = \begin{bmatrix} \theta \\ \omega \end{bmatrix}$ is 2-dimensional and satisfies a vector differential equation of the form $\dot{\mathbf{x}}(t) = \mathbf{f}(\mathbf{x}(t))$. Clearly

$$\mathbf{f}(\mathbf{x}(t)) = \begin{bmatrix} f_1(\mathbf{x}(t)) \\ f_2(\mathbf{x}(t)) \end{bmatrix} = \begin{bmatrix} \omega(t) \\ -\dfrac{g}{L} \sin \theta(t) \end{bmatrix} \qquad (19)$$

A Satellite As a second example let us consider the equations of motion of an asymmetrical satellite in space (Fig. 7.3.4). We define three body axes (labeled 1, 2, and 3) through the center of gravity. We let J_1, J_2, and J_3 denote the principal moments of inertia about axes 1, 2, and 3, respectively. We let ω_1, ω_2, and ω_3 denote the angular velocities about the three axes measured with respect to the body. We suppose that a reaction-wheel mechanism can provide torques T_1, T_2, T_3 about the three axes.

For this satellite the angular velocities ω_1, ω_2, ω_3 satisfy the so-called Euler equations†

$$J_1 \dot{\omega}_1 = (J_2 - J_3)\omega_2 \omega_3 + T_1$$
$$J_2 \dot{\omega}_2 = (J_3 - J_1)\omega_3 \omega_1 + T_2 \qquad (20)$$
$$J_3 \dot{\omega}_3 = (J_1 - J_2)\omega_1 \omega_2 + T_3$$

or

$$\dot{\omega}_1 = \frac{J_2 - J_3}{J_1} \omega_2 \omega_3 + \frac{T_1}{J_1}$$
$$\dot{\omega}_2 = \frac{J_3 - J_1}{J_2} \omega_3 \omega_1 + \frac{T_2}{J_2} \qquad (21)$$
$$\dot{\omega}_3 = \frac{J_1 - J_2}{J_3} \omega_1 \omega_2 + \frac{T_3}{J_3}$$

---

† See, for example, H. Goldstein, "Classical Mechanics," chap. 5, Addison-Wesley, Reading, Mass., 1959.

The system (21) is nonlinear due to the products $\omega_i \omega_j$ that appear in the equations (gyroscopic coupling).

In this example we can define the state vector $\mathbf{x}(t)$ to be the column 3-vector of the angular velocities, i.e.,

$$\mathbf{x}(t) = \begin{bmatrix} \omega_1(t) \\ \omega_2(t) \\ \omega_3(t) \end{bmatrix}, \tag{22}$$

and the input vector $\mathbf{u}(t)$ to be the torque vector

$$\mathbf{u}(t) = \begin{bmatrix} T_1(t) \\ T_2(t) \\ T_3(t) \end{bmatrix} \tag{23}$$

The state vector satisfies a differential equation of the form

$$\dot{\mathbf{x}}(t) = \mathbf{f}(\mathbf{x}(t)) + \mathbf{B}\mathbf{u}(t) \tag{24}$$

where

$$\mathbf{B} = \begin{bmatrix} \dfrac{1}{J_1} & 0 & 0 \\ 0 & \dfrac{1}{J_2} & 0 \\ 0 & 0 & \dfrac{1}{J_3} \end{bmatrix} \tag{25}$$

and

$$\mathbf{f}(\mathbf{x}(t)) = \begin{bmatrix} \dfrac{J_2 - J_3}{J_1} \omega_2(t)\omega_3(t) \\ \dfrac{J_3 - J_1}{J_2} \omega_3(t)\omega_1(t) \\ \dfrac{J_1 - J_2}{J_3} \omega_1(t)\omega_2(t) \end{bmatrix} \tag{26}$$

7.4 ECOLOGICAL SYSTEMS AND POPULATION DYNAMICS

The recent worldwide interest in ecology and pollution has encouraged us to present some mathematical models of population dynamics. These models are extremely interesting since they are described by systems of nonlinear differential equations, whose solutions can exhibit unstable and oscillatory behavior.

In this section we present a brief introduction to the types of mathematical ecological models used. We urge the reader to consult Ref. 1 for an excellent discussion of these and more complex models; see also Refs. 2 and 3.

7.4.1 Single-Species Dynamics

The simplest population equation for a single species whose population at time t is denoted by $x(t)$ is

$$\dot{x}(t) = ax(t) \tag{1}$$

where a is in general a positive constant. Basically, this equation states that the rate of growth is proportional to the population. The solution of Eq. (1) is

$$x(t) = e^{a(t-t_0)}x(t_0) \tag{2}$$

and since a is positive, the population undergoes an exponential growth.

Although exponential growth is realistic for small populations and over a limited amount of time, uninhibited growth does not occur in nature. Overcrowding and limited resources result in a saturation effect, which means that $x(t)$ must eventually reach a constant level. Equation (1) clearly does not do so. A modified mathematical model that does have this property is the so-called *Verlhust-Pearl logistic equation*

$$\dot{x}(t) = ax(t) - bx^2(t) \tag{3}$$

where a and b are positive constants. For small $x(t)$, the quadratic term $x^2(t)$ is negligible compared with $x(t)$, and hence exponential growth occurs. However, as $x(t)$ increases, the quadratic term in Eq. (3) slows down the growth.

Although Eq. (3) is a nonlinear differential equation, it is possible to obtain an analytical solution. If $x(0) > 0$ is known, the solution of (3) is given by

$$x(t) = \frac{ax(0)e^{at}}{a + bx(0)(e^{at} - 1)} \tag{4}$$

Note that as $t \to \infty$,

$$\lim_{t \to \infty} x(t) = \frac{a}{b} \tag{5}$$

which shows that the population eventually reaches a constant level.

7.4.2 Population Growth of Competing Species†

In the above development, it is assumed that the population dynamics of one species has a monopoly on the resources available. This is seldom the case. In general, two

---

† The discrete-time version of this problem was considered in Sec. 4.7.4.

or more coexistent species may compete for the same food and other natural resources. Thus it is reasonable to expect that the growth of each population may be further inhibited by the presence and magnitude of the other population. This effect leads to coupling between individual populations.

We start with the case of two competing species and then we generalize the equations to the case of n species. Suppose that $x_1(t)$ and $x_2(t)$ denote the populations of two species competing for the same resources. Then the following system of nonlinear equations has been used to model the population changes:

$$
\begin{aligned}
\dot{x}_1(t) &= a_1 x_1(t) - b_{11} x_1{}^2(t) - b_{12} x_1(t) x_2(t) \\
\dot{x}_2(t) &= a_2 x_2(t) - b_{22} x_2{}^2(t) - b_{21} x_2(t) x_1(t)
\end{aligned}
\tag{6}
$$

Note that in the absence of the other population, Eq. (6) reduces to the Verlhust-Pearl logistic equation (3) for the remaining species. Thus, if $x_2(t) = 0$, a_1/b_{11} represents the saturation level for the $x_1(t)$ population; if $x_1(t) = 0$, then a_2/b_{22} represents the saturation level for the $x_2(t)$ population.

This mathematical model can be readily generalized to the case of n species. Let $x_i(t)$, $i = 1, 2, \ldots, n$, denote the respective populations of n species competing for the same resources. Then the following system of simultaneous nonlinear differential equations can be used to describe the changes in population:

$$
\begin{aligned}
\dot{x}_1(t) &= a_1 x_1(t) - b_{11} x_1{}^2(t) - b_{12} x_1(t) x_2(t) - \cdots - b_{1n} x_1(t) x_n(t) \\
\dot{x}_2(t) &= a_2 x_2(t) - b_{22} x_2{}^2(t) - b_{21} x_2(t) x_1(t) - \cdots - b_{2n} x_2(t) x_n(t) \\
&\vdots \\
\dot{x}_n(t) &= a_n x_n(t) - b_{nn} x_n{}^2(t) - b_{n1} x_n(t) x_1(t) - \cdots - b_{n,n-1} x_n(t) x_{n-1}(t)
\end{aligned}
\tag{7}
$$

or in more compact form

$$
\dot{x}_i(t) = a_i x_i(t) - \sum_{j=1}^{n} b_{ij} x_i(t) x_j(t) \qquad i = 1, 2, \ldots, n
\tag{8}
$$

Clearly Eq. (8) is of the form

$$
\dot{\mathbf{x}}(t) = \mathbf{f}(\mathbf{x}(t))
\tag{9}
$$

where the state vector $\mathbf{x}(t)$ is simply the vector of the individual populations $x_i(t)$.

In general it is impossible to obtain analytical closed-form solutions for Eq. (6) or (7). One must use numerical integration, discussed in Chap. 9. Not all of the b_{ij}'s in Eq. (8) have to be positive. For example, if the population $x_\alpha(t)$ *does not* use the same resources as population $x_\beta(t)$, then

$$
b_{\alpha\beta} = b_{\beta\alpha} = 0
$$

However, their growth or decay is still coupled because of the other populations in the ecosystem.

7.4.3 Predator-Prey Population Dynamics

Often the food required for the survival of a particular species is another species in the ecosystem. The same type of behavior arises in the host-parasite situation. The simplest equations that can be used to describe this interchange are the so-called *Lotka-Volterra equations* (see Ref. 2). Let us suppose that $x_1(t)$ is the population of the prey and $x_2(t)$ is the population of the predator. The Lotka-Volterra equations are of the form

$$\dot{x}_1(t) = a_1 x_1(t) - b_1 x_1(t) x_2(t)$$
$$\dot{x}_2(t) = -a_2 x_2(t) + b_2 x_2(t) x_1(t) \tag{10}$$

with a_1, a_2, b_1, b_2 positive. This equation means that in the absence of prey $[x_1(t) = 0]$ the predator population will die in an exponential manner due to lack of food. In the absence of predators $[x_2(t) = 0]$, the prey population would undergo an exponential growth.

Of course, these dynamics can be further complicated in several ways. For example, one may wish to model the saturation effect on the prey population. In this case, Eq. (10) would look like

$$\dot{x}_1(t) = a_1 x_1(t) - c_1 x_1^2(t) - b_1 x_1(t) x_2(t)$$
$$\dot{x}_2(t) = -a_2 x_2(t) + b_2 x_2(t) x_1(t) \tag{11}$$

where c_1/a_1 is the prey saturation level in the absence of predators.

It is easy to see that one can construct the dynamics of complex ecosystems. For example, consider n species $i = 1, 2, \ldots, n$. Suppose that species 2 eats species 1, that species 3 eats species 2, and so on. In this case, one would obtain the following equations (not modeling saturation effects):

$$\dot{x}_1(t) = a_1 x_1(t) - b_1 x_1(t) x_2(t)$$
$$\dot{x}_2(t) = -a_2 x_2(t) + c_2 x_2(t) x_1(t) - b_2 x_2(t) x_3(t)$$
$$\dot{x}_3(t) = -a_3 x_3(t) + c_3 x_3(t) x_2(t) - b_3 x_3(t) x_4(t) \tag{12}$$
$$\vdots$$
$$\dot{x}_n(t) = -a_n x_n(t) + c_n x_n(t) x_{n-1}(t)$$

where all the constants are positive. The solution of such equations can exhibit extremely complex and interesting cyclic phenomena.

The above equations do not explicitly contain any external inputs. These usually can be introduced in several ways. For example, let us suppose that in Eq. (12) man harvests (or kills) a certain amount of one or more of the species in the ecosystem; at the same time he may intentionally breed and introduce in the ecosystem certain species. To model this let $u_i(t)$ denote the rate at which man introduces $[u_i(t) > 0]$ or removes $[u_i(t) < 0]$ species of the ith population. In this case, Eq. (12) would be modified to

$$\dot{x}_1(t) = a_1 x_1(t) - b_1 x_1(t) x_2(t) + u_1(t)$$
$$\dot{x}_2(t) = -a_2 x_2(t) + c_2 x_2(t) x_1(t) - b_2 x_2(t) x_3(t) + u_2(t)$$
$$\vdots$$
$$\dot{x}_n(t) = -a_n x_n(t) + c_n x_n(t) x_{n-1}(t) + u_n(t)$$

$$(13)$$

Other actions by man may be introduced indirectly by affecting the birth and death rates. For example, let us suppose that $u(t)$ denotes the amount of a certain chemical introduced in the ecosystem at time t. Suppose that only the basic growth rate a_1 of population 1 is instantaneously affected. It is not unreasonable to imagine that the growth rate a_1 decreases with increasing $u(t)$. Then we can write in Eq. (12)

$$a_1 = f(u(t))$$

and the time history of the chemical release $u(t)$ will have an influence in all population levels in this sequential prey-predator ecosystem.

7.5 BASIC MATHEMATICAL DEFINITIONS

We start our discussion of vector differential equations by presenting our notation and some fundamental definitions. The variable t is a real scalar-valued variable, often called the *independent variable*. In most problems of interest t has the units of *time*; however, in an abstract sense the units of t are immaterial. Thus, in many physical problems the units of t may be those of pressure, velocity, or some other physical quantity. We shall refer to t as the time variable since this corresponds to its physical meaning in the examples considered in Secs. 7.3 and 7.4 and Chap. 6.

Let R_n denote the n-dimensional euclidean space, and let R_m denote the m-dimensional euclidean space. For simplicity let us suppose that we have chosen the natural basis in both these spaces.

Throughout our discussion $\mathbf{x}(t)$ for each value of t will denote a time-varying vector in the n-dimensional euclidean space, i.e.,

$$\mathbf{x}(t) \in R_n$$

$$(1)$$

The components of $\mathbf{x}(t)$ with respect to the natural basis in R_n (or any other basis we may have chosen) are used to define the column n-vector

$$\mathbf{x}(t) = \begin{bmatrix} x_1(t) \\ x_2(t) \\ \vdots \\ x_n(t) \end{bmatrix} \qquad (2)$$

We shall often call $\mathbf{x}(t)$ the *state vector* to remind the reader of its physical significance.

Throughout our discussion $\mathbf{u}(t)$ will denote a time-varying vector in the m-dimensional euclidean space, i.e.,

$$\mathbf{u}(t) \in R_m \qquad \text{for each value of } t \qquad (3)$$

The components of $\mathbf{u}(t)$ with respect to the natural basis in R_m are used to define the column m-vector

$$\mathbf{u}(t) = \begin{bmatrix} u_1(t) \\ u_2(t) \\ \vdots \\ u_m(t) \end{bmatrix} \qquad (4)$$

We shall often call $\mathbf{u}(t)$ the *input vector*.

7.5.1   Linear Vector Differential Equations

Throughout this discussion $\mathbf{A}$ will denote an $n \times n$ time-invariant matrix with real constant elements $a_{ij}(i,j = 1, 2, \ldots, n)$. We shall often refer to $\mathbf{A}$ as the *system matrix*. In addition, $\mathbf{B}$ will denote an $n \times m$ time-invariant matrix with real constant elements $b_{ik}(i = 1, 2, \ldots, n; k = 1, 2, \ldots, m)$. We shall often refer to $\mathbf{B}$ as the *input-gain matrix*.

Definition 1   *The relation*

$$\dot{\mathbf{x}}(t) = \mathbf{A}\mathbf{x}(t) + \mathbf{B}\mathbf{u}(t) \qquad \dot{\mathbf{x}}(t) = \frac{d}{dt}\mathbf{x}(t) \qquad (5)$$

is called a forced LTI ordinary vector differential equation of the nth order.

Under our definitions, Eq. (5) is identical to the set of n first-order differential equations

$$\dot{x}_i(t) = \sum_{j=1}^{n} a_{ij} x_j(t) + \sum_{k=1}^{m} b_{ik} u_k(t) \qquad i = 1, 2, \ldots, n \qquad (6)$$

It should be clear that the derivative $\dot{\mathbf{x}}(t)$ of the state vector $\mathbf{x}(t)$ is also a vector in R_n, that is

$$\dot{\mathbf{x}}(t) \in R_n \tag{7}$$

Definition 2   *If in Eq. (5),* $\mathbf{u}(t) = \mathbf{0}$ *(for all t), then the resulting differential equation*

$$\dot{\mathbf{x}}(t) = \mathbf{A}\mathbf{x}(t) \tag{8}$$

is called unforced *or* homogeneous.

As we shall see later, the complete solution of (5) is made up of two parts; one involves terms arising from the homogeneous equation (8), and the other involves terms due to the input.

7.5.2   Boundary Conditions

We now discuss the notion of boundary conditions to a differential equation. Consider a differential equation of the form $\dot{\mathbf{x}} = \mathbf{f}(\mathbf{x}, \mathbf{u})$. Let t_0 denote some particular value of the time (independent variable) t. Let us suppose that at $t = t_0$ we know the value of the state vector, i.e., that

$$\boldsymbol{\xi} \triangleq \mathbf{x}(t_0) \qquad \text{is known} \tag{9}$$

It follows then that knowledge of the input $\mathbf{u}(t_0)$ defines the value of the derivative vector $\dot{\mathbf{x}}(t_0)$ because it is specified by the mapping $\mathbf{f}(.,.)$, that is,

$$\dot{\mathbf{x}}(t_0) = \mathbf{f}(\mathbf{x}(t_0), \mathbf{u}(t_0)) \tag{10}$$

Thus, we have a point $\mathbf{x}(t_0) \in R_n$ and a "direction" specified by $\dot{\mathbf{x}}(t_0)$. This information can be used to deduce the value of the state at $t = t_0 + \Delta t$, where Δt is small. The reason is that, for small Δt,

$$\mathbf{x}(t_0 + \Delta t) \approx \mathbf{x}(t_0) + \mathbf{f}(\mathbf{x}(t_0), \mathbf{u}(t_0)) \Delta t \tag{11}$$

This process can be repeated by starting at $t = t_0 + \Delta t$ to determine the motion of the state $\mathbf{x}(t)$ in R_n as t varies.

Thus, in an intuitive way the boundary condition or initial condition (9) provides a starting point $\boldsymbol{\xi} \in R_n$ and a starting time for the solution. For this reason, we shall always talk about the solution of a vector differential equation with a given boundary or initial condition.

7.5.3 Solution to a Differential Equation

Definition 3 *Consider the differential equation and boundary condition*

$$\dot{\mathbf{x}}(t) = \mathbf{f}(\mathbf{x}(t), \mathbf{u}(t)) \qquad \mathbf{x}(t_0) \triangleq \boldsymbol{\xi} \tag{12}$$

Suppose that $\mathbf{u}(t)$ is known for all $t \in [t_1, t_2]$. Suppose that $t_0 \in [t_1, t_2]$. Then, a vector-valued function $\boldsymbol{\phi}(t)$, such that $\boldsymbol{\phi}(t) \in R_n$ for each $t \in [t_1, t_2]$ is said to be a solution of (12), if

(a) $\quad \boldsymbol{\phi}(t_0) = \boldsymbol{\xi}$ *and* $\tag{13}$

(b) $\quad \dot{\boldsymbol{\phi}}(t) = \mathbf{f}(\boldsymbol{\phi}(t), \mathbf{u}(t))$ *for all $t \in [t_1, t_2]$.* $\tag{14}$

7.5.4 Existence and Uniqueness of Solutions

Our basic concern in this chapter is to obtain solutions (analytical if possible) to differential equations. In addition, we are interested in the properties of such solutions. Before we can attack the issue of solving a set of differential equations, we must investigate certain fundamental questions:

1 Under what conditions can we guarantee the existence of a solution? Is it possible to have a solution function $\boldsymbol{\phi}(t)$ blow up, i.e., go to infinity, in a finite time?

2 If a solution exists, is it unique? In other words, is it possible that a given differential equation with a given boundary condition has more than one solution according to Definition 3?

These fundamental issues of *existence* and *uniqueness* cannot be ignored. They are natural questions that must be answered. Indeed, it requires no mathematical sophistication to find examples of this type. For example, there exist *no real* solutions to the algebraic equation

$$x^2 + 1 = 0$$

On the other hand, the equation

$$x^2 - 1 = 0$$

has two real solutions; i.e., it does not have a unique solution.

If relatively innocent looking algebraic equations present problems with regard to existence and uniqueness, there is no reason to expect that more complicated mathematical ones, such as differential equations, would tend to behave nicely. Yet, the majority of introductory network and system courses ignore this question, the reason, as we shall prove in the sequel, being that *LTI differential equations do have unique solutions for all time.* As one may suspect, however, nonlinear differential equations present problems.

To illustrate these ideas consider the linear but time-varying differential equation

$$\frac{dx(t)}{dt} = \frac{1}{1-t} \qquad x(0) = 1 \tag{15}$$

As long as $t < 1$, this equation has a unique solution, as can be verified by direct integration

$$x(t) = x(0) + \int_0^t \frac{d\tau}{1-\tau} \qquad t < 1 \tag{16}$$

Hence, for $t < 1$, the time function

$$\phi(t) = 1 + \ln \frac{1}{1-t} \tag{17}$$

is the unique solution of (15) according to Definition 3. Note, however, that $\phi(t) \to +\infty$ as $t \to 1$; hence the solution cannot be extended past the time $t = 1$. Thus, the solution ceases to exist at $t = 1$.

As an example of nonunique solutions, consider the first-order nonlinear differential equation

$$\frac{dx}{dt} = \sqrt{|x|} \qquad x(0) = 0 \tag{18}$$

We claim that the time function $\phi_1(t)$

$$\phi_1(t) = 0 \qquad \text{for all } t \geq 0 \tag{19}$$

is a solution of (18), because $\phi_1(0) = 0$ and $\dot{\phi}_1(t) = \sqrt{|\phi_1(t)|}$. We also claim that the time function

$$\phi_2(t) = \tfrac{1}{4} t^2 \qquad \text{for all } t \geq 0 \tag{20}$$

is also a solution of (18) because $\phi_2(0) = 0$; and since $\dot{\phi}_2(t) = \tfrac{1}{2} t$, it is easy to see that $\dot{\phi}_2(t) = \sqrt{|\phi_2(t)|}$ holds. Thus, the nonlinear differential equation (17) does not have a unique solution.

7.5.5 Discussion

It should be evident from this discussion that questions of existence and uniqueness are important from a mathematical point of view. Such questions are also important from an engineering point of view.

Differential equations arise when one wishes to use mathematical relations to model a physical process, e.g., a network. In modeling dynamical processes, the differential equations provide us with the cause-and-effect relationship between several interrelated physical quantities. Our physical experience indicates that physical processes do not admit nonunique solutions, nor do they fail to have solutions. For this reason, one may seriously question the need for investigating the existence and uniqueness of solutions to differential equations which arise from the modeling of, say, electric networks. The argument may be roughly as follows: I know that physical processes admit unique responses for any given inputs. It therefore follows that the differential equations which I have written for this physical process must also have unique solutions, and so I do not have to worry about this topic. This seemingly plausible argument is nonetheless false. An engineer who writes the equations describing a real system or network always makes some idealizations. For example, he may have ignored the stray capacitance and resistance of an inductor. Therefore, since he can never be absolutely sure whether the mathematical relations represent a faithful model of the physical process, he *cannot* dismiss the issues of existence and uniqueness using physical reasoning.

One can carry this line of thought further. Suppose that an engineer has modeled a well-behaved physical network by a set of differential equations. Furthermore, suppose that these differential equations do not have a solution or have nonunique solutions. Then, since the physical network is well behaved but the differential equations are not, one can conclude that the mathematical model is *not* an accurate representation of the physical network. To say this in a different way: *a necessary condition that a set of differential equations represents faithfully a well-behaved physical dynamical process is that these equations do have unique solutions.*

On the other hand, the fact that a set of differential equations misbehaves, e.g., solutions go to infinity in finite time, may imply that the physical process or network will get in trouble. For example, transistors in a network may burn up or boilers in a chemical process may blow up. Thus, there is an intimate connection between pathological behavior of differential equations and bad design of a physical system.

With these remarks in mind, we are ready to start our detailed examination of classes of differential equations, their properties, and solutions.

7.6   THE SOLUTION OF THE LTI UNFORCED VECTOR DIFFERENTIAL EQUATION $\dot{\mathbf{x}}(t) = \mathbf{A}\mathbf{x}(t)$

If we consider a network containing LTI resistors, capacitors, and inductors but *no* independent sources, we know that the state variables, say $x_1(t)$, $x_2(t)$, $\ldots$, $x_n(t)$, satisfy a set of first-order differential equations of the form

$$\dot{x}_1(t) = a_{11} x_1(t) + a_{12} x_2(t) + \cdots + a_{1n} x_n(t)$$
$$\dot{x}_2(t) = a_{21} x_1(t) + a_{22} x_2(t) + \cdots + a_{2n} x_n(t)$$
$$\vdots \tag{1}$$
$$\dot{x}_n(t) = a_{n1} x_1(t) + a_{n2} x_2(t) + \cdots + a_{nn} x_n(t)$$

Obviously, Eq. (1) can be written

$$\boxed{\dot{\mathbf{x}}(t) = \mathbf{A}\mathbf{x}(t)} \tag{2}$$

where the $x_i(t)$ are the elements of the vector $\mathbf{x}(t)$ and the a_{ij} are the real constant elements of the matrix $\mathbf{A}$.

In this section we concentrate on the solution of the unforced vector differential equation (2). We shall assume that at $t = 0$ we know the value (initial condition) of $\mathbf{x}(0)$. So we suppose that the vector $\boldsymbol{\xi}$

$$\boldsymbol{\xi} \triangleq \mathbf{x}(0) \tag{3}$$

is known.

Let us now state precisely what we mean by the solution of

$$\dot{\mathbf{x}}(t) = \mathbf{A}\mathbf{x}(t) \qquad \mathbf{x}(0) = \boldsymbol{\xi} \tag{4}$$

We are looking for a time-varying vector $\boldsymbol{\phi}(t)$ such that

1   It satisfies the boundary conditions, i.e.,

$$\boldsymbol{\phi}(0) = \boldsymbol{\xi} \tag{5}$$

2   It satisfies the differential equation, i.e.,

$$\dot{\boldsymbol{\phi}}(t) = \mathbf{A}\boldsymbol{\phi}(t) \qquad \text{for all } t \tag{6}$$

We are immediately faced with the following questions:

1   Does a solution exist?
2   If a solution exists, is it unique?
3   If a unique solution exists, what is it?

7.6.1 The Matrix Exponential

Before proceeding with the answers to these questions let us recall that if we deal with the first-order differential equation

$$\dot{x}(t) = ax(t) \qquad x(0) = \xi \tag{7}$$

then a solution exists, is unique, and is given by

$$\phi(t) = e^{at}\xi \tag{8}$$

Let us also recall that one of the possible definitions of the exponential function, say e^z is by means of the infinite series

$$e^z \triangleq 1 + z + \frac{1}{2!}z^2 + \frac{1}{3!}z^3 + \cdots = \sum_{k=0}^{\infty} \frac{1}{k!}z^k \tag{9}$$

so that

$$e^{at} = 1 + at + \frac{1}{2!}(at)^2 + \frac{1}{3!}(at)^3 + \cdots = \sum_{k=0}^{\infty} \frac{1}{k!}a^k t^k \tag{10}$$

In a completely analogous manner we *define* the matrix exponential function $e^{\mathbf{Z}}$, where $\mathbf{Z}$ is a $n \times n$ (square) matrix, by[†]

$$\boxed{e^{\mathbf{Z}} \triangleq \mathbf{I} + \mathbf{Z} + \frac{1}{2!}\mathbf{Z}^2 + \frac{1}{3!}\mathbf{Z}^3 + \cdots = \sum_{k=0}^{\infty} \frac{1}{k!}\mathbf{Z}^k} \tag{11}$$

7.6.2 Existence of Solutions to $\dot{\mathbf{x}}(t) = \mathbf{A}\mathbf{x}(t)$; $\mathbf{x}(0) = \xi$

We shall show that Eq. (4) has a solution by using Picard's method of successive approximations.[‡] First we note that Eq. (4) is the same as the (vector) integral equation

$$\mathbf{x}(t) = \xi + \int_0^t \mathbf{A}\mathbf{x}(\tau)\,d\tau \tag{12}$$

[†] Note that $\mathbf{Z}^2 \triangleq \mathbf{ZZ}$, $\mathbf{Z}^3 \triangleq \mathbf{ZZZ}$, and so on.

[‡] Conceptually, this method is analogous to Picard's method for solving algebraic equations $\mathbf{x} = \mathbf{f}(\mathbf{x})$ discussed in Chap. 3. Here the same technique is used to solve integral equations.

Since **A** is a constant matrix, Eq. (12) can also be written

$$\mathbf{x}(t) = \boldsymbol{\xi} + \mathbf{A} \int_0^t \mathbf{x}(\tau)d\tau \tag{13}$$

We are seeking the solution to the integral equation (13); that is, we seek a function $\boldsymbol{\phi}(t)$ such that

$$\boldsymbol{\phi}(t) = \boldsymbol{\xi} + \mathbf{A} \int_0^t \boldsymbol{\phi}(\tau)\,d\tau \qquad \tau \in [0,t] \tag{14}$$

[Note that $\boldsymbol{\phi}(0) = \boldsymbol{\xi}$ as required by Eq. (5).] Consider the sequence $\{\boldsymbol{\phi}_n(t)\}$ of vector-valued time functions $\boldsymbol{\phi}_0(t), \boldsymbol{\phi}_1(t), \ldots$ constructed as follows:

$$\boldsymbol{\phi}_0(t) \triangleq \boldsymbol{\xi} \tag{15}$$

$$\boldsymbol{\phi}_1(t) \triangleq \boldsymbol{\xi} + \mathbf{A} \int_0^t \boldsymbol{\phi}_0(\tau)\,d\tau \tag{16}$$

$$\boldsymbol{\phi}_2(t) \triangleq \boldsymbol{\xi} + \mathbf{A} \int_0^t \boldsymbol{\phi}_1(\tau)\,d\tau \tag{17}$$

$$\vdots$$

$$\boldsymbol{\phi}_n(t) \triangleq \boldsymbol{\xi} + \mathbf{A} \int_0^t \boldsymbol{\phi}_{n-1}(\tau)\,d\tau \tag{18}$$

$$\vdots$$

It is easy to see that

$$\boldsymbol{\phi}_0(t) = \boldsymbol{\xi} \tag{19}$$

$$\boldsymbol{\phi}_1(t) = \boldsymbol{\xi} + \mathbf{A} \int_0^t \boldsymbol{\xi}\,d\tau = (\mathbf{I} + \mathbf{A}t)\boldsymbol{\xi} \tag{20}$$

$$\boldsymbol{\phi}_2(t) = \boldsymbol{\xi} + \mathbf{A} \int_0^t (\mathbf{I} + \mathbf{A}t)\boldsymbol{\xi}\,d\tau = \left(\mathbf{I} + \mathbf{A}t + \frac{1}{2!}\mathbf{A}^2 t^2\right)\boldsymbol{\xi} \tag{21}$$

$$\vdots$$

$$\boldsymbol{\phi}_n(t) = \left(\sum_{k=0}^{n} \frac{1}{k!}\mathbf{A}^k t^k\right)\boldsymbol{\xi} \tag{22}$$

In view of Eq. (11), $\lim_{n\to\infty} \boldsymbol{\phi}_n(t)$ exists, and in fact

$$\lim_{n\to\infty} \boldsymbol{\phi}_n(t) = \left(\sum_{k=0}^{\infty} \frac{1}{k!}\mathbf{A}^k t^k\right)\boldsymbol{\xi} = e^{\mathbf{A}t}\boldsymbol{\xi} \tag{23}$$

We shall now show that $\boldsymbol{\phi}(t) = e^{\mathbf{A}t}\boldsymbol{\xi}$ satisfies Eq. (13). To see this we note that

$$\xi + A \int_0^t e^{A t} \xi \, d\tau = \left(I + A \int_0^t \sum_{k=0}^\infty \frac{1}{k!} A^k \tau^k \, d\tau \right) \xi$$

$$= \left(I + A \sum_{k=0}^\infty \frac{1}{k!} A^k \int_0^t \tau^k \, d\tau \right) \xi$$

$$= \left(I + \sum_{k=0}^\infty \frac{1}{k!} A^{k+1} \frac{1}{k+1} t^{k+1} \right) \xi$$

$$= \left[I + \sum_{k=0}^\infty \frac{1}{(k+1)!} A^{k+1} t^{k+1} \right] \xi$$

$$= \left(I + \sum_{\beta=1}^\infty \frac{1}{\beta!} A^\beta t^\beta \right) \xi$$

$$= \left(\sum_{\beta=0}^\infty \frac{1}{\beta!} A^\beta t^\beta \right) \xi = e^{A t} \xi$$

We now claim that

$$\phi(t) = e^{A t} \xi \tag{24}$$

is a solution to our differential equation. To see this we must show that Eqs. (5) and (6) hold. We note that

$$\phi(0) = e^{A 0} \xi = \left(I + A \cdot 0 + \frac{1}{2!} A^2 \cdot 0 + \cdots \right) \xi = I \xi = \xi \tag{25}$$

so that the boundary condition (5) is satisfied. Next, we compute the time derivative of the vector $\phi(t)$ to obtain

$$\frac{d}{dt} \phi(t) = \frac{d}{dt} (e^{A t} \xi)$$

$$= \frac{d}{dt} \left(I + A t + \frac{1}{2!} A^2 t^2 + \frac{1}{3!} A^3 t^3 + \cdots \right) \xi$$

$$= \left(0 + A + \frac{2}{2!} A^2 t + \frac{3}{3!} A^3 t^2 + \cdots \right) \xi \tag{26}$$

$$= A \left(I + A t + \frac{1}{2!} A^2 t^2 + \cdots \right) \xi$$

$$= A e^{A t} \xi$$

$$= A \phi(t)$$

which implies that Eq. (6) holds. Thus, we proved the following theorem.

Theorem 1 *The time function (vector-valued)*

$$\boldsymbol{\phi}(t) = e^{\mathbf{A}t}\boldsymbol{\xi} \tag{27}$$

is a solution to the unforced LTI vector differential equation

$$\dot{\mathbf{x}}(t) = \mathbf{A}\mathbf{x}(t) \qquad \mathbf{x}(0) = \boldsymbol{\xi} \tag{28}$$

where $e^{\mathbf{A}t}$ is the matrix exponential defined by

$$e^{\mathbf{A}t} \triangleq \sum_{k=0}^{\infty} \frac{1}{k!}\mathbf{A}^k t^k \tag{29}$$

7.6.3 Uniqueness of Solutions to $\dot{\mathbf{x}}(t) = \mathbf{A}\mathbf{x}(t)$; $\mathbf{x}(0) = \boldsymbol{\xi}$

We have shown above that $\boldsymbol{\phi}(t) = e^{\mathbf{A}t}\boldsymbol{\xi}$ is *a* solution of Eq. (4) or, equivalently, of (14). *We shall now show that $\boldsymbol{\phi}(t)$ is the unique solution.* In other words, we shall show that there are no other solutions. The proof will be by contradiction. We suppose that $\mathbf{z}(t)$ is yet another solution of Eq. (14) and that $\mathbf{z}(t)$ and $\boldsymbol{\phi}(t)$ are *distinct* [this means that there is a finite interval of time, say T, such that $\mathbf{z}(t) \neq \boldsymbol{\phi}(t)$ for $t \in T$].

The assumption that $\mathbf{z}(t)$ is also a solution of Eq. (14) implies that $\mathbf{z}(t)$ must satisfy the integral equation

$$\mathbf{z}(t) = \boldsymbol{\xi} + \mathbf{A}\int_0^t \mathbf{z}(\tau)\,d\tau \tag{30}$$

From Eqs. (30) and (16) we have

$$\mathbf{z}(t) - \boldsymbol{\phi}_1(t) = \mathbf{A}\int_0^t [\mathbf{z}(\tau) - \boldsymbol{\phi}_0(\tau)]\,d\tau \tag{31}$$

Taking norms,† and since $\boldsymbol{\phi}_0(\tau) = \boldsymbol{\xi}$, we have

$$
\begin{aligned}
\|\mathbf{z}(t) - \boldsymbol{\phi}_1(t)\| &= \left\|\mathbf{A}\int_0^t [\mathbf{z}(\tau) - \boldsymbol{\xi}]\,d\tau\right\| \\[2mm]
&\leq \|\mathbf{A}\| \cdot \left\|\int_0^t [\mathbf{z}(\tau) - \boldsymbol{\xi}]\,d\tau\right\| \\[2mm]
&\leq \|\mathbf{A}\| \int_0^t \|\mathbf{z}(\tau) - \boldsymbol{\xi}\|\,d\tau \\[2mm]
&\leq \|\mathbf{A}\| \int_0^t (\|\mathbf{z}(\tau)\| + \|\boldsymbol{\xi}\|)\,d\tau
\end{aligned}
\tag{32}
$$

Let

---

† Any norm can be used; see Appendix A, Sec. A.15

$$c = \max_{\tau \in [0,t]} \|\mathbf{z}(\tau)\| \tag{33}$$

Then

$$\|\mathbf{z}(t) - \boldsymbol{\phi}_1(t)\| \leq \|\mathbf{A}\| \int_0^t (c + \|\boldsymbol{\xi}\|) \, d\tau \tag{34}$$

Therefore,

$$\|\mathbf{z}(t) - \boldsymbol{\phi}_1(t)\| \leq \|\mathbf{A}\| (c + \|\boldsymbol{\xi}\|) t \tag{35}$$

Using similar arguments, we find that

$$\|\mathbf{z}(t) - \boldsymbol{\phi}_2(t)\| \leq \|\mathbf{A}\| \int_0^t \|\mathbf{z}(\tau) - \boldsymbol{\phi}_1(\tau)\| \, d\tau \tag{36}$$

From Eqs. (35) and (36) we deduce that

$$\|\mathbf{z}(t) - \boldsymbol{\phi}_2(t)\| \leq \|\mathbf{A}\| \int_0^t \|\mathbf{A}\| (c + \|\boldsymbol{\xi}\|) \tau \, d\tau \tag{37}$$

or

$$\|\mathbf{z}(t) - \boldsymbol{\phi}_2(t)\| \leq \frac{1}{2!} \|\mathbf{A}\|^2 (c + \|\boldsymbol{\xi}\|) t^2 \tag{38}$$

Using similar arguments, we establish that

$$\|\mathbf{z}(t) - \boldsymbol{\phi}_n(t)\| \leq \|\mathbf{A}\|^n (c + \|\boldsymbol{\xi}\|) \frac{t^n}{n!} \tag{39}$$

Therefore,

$$\lim_{n \to \infty} \|\mathbf{z}(t) - \boldsymbol{\phi}_n(t)\| \leq (c + \|\boldsymbol{\xi}\|) \lim_{n \to \infty} \frac{\|\mathbf{A}\|^n t^n}{n!} = 0 \tag{40}$$

Hence,

$$\|\mathbf{z}(t) - \boldsymbol{\phi}(t)\| \leq 0 \tag{41}$$

which implies that

$$\mathbf{z}(t) = \boldsymbol{\phi}(t) \tag{42}$$

contradicting our assumption that $\mathbf{z}(t)$ and $\boldsymbol{\phi}(t)$ are distinct. Therefore, $\boldsymbol{\phi}(t)$ is the unique solution.

7.6.4   Recapitulation

We have shown that the function $\boldsymbol{\phi}(t) = e^{\mathbf{A}t} \boldsymbol{\xi}$ is the unique solution of the vector differential equation $\dot{\mathbf{x}}(t) = \mathbf{A}\mathbf{x}(t)$ subject to the boundary condition $\mathbf{x}(0) = \boldsymbol{\xi}$. The same ideas can be used to prove the following theorem.

Theorem 2 *The solution of*

$$\dot{\mathbf{x}}(t) = \mathbf{A}\mathbf{x}(t) \qquad \mathbf{x}(t_0) = \boldsymbol{\xi} \tag{43}$$

exists, is unique, and is given by

$$\boldsymbol{\phi}(t) = e^{\mathbf{A}(t-t_0)}\boldsymbol{\xi} = \left[\sum_{k=0}^{\infty} \frac{1}{k!}\mathbf{A}^k(t-t_0)^k\right]\boldsymbol{\xi} \tag{44}$$

The matrix $e^{\mathbf{A}t}$ or $e^{\mathbf{A}(t-t_0)}$ is the *matrix exponential*. We discuss its evaluation in a following section. In the remainder of this section we state some of its properties, most of which follow from the definition of the matrix exponential by its infinite-series representation. We therefore leave it to the reader to prove that

$$e^{\mathbf{A}t}e^{\mathbf{A}\tau} = e^{\mathbf{A}(t+\tau)} \tag{45}$$

$$e^{(\mathbf{A}+\mathbf{B})t} = e^{\mathbf{A}t}e^{\mathbf{B}t} \qquad \text{for all } t \text{ if and only if } \mathbf{A}\mathbf{B} = \mathbf{B}\mathbf{A} \tag{46}$$

By setting $\tau = -t$ we obtain the very important result

$$e^{\mathbf{A}t}e^{-\mathbf{A}t} = \mathbf{I} \tag{47}$$

which means that $e^{\mathbf{A}t}$ is nonsingular and that its inverse is given by

$$(e^{\mathbf{A}t})^{-1} = e^{-\mathbf{A}t} \qquad \text{for all } t \tag{48}$$

We remark that $e^{\mathbf{A}t}$ is nonsingular even though $\mathbf{A}$ may be a singular matrix.

Again through the use of the infinite series we can show that

$$\frac{d}{dt}e^{\mathbf{A}t} = \mathbf{A}e^{\mathbf{A}t} = e^{\mathbf{A}t}\mathbf{A} \tag{49}$$

Finally, if $\mathbf{A}$ is a nonsingular matrix,

$$\int_{t_0}^{t} e^{\mathbf{A}\tau}\,d\tau = \mathbf{A}^{-1}(e^{\mathbf{A}t} - e^{\mathbf{A}t_0}) \tag{50}$$

7.6.5 Superposition Properties

We have seen that we can readily compute the solution of $\dot{\mathbf{x}}(t) = \mathbf{A}\mathbf{x}(t)$ for any given initial condition. However, the linearity of the equation has some interesting implications. For example, suppose we have computed the solution for a particular initial condition $\mathbf{x}(t_0) = \boldsymbol{\xi}$. Suppose now we double the initial condition; that is, $\mathbf{x}(t_0) = 2\boldsymbol{\xi}$. Do we have to recompute the solution, or can we write the solution by inspection?

The linearity of the differential equation allows us to develop the so-called *superposition with respect to initial-state* properties. We give the main result in the theorem below.

Theorem 3 (Superpositon property)   *Consider the linear system*

$$\dot{\mathbf{x}}(t) = \mathbf{A}\mathbf{x}(t) \tag{51}$$

Let $\boldsymbol{\gamma}_1(t)$ denote the solution when the initial condition is $\mathbf{x}(t_0) = \boldsymbol{\xi}_1$. Let $\boldsymbol{\gamma}_2(t)$ denote the solution when the initial condition is $\mathbf{x}(t_0) = \boldsymbol{\xi}_2$. Now consider the initial condition

$$\mathbf{x}(t_0) = \boldsymbol{\xi} = \alpha_1 \boldsymbol{\xi}_1 + \alpha_2 \boldsymbol{\xi}_2 \tag{52}$$

where α_1 and α_2 are arbitrary scalars; let $\boldsymbol{\gamma}(t)$ denote the solution when the initial condition is given by (52). Then

$$\boldsymbol{\gamma}(t) = \alpha_1 \boldsymbol{\gamma}_1(t) + \alpha_2 \boldsymbol{\gamma}_2(t) \qquad \text{for all } t \tag{53}$$

PROOF   Since $\boldsymbol{\gamma}_1(t)$ is a solution, then from Theorem 2 we have

$$\boldsymbol{\gamma}_1(t) = e^{\mathbf{A}(t-t_0)}\boldsymbol{\xi}_1 \tag{54}$$

Similarly

$$\boldsymbol{\gamma}_2(t) = e^{\mathbf{A}(t-t_0)}\boldsymbol{\xi}_2 \tag{55}$$

and

$$\boldsymbol{\gamma}(t) = e^{\mathbf{A}(t-t_0)}\boldsymbol{\xi} \tag{56}$$

Substituting Eq. (52) into (56), we obtain

$$\boldsymbol{\gamma}(t) = e^{\mathbf{A}(t-t_0)}\boldsymbol{\xi} = e^{\mathbf{A}(t-t_0)}(\alpha_1 \boldsymbol{\xi}_1 + \alpha_2 \boldsymbol{\xi}_2)$$

$$= \alpha_1 \underbrace{e^{\mathbf{A}(t-t_0)}\boldsymbol{\xi}_1}_{\boldsymbol{\gamma}_1(t)} + \alpha_2 \underbrace{e^{\mathbf{A}(t-t_0)}\boldsymbol{\xi}_2}_{\boldsymbol{\gamma}_2(t)} \tag{57}$$

and so in view of Eqs. (48) and (49) we conclude that

$$\gamma(t) = \alpha_1 \gamma_1(t) + \alpha_2 \gamma_2(t) \tag{58}$$

The use of the superposition principle allows us to save computation when we wish to compute solutions of the same equation for many initial conditions. To see further how superposition helps, let us suppose that we select the natural basis vectors $e_1, e_2, \ldots, e_n$ in R_n, that is,

$$e_1 \triangleq \begin{bmatrix} 1 \\ 0 \\ \vdots \\ 0 \end{bmatrix} \qquad e_2 \triangleq \begin{bmatrix} 0 \\ 1 \\ \vdots \\ 0 \end{bmatrix}, \ldots, e_n \triangleq \begin{bmatrix} 0 \\ 0 \\ \vdots \\ 1 \end{bmatrix} \tag{59}$$

Let $\psi_i(t)$ denote the solution of (51) when $x(t_0) = e_i$, that is,

$$\psi_i(t) = e^{A(t-t_0)} e_i \qquad i = 1, 2, \ldots, n \tag{60}$$

Now let ξ be an arbitrary vector in R_n; let us express ξ as a linear combination of the basis vectors

$$\xi = \sum_{i=1}^{n} \alpha_i e_i \tag{61}$$

where the α_i are the coordinates of ξ with respect to the natural basis coordinate system. Then the solution $\gamma(t)$ of (45) with initial condition

$$x(t_0) = \xi \tag{62}$$

is given, using the superposition theorem, by

$$\gamma(t) = \sum_{i=1}^{n} \alpha_i \psi_i(t) \tag{63}$$

7.6.6 The Time-shifting property

Another useful property of LTI differential equations is related to the problem of computing the solution of the differential equation $\dot{x}(t) = Ax(t)$ starting at the *same* initial state ξ but at *different* initial values of time. Once more we state this as a theorem.

Theorem 4 (Time-shifting property)   *Consider the LTI differential equation*

$$\dot{x}(t) = Ax(t) \tag{64}$$

Let $\gamma_1(t)$ denote the solution when $\mathbf{x}(t_1) = \xi$, and let $\gamma_2(t)$ denote the solution when $\mathbf{x}(t_2) = \xi$. Let

$$\tau \triangleq t_2 - t_1 \tag{65}$$

Then

$$\gamma_2(t) = \gamma_1(t - \tau) \tag{66}$$

PROOF   The solution $\gamma_1(t)$ is given by

$$\gamma_1(t) = e^{\mathbf{A}(t-t_1)}\xi \tag{67}$$

according to Theorem 2; $\gamma_2(t)$ is given by

$$\gamma_2(t) = e^{\mathbf{A}(t-t_2)}\xi \tag{68}$$

Substituting $t_2 = t_1 + \tau$ into Eq. (68), we obtain

$$\gamma_2(t) = e^{\mathbf{A}(t-t_1-\tau)}\xi \tag{69}$$

But from (67) we also have

$$\gamma_1(t - \tau) = e^{\mathbf{A}(t-\tau-t_1)}\xi \tag{70}$$

Since Eqs. (69) and (70) are identical,

$$\gamma_2(t) = \gamma_1(t - \tau) \tag{71}$$

7.7   THE EVALUATION OF THE MATRIX EXPONENTIAL $e^{\mathbf{A}t}$

We have seen (Sec. 7.6) that the solution of

$$\dot{\mathbf{x}}(t) = \mathbf{A}\mathbf{x}(t) \qquad \mathbf{x}(0) = \xi \tag{1}$$

is given by

$$\mathbf{x}(t) = e^{\mathbf{A}t}\xi \tag{2}$$

[Here, by abuse of notation we use the same vector $\mathbf{x}(t)$ both as the variable and the solution.] Thus, the problem reduces to determination of the matrix $e^{\mathbf{A}t}$.

7.7.1 Series Evaluation

One straightforward way is to evaluate the exponential matrix using a digital computer. This method simply approximates e^{At} by evaluating only the first, say, N terms in the series expansion. In other words, we use

$$e^{At} \approx I + At + \frac{1}{2!}A^2 t^2 + \cdots + \frac{1}{N!}A^N t^N \tag{3}$$

The larger the N, the better the approximation.

In a practical sense, it is difficult to predict the number of terms *that* one needs to calculate. From the viewpoint of digital computation it is best to let the computer decide when to stop. To do this one prespecifies an accuracy constant, $\epsilon > 0$, and if

$$\left\| \frac{1}{n!}A^n t^n \right\| < \epsilon$$

then stop: $n = N$; if not, let $n = n + 1$ and continue.

7.7.2 Evaluation via Eigenvalues and the Similarity Transformation

It is possible to evaluate the exponential matrix e^{At} analytically. This is easily demonstrated if the matrix A has distinct eigenvalues.

Suppose that A is a square matrix with distinct eigenvalues $\lambda_1, \lambda_2, \ldots, \lambda_n$. These may be real or complex.[†] We define the diagonal matrix Λ by

$$\Lambda = \begin{bmatrix} \lambda_1 & 0 & \cdots & 0 \\ 0 & \lambda_2 & \cdots & 0 \\ \vdots & \vdots & \vdots & \vdots \\ 0 & 0 & \cdots & \lambda_n \end{bmatrix} \tag{4}$$

We know that A and Λ have the same eigenvalues.[‡] Therefore, A and Λ are similar matrices. Hence, there is a nonsingular matrix P such that

$$\Lambda = P^{-1}AP \tag{5}$$

Now

$$e^A = I + A + \frac{1}{2!}A^2 + \cdots \tag{6}$$

$$e^\Lambda = I + \Lambda + \frac{1}{2!}\Lambda^2 + \cdots \tag{7}$$

[†] Recall that complex eigenvalues of real matrices appear in complex-conjugate pairs.
[‡] See Appendix A, Sec. A.8 to A.10.

We form the matrix $\mathbf{P}^{-1}e^{\mathbf{A}}\mathbf{P}$ to find

$$
\begin{aligned}
\mathbf{P}^{-1}e^{\mathbf{A}}\mathbf{P} &= \mathbf{P}^{-1}(\mathbf{I} + \mathbf{A} + \frac{1}{2!}\mathbf{A}^2 + \cdots)\mathbf{P} \\
&= \mathbf{I} + \mathbf{P}^{-1}\mathbf{A}\mathbf{P} + \frac{1}{2!}\mathbf{P}^{-1}\mathbf{A}^2\mathbf{P} + \cdots \\
&= \mathbf{I} + \mathbf{P}^{-1}\mathbf{A}\mathbf{P} + \frac{1}{2!}\mathbf{P}^{-1}\mathbf{A}\mathbf{P}\mathbf{P}^{-1}\mathbf{A}\mathbf{P} + \cdots \\
&= \mathbf{I} + \mathbf{\Lambda} + \frac{1}{2!}\mathbf{\Lambda}^2 + \cdots
\end{aligned}
\tag{8}
$$

From Eqs. (8) and (7) we conclude that

$$
\boxed{e^{\mathbf{\Lambda}} = \mathbf{P}^{-1}e^{\mathbf{A}}\mathbf{P}}
\tag{9}
$$

Therefore, the matrices $e^{\mathbf{\Lambda}}$ and $e^{\mathbf{A}}$ are similar.

Next we turn our attention to Eq. (7). Since $\mathbf{\Lambda}$ is a diagonal matrix, it is easy to see that $\mathbf{\Lambda}^k$ can be written immediately as

$$
\mathbf{\Lambda}^k = \begin{bmatrix} \lambda_1 & 0 & \cdots & 0 \\ 0 & \lambda_2 & \cdots & 0 \\ \vdots & \vdots & \vdots & \vdots \\ 0 & 0 & \cdots & \lambda_n \end{bmatrix}^k = \begin{bmatrix} \lambda_1^k & 0 & \cdots & 0 \\ 0 & \lambda_2^k & \cdots & 0 \\ \vdots & \vdots & \vdots & \vdots \\ 0 & 0 & \cdots & \lambda_n^k \end{bmatrix}
\tag{10}
$$

For this reason we can readily see that

$$
e^{\mathbf{\Lambda}} = \begin{bmatrix} \sum_{k=0}^{\infty} \frac{\lambda_1^k}{k!} & 0 & \cdots & 0 \\ 0 & \sum_{k=0}^{\infty} \frac{\lambda_2^k}{k!} & \cdots & 0 \\ \vdots & \vdots & \vdots & \vdots \\ 0 & 0 & \cdots & \sum_{k=0}^{\infty} \frac{\lambda_n^k}{k!} \end{bmatrix} = \begin{bmatrix} e^{\lambda_1} & 0 & \cdots & 0 \\ 0 & e^{\lambda_2} & \cdots & 0 \\ \vdots & \vdots & \vdots & \vdots \\ 0 & 0 & \cdots & e^{\lambda_n} \end{bmatrix}
\tag{11}
$$

We have demonstrated that the fact that $\mathbf{\Lambda}$ is a diagonal matrix implies that $e^{\mathbf{\Lambda}}$ is also diagonal. In fact, if $\lambda_1, \lambda_2, \ldots, \lambda_n$ are the eigenvalues of $\mathbf{\Lambda}$, then $e^{\lambda_1}, e^{\lambda_2}, \ldots, e^{\lambda_n}$ are the eigenvalues of the exponential matrix $e^{\mathbf{\Lambda}}$. Furthermore, in view of Eq. (9), we know that $e^{\mathbf{\Lambda}}$ and $e^{\mathbf{A}}$ are similar matrices. Therefore, the eigenvalues of $e^{\mathbf{A}}$ are $e^{\lambda_1}, e^{\lambda_2}, \ldots, e^{\lambda_n}$.

In view of these considerations it should be clear how one can evaluate the exponential matrix $e^{\mathbf{A}t}$ that governs the solution of the vector differential equation $\dot{\mathbf{x}}(t) = \mathbf{A}\mathbf{x}(t)$. The procedure is as follows.

STEP 1 Evaluate the eigenvalues $\lambda_1, \lambda_2, \ldots, \lambda_n$ of $\mathbf{A}$ by finding the roots of its characteristic polynomial

$$p(\lambda) = \det(\lambda \mathbf{I} - \mathbf{A}) \tag{12}$$

(We suppose that the eigenvalues are distinct.)

STEP 2 Form the diagonal matrix $\boldsymbol{\Lambda}$ of the eigenvalues [see Eq. (4)].

STEP 3 Evaluate the matrix $\mathbf{P}$ and its inverse $\mathbf{P}^{-1}$ such that

$$\boldsymbol{\Lambda} = \mathbf{P}^{-1}\mathbf{A}\mathbf{P} \quad \text{or} \quad \mathbf{A}\mathbf{P} = \mathbf{P}\boldsymbol{\Lambda} \tag{13}$$

STEP 4 Since the matrices $e^{\Lambda t}$ and e^{At} are similar, find e^{At} by the equation

$$e^{At} = \mathbf{P}e^{\Lambda t}\mathbf{P}^{-1} \tag{14}$$

where

$$e^{\Lambda t} = \begin{bmatrix} e^{\lambda_1 t} & 0 & \cdots & 0 \\ 0 & e^{\lambda_2 t} & \cdots & 0 \\ \vdots & \vdots & \vdots & \vdots \\ 0 & 0 & \cdots & e^{\lambda_n t} \end{bmatrix} \tag{15}$$

STEP 5 Then the (unique) solution of

$$\dot{\mathbf{x}}(t) = \mathbf{A}\mathbf{x}(t) \qquad \mathbf{x}(0) = \boldsymbol{\xi} \tag{16}$$

is given by

$$\mathbf{x}(t) = e^{At}\boldsymbol{\xi} = \mathbf{P}e^{\Lambda t}\mathbf{P}^{-1}\boldsymbol{\xi} \tag{17}$$

for all t.

This method of evaluating the solutions of differential equations illustrates how the eigenvalues of a matrix enter into the character of the solution of a differential equation. It should be clear now why it is important to know the meaning of the eigenvalues of a matrix and numerical means for their evaluation.

In general the elements of the matrix $\mathbf{P}e^{\Lambda t}\mathbf{P}^{-1}$ are linear combinations of $e^{\lambda_i t}$. Thus, the general form of the jth component of $\mathbf{x}(t)$ is

$$x_j(t) = \sum_{k=1}^{n} \beta_{jk} e^{\lambda_k t} \tag{18}$$

where the β_{jk} involve the elements of $\mathbf{P}$, $\mathbf{P}^{-1}$, and the initial condition vector $\boldsymbol{\xi}$.

7.7.3 Solution Using the Eigenvalues and the Eigenvectors of the A Matrix

Although Eq. (18) shows the explicit dependence of the solution on the eigenvalues of the **A** matrix, it masks some of the interesting geometric issues of the solution of $\dot{x}(t) = Ax(t)$. An alternate procedure clearly shows the role of the eigenvalues and the eigenvectors of the **A** matrix. This process is completely analogous to that carried out for the discrete-time systems in Sec. 4.5.

Once more let us assume that $\lambda_1, \lambda_2, \ldots, \lambda_n$ are the *assumed distinct* eigenvalues of the matrix **A**. Let $v_1, v_2, \ldots, v_n$ denote the corresponding eigenvectors. Remember that this means

$$Av_i = \lambda_i v_i \qquad i = 1, 2, \ldots, n \tag{19}$$

First let us suppose that the initial state of the system

$$\dot{x}(t) = Ax(t) \tag{20}$$

is one of the eigenvectors, i.e., suppose that

$$x(0) = v_i \tag{21}$$

Let us denote the solution of (20) with initial condition (21) by $x_i(t)$. Then

$$x_i(t) = e^{At} v_i \tag{22}$$

Using the infinite-series expansion for the matrix exponential, we obtain

$$x_i(t) = \left(I + At + \frac{1}{2!} A^2 t^2 + \frac{1}{3!} A^3 t^3 + \cdots \right) v_i \tag{23}$$

$$= v_i + tAv_i + \frac{1}{2!} t^2 A^2 v_i + \frac{1}{3!} t^3 A^3 v_i + \cdots$$

Now let us recall (see Theorem 15, Sec. A.10 of Appendix A) that if λ_i is an eigenvalue of **A** and v_i is the corresponding eigenvector, then

$$A^k v_i = (\lambda_i)^k v_i \tag{24}$$

Substituting Eq. (24) into Eq. (23), we see that

$$x_i(t) = v_i + t\lambda_i v_i + \frac{1}{2!} t^2 \lambda_i^2 v_i + \frac{1}{3!} t^3 \lambda_i^3 v_i + \cdots \tag{25}$$

or

$$x_i(t) = \sum_{k=0}^{\infty} \frac{\lambda_i^k t^k}{k!} v_i \tag{26}$$

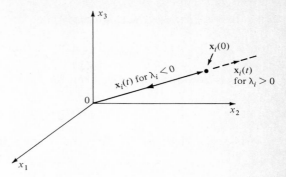

FIGURE 7.7.1
Geometric interpretation of the solution
$\mathbf{x}_i(t)$ of $\dot{\mathbf{x}}(t) = \mathbf{A}\mathbf{x}(t)$, $\mathbf{x}(0) = \mathbf{v}_i$.

But the infinite sum is simply $\exp(\lambda_i t)$; that is,

$$\sum_{k=0}^{\infty} \frac{\lambda_i^k t^k}{k!} = e^{\lambda_i t} \tag{27}$$

Hence,

$$\mathbf{x}_i(t) = e^{\lambda_i t}\mathbf{v}_i \tag{28}$$

The geometric interpretation of Eq. (28) is that if the initial state $\mathbf{x}(0)$ coincides with one of the eigenvectors, the solution remains for all time along the straight line defined by this eigenvector; also, the time variation is determined only by the associated eigenvalue λ_i.

Note that if λ_i is real and positive, $\mathbf{x}_i(t)$ will grow along the direction of $\mathbf{v}_i$. If λ_i is real and negative, $\mathbf{x}_i(t)$ will go toward the origin along the direction of $\mathbf{v}_i$. If λ_i is zero, $\mathbf{x}_i(t) = \mathbf{v}_i$ for all t; that is, the system stays at its initial condition. These geometric ideas are illustrated in Fig. 7.7.1.

Once we understand the nature of the solution of the differential equation when its initial state coincides with any one of the eigenvectors of the $\mathbf{A}$ matrix, we can use the superposition theorem to advantage.

If $\mathbf{A}$ has distinct eigenvalues, its eigenvectors $\mathbf{v}_i, \mathbf{v}_2, \ldots, \mathbf{v}_n$ form a linearly independent set and hence they can be used to define a (skewed) coordinate system in R_n (see Theorem 17, Sec. A.10 of Appendix A). This means that any vector $\boldsymbol{\xi}$ in R_n can be written as a linear combination of the eigenvectors, i.e.,

$$\boldsymbol{\xi} = \beta_1 \mathbf{v}_1 + \beta_2 \mathbf{v}_2 + \cdots + \beta_n \mathbf{v}_n \tag{29}$$

where the scalars β_i (real or complex) are the coordinates of $\boldsymbol{\xi}$ with respect to the coordinate system defined by the eigenvectors.

Let us consider the solution $\mathbf{x}(t)$ of

$$\dot{\mathbf{x}}(t) = \mathbf{A}\mathbf{x}(t) \qquad \mathbf{x}(0) = \boldsymbol{\xi} = \sum_{i=1}^{n} \beta_i \mathbf{v}_i \tag{30}$$

where $\boldsymbol{\xi}$ is now given by (29). Then by the superposition property we have

$$\mathbf{x}(t) = \sum_{i=1}^{n} \beta_i \mathbf{x}_i(t) \tag{31}$$

where $\mathbf{x}_i(t)$ is given by Eq. (28), and so the solution of (30) is

$$\boxed{\mathbf{x}(t) = \beta_1 e^{\lambda_1 t} \mathbf{v}_1 + \beta_2 e^{\lambda_2 t} \mathbf{v}_2 + \cdots + \beta_n e^{\lambda_n t} \mathbf{v}_n} \tag{32}$$

Equation (32) is identical to (17) except that it contains more information about the role of the eigenvectors of the $\mathbf{A}$ matrix.

7.8 AN EXAMPLE

To illustrate some of the ideas presented in Sec. 7.7 we shall solve a very simple example. Consider the system of the two first-order differential equations

$$\begin{aligned}
\dot{x}_1(t) &= 7x_1(t) - 2x_2(t) & x_1(0) &= 1 \\
\dot{x}_2(t) &= 15x_1(t) - 4x_2(t) & x_2(0) &= 1
\end{aligned} \tag{1}$$

We let

$$\mathbf{x}(t) = \begin{bmatrix} x_1(t) \\ x_2(t) \end{bmatrix} \qquad \mathbf{x}(0) = \boldsymbol{\xi} = \begin{bmatrix} 1 \\ 1 \end{bmatrix} \qquad \mathbf{A} = \begin{bmatrix} 7 & -2 \\ 15 & -4 \end{bmatrix} \tag{2}$$

so that Eq. (1) is in the standard form

$$\dot{\mathbf{x}}(t) = \mathbf{A}\mathbf{x}(t) \qquad \mathbf{x}(0) = \boldsymbol{\xi} \tag{3}$$

To find the solution of this vector differential equation we follow the step-by-step procedure of Sec. 7.7.2.

STEP 1 We want to determine the two eigenvalues λ_1, λ_2 of the matrix $\mathbf{A}$. The characteristic polynomial is

$$\begin{aligned}
p(\lambda) = \det(\lambda\mathbf{I} - \mathbf{A}) &= \det \begin{bmatrix} \lambda - 7 & 2 \\ -15 & \lambda + 4 \end{bmatrix} \\
&= (\lambda - 7)(\lambda + 4) + 30 = \lambda^2 - 3\lambda + 2 = (\lambda - 1)(\lambda - 2)
\end{aligned} \tag{4}$$

Thus the eigenvalues of **A** are

$$\lambda_1 = 1 \qquad \lambda_2 = 2 \tag{5}$$

STEP 2 The diagonal matrix of the eigenvalues Λ is

$$\Lambda \triangleq \begin{bmatrix} \lambda_1 & 0 \\ 0 & \lambda_2 \end{bmatrix} = \begin{bmatrix} 1 & 0 \\ 0 & 2 \end{bmatrix} \tag{6}$$

STEP 3 To find the matrix **P** we start from

$$\mathbf{AP} = \mathbf{P\Lambda} \tag{7}$$

which is

$$\begin{bmatrix} 7 & -2 \\ 15 & -4 \end{bmatrix}\begin{bmatrix} P_{11} & P_{12} \\ P_{21} & P_{22} \end{bmatrix} = \begin{bmatrix} P_{11} & P_{12} \\ P_{21} & P_{22} \end{bmatrix}\begin{bmatrix} 1 & 0 \\ 0 & 2 \end{bmatrix} \tag{8}$$

Equation (8) yields the simultaneous equations

$$7P_{11} - 2P_{21} = P_{11}$$
$$7P_{12} - 2P_{22} = 2P_{12}$$
$$15P_{11} - 4P_{21} = P_{21} \tag{9}$$
$$15P_{12} - 4P_{22} = 2P_{22}$$

or

$$6P_{11} - 2P_{21} = 0$$
$$15P_{11} - 5P_{21} = 0$$
$$5P_{12} - 2P_{22} = 0 \tag{10}$$
$$15P_{12} - 6P_{22} = 0$$

The first two equations are the same, and the last two equations are also the same.[†]
Thus, we have

$$3P_{11} - P_{21} = 0 \tag{11}$$
$$5P_{12} - 2P_{22} = 0 \tag{12}$$

[†] The fact that Eq. (10) does not have a unique solution should not surprise the reader. The reason is that if $\mathbf{AP} = \mathbf{P\Lambda}$, then $\mathbf{A}(\beta\mathbf{P}) = (\beta\mathbf{P})\Lambda$; so there is nonuniqueness in the elements of **P**.

and we can set

$$
\begin{array}{cc}
P_{11} = 1 & P_{21} = 3 \\
P_{12} = 2 & P_{22} = 5
\end{array}
\tag{13}
$$

so that we work with small integers. Thus

$$
\mathbf{P} = \begin{bmatrix} 1 & 2 \\ 3 & 5 \end{bmatrix}
\tag{14}
$$

Since det $\mathbf{P} = -1$

$$
\mathbf{P}^{-1} = \frac{1}{-1} \begin{bmatrix} 5 & -2 \\ -3 & 1 \end{bmatrix} = \begin{bmatrix} -5 & 2 \\ 3 & -1 \end{bmatrix}
\tag{15}
$$

STEP 4   The matrix $e^{\Lambda t}$ is given by

$$
e^{\Lambda t} = \begin{bmatrix} e^{\lambda_1 t} & 0 \\ 0 & e^{\lambda_2 t} \end{bmatrix} = \begin{bmatrix} e^t & 0 \\ 0 & e^{2t} \end{bmatrix}
\tag{16}
$$

The matrix $e^{\mathbf{A}t}$ is found from the equation $e^{\mathbf{A}t} = \mathbf{P} e^{\Lambda t} \mathbf{P}^{-1}$. Thus

$$
\begin{aligned}
e^{\mathbf{A}t} &= \begin{bmatrix} 1 & 2 \\ 3 & 5 \end{bmatrix} \begin{bmatrix} e^t & 0 \\ 0 & e^{2t} \end{bmatrix} \begin{bmatrix} -5 & 2 \\ 3 & -1 \end{bmatrix} \\
&= \begin{bmatrix} -5e^t + 6e^{2t} & 2e^t - 2e^{2t} \\ -15e^t + 15e^{2t} & 6e^t - 5e^{2t} \end{bmatrix}
\end{aligned}
\tag{17}
$$

STEP 5   The solution is $\mathbf{x}(t) = e^{\mathbf{A}t}\boldsymbol{\xi}$. Since $\boldsymbol{\xi} = \begin{bmatrix} 1 \\ 1 \end{bmatrix}$ we have (in terms of components) the solution

$$
\begin{aligned}
x_1(t) &= -3e^t + 4e^{2t} \\
x_2(t) &= -9e^t + 10e^{2t}
\end{aligned}
\tag{18}
$$

This completes the derivation of the analytical solution of Eq. (1).

Next we shall use the eigenvalue-eigenvector method of Sec. 7.7.2 explicitly. The eigenvector $\mathbf{v}_1$ associated with the eigenvalue $\lambda_1 = 1$ can be shown to be

$$
\mathbf{v}_1 = \begin{bmatrix} 1 \\ 3 \end{bmatrix}
\tag{19}
$$

The eigenvector $\mathbf{v}_2$ associated with the eigenvalue $\lambda_2 = 2$ can be shown to be

$$\mathbf{v}_2 = \begin{bmatrix} 2 \\ 5 \end{bmatrix} \tag{20}$$

Let us now write the intial state $\boldsymbol{\xi}$ as a linear combination of $\mathbf{v}_1$ and $\mathbf{v}_2$, that is,

$$\boldsymbol{\xi} = \beta_1 \mathbf{v}_1 + \beta_2 \mathbf{v}_2 \tag{21}$$

Hence,

$$\underbrace{\begin{bmatrix} 1 \\ 1 \end{bmatrix}}_{\boldsymbol{\xi}} = \beta_1 \underbrace{\begin{bmatrix} 1 \\ 3 \end{bmatrix}}_{\mathbf{v}_1} + \beta_2 \underbrace{\begin{bmatrix} 2 \\ 5 \end{bmatrix}}_{\mathbf{v}_2} \tag{22}$$

Therefore,

$$\begin{aligned} 1 &= \beta_1 + 2\beta_2 \\ 1 &= 3\beta_1 + 5\beta_2 \end{aligned} \tag{23}$$

from which we find

$$\beta_1 = -3 \qquad \beta_2 = 2 \tag{24}$$

Hence, the solution is

$$\begin{aligned} \mathbf{x}(t) &= \beta_1 e^{\lambda_1 t} \mathbf{v}_1 + \beta_2 e^{\lambda_2 t} \mathbf{v}_2 \\ &= -3e^t \begin{bmatrix} 1 \\ 3 \end{bmatrix} + 2e^{2t} \begin{bmatrix} 2 \\ 5 \end{bmatrix} \end{aligned} \tag{25}$$

Hence,

$$\begin{aligned} x_1(t) &= -3e^t + 4e^{2t} \\ x_2(t) &= -9e^t + 10e^{2t} \end{aligned} \tag{26}$$

Of course, (26) is identical to (18).

7.9 AN INTERESTING EXAMPLE OF LINEAR SYSTEMS: BEAT GENERATION

Many examples of physical systems consist of almost identical components coupled weakly to each other. This link between otherwise similar systems serves as a bridge which can transfer energy from one system to another again and again.

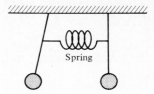

FIGURE 7.9.1
Two pendulums coupled by a spring.

The most intuitive example is the case of two pendulums, otherwise identical, which are interconnected by a very "soft" spring as shown in Fig. 7.9.1. If such a device is constructed and the left pendulum is initially displaced one observes that the amplitude of the oscillations of the left pendulum slowly decrease while the right pendulum starts oscillating at an increasing amplitude. Eventually, motion in the left pendulum almost ceases, while the right pendulum exhibits full oscillation. Next, the reverse action takes place; i.e., the energy from the right pendulum is transferred to the left, and so on.

This is a common phenomenon. For example, in propeller aircraft passengers often perceive a slowly varying increase and then decrease of vibration frequency, due to the coupling, through the air frame, of vibrations between the almost identical propellers.

In this section we present the electrical counterpart to this phenomenon. Such a system consists of two identical LC networks connected by a relatively large capacitor. The specific network is shown in Fig. 7.9.2. Following the material presented in Chap. 6, we define the state variables as follows:

$x_1(t)$ = current through left inductor
$x_2(t)$ = current through right inductor
$x_3(t)$ = voltage across middle capacitor
$x_4(t)$ = voltage across left capacitor
$x_5(t)$ = voltage across right capacitor

The differential equations are

$$\dot{x}_1(t) = x_3(t) - x_4(t)$$
$$\dot{x}_2(t) = x_3(t) - x_5(t)$$
$$20\dot{x}_3(t) = -x_1(t) - x_2(t) \tag{1}$$
$$\dot{x}_4(t) = x_1(t)$$
$$\dot{x}_5(t) = x_2(t)$$

In vector form Eq. (1) becomes

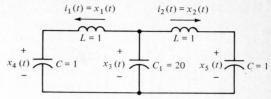

FIGURE 7.9.2
An *LC* network that exhibits beat generation. The values of the capacitors are in farads and of the inductors in henrys.

$$
\frac{d}{dt}
\begin{bmatrix}
x_1(t) \\
x_2(t) \\
x_3(t) \\
x_4(t) \\
x_5(t)
\end{bmatrix}
=
\begin{bmatrix}
0 & 0 & 1 & -1 & 0 \\
0 & 0 & 1 & 0 & -1 \\
-\frac{1}{20} & -\frac{1}{20} & 0 & 0 & 0 \\
1 & 0 & 0 & 0 & 0 \\
0 & 1 & 0 & 0 & 0
\end{bmatrix}
\begin{bmatrix}
x_1(t) \\
x_2(t) \\
x_3(t) \\
x_4(t) \\
x_5(t)
\end{bmatrix}
\qquad (2)
$$

$$\underbrace{\dot{\mathbf{x}}(t)} \qquad \underbrace{\mathbf{A}} \qquad \underbrace{\mathbf{x}(t)}$$

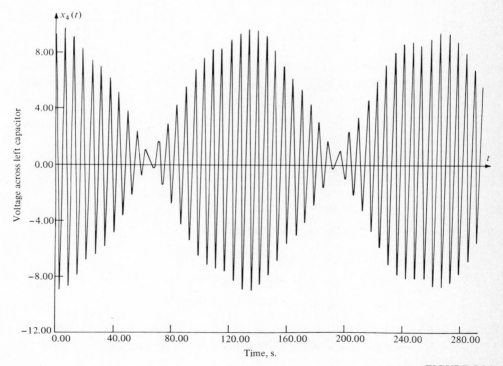

FIGURE 7.9.3
$x_4(t)$ vs. time.

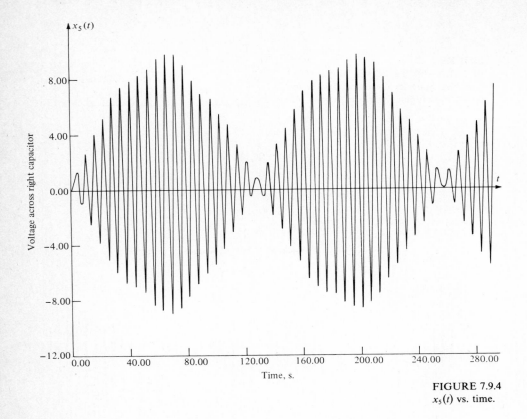

FIGURE 7.9.4
$x_5(t)$ vs. time.

For this example, the eigenvalues of the **A** matrix are found (from a computer program) at

$$\lambda_1 = 0$$
$$\lambda_2 = j0.999996185302734$$
$$\lambda_3 = -j0.999996185302734 \qquad\qquad (3)$$
$$\lambda_4 = j1.048796653747559$$
$$\lambda_5 = -j1.048796653747559$$

Note that they all lie on the imaginary $j\omega$ axis in the complex plane. The presence of the 20-F capacitor has slightly shifted the natural frequencies ($= 1$ rad/s) of each individual LC network.

Figures 7.9.3 to 7.9.5 show the response of some of the state variables when the initial condition is a 10-V initial voltage across the left capacitor. Thus, the initial state vector is

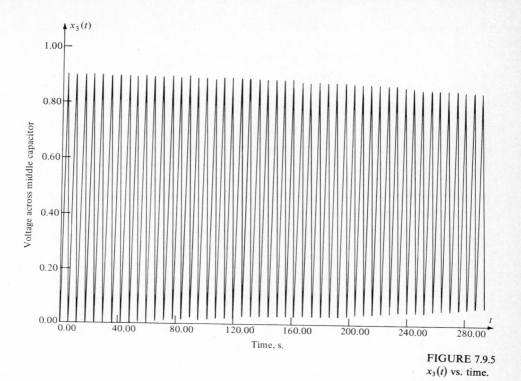

FIGURE 7.9.5
$x_3(t)$ vs. time.

$$\mathbf{x}(0) = \begin{bmatrix} 0 \\ 0 \\ 0 \\ 10 \\ 0 \end{bmatrix} \qquad (4)$$

As can be evidenced by Fig. 7.9.3, the energy from the left LC network decreases until, at about 60 s, most of the energy has been transferred to the right LC network; as shown by Fig. 7.9.4, where the capacitor voltage of the right LC network peaks up at 60 s. This energy transfer is then reversed, so that at about 180 s the energy has been retransferred to the left LC network. The main oscillation frequency is about 1 rad/s. The period of the envelop is about 130 s. It can be shown by carrying out the solution analytically that the fast frequency is the numerical average of the eigenvalue frequency, i.e.,

$$\tfrac{1}{2}(0.999996 \cdots + 1.048796 \cdots)$$

while the frequency of the slow oscillation, the beat frequency, is the difference

$$1.048796 - 0.99996 \approx 0.0488 \text{ rad}/s$$

This leads to a period

$$T = \frac{2\pi}{0.0488} \approx 130\,s$$

which agrees with that given by the computer solution.

7.10 THE SOLUTION OF LTI FORCED VECTOR DIFFERENTIAL EQUATIONS

In this section we examine the solution of differential equations of the form

$$\dot{\mathbf{x}}(t) = \mathbf{A}\mathbf{x}(t) + \mathbf{v}(t) \qquad \mathbf{x}(0) = \boldsymbol{\xi} \tag{1}$$

where $\mathbf{x}(t)$ = column n-vector
$\mathbf{v}(t)$ = column n-vector
$\mathbf{A}$ = $n \times n$ constant matrix

By way of motivation, let us recall that the analysis of a LTI network with dependent and independent sources led to a vector differential equation of the form (see Sec. 6.4)

$$\dot{\mathbf{x}}(t) = \mathbf{A}\mathbf{x}(t) + \mathbf{B}\mathbf{u}(t) \tag{2}$$

where the components of the state vector $\mathbf{x}(t)$ were the inductor currents and the capacitor voltages and the components of the vector $\mathbf{u}(t)$ were the *independent* current and voltage sources in the network. The matrices $\mathbf{A}$ and $\mathbf{B}$ are fixed by the network topology and the elements in the network branches. If the matrix $\mathbf{B}$ is known, and if the vector $\mathbf{u}(t)$ is also known for all t, setting $\mathbf{v}(t) = \mathbf{B}\mathbf{u}(t)$ reduces the vector differential equation (2) to that given by Eq. (1).

7.10.1. The Nature of the Solutions

In discussing the structure of the solution of Eq. (1) [or (2)] we are interested in finding a formula that works for all initial conditions $\boldsymbol{\xi}$ and all vector time functions $\mathbf{v}(t)$. We note that if we set $\mathbf{v}(t) = \mathbf{0}$ for all t, then Eq. (1) reduces to the unforced differential equation $\dot{\mathbf{x}}(t) = \mathbf{A}\mathbf{x}(t)$, $\mathbf{x}(0) = \boldsymbol{\xi}$, which we studied earlier in this chapter. Thus, we should expect that the vector $e^{\mathbf{A}t}\boldsymbol{\xi}$ will appear in our solution.

[Note that if we think of a network, setting $\mathbf{v}(t) = \mathbf{0}$, for all t, means that we are short-circuiting all independent voltage sources and removing all independent current sources. In this case, the network responds due to the initially stored capacitor and/or inductor energy.] On the other hand, if $\mathbf{v}(t) \neq \mathbf{0}$, we expect our solution to depend also on the specific function $\mathbf{v}(t)$ used to drive the system.

The following theorem yields a formula for computing the unique solution of Eq. (1).

Theorem 1 *Consider the LTI vector differential equation*

$$\dot{\mathbf{x}}(t) = \mathbf{A}\mathbf{x}(t) + \mathbf{v}(t) \tag{3}$$

subject to the initial condition

$$\mathbf{x}(0) = \boldsymbol{\xi} \tag{4}$$

Suppose that $\mathbf{v}(\tau)$, $0 < \tau < t$, has a finite number of discontinuities in the time interval $0 < \tau < t$. Moreover, suppose that $\mathbf{v}(\tau)$ is bounded for all $\tau \in [0, t]$; that is, there exists a finite number M such that

$$\|\mathbf{v}(\tau)\| < M \quad \text{for all } \tau \in [0, t] \tag{5}$$

Then there exists a unique solution $\boldsymbol{\phi}(t)$ OF Eq. (3) subject to the initial condition (4), and it is given by

$$\boldsymbol{\phi}(t) = e^{\mathbf{A}t}\left[\boldsymbol{\xi} + \int_0^t e^{-\mathbf{A}\tau}\mathbf{v}(\tau)\,d\tau\right] \tag{6}$$

where the matrix exponential is defined as before by the infinite series

$$e^{\mathbf{A}t} \triangleq \sum_{k=0}^{\infty} \frac{1}{k!}\mathbf{A}^k t^k \tag{7}$$

PROOF   First we prove that $\boldsymbol{\phi}(t)$ is a solution. To do so we must prove that

$$\boldsymbol{\phi}(0) = \boldsymbol{\xi} \tag{8}$$

and

$$\dot{\boldsymbol{\phi}}(t) = \mathbf{A}\boldsymbol{\phi}(t) + \mathbf{v}(t) \tag{9}$$

To prove Eq. (8) we note that by evaluating Eq. (6) at $t = 0$ we have

$$\boldsymbol{\phi}(0) = e^{\mathbf{A}0}\left[\boldsymbol{\xi} + \int_0^0 e^{-\mathbf{A}\tau}\mathbf{v}(\tau)\,d\tau\right] \tag{10}$$

But in view of Eq. (7)

$$e^{\mathbf{A}0} = \mathbf{I} \tag{11}$$

and in view of Eq. (5) (no impulses are allowed)

$$\int_0^0 e^{-\mathbf{A}\tau} \mathbf{v}(\tau)\, d\tau = \mathbf{0} \tag{12}$$

Thus, $\boldsymbol{\phi}(0) = \mathbf{0}$ as required.

To show that Eq. (9) holds, we differentiate Eq. (6) with respect to t. We use the relation

$$\frac{d}{dt} e^{\mathbf{A}t} = \mathbf{A}e^{\mathbf{A}t} \tag{13}$$

and

$$\frac{d}{dt} \int_0^t e^{-\mathbf{A}\tau} \mathbf{v}(\tau)\, d\tau = e^{-\mathbf{A}t} \mathbf{v}(t) \tag{14}$$

to deduce from Eq. (6) that

$$\frac{d}{dt}\boldsymbol{\phi}(t) = \frac{d}{dt}\left[e^{\mathbf{A}t}\left(\boldsymbol{\xi} + \int_0^t e^{-\mathbf{A}\tau}\mathbf{v}(\tau)\,d\tau \right) \right]$$

$$= \mathbf{A}e^{\mathbf{A}t}\underbrace{\left[\boldsymbol{\xi} + \int_0^t e^{-\mathbf{A}\tau}\mathbf{v}(\tau)\,d\tau \right]}_{\boldsymbol{\phi}(t)} + e^{\mathbf{A}t}\frac{d}{dt}\int_0^t e^{-\mathbf{A}\tau}\mathbf{v}(\tau)\,d\tau \tag{15}$$

$$= \mathbf{A}\boldsymbol{\phi}(t) + e^{\mathbf{A}t}e^{-\mathbf{A}t}\mathbf{v}(t)$$

$$= \mathbf{A}\boldsymbol{\phi}(t) + \mathbf{v}(\tau)$$

so that Eq. (9) holds. Thus, by proving that $\boldsymbol{\phi}(t)$ is a solution, we have proved the existence of a solution.

Next, we shall prove that the solution $\boldsymbol{\phi}(t)$ is *unique*. This proof (like most proofs regarding uniqueness) is by contradiction. Suppose that $\mathbf{z}(t)$ is also a solution and that $\boldsymbol{\phi}(t)$ and $\mathbf{z}(t)$ are distinct. Since we have assumed that both $\boldsymbol{\phi}(t)$ and $\mathbf{z}(t)$ are solutions, we must have

$$\dot{\boldsymbol{\phi}}(t) = \mathbf{A}\boldsymbol{\phi}(t) + \mathbf{v}(t) \qquad \boldsymbol{\phi}(0) = \boldsymbol{\xi} \tag{16}$$

$$\dot{\mathbf{z}}(t) = \mathbf{A}\mathbf{z}(t) + \mathbf{v}(t) \qquad \mathbf{z}(0) = \boldsymbol{\xi} \tag{17}$$

Thus,

$$\frac{d}{dt}[\phi(t) - \mathbf{z}(t)] = \mathbf{A}[\phi(t) - \mathbf{z}(t)] \tag{18}$$

Define the vector $\boldsymbol{\beta}(t)$ by

$$\boldsymbol{\beta}(t) \triangleq \phi(t) - \mathbf{z}(t) \qquad \text{for all } t \tag{19}$$

$$\boldsymbol{\beta}(0) = \phi(0) - \mathbf{z}(0) = \boldsymbol{\xi} - \boldsymbol{\xi} = \mathbf{0} \tag{20}$$

Then Eqs. (18) to (20) imply that the vector $\boldsymbol{\beta}(t)$ must satisfy the unforced equation

Then

$$\dot{\boldsymbol{\beta}}(t) = \mathbf{A}\boldsymbol{\beta}(t) \qquad \boldsymbol{\beta}(0) = \mathbf{0} \tag{21}$$

whose solution we know exists, is unique, and is given by

$$\boldsymbol{\beta}(t) = e^{\mathbf{A}t}\mathbf{0} = \mathbf{0} \qquad \text{for all } t \tag{22}$$

Hence,

$$\boldsymbol{\beta}(t) = \mathbf{0} = \phi(t) - \mathbf{z}(t) \qquad \text{for all } t \tag{23}$$

which *contradicts* the assumption that $\mathbf{z}(t)$ and $\phi(t)$ are distinct. Thus $\phi(t) = \mathbf{z}(t)$, and so $\phi(t)$ represents the unique solution.

Theorem 1 provides us with the solution to Eq. (1) when the boundary condition holds at $t = 0$. It is easy to verify that the vector differential equation

$$\dot{\mathbf{x}}(t) = \mathbf{A}\mathbf{x}(t) + \mathbf{v}(t) \qquad \mathbf{x}(t_0) = \boldsymbol{\xi} \tag{24}$$

has the unique solution

$$\phi(t) = e^{\mathbf{A}(t-t_0)}\left[\boldsymbol{\xi} + \int_{t_0}^{t} e^{-\mathbf{A}(\tau-t_0)}\mathbf{v}(\tau)\,d\tau\right] \tag{25}$$

There are several ways that one can write Eq. (25) by using the fact that

$$e^{\mathbf{A}(t+\tau)} = e^{\mathbf{A}t}e^{\mathbf{A}\tau} \tag{26}$$

Thus, from Eq. (25) we obtain the several alternate forms

$$\phi(t) = e^{A(t-t_0)}\boldsymbol{\xi} + e^{A(t-t_0)} \int_{t_0}^{t} e^{-A(\tau-t_0)} \mathbf{v}(\tau)\, d\tau \tag{27}$$

or

$$\phi(t) = e^{A(t-t_0)}\boldsymbol{\xi} + e^{At} \int_{t_0}^{t} e^{-A\tau} \mathbf{v}(\tau)\, d\tau \tag{28}$$

or

$$\phi(t) = e^{A(t-t_0)}\boldsymbol{\xi} + \int_{t_0}^{t} e^{A(t-\tau)} \mathbf{v}(\tau)\, d\tau \tag{29}$$

Equation (29) clearly illustrates that the complete solution $\phi(t)$ is composed of two parts, *the zero-input response*

$$e^{A(t-t_0)}\boldsymbol{\xi} \tag{30}$$

and *the zero-state response*

$$\int_{t_0}^{t} e^{A(t-\tau)} \mathbf{v}(\tau)\, d\tau \tag{31}$$

Clearly, the *zero-input response* is obtained by setting $\mathbf{v}(\tau) = \mathbf{0}$ for all $\tau \in [t_0, t]$; the *zero-state response* is obtained by setting the initial-state vector $\boldsymbol{\xi} = \mathbf{0}$. Observe that the matrix exponential $e^{A(t-t_0)}$ appears in both the zero-input and the zero-state response.

7.10.2 Superposition Properties

We shall now examine under what circumstances a system described by

$$\dot{\mathbf{x}}(t) = \mathbf{A}\mathbf{x}(t) + \mathbf{v}(t) \qquad \mathbf{x}(t_0) = \boldsymbol{\xi} \tag{32}$$

obeys the principle of superposition. First we shall prove that the system (32) obeys the principle of superposition for the zero-input case. Let $\phi_1(t)$ be the *zero-input response* when $\mathbf{x}(t_0) = \boldsymbol{\xi}_1$, and let $\phi_2(t)$ be the zero-input response when $\mathbf{x}(t_0) = \boldsymbol{\xi}_2$. In other words,

$$\phi_1(t) = e^{A(t-t_0)}\boldsymbol{\xi}_1 \tag{33}$$

$$\phi_2(t) = e^{A(t-t_0)}\boldsymbol{\xi}_2 \tag{34}$$

If we choose for an initial condition a linear combination of ξ_1 and ξ_2

$$\mathbf{x}(t_0) = \alpha_1 \xi_1 + \alpha_2 \xi_2 \tag{35}$$

then the zero-input response is given by

$$
\begin{aligned}
\phi(t) &= e^{\mathbf{A}(t-t_0)}(\alpha_1 \xi_1 + \alpha_2 \xi_2) \\
&= \alpha_1 e^{\mathbf{A}(t-t_0)}\xi_1 + \alpha_2 e^{\mathbf{A}(t-t_0)}\xi_2 \\
&= \alpha_1 \phi_1(t) + \alpha_2 \phi_2(t)
\end{aligned}
\tag{36}
$$

Therefore, *the zero-input response obeys the superposition principle.*

Second, we shall prove that the *zero-state response* also obeys the superposition principle. Let $\phi_1(t)$ be the *zero-state response* for the input $\mathbf{v}_1(\tau)$, $\tau \in [t_0, t]$, and let $\phi_2(t)$ be the *zero-state response* for the input $\mathbf{v}_2(\tau)$, $\tau \in [t_0, t]$. More precisely,

$$\phi_1(t) = \int_{t_0}^{t} e^{\mathbf{A}(t-\tau)}\mathbf{v}_1(\tau)\,d\tau \tag{37}$$

$$\phi_2(t) = \int_{t_0}^{t} e^{\mathbf{A}(t-\tau)}\mathbf{v}_2(\tau)\,d\tau \tag{38}$$

If the input is a linear combination of $\mathbf{v}_1(t)$ and $\mathbf{v}_2(t)$, that is,

$$\mathbf{v}(\tau) = \beta_1 \mathbf{v}_1(\tau) + \beta_2 \mathbf{v}_2(\tau) \tag{39}$$

then the *zero-state response* is

$$
\begin{aligned}
\phi(t) &= \int_{t_0}^{t} e^{\mathbf{A}(t-\tau)}\mathbf{v}(\tau)\,d\tau \\
&= \int_{t_0}^{t} e^{\mathbf{A}(t-\tau)}[\beta_1 \mathbf{v}_1(\tau) + \beta_2 \mathbf{v}_2(\tau)]\,d\tau \\
&= \beta_1 \int_{t_0}^{t} e^{\mathbf{A}(t-\tau)}\mathbf{v}_1(\tau)\,d\tau + \beta_2 \int_{t_0}^{t} e^{\mathbf{A}(t-\tau)}\mathbf{v}_2(\tau)\,d\tau \\
&= \beta_1 \phi_1(t) + \beta_2 \phi_2(t)
\end{aligned}
\tag{40}
$$

Thus, *the zero-state response obeys the superposition principle with respect to the inputs.*

We shall leave it to the reader to show that the complete solution $\phi(t)$ [see Eq. (29)] does *not* obey the principle of superposition.

★ 7.10.3 The Time-shifting Property

We shall now prove that a system described by the vector differential equation

$$\dot{\mathbf{x}}(t) = \mathbf{A}\mathbf{x}(t) + \mathbf{v}(t) \qquad \mathbf{x}(t_0) = \xi \tag{41}$$

obeys a time-shifting property. More precisely, let $\phi(t)$ be the solution of (41), that is,

$$\phi(t) = e^{\mathbf{A}(t-t_0)}\xi + \int_{t_0}^{t} e^{\mathbf{A}(t-\tau)}\mathbf{v}(\tau)\,d\tau \qquad (42)$$

Now we consider the system

$$\dot{\mathbf{x}}(t) = \mathbf{A}\mathbf{x}(t) + \mathbf{u}(t) \qquad \mathbf{x}(t_1) = \xi \qquad (43)$$

with

$$t_1 = t_0 + \Delta \qquad (44)$$

and suppose that $\mathbf{u}(t)$ is a time-shifted replica of $\mathbf{v}(t)$, that is,

$$\mathbf{u}(t) = \mathbf{v}(t - \Delta) \qquad (45)$$

Let $\psi(t)$ be the solution of (43). We want to prove that

$$\psi(t) = \phi(t - \Delta) \qquad (46)$$

To do this we have

$$\psi(t) = e^{\mathbf{A}(t-t_1)}\xi + \int_{t_1}^{t} e^{\mathbf{A}(t-\sigma)}\mathbf{u}(\sigma)\,d\sigma \qquad (47)$$

$$= e^{\mathbf{A}(t-t_0-\Delta)}\xi + \int_{t_0+\Delta}^{t} e^{\mathbf{A}(t-\sigma)}\mathbf{v}(\sigma - \Delta)\,d\sigma \qquad (48)$$

Evaluating Eq. (42) at $t - \Delta$, we have

$$\phi(t - \Delta) = e^{\mathbf{A}(t-t_0-\Delta)}\xi + \int_{t_0}^{t-\Delta} e^{\mathbf{A}(t-\tau-\Delta)}\mathbf{v}(\tau)\,d\tau \qquad (49)$$

In Eq. (49) we have

$$t_0 \le \tau \le t - \Delta \qquad (50)$$

Define a new (dummy) variable μ by

$$\mu = \tau + \Delta \qquad \tau = \mu - \Delta \qquad (51)$$

Then in view of Eq. (50)

$$t_0 + \Delta \le \mu \le t \qquad (52)$$

From Eqs. (49) to (52) we deduce

$$\phi(t - \Delta) = e^{\mathbf{A}(t-t_0-\Delta)}\xi + \int_{t_0+\Delta}^{t} e^{\mathbf{A}(t-\mu)}\mathbf{v}(\mu - \Delta)\,d\mu \qquad (53)$$

Comparing Eqs. (48) and (53), since both σ and μ are dummy variables, we conclude that $\psi(t) = \phi(t - \Delta)$, which is the implication of the time invariance of the system.

EXAMPLE 1 Let us illustrate the concepts by a simple numerical example. Consider the second-order system

$$\underbrace{\begin{bmatrix} \dot{x}_1(t) \\ \dot{x}_2(t) \end{bmatrix}}_{\dot{\mathbf{x}}(t)} = \underbrace{\begin{bmatrix} 0 & 1 \\ 0 & 0 \end{bmatrix}}_{\mathbf{A}} \underbrace{\begin{bmatrix} x_1(t) \\ x_2(t) \end{bmatrix}}_{\mathbf{x}(t)} + \underbrace{\begin{bmatrix} 0 \\ 1 \end{bmatrix} u(t)}_{\mathbf{B} \quad u(t)} \tag{54}$$

with the initial condition at $t = 0$,

$$\mathbf{x}(0) = \boldsymbol{\xi} = \begin{bmatrix} 2 \\ 3 \end{bmatrix} \tag{55}$$

and the specific input

$$u(t) = t \qquad 0 \le t \tag{56}$$

We shall use the solution formula (6), that is,

$$\mathbf{x}(t) = e^{\mathbf{A}t}\left[\boldsymbol{\xi} + \int_0^t e^{-\mathbf{A}\tau}\mathbf{v}(\tau)\,d\tau\right] \tag{57}$$

In our example

$$\mathbf{v}(t) = \mathbf{B}u(t) = \begin{bmatrix} 0 \\ u(t) \end{bmatrix} = \begin{bmatrix} 0 \\ t \end{bmatrix} \tag{58}$$

First we must calculate the matrix exponential $e^{\mathbf{A}t}$, where $\mathbf{A}$ is the 2×2 matrix defined by Eq. (54). It is easy to show by summing the infinite series that

$$e^{\mathbf{A}t} = \begin{bmatrix} 1 & t \\ 0 & 1 \end{bmatrix} \tag{59}$$

Next we must find $e^{-\mathbf{A}\tau}$. We use the property

$$e^{-\mathbf{A}\tau} = (e^{\mathbf{A}\tau})^{-1} \tag{60}$$

and the fact that (59) is a simple 2×2 matrix to deduce

$$e^{-\mathbf{A}\tau} = \begin{bmatrix} 1 & -\tau \\ 0 & 1 \end{bmatrix} \tag{61}$$

The evaluation of the integral (57) requires $e^{-\mathbf{A}\tau}\mathbf{v}(\tau)$. From (61) and (58) we find

$$e^{-\mathbf{A}\tau}\mathbf{v}(\tau) = \begin{bmatrix} 1 & -\tau \\ 0 & 1 \end{bmatrix}\begin{bmatrix} 0 \\ \tau \end{bmatrix} = \begin{bmatrix} -\tau^2 \\ \tau \end{bmatrix} \tag{62}$$

Hence,

$$\int_0^t e^{-\mathbf{A}\tau}\mathbf{v}(\tau)\,d\tau = \int_0^t \begin{bmatrix} -\tau^2 \\ \tau \end{bmatrix} d\tau = \begin{bmatrix} -\frac{1}{3}t^3 \\ \frac{1}{2}t^2 \end{bmatrix} \tag{63}$$

Substituting Eqs. (55), (59), and (63), into Eq. (57), we find

$$\mathbf{x}(t) = \begin{bmatrix} 1 & t \\ 0 & 1 \end{bmatrix}\left[\begin{bmatrix} 2 \\ 3 \end{bmatrix} + \begin{bmatrix} -\frac{1}{3}t^3 \\ \frac{1}{2}t^2 \end{bmatrix}\right] = \begin{bmatrix} 2 + 3t - \frac{1}{3}t^3 + \frac{1}{2}t^3 \\ 3 + \frac{1}{2}t^2 \end{bmatrix} \tag{64}$$

which simplifies to

$$\begin{aligned} x_1(t) &= 2 + 3t + \tfrac{1}{6}t^3 \\ x_2(t) &= 3 + \tfrac{1}{2}t^2 \end{aligned} \tag{65}$$

7.11   STABILITY OF LTI SYSTEMS

In this section we shall investigate the stability of networks and systems described by the vector differential equation

$$\dot{\mathbf{x}}(t) = \mathbf{A}\mathbf{x}(t) + \mathbf{B}\mathbf{u}(t) \qquad \mathbf{x}(0) = \boldsymbol{\xi} \tag{1}$$

Although stability is a very intuitive notion, we must be very precise in its meaning. In order to tie our intuition to mathematics we shall think of Eq. (1) as describing a network containing LTI resistors, capacitors, and inductors as well as independent and dependent current and voltage sources. Thus

$\mathbf{x}(t)$ (the state) is the vector of all inductor currents and capacitor voltages.
$\mathbf{u}(t)$ (the input) is the vector of all independent current and voltage sources.

Suppose that

$$\mathbf{u}(t) = \mathbf{0} \quad \text{for all } t \geq 0 \tag{2}$$

so that we deal with the unforced system

$$\dot{\mathbf{x}}(t) = \mathbf{A}\mathbf{x}(t) \qquad \mathbf{x}(0) = \boldsymbol{\xi} \tag{3}$$

In terms of a network, this case corresponds to setting to zero all independent sources; i.e., voltage sources are short-circuited, and current sources are disconnected. We assume that $\xi \neq \mathbf{0}$, which means that our network is initially excited. Thus, at time $t = 0$ a certain amount of energy is stored in the network capacitors and inductors. If the network contained only positive R, L, and C elements, we would expect this initial stored energy to be dissipated (in the form of heat) in the network resistors and all currents and voltages to tend to zero, i.e.,

$$\lim_{t\to\infty} x_i(t) = 0 \qquad \text{for all } i = 1, 2, \ldots, n \tag{4}$$

or, equivalently,

$$\lim_{t\to\infty} \mathbf{x}(t) = \mathbf{0} \tag{5}$$

which is true if and only if

$$\lim_{t\to\infty} \|\mathbf{x}(t)\| = 0 \tag{6}$$

where $\|\mathbf{x}(t)\|$ is any norm of the state vector $\mathbf{x}(t)$. In this case, we deal with a stable network.

If our network contains, in addition, dependent sources (and this can arise if we have transistors in our network), these dependent sources may be delivering external power to the network so that the currents and voltages may not tend to zero. For this reason, it becomes important to investigate under what precise conditions we have stability.

Definition 1 *The system $\dot{\mathbf{x}}(t) = \mathbf{A}\mathbf{x}(t)$; $\mathbf{x}(0) = \xi$ is called asymptotically stable in the large (ASIL) if*

$$\lim_{t\to\infty} \|\mathbf{x}(t)\| = 0 \qquad \text{for all } \xi \in R_n \tag{7}$$

Equation (7) implies that Eq. (4) or (5) holds and is implied by them.

To study the stability of the system (3) we write the solution for $t \geq 0$

$$\mathbf{x}(t) = e^{\mathbf{A}t}\xi \tag{8}$$

But[†]

$$\|\mathbf{x}(t)\| = \|e^{\mathbf{A}t}\xi\| \leq \|e^{\mathbf{A}t}\| \cdot \|\xi\| \tag{9}$$

It is clear that the system (1) is ASIL *if and only if*

$$\lim_{t\to\infty} \|e^{\mathbf{A}t}\| = 0 \tag{10}$$

[†] It may be a good idea to review norms at this point (Appendix A, Sec. A.15).

since $\|\boldsymbol{\xi}\|$ is some constant. Thus, we must investigate under what conditions Eq. (10) holds.

We suppose that the matrix $\mathbf{A}$ has *distinct* eigenvalues (real or complex) $\lambda_1, \lambda_2, \ldots, \lambda_n$. Let

$$e^{\Lambda t} = \begin{bmatrix} e^{\lambda_1 t} & 0 & \cdots & 0 \\ 0 & e^{\lambda_2 t} & \cdots & 0 \\ \vdots & \vdots & \vdots & \vdots \\ 0 & 0 & \cdots & e^{\lambda_n t} \end{bmatrix} \tag{11}$$

be the diagonal exponential matrix. We know from Sec. 7.5 that

$$e^{\mathbf{A}t} = \mathbf{P}e^{\Lambda t}\mathbf{P}^{-1} \tag{12}$$

Using the matrix norm property $\|\mathbf{AB}\| \leq \|\mathbf{A}\| \cdot \|\mathbf{B}\|$, we have

$$\|e^{\mathbf{A}t}\| = \|\mathbf{P}e^{\Lambda t}\mathbf{P}^{-1}\| \leq \|\mathbf{P}\| \cdot \|\mathbf{P}^{-1}\| \cdot \|e^{\Lambda t}\| = M\|e^{\Lambda t}\| \tag{13}$$

where $M = \|\mathbf{P}\| \cdot \|\mathbf{P}^{-1}\|$ is a bounded positive number. Let us choose the matrix norm

$$\|\mathbf{A}\| = \sum_{i=1}^{n} \sum_{j=1}^{n} |a_{ij}|$$

then

$$\|e^{\Lambda t}\| = |e^{\lambda_1 t}| + |e^{\lambda_2 t}| + \cdots + |e^{\lambda_n t}| \tag{14}$$

Let us decompose each eigenvalue into its real and imaginary parts

$$\lambda_i = \sigma_i + j\omega_i \qquad i = 1, 2, \ldots, n \tag{15}$$

so that

$$\sigma_i = \mathrm{Re}(\lambda_i) \qquad \omega_i = \mathrm{Im}(\lambda_i) \tag{16}$$

Then

$$|e^{\lambda_i t}| = |e^{\sigma_i t + j\omega_i t}| = |e^{\sigma_i t}| \cdot |e^{j\omega_i t}| = e^{\sigma_i t}|e^{j\omega_i t}| = e^{\sigma_i t} \tag{17}$$

Therefore

$$\|e^{\Lambda t}\| = \sum_{i=1}^{n} e^{\sigma_i t} = \sum_{i=1}^{n} e^{\mathrm{Re}(\lambda_i)t} \tag{18}$$

From Eq. (18) we deduce that a sufficient condition for

$$\lim_{t \to \infty} \|e^{\Lambda t}\| \to 0 \tag{19}$$

is that

$$\mathrm{Re}(\lambda_1) < 0, \; \mathrm{Re}(\lambda_2) < 0, \; \ldots, \; \mathrm{Re}(\lambda_n) < 0 \tag{20}$$

Since $\|e^{\mathbf{A}t}\| < M\|e^{\mathbf{\Lambda}t}\|$, we have proved the sufficiency part of the following theorem (it is easy also to prove the necessity part).

Theorem 1 *Consider the system* $\dot{\mathbf{x}}(t) = \mathbf{A}\mathbf{x}(t); \; \mathbf{x}(0) = \boldsymbol{\xi}$. *Suppose that* $\lambda_1, \lambda_2,$ $\ldots, \lambda_n$ *are the distinct eigenvalues of* $\mathbf{A}$. *Then the system is asymptotically stable in the large if and only if all the eigenvalues have negative real parts, i.e.,*

$$\mathrm{Re}(\lambda_i) < 0 \quad \text{for all } i = 1, 2, \ldots, n \tag{21}$$

This theorem is true even for repeated (nondistinct) eigenvalues, but the proof falls outside the scope of this book.

We can see more and more why the concept of the eigenvalues of a matrix is so central in the study of linear systems; the eigenvalues govern the general solution of linear vector differential equations as well as the stability of these solutions.

Of course, we could also obtain the same result once we recall that the solution of (3) can also be exhibited using the formula

$$\mathbf{x}(t) = \beta_1 e^{\lambda_1 t} \mathbf{v}_1 + \beta_2 e^{\lambda_2 t} \mathbf{v}_2 + \cdots + \beta_n e^{\lambda_n t} \mathbf{v}_n \tag{22}$$

where $\mathbf{v}_1, \mathbf{v}_2, \ldots, \mathbf{v}_n$ are the eigenvectors of the $\mathbf{A}$ matrix and $\beta_1, \beta_2, \ldots, \beta_n$ are the coordinates of $\boldsymbol{\xi}$ with respect to the basis defined by the eigenvectors, i.e.,

$$\boldsymbol{\xi} = \beta_1 \mathbf{v}_1 + \beta_2 \mathbf{v}_2 + \cdots + \beta_n \mathbf{v}_n \tag{23}$$

Once more using the decomposition (16), we see that Eq. (22) yields

$$\mathbf{x}(t) = \sum_{i=1}^{n} \beta_i e^{\sigma_i t} e^{j\omega_i t} \mathbf{v}_i \tag{24}$$

If all the σ_i's are negative, each term in the series goes to zero as $t \to \infty$ and hence $\mathbf{x}(t)$ tends to the origin.

On the other hand, if any one of the σ_i's is positive, i.e., if at least one eigenvalue of the $\mathbf{A}$ matrix is in the right half of the complex plane, then the exponential term $e^{\sigma_i t}$ with $\sigma_i > 0$ will grow without bound as $t \to \infty$. Hence, at least one component of $\mathbf{x}(t)$ will also grow without bound, and so $\|\mathbf{x}(t)\| \to \infty$ as $t \to \infty$. In this case, the system is *unstable*.

A dividing case occurs when one or more of the eigenvalues of the system have zero real parts, i.e., they lie on the $j\omega$ axis of the complex plane. In this case, the $j\omega$ axis acts as the dividing line between asymptotic stability and instability. We shall distinguish some cases.

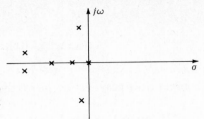

FIGURE 7.11.1
Special case of a system with zero
eigenvalue. The system is stable but not
asymtotically stable.

CASE 1: *Single eigenvalue at zero, all others in left half plane.* Suppose (see Fig. 7.11.1) that

$$\lambda_1 = 0 \quad \text{and that} \quad \text{Re}(\lambda_i) < 0 \quad i = 2, 3, \ldots, n \quad (25)$$

In this case Eq. (22) reduces to

$$\mathbf{x}(t) = \beta_1 \mathbf{v}_1 + \sum_{i=2}^{n} \beta_i e^{\lambda_i t} \mathbf{v}_i \quad (26)$$

Clearly under our assumptions

$$\lim_{t \to \infty} \sum_{i=2}^{n} \beta_i e^{\lambda_i t} \mathbf{v}_i = \mathbf{0} \quad (27)$$

Hence,

$$\lim_{t \to \infty} \mathbf{x}(t) = \beta_1 \mathbf{v}_1 = \text{const} \quad (28)$$

Thus $\mathbf{x}(t)$ neither goes to zero nor to infinity; it simply approaches a constant value determined by the component of the initial state $\boldsymbol{\xi}$ upon the eigenvector $\mathbf{v}_1$ associated with the zero eigenvalue.

Under these conditions the system is said to be *stable* because the state vector does not blow up.

CASE 2: *A pair of complex conjugate eigenvalues on the $j\omega$ axis.* Suppose (see Fig. 7.11.2) that

$$\lambda_1 = j\omega_1 \quad \lambda_2 = \lambda_1^* = -j\omega_1 \quad (29)$$

and that

$$\text{Re}(\lambda_i) < 0 \quad i = 3, \ldots, n \quad (30)$$

In this case $\mathbf{x}(t)$ is given by

$$\mathbf{x}(t) = \beta_1 e^{j\omega_1 t} \mathbf{v}_1 + \beta_2 e^{-j\omega_1 t} \mathbf{v}_2 + \sum_{i=3}^{n} \beta_i e^{\lambda_i t} \mathbf{v}_i \quad (31)$$

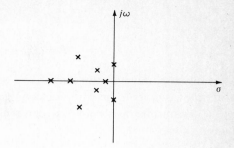

FIGURE 7.11.2
Special case of a system with two purely imaginary eigenvalues. The system is stable but not asymptotically stable.

Once more, in view of our assumption (30)

$$\lim_{t \to \infty} \sum_{i=3}^{n} \beta_i e^{\lambda_i t} \mathbf{v}_i = \mathbf{0} \tag{32}$$

Hence, the limiting behavior of $\mathbf{x}(t)$ is governed by

$$\beta_1 e^{j\omega_1 t} \mathbf{v}_1 + \beta_2 e^{-j\omega_1 t} \mathbf{v}_2 \tag{33}$$

Now since $\lambda_1 = j\omega_1$ is an eigenvalue of $\mathbf{A}$,

$$\mathbf{A}\mathbf{v}_1 = \lambda_1 \mathbf{v}_1 = j\omega_1 \mathbf{v}_1 \tag{34}$$

and since $\lambda_2 = -j\omega_1$ is also an eigenvalue of $\mathbf{A}$,

$$\mathbf{A}\mathbf{v}_2 = \lambda_2 \mathbf{v}_2 = -j\omega_1 \mathbf{v}_2 \tag{35}$$

Since $\mathbf{A}$ is a real matrix, by computing the complex conjugate of both sides of (35) we obtain

$$\mathbf{A}\mathbf{v}_2^* = j\omega_1 v_2^* \tag{36}$$

Comparing Eqs. (34) and (36), we deduce that

$$\mathbf{v}_1 = \mathbf{v}_2^*$$

i.e., the eigenvectors that correspond to complex conjugate eigenvalue pairs are themselves complex conjugate.

Now since the initial state $\boldsymbol{\xi}$ is a real vector, then

$$\boldsymbol{\xi} = \beta_1 \mathbf{v}_1 + \beta_2 \mathbf{v}_1^* + \sum_{i=3}^{n} \beta_i \mathbf{v}_i \tag{37}$$

This implies that

$$\beta_2 = \beta_1^* \tag{38}$$

so that

$$\xi = \beta_1 \mathbf{v}_1 + \beta_1^* \mathbf{v}_1^* + \sum_{i=3}^{n} \beta_1 \mathbf{v}_i \tag{39}$$

because the only way the sum of two complex vectors can be a real vector is for the complex vectors to be complex conjugate.

From this discussion one can see that Eq. (33) leads to

$$\beta_1 e^{j\omega_1 t} \mathbf{v}_1 + \beta_2 e^{-j\omega_1 t} \mathbf{v}_2 = \beta_1 e^{j\omega_1 t} \mathbf{v}_1 + \beta_1^* e^{-j\omega_1 t} \mathbf{v}_1^* \tag{40}$$

Hence, as $t \to \infty$, the state vector $\mathbf{x}(t)$ is characterized by

$$\mathbf{x}(t) \to \beta_1 e^{j\omega_1 t} \mathbf{v}_1 + \beta_1^* e^{-j\omega_1 t} \mathbf{v}_1^* \tag{41}$$

Let us write the components of the vector $\beta_1 \mathbf{v}_1$ explicitly

$$\beta_1 \mathbf{v}_1 = \begin{bmatrix} a_1 + jb_1 \\ a_2 + jb_2 \\ \vdots \\ a_n + jb_n \end{bmatrix} \tag{42}$$

Then

$$\beta_1^* \mathbf{v}_1^* = (\beta_1 \mathbf{v}_1)^* = \begin{bmatrix} a_1 - jb_1 \\ a_2 - jb_2 \\ \vdots \\ a_n - jb_n \end{bmatrix} \tag{43}$$

Then the ith component $x_i(t)$ of the vector $\mathbf{x}(t)$, is given by

$$x_i(t) \to (a_i + jb_i)e^{j\omega_1 t} + (a_i - jb_i)e^{-j\omega_1 t} \tag{44}$$

$$= \sqrt{a_i^2 + b_i^2} \, \cos(\omega_1 t + \gamma_i) \tag{45}$$

where

$$\gamma_i = \tan^{-1} \frac{b_i}{a_i} \tag{46}$$

Hence, as $t \to \infty$, each component of the state vector tends to a sinusoid of the *same* frequency ω_1. The amplitude and phase of each sinusoid depend, of course, upon the initial conditions.

Since $\mathbf{x}(t)$ is bounded, the system is said to be *stable* but not asymptotically stable.

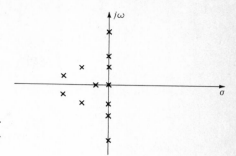

FIGURE 7.11.3
Special case of a system with many distinct eigenvalues on the imaginary axis. This type of system is stable.

CASE 3: *Many distinct eigenvalues on the $j\omega$ axis.* (see Fig. 7.11.3). As long as the eigenvalues are distinct, the state vector $\mathbf{x}(t)$ may have a constant component as well as the sum of sinusoids at the frequencies of the imaginary eigenvalues. Hence $\mathbf{x}(t)$ remains finite and the system is *stable*.

CASE 4: *Repeated eigenvalues on the $j\omega$ axis.* We have not analyzed this case in detail since our whole derivation was carried out using the assumption of distinct eigenvalues. It suffices to state that in this case $\mathbf{x}(t)$ grows without bound and the system is *unstable* provided that $\mathbf{A}$ is in companion form (see Appendix A, Sec. A.10, Definition 1).

A summary of definitions for stability considerations for LTI systems is given in Table 7.1 in terms of the asymptotic behavior of the state vector. Table 7.2 summarizes tests for stability, asymptotic stability, and instability in terms of the eigenvalues of the $\mathbf{A}$ matrix.

The concept of asymptotic stability is not solely restricted to the notion that all zero-input solutions tend to zero as $t \to \infty$. It can be extended to an investigation of whether solutions of the same differential equation with identical inputs but with different initial conditions do indeed approach each other. To make this notion more precise, consider the differential equation

$$\dot{\mathbf{x}}(t) = \mathbf{A}\mathbf{x}(t) + \mathbf{v}(t) \tag{47}$$

Let $\mathbf{x}_1(t)$ denote the solution of (47) when $\mathbf{x}(0) = \xi_1$. Let $\mathbf{x}_2(t)$ denote the solution of (47) when $\mathbf{x}(0) = \xi_2$. We are interested in the asymptotic behavior of these two solutions, namely, whether or not they approach each other. Let us define an error vector $\mathbf{e}(t)$

$$\mathbf{e}(t) \triangleq \mathbf{x}_1(t) - \mathbf{x}_2(t) \tag{48}$$

which simply measures at any time t the difference between these two solutions. By

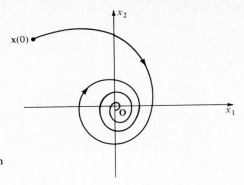

FIGURE 7.11.4
Motion of $\mathbf{x}(t)$ in state plane for an asymptotically stable system.

differentiating (48) and using the fact [since both $\mathbf{x}_1(t)$ and $\mathbf{x}_2(t)$ are solutions] that

$$\dot{\mathbf{x}}_1(t) = \mathbf{A}\mathbf{x}_1(t) + \mathbf{v}(t)$$
$$\dot{\mathbf{x}}_2(t) = \mathbf{A}\mathbf{x}_2(t) + \mathbf{v}(t)$$

$$(49)$$

Table 7.1   SUMMARY OF STABILITY DEFINITIONS FOR $\dot{\mathbf{x}}(t)$ $= \mathbf{A}\mathbf{x}(t)$; $\mathbf{x}(0) = \boldsymbol{\xi}$ (ARBITRARY)

Asymptotic stability	$\|\mathbf{x}(t)\| \to 0$	as $t \to \infty$
Stability	$\|\mathbf{x}(t)\| < \infty$	as $t \to \infty$
Instability	$\|\mathbf{x}(t)\| \to \infty$	as $t \to \infty$

Table 7.2   TESTS FOR STABILITY AND INSTABILITY OF $\dot{\mathbf{x}}(t) = \mathbf{A}\mathbf{x}(t)$ IN TERMS OF THE EIGENVALUES $\lambda_i(\mathbf{A})$, $i = 1, 2, \ldots, n$, OF THE MATRIX A

Asymptotic stability (see Fig. 7.11.4)	All eigenvalues of $\mathbf{A}$, distinct or repeated, strictly in the left half complex plane, i.e., $\mathrm{Re}[\lambda_i(\mathbf{A})] < 0$ for all $i = 1, 2, \ldots, n$
Stability (see Fig. 7.11.5)	All eigenvalues either in left half complex plane or imaginary axis; those strictly in the left half plane may be repeated; those on the imaginary axis must be distinct, i.e., $\mathrm{Re}[\lambda_i(\mathbf{A})] \leq 0$ for all $1 = 1, 2, \ldots, n$ and if $\mathrm{Re}[\lambda_k(\mathbf{A})] = 0$, then $\lambda_j(\mathbf{A}) \neq \lambda_k(\mathbf{A})$ for all $j \neq k$
Instability (see Fig. 7.11.6)	One or more eigenvalues strictly in the right half complex plane or repeated eigenvalues on the imaginary axis and $\mathbf{A}$ in companion form $\mathrm{Re}[\lambda_i(\mathbf{A})] > 0$ for some i or for some j and k $\mathrm{Re}[\lambda_j(\mathbf{A})] = \mathrm{Re}[\lambda_k(\mathbf{A})] = 0$ and $\mathrm{Im}[\lambda_j(\mathbf{A})] = \mathrm{Im}[\lambda_k(\mathbf{A})]$

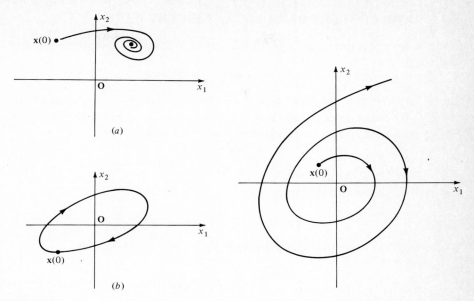

FIGURE 7.11.5
Examples of stable by not asymptotically
stable systems.

FIGURE 7.11.6
Example of unstable system.

we can easily deduce that the error vector $\mathbf{e}(t)$ satisfies the unforced LTI differential
equation

$$\dot{\mathbf{e}}(t) = \mathbf{A}\mathbf{e}(t) \qquad \mathbf{e}(0) = \boldsymbol{\xi}_1 - \boldsymbol{\xi}_2 \tag{50}$$

It therefore follows that if all the eigenvalues of $\mathbf{A}$ have negative real parts, then

$$\lim_{t \to \infty} \mathbf{e}(t) = \mathbf{0} \tag{51}$$

which in turn implies that $\mathbf{x}_1(t)$ and $\mathbf{x}_2(t)$ approach each other as $t \to \infty$. Since the
argument holds for any initial conditions, we can conclude that if we compute the
zero-state solution $\boldsymbol{\phi}_{zs}(t)$ of (47), that is,

$$\boldsymbol{\phi}_{zs}(t) = \int_0^t e^{\mathbf{A}(t-\tau)} \mathbf{v}(\tau)\, d\tau \tag{52}$$

and if we denote by $\boldsymbol{\phi}_{ss}(t)$ (called the steady-state solution) the function of time to
which $\boldsymbol{\phi}_{zs}(t)$ tends to as $t \to \infty$, then all solutions of (47), for any initial conditions,
tend to the steady-state solution $\boldsymbol{\phi}_{ss}(t)$.

⋆ 7.12   THE CONCEPT OF DOMINANT EIGENVALUES

We have seen that the eigenvalues of the matrix $\mathbf{A}$ govern the response of the system

$$\dot{\mathbf{x}}(t) = \mathbf{A}\mathbf{x}(t) \qquad \mathbf{x}(0) = \boldsymbol{\xi} \tag{1}$$

Let us suppose that we have computed the eigenvalues of $\mathbf{A}$ and that they are distinct. We remind the reader that if $\mathbf{A}$ is a real matrix, its eigenvalues are either real or consist of complex conjugate pairs. At any rate, we set

$$\lambda_i = \sigma_i + j\omega_i \qquad \sigma_i \triangleq \operatorname{Re}(\lambda_i) \qquad \omega_i \triangleq \operatorname{Im}(\lambda_i) \qquad \text{for } i = 1, 2, \ldots, n \tag{2}$$

Let us arrange the eigenvalues so that

$$\sigma_1 \geq \sigma_2 \geq \sigma_2 \geq \cdots \geq \sigma_n \tag{3}$$

If λ_1 is real, our assumption that the eigenvalues are distinct yields

$$\sigma_1 > \sigma_2 \geq \sigma_3 \geq \cdots \geq \sigma_n \tag{4}$$

If λ_1 and λ_2 represent a complex conjugate pair, then $\sigma_1 + j\omega_1 = \sigma_2 - j\omega_2$, so that $\sigma_1 = \sigma_2$ and $\omega_1 = -\omega_2$. In this case we have

$$\sigma_1 = \sigma_2 > \sigma_3 \geq \cdots \geq \sigma_n \tag{5}$$

In either case, the eigenvalue λ_1 is called the dominant eigenvalue of the matrix $\mathbf{A}$.

We shall now demonstrate that we can use the dominant eigenvalue in some calculations pertaining to a gross evaluation of the system response. The basic idea is to investigate the time evolution of an appropriate norm of the state vector. In this case we look at a scalar-valued function of the state variables rather than at the time response of each individual state variable.

We know that the solution of Eq. (1) is given by

$$\mathbf{x}(t) = e^{\mathbf{A}t}\boldsymbol{\xi} \tag{6}$$

Thus, by taking norms, we have

$$\|\mathbf{x}(t)\| = \|e^{\mathbf{A}t}\boldsymbol{\xi}\| \leq \|e^{\mathbf{A}t}\| \cdot \|\boldsymbol{\xi}\| \tag{7}$$

Let us recall that

$$e^{\mathbf{A}t} = \mathbf{P}e^{\boldsymbol{\Lambda}t}\mathbf{P}^{-1} \tag{8}$$

Thus,

$$\|e^{\mathbf{A}t}\| \leq \|\mathbf{P}\| \cdot \|\mathbf{P}^{-1}\| \cdot \|e^{\boldsymbol{\Lambda}t}\| \tag{9}$$

In particular, if we choose the norm $\|\cdot\|_\infty$, that is,

$$\|e^{\Lambda t}\| = \max_i \left\{ \sum_{i=1}^n |e^{\lambda_i t} \delta_{ij}| \right\} \tag{10}$$

where δ_{ij} is the Kronecker delta, then

$$\|e^{\Lambda t}\| = e^{\sigma_1 t} \tag{11}$$

Substituting Eqs. (9) and (11) into Eq. (7), we obtain the desired relationship

$$\|\mathbf{x}(t)\| \leq \|\mathbf{P}\| \cdot \|\mathbf{P}^{-1}\| \cdot \|\boldsymbol{\xi}\| e^{\sigma_1 t} \tag{12}$$

This equation indicates that the dominant eigenvalue λ_1 provides an upper bound on the time evolution of $\|\mathbf{x}(t)\|$. Of course, if the matrix $\mathbf{A}$ is stable, then $\sigma_1 < 0$ and (12) gives a decaying exponential which bounds the decay norm $\|\mathbf{x}(t)\|$ from above.

This method of computing an upper bound is more suitable than the following one. If we use our series representation of the matrix exponential, we have

$$\|e^{\mathbf{A}t}\| = \left\| \mathbf{I} + \mathbf{A}t + \frac{1}{2!}\mathbf{A}^2 t^2 + \cdots \right\| \leq 1 + \|\mathbf{A}\|t + \frac{1}{2!}\|\mathbf{A}\|^2 t^2 + \cdots = e^{\|\mathbf{A}\|t} \tag{13}$$

From Eqs. (13) and (7) we now deduce that

$$\|\mathbf{x}(t)\| \leq e^{\|\mathbf{A}\|t} \|\boldsymbol{\xi}\| \tag{14}$$

Note that $\|\mathbf{A}\| > 0$ even though $\mathbf{A}$ may be a stable matrix. Thus (14) bounds $\|\mathbf{x}(t)\|$ by a growing exponential while (12) bounds $\mathbf{x}(t)$ by a decaying exponential if $\mathbf{A}$ is a stable matrix.

⋆ 7.13 NONLINEAR VECTOR DIFFERENTIAL EQUATIONS: GENERAL CONCEPTS AND ASSUMPTIONS

In Chap. 6 we analyzed classes of electric networks containing nonlinear resistors, capacitors, and inductors, and we saw that the network state variables satisfy nonlinear differential equations. Sections 7.3 and 7.4 included examples of other systems whose state variables also satisfy nonlinear differential equations. The remainder of this chapter presents ideas related to such nonlinear dynamical systems.

In a nonlinear dynamical system its state variables $x_1(t), x_2(t), \ldots, x_n(t)$ are related to the input variables $u_1(t), u_2(t), \ldots, u_m(t)$ by first-order differential equations of the form

$$\dot{x}_1 = f_1(x_1, x_2, \ldots, x_n; u_1, u_2, \ldots, u_m)$$
$$\dot{x}_2 = f_2(x_1, x_2, \ldots, x_n; u_1, u_2, \ldots, u_m)$$
$$\vdots \tag{1}$$
$$\dot{x}_n = f_n(x_1, x_2, \ldots, x_n; u_1, u_2, \ldots, u_m)$$

This set of equations can be written in vector form by defining the state vector $\mathbf{x}(t) \in R_n$ and the input vector $\mathbf{u}(t) \in R_m$ in the form

$$\dot{\mathbf{x}}(t) = \mathbf{f}(\mathbf{x}(t), \mathbf{u}(t)) \tag{2}$$

where $\mathbf{f}(.,.)$ is a vector-valued nonlinear function whose components are $f_1(\mathbf{x}, \mathbf{u})$, $f_2(\mathbf{x}, \mathbf{u}), \ldots, f_n(\mathbf{x}, \mathbf{u})$.

We shall be concerned with the solution of differential equations of type (2) together with some initial boundary conditions. In other words, we consider equations of the form

$$\dot{\mathbf{x}}(t) = \mathbf{f}(\mathbf{x}(t), \mathbf{u}(t)) \qquad \mathbf{x}(t_0) = \xi \tag{3}$$

and ask: *Suppose we specify the input vector* $\mathbf{u}$ *over some time interval, i.e. we specify* $\mathbf{u}(\tau)$ *for* $t_0 \le \tau \le t$; *then what is the value of the state vector* $\mathbf{x}(t)$ *at time t?*

From the start we make it perfectly clear that in general we cannot obtain an analytical closed-form formula which specifies $\mathbf{x}(t)$ for nonlinear systems, in contradistinction to linear ones. One can only evaluate $\mathbf{x}(t)$ numerically. Numerical techniques for the solution of nonlinear differential equations are examined in Chap. 9.

Thus, our main interest in this chapter is not numerical techniques for the solution of differential equations but some of their general properties. Even these general properties can be obtained only under certain general assumptions about the mathematical properties of the function $\mathbf{f}(.,.)$ and of the input vector $\mathbf{u}(t)$. Fortunately, these mathematical assumptions are satisfied by well-behaved physical networks and systems. For this reason, the assumptions we make should not worry the engineering-oriented reader. Since we have examined nonlinear networks, we can always go back to them and analyze the meaning of these assumptions in terms of the nature of the physical variables involved. The assumptions which will be in force for the remainder of the chapter are summarized in the remainder of this section.

7.13.1 Assumptions on the Input Vector u(t)

Let us consider the vector differential equation

$$\dot{\mathbf{x}}(t) = \mathbf{f}(\mathbf{x}(t), \mathbf{u}(t)) \qquad \mathbf{x}(t_0) = \xi \tag{4}$$

and the input vector

$$\mathbf{u}(t) = \begin{bmatrix} u_1(t) \\ u_2(t) \\ \vdots \\ u_m(t) \end{bmatrix} \tag{5}$$

Our physical interpretation of the input vector $\mathbf{u}(t)$ is that its components represent the various voltage and current waveforms generated by the *independent* voltage and current sources in our network.

The first of our basic assumptions is

$$\|\mathbf{u}(t)\| \leq U \qquad \text{for all finite } t \tag{6}$$

This means that some suitable norm of the input vector is bounded. Hence, each and every component $u_j(t)$ must be bounded. Physically, this means that the independent voltage and current sources generate finite signals, certainly a reasonable assumption from a physical point of view. Note that *no impulses, doublets, etc., are allowed.*

The second of our basic assumptions is that the components of $\mathbf{u}(t)$ are allowed to have a finite number of discontinuities over a finite interval of time and that between the discontinuities the components of $\mathbf{u}(t)$ are continuous functions of t. Mathematically this is stated as

$$\mathbf{u}(t) = \text{piecewise continuous function of time} \tag{7}$$

From a physical point of view, this assumption means that our voltage and current sources are allowed to produce finite step-function changes a finite number of times, again a good physical assumption.

7.13.2 Assumptions on the Function f(x, u)

Once more we consider the vector differential equation

$$\dot{\mathbf{x}}(t) = \mathbf{f}(\mathbf{x}(t), \mathbf{u}(t)) \qquad \mathbf{x}(t_0) = \boldsymbol{\xi} \tag{8}$$

At each instant of time $\mathbf{x} \in R_n$, $\mathbf{u} \in R_m$. The vector-valued function $\mathbf{f}(.,.)$ depends both on the state $\mathbf{x}$ and the input $\mathbf{u}$.

Continuity Our first assumption is that $\mathbf{f}(.,.)$ is a continuous function of both its arguments $\mathbf{x}$ and $\mathbf{u}$ for all $\mathbf{x} \in R_n$ and all $\mathbf{u} \in R_m$.

Boundedness Our second assumption is that $\mathbf{f}(\mathbf{x},\mathbf{u})$ is a bounded function. Let us recall that we have already assumed that $\mathbf{u}$ is bounded by $\|\mathbf{u}\| \leq U$. Our assumption then is that there exists an F such that for all finite values of $\mathbf{x}$

$$\|\mathbf{f}(\mathbf{x},\mathbf{u})\| \leq F \tag{9}$$

Intuitively, this means that the right-hand side of Eq. (8) does not blow up for finite values of its arguments. Furthermore, this assumption means that the rates of change of the state variables are finite.

Lipschitz conditions Our final assumption for the function $\mathbf{f}(\mathbf{x},\mathbf{u})$ is that it satisfies a Lipschitz condition uniformly in $\mathbf{u}$. To explain this, suppose that we fix the vector $\mathbf{u}$ at some value, say $\hat{\mathbf{u}}$, and we examine the function $\mathbf{f}(\mathbf{x},\hat{\mathbf{u}})$ as we change the vector $\mathbf{x}$. In particular, let us select any two values $\mathbf{x}_\alpha$ and $\mathbf{x}_\beta$ for the vector $\mathbf{x}$. When $\mathbf{x} = \mathbf{x}_\alpha$, the value of the function is $\mathbf{f}(\mathbf{x}_\alpha,\hat{\mathbf{u}})$; when $\mathbf{x} = \mathbf{x}_\beta$, the value of the function is $\mathbf{f}(\mathbf{x}_\beta,\hat{\mathbf{u}})$.

The Lipschitz condition relates the norm of the difference of the vectors $\mathbf{x}_\alpha$ and $\mathbf{x}_\beta$. More precisely, we say that $\mathbf{f}(\mathbf{x},\mathbf{u})$ satisfies a Lipschitz condition if the relation

$$\|\mathbf{f}(\mathbf{x}_\alpha,\hat{\mathbf{u}}) - \mathbf{f}(\mathbf{x}_\beta,\hat{\mathbf{u}})\| \leq L\|\mathbf{x}_\alpha - \mathbf{x}_\beta\| \tag{10}$$

holds, where L (the Lipschitz constant) is constant and finite. Relation (10) must holds for any value of $\hat{\mathbf{u}}$ and for any two vectors $\mathbf{x}_\alpha$ and $\mathbf{x}_\beta$ in R_n.

To simplify our subsequent development we shall further assume (with no loss in generality) that

$$L = F \tag{11}$$

where F is the bound on $\mathbf{f}(\mathbf{x},\mathbf{u})$ [see Eq. (9)]. Thus, we shall, henceforth assume that

$$\begin{aligned}\|\mathbf{f}(\mathbf{x}_\alpha,\hat{\mathbf{u}}) - \mathbf{f}(\mathbf{x}_\beta,\hat{\mathbf{u}})\| &\leq F\|\mathbf{x}_\alpha - \mathbf{x}_\beta\| \\ \|\mathbf{f}(\mathbf{x},\hat{\mathbf{u}})\| &\leq F\end{aligned} \tag{12}$$

A special class of functions satisfying the Lipschitz condition (13) is the class of functions $\mathbf{f}(\mathbf{x},\mathbf{u})$ which are differentiable with respect to $\mathbf{x}$ where the gradient matrix

$$\frac{\partial \mathbf{f}(\mathbf{x},\mathbf{u})}{\partial \mathbf{x}} \tag{13}$$

is continuous with respect to $\mathbf{x}$. In simpler terms this means that if we examine Eq. (1), all the different partial derivatives

$$\frac{\partial f_i(x_1, x_2, \ldots, x_n; u_1, u_2, \ldots, u_m)}{\partial x_j} \tag{14}$$

must exist for all $i,j = 1, 2, \ldots, n$ and must be continuous functions of the x_i's.

To prove that this class of functions automatically satisfies the Lipschitz conditions, let us expand $\mathbf{f}(\mathbf{x}, \hat{\mathbf{u}})$ about $\mathbf{x} = \mathbf{x}_\beta$ using a Taylor series (see Appendix B)

$$\mathbf{f}(\mathbf{x}, \hat{\mathbf{u}}) = \mathbf{f}(\mathbf{x}_\beta, \hat{\mathbf{u}}) + \left.\frac{\partial \mathbf{f}(\mathbf{x}, \hat{\mathbf{u}})}{\partial \mathbf{x}}\right|_{\mathbf{x}=\mathbf{x}_o} (\mathbf{x} - \mathbf{x}_\beta) \tag{15}$$

where $\|\mathbf{x}_o\| \leq X$. In particular if we set $\mathbf{x} = \mathbf{x}_\alpha$ in Eq. (15), we obtain

$$\mathbf{f}(\mathbf{x}_\alpha, \hat{\mathbf{u}}) - \mathbf{f}(\mathbf{x}_\beta, \hat{\mathbf{u}}) = \left[\left.\frac{\partial \mathbf{f}(\mathbf{x}, \hat{\mathbf{u}})}{\partial \mathbf{x}}\right|_{\mathbf{x}=\mathbf{x}_o} \right] (\mathbf{x}_\alpha - \mathbf{x}_\beta) \tag{16}$$

Hence

$$\|\mathbf{f}(\mathbf{x}_\alpha, \hat{\mathbf{u}}) - \mathbf{f}(\mathbf{x}_\beta, \hat{\mathbf{u}})\| \leq \left\| \left.\frac{\partial \mathbf{f}(\mathbf{x}, \hat{\mathbf{u}})}{\partial \mathbf{x}}\right|_{\mathbf{x}=\mathbf{x}_o} \right\| \cdot \|\mathbf{x}_\alpha - \mathbf{x}_\beta\| \tag{17}$$

If the Jacobian matrix $\partial \mathbf{f}(\mathbf{x}, \hat{\mathbf{u}})/\partial \mathbf{x}$ exists and is continuous, its elements are bounded and hence

$$\left\| \left.\frac{\partial \mathbf{f}(\mathbf{x}, \hat{\mathbf{u}})}{\partial \mathbf{x}}\right|_{\mathbf{x}=\mathbf{x}_o} \right\| \leq L \tag{18}$$

Hence, from Eqs. (17) and (18) we conclude that this class of functions satisfies the Lipschitz conditions.

7.13.3   Discussion

In the remainder of this chapter we shall assume that the above assumptions hold, thus establishing precise conditions for the existence and uniqueness of solutions to nonlinear vector differential equations.

★ 7.14   EXISTENCE AND UNIQUENESS

The vector differential equation

$$\dot{\mathbf{x}}(t) = \mathbf{f}(\mathbf{x}(t), \mathbf{u}(t)) \qquad \mathbf{x}(t_0) = \boldsymbol{\xi} \tag{1}$$

is completely equivalent to the vector integral equation

$$\mathbf{x}(t) = \boldsymbol{\xi} + \int_{t_0}^{t} \mathbf{f}(\mathbf{x}(\tau), \mathbf{u}(\tau))\, d\tau \tag{2}$$

For this reason, our treatment of existence and uniqueness of solutions of (1) will be carried out in terms of the integral equation (2).

Before we start our detailed exposition, we remark that Eq. (2) is of the form

$$\mathbf{x} = \mathbf{K}(\mathbf{x}) \tag{3}$$

where the mapping $\mathbf{K}$ is defined by the operation of integration and by the function $\mathbf{f}(\mathbf{x}, \mathbf{u})$, which is the integrand. The reader may guess that we shall use some form of Picard's method to attack the problem of existence and uniqueness, just as we did in Chap. 3. This is precisely what we shall do; of course, our guesses involve vector-valued *time functions* rather than constant vectors.

7.14.1 The Basic Philosophy

To establish the existence of solutions to the integral equation (2) we use the assumptions on $\mathbf{u}(t)$ and $\mathbf{f}(\mathbf{x}, \mathbf{u})$ stated in Sec. 7.13. We want to establish the existence of a function $\phi(t)$ such that when we plug it in Eq. (2), we obtain

$$\phi(t) = \xi + \int_{t_0}^{t} \mathbf{f}(\phi(\tau), \mathbf{u}(\tau)) \, d\tau \tag{4}$$

in some finite time interval of interest.

To establish this we shall construct a sequence of time functions $\phi_0(t)$, $\phi_1(t)$, $\phi_2(t)$, ... and (we hope) establish that the limit

$$\lim_{K \to \infty} \phi_K(t) \qquad t \text{ finite}$$

exists and that the limit function indeed has the property (3). We recall that this is really the same procedure as that used to prove that the linear differential equation $\dot{\mathbf{x}} = \mathbf{A}\mathbf{x}$ has a solution (see Sec. 7.6). In the linear case, this trick also gave us an analytical expression for the solution; this will *not* happen here.

As we shall see, this sequence of functions $\{\phi_K(t)\}$ will also come in handy in proving uniqueness of solutions, once we have taken care of existence.

As a final comment, we stress that the ideas involved are really simple. The apparent mathematical formalism has to do with establishing convergence of this sequence. The reader should be aware that convergence proofs tend to be sticky; this also happens here.

7.14.2 Existence of Solutions

We define the sequence of time functions $\phi_0(t)$, $\phi_1(t)$, $\phi_2(t)$, ... in the following manner:

$$\phi_0(t) \triangleq \xi$$

$$\phi_1(t) \triangleq \xi + \int_{t_0}^{t} \mathbf{f}(\phi_0(\tau), \mathbf{u}(\tau)) \, d\tau$$

$$\phi_2(t) \triangleq \xi + \int_{t_0}^{t} \mathbf{f}(\phi_1(\tau), \mathbf{u}(\tau)) \, d\tau \qquad (5)$$

$$\vdots$$

$$\phi_{K+1}(t) \triangleq \xi + \int_{t_0}^{t} \mathbf{f}(\phi_K(\tau), \mathbf{u}(\tau)) \, d\tau$$

The recursive nature of these functions should be transparent.

First we can see that this sequence of functions is well defined because the fact that $\phi_0(t)$ is bounded and $\mathbf{u}(\tau)$ is bounded, together with our assumptions that $\mathbf{f}(\mathbf{x}, \mathbf{u})$ is continuous and bounded, implies that the integrals are well defined for finite values of t. Hence, for finite values of the index K these functions are well behaved, and, in fact, each $\phi_K(t)$ is continuous in t.

The next question is: Does the limit

$$\lim_{K \to \infty} \phi_K(t)$$

exist for finite values of t? To answer we must use standard arguments on convergence of sequences. Let us consider the sequence

$$\{\phi_K(t)\} \qquad \text{for all finite } t \qquad (6)$$

We want to see whether this sequence converges uniformly to a limit function $\phi(t)$, that is,

$$\phi(t) = \lim_{K \to \infty} \phi_K(t) \qquad \text{for all finite } t \qquad (7)$$

At this point we use a trick; we note that we can write

$$\phi_K(t) = \phi_0(t) - \phi_0(t) + \phi_1(t) - \phi_1(t) + \cdots + \phi_{K-1}(t) - \phi_{K-1}(t) + \phi_K(t) \qquad (8)$$

or

$$\phi_K(t) = \phi_0(t) + [\phi_1(t) - \phi_0(t)] + [\phi_2(t) - \phi_1(t)] + \cdots + [\phi_K(t) - \phi_{K-1}(t)] \qquad (9)$$

This means that if we can prove that the series

$$\sum_{K=0}^{n} [\phi_{K+1}(t) - \phi_K(t)] \qquad (10)$$

converges, then $\phi_K(t)$ also converges to a limit. But to prove that the series (10) converges it is sufficient to prove that the series

$$\sum_{K=0}^{n} \|\phi_{K+1}(t) - \phi_K(t)\| \qquad (11)$$

converges for all finite values of t. For this reason we shall now examine the individual terms of the series of Eq. (11).

We start by computing

$$\phi_1(t) - \phi_0(t) = \int_{t_0}^{t} \mathbf{f}(\phi_0(\tau), \mathbf{u}(\tau)) \, d\tau \tag{12}$$

Hence, since $\|\mathbf{f}(\xi, \mathbf{u})\| = \|\mathbf{f}(\phi_0(\tau), \mathbf{u}(\tau))\|$,

$$\|\phi_1(t) - \phi_0(t)\| \leq \int_{t_0}^{t} \|\mathbf{f}(\phi_0(\tau), \mathbf{u}(\tau))\| \, d\tau \leq \int_{t_0}^{t} F \, d\tau = F|t - t_0| \tag{13}$$

Next, we compute

$$\phi_2(t) - \phi_1(t) = \int_{t_0}^{t} [\mathbf{f}(\phi_1(\tau), \mathbf{u}(\tau)) - \mathbf{f}(\phi_0(\tau), \mathbf{u}(\tau))] \, d\tau \tag{14}$$

Using the Lipschitz condition, we deduce that

$$\|\phi_2(t) - \phi_1(t)\| \leq \int_{t_0}^{t} F\|\phi_1(\tau) - \phi_0(\tau)\| \, d\tau \tag{15}$$

From Eq. (13) we can readily see that

$$\|\phi_2(t) - \phi_1(t)\| \leq \int_{t_0}^{t} F^2 |\tau - t_0| \, d\tau = \frac{1}{2!} F^2 |t - t_0|^2 \tag{16}$$

The form of inequalities (13) and (16) immediately suggests that we can utilize an *inductive* proof. In other words, we *assume* that

$$\|\phi_K(t) - \phi_{K-1}(t)\| \leq \frac{1}{K!} F^K |t - t_0|^K \tag{17}$$

and we try to show that (17) implies that

$$\|\phi_{K+1}(t) - \phi_K(t)\| \leq \frac{1}{(K+1)!} F^{K+1} |t - t_0|^{k+1} \tag{18}$$

To prove that (18) is true we note that

$$\phi_{K+1}(t) - \phi_K(t) = \int_{t_0}^{t} [\mathbf{f}(\phi_K(\tau), \mathbf{u}(\tau)) - \mathbf{f}(\phi_{K-1}(\tau), \mathbf{u}(\tau))] \, d\tau \tag{19}$$

Use of the Lipschitz condition yields

$$\|\phi_{K+1}(t) - \phi_K(t)\| \leq \int_{t_0}^{t} F\|\phi_K(\tau) - \phi_{K-1}(\tau)\| \, d\tau \tag{20}$$

Since we assumed that (17) is true, we use this in Eq. (20) to deduce that

$$\|\phi_{K+1}(t) - \phi_K(t)\| \leq \int_{t_0}^{t} \frac{1}{K!} F^{K+1} |\tau - t_0|^K \, d\tau = \frac{1}{(K+1)!} F^{K+1} |t - t_0|^{K+1} \tag{21}$$

which is precisely what we wanted to prove [see Eq. (18)].

The established inequalities (13), (16), (18), and (21) can be used to bound the elements of the series (11), that is

$$\sum_{K=0}^{n} \|\phi_{K+1}(t) - \phi_K(t)\| \leq \sum_{K=0}^{n} \frac{1}{(K+1)!} F^{K+1} |t - t_0|^{K+1} \tag{22}$$

But the series in the right-hand side of Eq. (22) converges uniformly to the function

$$e^{F|t-t_0|} - 1$$

for all finite values of $|t - t_0|$. Hence we conclude that the functions $\phi_K(t)$ converge uniformly to a limit function $\phi(t)$ in any finite time interval.

The final question to be answered is: How do we know that this limit function is a solution of the integral equation (2)? This is easily proved as follows.

From Eq. (5) we see that

$$\lim_{K \to \infty} \phi_{K+1}(t) = \lim_{K \to \infty} \left[\xi + \int_{t_0}^{t} \mathbf{f}(\phi_K(\tau), \mathbf{u}(\tau)) \, d\tau \right] \tag{23}$$

Since $\phi_K(t)$ converges to $\phi(t)$, and since ξ is a constant vector, we have

$$\phi(t) = \xi + \lim_{K \to \infty} \int_{t_0}^{t} \mathbf{f}(\phi_K(\tau), \mathbf{u}(\tau)) \, d\tau \tag{24}$$

Our previous assumptions that

1 $\mathbf{f}$ is continuous in both its arguments
2 $\mathbf{u}(\tau)$ is bounded and piecewise continuous

and the fact that $\phi_K(\tau)$ converges uniformly to $\phi(t)$ guarantee that we can take the limit inside the integral and the function, i.e., that

$$\phi(t) = \xi + \int_{t_0}^{t} \lim_{K \to \infty} \mathbf{f}(\phi_K(\tau), \mathbf{u}(\tau)) \, d\tau$$

$$= \xi + \int_{t_0}^{t} \mathbf{f}(\lim_{K \to \infty} \phi_K(\tau), \mathbf{u}(\tau)) \, d\tau \tag{25}$$

$$= \xi + \int_{t_0}^{t} \mathbf{f}(\phi(\tau), \mathbf{u}(\tau)) \, d\tau$$

Thus, existence of a solution $\phi(t)$ has been established for all finite values of $t - t_0$.

7.14.3   Uniqueness of Solutions

We shall now prove that under our assumptions the solution $\phi(t)$ is unique. Since most uniqueness proofs are *by contradiction*, we shall follow this route. Suppose that in addition to the solution $\phi(t)$, whose existence we demonstrated above, there is *another* solution $\psi(t)$, which is distinct from $\phi(t)$.

Since $\psi(t)$ is assumed to be a solution, then by definition it must satisfy the integral equation

$$\psi(t) = \xi + \int_{t_0}^{t} \mathbf{f}(\psi(\tau), \mathbf{u}(\tau)) \, d\tau \tag{26}$$

Now we note [see Eq. (5)] that

$$\psi(t) - \phi_0(t) = \int_{t_0}^{t} \mathbf{f}(\psi(\tau), \mathbf{u}(\tau)) \, d\tau \tag{27}$$

and so

$$\|\psi(t) - \phi_0(t)\| \le F|t - t_0| \tag{28}$$

Similarly

$$\psi(t) - \phi_1(t) = \int_{t_0}^{t} [\mathbf{f}(\psi(\tau), \mathbf{u}(\tau)) - \mathbf{f}(\phi_0(\tau), \mathbf{u}(\tau))] \, d\tau \tag{29}$$

Use of the Lipschitz condition and of Eq. (18) yields

$$\|\psi(t) - \phi_1(t)\| \le \int_{t_0}^{t} F\|\psi(\tau) - \phi_0(\tau)\| \, d\tau \le \frac{1}{2!} F^2 |t - t_0|^2 \tag{30}$$

A straightforward inductive argument yields

$$\|\psi(t) - \phi_K(t)\| \le \frac{1}{K!} F^K |t - t_0|^K \tag{31}$$

Hence, since $\phi_K(t)$ converges to $\phi(t)$,

$$\|\psi(t) - \phi(t)\| = \lim_{K \to \infty} \|\psi(t) - \phi_K(t)\| \le \lim_{K \to \infty} \frac{1}{K!} F^K |t - t_0|^K \tag{32}$$

But for any finite value of F and $|t - t_0|$

$$\lim_{K \to \infty} \frac{1}{K!} F^K |t - t_0|^K = 0 \tag{33}$$

Hence,

$$\|\psi(t) - \phi(t)\| \le 0 \tag{34}$$

which implies

$$\psi(t) = \phi(t) \tag{35}$$

a contradiction to our original assumption that there were two different solutions. Therefore, the solution is unique.

7.15 STATIC LINEARIZATION

We have already noted that there is very little one can say about properties of the solution of the vector differential equation $\dot{\mathbf{x}} = \mathbf{f}(\mathbf{x}, \mathbf{u})$. The situation is analogous to the one occurring in our analysis of nonlinear resistive networks (Chapt. 2) because we are faced with the solution of a nonlinear algebraic equation of the form $\mathbf{x} = \mathbf{f}(\mathbf{x})$. There, however, we were able to carry out a *small-signal analysis* of our network about a given operating point and describe how small-signal variations about this operating point can be described by linear equations. We also used the same technique in Chap. 4, for nonlinear difference equations.

In this section we shall show that the same conceptual technique can be extended to networks and systems described by nonlinear differential equations. This is an *extremely* powerful engineering technique which is widely utilized in the analysis and control of many dynamical systems. We shall describe the basic ideas behind the general method and then illustrate its application by examining (Sec. 7.16) a simple nonlinear network which can have both stable and unstable operating points.

The material in this section follows a development almost identical to that given for linearization methods for nonlinear discrete-time systems. We urge the reader to review the material of Sec. 4.7 to see this strong interrelationship. Of course since in Chap. 4 we dealt with systems described by difference equations while in this chapter we deal with systems described by differential equations, the implications of the linearization analysis will be somewhat different. In particular, in Chap. 4 we showed that the linearization method leads to a set of LTI difference equations describing the state and input perturbation vectors about their constant equilibrium values; in this chapter we shall demonstrate that the state and input perturbation vectors are related by LTI differential equations.

7.15.1 Equilibrium Pairs

Let us consider a nonlinear system or network whose state vector $\mathbf{x}(t)$ and input vector $\mathbf{u}(t)$ are related by the vector differential equation

$$\dot{\mathbf{x}}(t) = \mathbf{f}(\mathbf{x}(t), \mathbf{u}(t)) \qquad \begin{matrix} \mathbf{x}(t) \in R_n \\ \mathbf{u}(t) \in R_m \end{matrix} \qquad (1)$$

Let us suppose that for the function $\mathbf{f}(.,.)$ we can find two *constant* vectors $\mathbf{x}^*$ and $\mathbf{u}^*$ such that

$$\mathbf{f}(\mathbf{x}^*, \mathbf{u}^*) = 0 \qquad (2)$$

What is the meaning of this? It means that if the system (1) starts at the initial condition

$$\mathbf{x}(t_0) = \mathbf{x}^* \tag{3}$$

and if we apply the *constant* input

$$\mathbf{u}(t) = \mathbf{u}^* \qquad \text{for all } t \geq t_0 \tag{4}$$

then the solution of (1) will be

$$\mathbf{x}(t) = \mathbf{x}^* \qquad \text{for all } t \geq t_0 \tag{5}$$

i.e., it will stay where it started. To prove this, let us recall that the solution of (1) is given by

$$\mathbf{x}(t) = \mathbf{x}(t_0) + \int_{t_0}^{t} \mathbf{f}(\mathbf{x}(\tau), \mathbf{u}(\tau)) \, d\tau \tag{6}$$

and so if we use Eqs. (3) and (4), we have

$$\mathbf{x}(t) = \mathbf{x}^* + \int_{t_0}^{t} \mathbf{f}(\mathbf{x}(\tau), \mathbf{u}^*) \, d\tau \tag{7}$$

If we assume existence and uniqueness of solutions, we see that in view of Eq. (2), $\mathbf{x}(t) = \mathbf{x}^*$ is a unique solution.

The pair $(\mathbf{x}^*, \mathbf{u}^*)$ will be denoted as an *equilibrium state and input pair* for the differential equation (1). Differential equations may have more than one state and input equilibrium pair; e.g., the scalar nonlinear differential equation

$$\dot{x} = (x - 1)(x - 2)(x - 3) + u$$

has three equilibrium pairs for the value $u(t) = 0$, namely,

$$x^* = 1 \qquad u^* = 0$$
$$x^* = 2 \qquad u^* = 0$$
$$x^* = 3 \qquad u^* = 0$$

7.15.2 Small-Signal Analysis

Let us suppose that $(\mathbf{x}^*, \mathbf{u}^*)$ form an equilibrium pair for a nonlinear dynamical system or network. At this stage it is helpful to visualize a nonlinear network in which the current and voltage sources generate constant (bias) currents and voltages and yield the constant state vector $\mathbf{x}^*$, which represents the constant inductor currents and capacitor voltages in the network.

Now let us suppose that we allow one or more of the independent sources to generate a *small* signal about its previously constant value. We can express this fact mathematically by writing the input vector $\mathbf{u}(t)$ as

$$\mathbf{u}(t) = \mathbf{u}^* + \boldsymbol{\delta}\mathbf{u}(t) \tag{8}$$

Our assumption that the components of the *input-perturbation vector* $\boldsymbol{\delta}\mathbf{u}(t)$ are small compared to those of the constant vector $\mathbf{u}^*$ implies that

$$\|\boldsymbol{\delta}\mathbf{u}(t)\| \ll \|\mathbf{u}^*\| \qquad \text{for all } t \tag{9}$$

The time variations of the input vector will now cause a time variation in the state vector $\mathbf{x}(t)$. We express the time variation of $\mathbf{x}(t)$ as

$$\mathbf{x}(t) = \mathbf{x}^* + \boldsymbol{\delta}\mathbf{x}(t) \tag{10}$$

where $\boldsymbol{\delta}\mathbf{x}(t)$ represents the *perturbation state vector* about its constant value $\mathbf{x}^*$.

There are many questions one can pose at this point:

1 How are the state perturbations $\boldsymbol{\delta}\mathbf{x}(t)$ related to the input perturbations $\boldsymbol{\delta}\mathbf{u}(t)$?

2 Can we guarantee that "small" $\boldsymbol{\delta}\mathbf{u}(t)$'s generate "small" $\boldsymbol{\delta}\mathbf{x}(t)$'s? Can we be sure of this fact without solving the nonlinear differential equations?

Since such questions are intimately related to small-signal analysis, or perturbation analysis, a very useful engineering technique, we shall indicate how one goes about answering such questions for continuous-time dynamical systems.

7.15.3 Taylor Series

The mathematical tool of small-signal analysis is Taylor series expansions about the (operating) equilibrium conditions (see Appendix B). We deal with the nonlinear vector differential equation

$$\dot{\mathbf{x}}(t) = \mathbf{f}(\mathbf{x}(t), \mathbf{u}(t)) \tag{11}$$

We suppose that $\mathbf{x}^*$ and $\mathbf{u}^*$ form an equilibrium pair, i.e.,

$$\mathbf{0} = \mathbf{f}(\mathbf{x}^*, \mathbf{u}^*) \tag{12}$$

Let

$$\mathbf{u}(t) \stackrel{\triangle}{=} \mathbf{u}^* + \boldsymbol{\delta}\mathbf{u}(t) \tag{13}$$

and

$$\mathbf{x}(t) \triangleq \mathbf{x}^* + \boldsymbol{\delta}\mathbf{x}(t) \tag{14}$$

Clearly

$$\dot{\mathbf{x}}(t) \triangleq \mathbf{f}(\mathbf{x}(t), \mathbf{u}(t)) \tag{15}$$

Next we expand $\mathbf{f}(\mathbf{x}(t), \mathbf{u}(t))$ about the values $\mathbf{x}(t) = \mathbf{x}^*$, $\mathbf{u}(t) = \mathbf{u}^*$ to obtain

$$\mathbf{f}(\mathbf{x}(t), \mathbf{u}(t)) = \mathbf{f}(\mathbf{x}^*, \mathbf{u}^*) + \frac{\partial \mathbf{f}}{\partial \mathbf{x}}\bigg|_* (\mathbf{x}(t) - \mathbf{x}^*) + \frac{\partial \mathbf{f}}{\partial \mathbf{u}}\bigg|_* (\mathbf{u}(t) - \mathbf{u}^*)$$
$$+ \text{ higher-order terms} \tag{16}$$

where $\partial \mathbf{f}/\partial \mathbf{x}$ is the $n \times n$ Jacobian matrix of $\mathbf{f}(\mathbf{x}, \mathbf{u})$ with respect to the state vector $\mathbf{x}$ and $\partial \mathbf{f}/\partial \mathbf{u}$ is the $n \times m$ Jacobian matrix of $\mathbf{f}(\mathbf{x}, \mathbf{u})$ with respect to the input vector $\mathbf{u}$. The notation $\partial \mathbf{f}/\partial \mathbf{x}|_*$ and $\partial \mathbf{f}/\partial \mathbf{u}|_*$ means that we evaluate the elements of these two matrices at the constant equilibrium values $\mathbf{x} = \mathbf{x}^*$ and $\mathbf{u} = \mathbf{u}^*$. Thus if we know

1 The analytical expression for $\mathbf{f}(\mathbf{x}, \mathbf{u})$ and
2 The constant values of $\mathbf{x}^*$ and $\mathbf{u}^*$

we can certainly compute these two matrices. To stress this fact we shall write

$$\mathbf{A}_* \triangleq \frac{\partial \mathbf{f}}{\partial \mathbf{x}}\bigg|_* \qquad \text{an } n \times n \text{ } constant \text{ matrix} \tag{17}$$

$$\mathbf{B}_* \triangleq \frac{\partial \mathbf{f}}{\partial \mathbf{u}}\bigg|_* \qquad \text{an } n \times m \text{ } constant \text{ matrix} \tag{18}$$

Using this notation together with Eqs. (12) to (14), we can see that Eq. (16) yields

$$\mathbf{f}(\mathbf{x}(t), \mathbf{u}(t)) = \mathbf{A}_* \boldsymbol{\delta}\mathbf{x}(t) + \mathbf{B}_* \boldsymbol{\delta}\mathbf{u}(t) + \text{higher-order terms} \tag{19}$$

But since $\mathbf{x}^*$ is a constant vector,

$$\dot{\mathbf{x}}(t) = \frac{d}{dt}\boldsymbol{\delta}\mathbf{x}(t) = \mathbf{f}(\mathbf{x}(t), \mathbf{u}(t)) \tag{20}$$

in view of Eqs. (14) and (15). Therefore, from Eqs. (19) and (20) we can readily deduce that the state perturbation vector $\boldsymbol{\delta}\mathbf{x}(t)$ and the input-perturbation vector $\boldsymbol{\delta}\mathbf{u}(t)$ are related by

$$\boxed{\frac{d}{dt}\boldsymbol{\delta}\mathbf{x}(t) = \mathbf{A}_* \boldsymbol{\delta}\mathbf{x}(t) + \mathbf{B}_* \boldsymbol{\delta}\mathbf{u}(t) + \text{higher-order terms}} \tag{21}$$

Now let us recall that we assumed that the input perturbations were small, i.e.,

$$\|\delta\mathbf{u}(t)\| = \text{``small''} \tag{22}$$

If it turns out that the resultant state perturbations are also small, i.e.,

$$\|\delta\mathbf{x}(t)\| = \text{``small''} \tag{23}$$

then we are justified in neglecting† the higher-order terms in Eq. (21) to deduce the linear differential equation

$$\boxed{\frac{d}{dt}\delta\mathbf{x}(t) = \mathbf{A}_* \, \delta\mathbf{x}(t) + \mathbf{B}_* \, \delta\mathbf{u}(t)} \tag{24}$$

7.15.4 Conclusions

What is the net outcome of all this? We have demonstrated that the state perturbation vector $\delta\mathbf{x}(t)$ is related to the input-perturbation vector $\delta\mathbf{u}(t)$ by means of a LTI vector differential equation. The linearity is due to the assumption that the higher-order terms can be neglected. The time invariance is due to the fact that the matrices $\mathbf{A}_*$ and $\mathbf{B}_*$ are constant matrices, which in turn is a consequence of the fact that

1 The nonlinear function $\mathbf{f}(\mathbf{x}, \mathbf{u})$ does not explicitly depend on the time t.
2 The equilibrium values $\mathbf{x}^*$, $\mathbf{u}^*$ are constant.

We wish to stress the fact that the matrices $\mathbf{A}_*$ and $\mathbf{B}_*$ are *precomputable.* Hence, we can evaluate the transition matrix $e^{\mathbf{A}_* t}$ and write the solution of (24)

$$\delta\mathbf{x}(t) = e^{\mathbf{A}_* t}\left[\delta\mathbf{x}(0) + \int_0^t e^{-\mathbf{A}_* \tau}\mathbf{B}_* \, \delta\mathbf{u}(\tau)\, d\tau\right] \tag{25}$$

It is easy to see that if the eigenvalues of $\mathbf{A}_*$ have negative real parts, the system (24) is *stable* (see Sec. 7.10) and small $\delta\mathbf{u}(t)$'s will generate small $\delta\mathbf{x}(t)$'s. In this case, the pair $(\mathbf{x}^*, \mathbf{u}^*)$ is called a *stable equilibrium pair* (this is analogous to the fact that the bottom position of a pendulum is stable).

If, on the other hand, one or more of the eigenvalues of $\mathbf{A}_*$ have a positive real part, the system (24) is *unstable* and hence *small* $\delta\mathbf{u}(t)$'s will generate large $\delta\mathbf{x}(t)$'s. In this case, the pair $(\mathbf{x}^*, \mathbf{u}^*)$ is called an *unstable equilibrium pair.* (This is analogous to the fact that the top position of an inverted pendulum is an unstable equilibrium point.) In cases of unstable equilibrium, our assumption that the $\delta\mathbf{x}(t)$'s are small

---

† The higher-order terms contain quadratic, cubic, quartic, etc., terms in the components of $\delta\mathbf{x}(t)$, $\delta\mathbf{u}(t)$ and their cross products. Since the square, cube, etc., of a small number is much smaller than the small number itself, they can justifiably be neglected.

will eventually be violated; this means that we cannot ignore the higher-order terms in the Taylor series expansion and that the linear model (24) is *not* valid.

In the next section we shall illustrate these ideas by considering a nonlinear *RLC* network with three equilibrium conditions. This type of analysis is useful not only in the analysis of nonlinear networks but many other types of engineering systems. For example, the motion of an airplane is in general described by nonlinear differential equations; but for any given flight condition, i.e.,

1 Flight at constant altitude
2 Flight at constant speed
3 Flight at zero pitch, roll, and yaw angle

one can linearize the nonlinear differential equations of motion and describe the perturbations of the aircraft about this flight condition by means of LTI differential equations. In point of fact, the autopilots for most conventional aircraft are *designed to stabilize the aircraft under these assumptions*. This is why when a severe gust or turbulence hits the aircraft, resulting in *large* perturbations of the state vector, the linearized equations become invalid, the autopilot may not be able to function effectively, and it is up to the pilot to take appropriate action.

7.16 SMALL-SIGNAL ANALYSIS OF A NONLINEAR *RLC* NETWORK

In this section we illustrate the general principles of small-signal analysis by considering a simple *RLC* network containing a single current-controlled nonlinear resistor. Our objective is to show that this specific network has three equilibrium conditions, two stable and one unstable. We hope that the detailed analysis of this specific network will enhance the reader's understanding and appreciation of the general theory presented in Sec. 7.15.

7.16.1. The Nonlinear Network

The network shown in Fig. 7.16.1*a* contains a constant voltage source $u^* = 1\text{V}$, a linear 1-F capacitor, a linear 1-H inductor, a linear 1-Ω resistor, and a nonlinear voltage-controlled conductance G. We know that we can select the inductor current $i(t)$ and the capacitor voltage $v(t)$ as the two state variables for the network. As shown in Fig. 7.16.1*a*, we let v_G be the voltage across the nonlinear conductance, and we let i_G be the current through it. The voltage-current relationship is assumed to be

$$i_G = g(v_G) = v_G(v_G - 1)(v_G - 4) \qquad \textbf{A} \qquad\qquad (1)$$

The graph of $g(\,.\,)$ is shown in Fig. 7.16.1*b*.

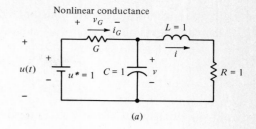

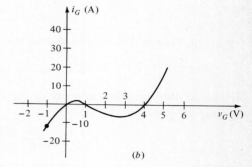

FIGURE 7.16.1
(a) A nonlinear network; (b) the current-voltage characteristics of the voltage-controlled nonlinear conductance G.

Next let us obtain the state differential equations for this network. We shall leave it as an exercise to the reader to deduce that

$$\frac{d}{dt}i(t) = -i(t) + v(t) \tag{2}$$

$$\frac{d}{dt}v(t) = -i(t) + g(u(t) - v(t)) \tag{3}$$

where the function $g(.)$ is analytically specified by Eq. (1).

Clearly, the state differential equations have the form

$$\begin{aligned}
\frac{d}{dt}i(t) &= f_1(i(t), v(t), u(t)) = -i(t) + v(t) \\
\frac{d}{dt}v(t) &= f_2(i(t), v(t), u(t)) = -i(t) + g(u(t) - v(t))
\end{aligned} \tag{4}$$

7.16.2 The Equilibrium Values

Since according to Fig. 7.16.1a, the voltage $u(t)$ is a constant

$$u(t) = u^* = 1V$$

we inquire whether there are constant values of the inductor current $i(t) = i^*$ and of the capacitor voltage $v(t) = v^*$ such that the equations

$$0 = f_1(i^*, v^*, u^*) = -i^* + v^* \tag{5}$$

$$0 = f_2(i^*, v^*, u^*) = -i^* + g(1 - v^*) \tag{6}$$

hold simultaneously. From Eq. (5) we deduce that

$$i^* = v^* \tag{7}$$

and from Eqs. (6) and (7) we deduce that

$$v^* = g(1 - v^*) \tag{8}$$

If we use the analytical expression for $g(.)$ given in Eq. (1), we see that Eq. (8) reduces to

$$v^* = (1 - v^*)(1 - v^* - 1)(1 - v^* - 4) \tag{9}$$

This is a nonlinear cubic equation in v^*, and it has the following three roots (as the reader can readily verify by direct substitution):

$$v_1^* = 0\,\text{V} \qquad v_2^* = -1 + \sqrt{3}\,\text{V} \qquad v_3^* = -1 - \sqrt{3}\,\text{V} \tag{10}$$

From Eqs. (10) and (7) we can then conclude that there are *three* possible equilibrium conditions

Case 1:	$i_1^* = 0\,\text{A}$	$v_1^* = 0\,\text{V}$	$u^* = 1\,\text{V}$	
Case 2:	$i_2^* = -1 + \sqrt{3}\,\text{A}$	$v_2^* = -1 + \sqrt{3}\,\text{V}$	$u^* = 1\,\text{V}$	(11)
Case 3:	$i_3^* = -1 - \sqrt{3}\,\text{A}$	$v_3^* = -1 - \sqrt{3}\,\text{V}$	$u^* = 1\,\text{V}$	

What is the interpretation of these equilibrium conditions? From a theoretical point of view, if the initial values of the inductor current and capacitor value correspond to the same equilibrium pair, they will not change. In practice, this will be true only for the stable equilibrium conditions. If there are any unstable equilibrium conditions, then the network will "get out" of them. For this reason, we shall carry out a small-signal analysis to see what happens.

7.16.3 Small-Signal Analysis

The easiest way of visualizing what we mean by small-signal analysis is to add a voltage source $\delta u(t)$ in series with the 1-V battery, as shown in Fig. 7.16.2. We shall

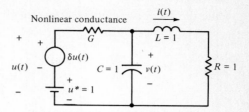

FIGURE 7.16.2
The network of Fig. 7.16.1a with a small
voltage source $\delta u(t)$ added.

assume that the voltage $\delta u(t)$ is small, i.e.,

$$|\delta u(t)| \ll 1\text{V} \tag{12}$$

The small voltage variation $\delta u(t)$ will now cause variations in both the state
variables $i(t)$ and $v(t)$ about their equilibrium values. For this reason, we write

$$i(t) = i^* + \delta i(t)$$
$$v(t) = v^* + \delta v(t) \tag{13}$$

The column 2-vector

$$\delta \mathbf{x}(t) = \begin{bmatrix} \delta i(t) \\ \delta v(t) \end{bmatrix} \tag{14}$$

plays the role of the state perturbation vector, and the scalar $\delta u(t)$ plays the role of
the input-perturbation vector. According to the results of Sec. 7.15, we expect these
perturbations to be related by an LTI vector differential equation of the form

$$\frac{d}{dt}\delta \mathbf{x}(t) = \mathbf{A}_* \,\delta \mathbf{x}(t) + \mathbf{B}_* \,\delta u(t) \tag{15}$$

In our example $\mathbf{A}_*$ is a 2×2 matrix and $\mathbf{B}_*$ is a 2×1 matrix (or column 2-vector).
Let us recall that these are evaluated by computing the two Jacobian matrices $\partial \mathbf{f}/\partial \mathbf{x}$
and $\partial \mathbf{f}/\partial \mathbf{u}$ and evaluating them at a given equilibrium pair. In our example

$$\mathbf{f}(\mathbf{x}, \mathbf{u}) = \begin{bmatrix} f_1(i, v, u) \\ f_2(i, v, u) \end{bmatrix} = \begin{bmatrix} -i + v \\ -i + (u - v)(u - v - 1)(u - v - 4) \end{bmatrix} \tag{16}$$

Hence, the Jacobian matrix $\partial \mathbf{f}/\partial \mathbf{x}$ is given by

$$\frac{\partial \mathbf{f}}{\partial \mathbf{x}} = \begin{bmatrix} \dfrac{\partial f_1}{\partial i} & \dfrac{\partial f_1}{\partial v} \\ \dfrac{\partial f_2}{\partial i} & \dfrac{\partial f_2}{\partial v} \end{bmatrix} = \begin{bmatrix} -1 & 1 \\ -1 & -3(u - v)^2 + 10(u - v) - 4 \end{bmatrix} \tag{17}$$

while the Jacobian matrix $\partial \mathbf{f}/\partial u$ (which is now a column vector) is

$$\frac{\partial \mathbf{f}}{\partial u} = \begin{bmatrix} \dfrac{\partial f_1}{\partial u} \\ \dfrac{\partial f_2}{\partial u} \end{bmatrix} = \begin{bmatrix} 0 \\ 3(u-v)^2 - 10(u-v) + 4 \end{bmatrix} \tag{18}$$

In our example we have three equilibrium pairs [see Eq. (11)]. Hence, we shall get three distinct linear systems depending upon which equilibrium pair we conduct our small-signal analysis for. We shall do it for all three.

CASE 1: In this case the equilibrium occurs at

$$i^* = 0 \qquad v^* = 0 \qquad u^* = 1 \tag{19}$$

Evaluating the gradient matrix (17) at these values, we obtain the matrix

$$\mathbf{A}_{*1} = \begin{bmatrix} -1 & 1 \\ -1 & 3 \end{bmatrix} \tag{20}$$

Evaluating the gradient matrix (column vector) (18) at these values, we obtain

$$\mathbf{B}_{*1} = \begin{bmatrix} 0 \\ -3 \end{bmatrix} \tag{21}$$

Hence the small-signal analysis yields the vector differential equation

$$\underbrace{\frac{d}{dt}\begin{bmatrix} \delta i(t) \\ \delta v(t) \end{bmatrix}}_{\dfrac{d}{dt}\,\delta \mathbf{x}(t)} = \underbrace{\begin{bmatrix} -1 & 1 \\ -1 & 3 \end{bmatrix}}_{\mathbf{A}_{*1}}\underbrace{\begin{bmatrix} \delta i(t) \\ \delta v(t) \end{bmatrix}}_{\delta \mathbf{x}(t)} + \underbrace{\begin{bmatrix} 0 \\ 3 \end{bmatrix}}_{\mathbf{B}_{*1}}\underbrace{\delta u(t)}_{\delta \mathbf{u}(t)} \tag{22}$$

or, equivalently,

$$\frac{d}{dt}\delta i(t) = -\delta i(t) + \delta v(t)$$

$$\frac{d}{dt}\delta v(t) = -\delta i(t) + 3\delta v(t) + 3\delta u(t) \tag{23}$$

Is this a stable equilibrium point? To check this we need only evaluate the eigenvalues of the matrix $\mathbf{A}_{*1}$ given by Eq. (20); it is easy to verify that its eigenvalues are at

$$1 \pm \sqrt{3} \tag{24}$$

and since one of the eigenvalues is positive, we readily conclude that this corresponds to an unstable equilibrium condition.

CASE 2:   In this case the equilibrium occurs [see Eq. (11)] at

$$i_2^* = -1 + \sqrt{3} \qquad v_2^* = -1 + \sqrt{3} \qquad u^* = 1 \tag{25}$$

If we evaluate the gradient matrix (17) at these values, we obtain the matrix

$$\mathbf{A}_{*2} = \begin{bmatrix} -1 & 1 \\ -1 & -5 + 2\sqrt{3} \end{bmatrix} \tag{26}$$

and evaluating the gradient matrix (18) at these values, we obtain

$$\mathbf{B}_{*2} = \begin{bmatrix} 0 \\ 5 - 2\sqrt{3} \end{bmatrix} \tag{27}$$

Hence, the small-signal analysis yields the vector differential equation

$$\frac{d}{dt} \underbrace{\begin{bmatrix} \delta i(t) \\ \delta v(t) \end{bmatrix}}_{} = \underbrace{\begin{bmatrix} -1 & 1 \\ -1 & -5 + 2\sqrt{3} \end{bmatrix}}_{} \underbrace{\begin{bmatrix} \delta i(t) \\ \delta v(t) \end{bmatrix}}_{} + \underbrace{\begin{bmatrix} 0 \\ 5 - 2\sqrt{3} \end{bmatrix}}_{} \underbrace{\delta u(t)}_{} \tag{28}$$

$$\frac{d}{dt}\ \delta\mathbf{x}(t) \qquad\qquad \mathbf{A}_{*2} \qquad\qquad \delta\mathbf{x}(t) \qquad\qquad \mathbf{B}_{*2} \qquad \delta\mathbf{u}(t)$$

or equivalently,

$$\frac{d}{dt}\delta i(t) = -\delta i(t) + \delta v(t)$$

$$\frac{d}{dt}\delta v(t) = -\delta i(t) + (-5 + 2\sqrt{3})\,\delta v(t) + (5 - 2\sqrt{3})\,\delta u(t) \tag{29}$$

To check the stability of this equilibrium point, we compute the eigenvalues of the matrix $\mathbf{A}_{*2}$ given by Eq. (26); its eigenvalues are at

$$-(3 - \sqrt{3}) \pm \sqrt{6 - 4\sqrt{3}} \tag{30}$$

Both the eigenvalues are negative, and therefore we conclude that this represents a stable equilibrium condition.

CASE 3:   In this case the equilibrium occurs [see Eq. (11)] at

$$i_3^* = -1 - \sqrt{3} \qquad v_3^* = -1 - \sqrt{3} \qquad u^* = 1 \tag{31}$$

If we evaluate the gradient matrix (17) at these values, we obtain the matrix

$$\mathbf{A}_{*3} = \begin{bmatrix} -1 & 1 \\ -1 & -5 - 2\sqrt{3} \end{bmatrix} \tag{32}$$

If we evaluate the gradient matrix (18) at these values, we obtain

$$\mathbf{B}_{*3} = \begin{bmatrix} 0 \\ 5 + 2\sqrt{3} \end{bmatrix} \tag{33}$$

Hence the small-signal analysis yields the vector differential equation

$$\underbrace{\frac{d}{dt}\begin{bmatrix} \delta i(t) \\ \delta v(t) \end{bmatrix}}_{\frac{d}{dt}\ \delta\mathbf{x}(t)} = \underbrace{\begin{bmatrix} -1 & 1 \\ -1 & -5 - 2\sqrt{3} \end{bmatrix}}_{\mathbf{A}_{*3}} \underbrace{\begin{bmatrix} \delta i(t) \\ \delta v(t) \end{bmatrix}}_{\delta\mathbf{x}(t)} + \underbrace{\begin{bmatrix} 0 \\ 5 + 2\sqrt{3} \end{bmatrix}}_{\mathbf{B}_{*3}} \underbrace{\delta u(t)}_{\delta\mathbf{u}(t)} \tag{34}$$

or, equivalently,

$$\frac{d}{dt}\delta i(t) = -\delta i(t) + \delta u(t)$$

$$\frac{d}{dt}\delta v(t) = -\delta i(t) - (5 + 2\sqrt{3})\,\delta v(t) + (5 + 2\sqrt{3})\delta u(t) \tag{35}$$

The eigenvalues of the matrix $\mathbf{A}_{*3}$, given by Eq. (32), are at

$$-(3 + \sqrt{3}) \pm \sqrt{6 + 4\sqrt{3}} \tag{36}$$

Since both eigenvalues have negative real parts, we conclude that this set of values corresponds to a stable equilibrium condition.

7.17 ANOTHER EXAMPLE OF LINEARIZATION ANALYSIS: A SPINNING SPACE VEHICLE

In this section we shall illustrate the concepts of linearization analysis once more by considering the rotational motion of a body in space. We presented the equations of motion for this system in Sec. 7.3:

$$\dot{\omega}_1(t) = \frac{I_2 - I_3}{I_1}\omega_2(t)\omega_3(t)$$

$$\dot{\omega}_2(t) = \frac{I_3 - I_1}{I_2}\omega_1(t)\omega_3(t) \tag{1}$$

$$\dot{\omega}_3(t) = \frac{I_1 - I_2}{I_3}\omega_1(t)\omega_2(t)$$

where $\omega_1(t)$, $\omega_2(t)$, and $\omega_3(t)$ denote the three angular velocities about the three principal axes and I_1, I_2, and I_3 are the corresponding moments of inertia. This set of equations represents the natural motion of this system in the absence of any external forces or torques. If the body is completely asymmetrical, then no two of the three moments of intertia are equal. Let us rank them according to the inequality

$$I_1 < I_2 < I_3 \tag{2}$$

so that $\omega_1(t)$ denotes angular rotation about the smallest moment of inertia axis, $\omega_2(t)$ denotes angular rotation about the intermediate moment of inertia axis, and $\omega_3(t)$ denotes angular rotation about the largest moment of inertia axis.

A problem of practical interest is as follows: Is it possible to maintain rotation of the space body, at a constant angular velocity, about only one axis while there is no rotation whatsoever about the other two axes? This problem is of practical interest since one may wish to establish artificial gravity by rotating the space vehicle about a single axis; rotation about the two other axes would be undesirable since it would correspond to tumbling of the space vehicle and of any people and equipment inside.

To analyze this problem we must check whether constant nonzero angular velocity about one axis and zero angular velocities about the other two axes correspond to an equilibrium state of the system (1). We shall distinguish three cases.

CASE 1: For rotation about the smallest moment of inertia axis we wish to check whether the constant state defined by

$$\omega_1^* \neq 0 \qquad \omega_2^* = 0 \qquad \omega_3^* = 0 \tag{3}$$

can be an equilibrium state of (1). The answer is yes because substitution of (3) into (1) yields three $0 = 0$ identities.

CASE 2: For rotation about the intermediate moment of inertia axis we can once more verify that the state variables

$$\omega_1^* = 0 \qquad \omega_2^* \neq 0 \qquad \omega_3^* = 0 \tag{4}$$

define an equilibrium state of (1).

CASE 3: For rotation about the largest moment of inertia axis once more we can show that the state variables

$$\omega_1^* = 0 \qquad \omega_2^* = 0 \qquad \omega_3^* \neq 0 \tag{5}$$

define an equilibrium state of (1).

The implication of the above analysis is that *theoretically* it is possible to maintain pure rotation about any one of the three axes of the space vehicle, since all three conditions correspond to equilibrium states of the nonlinear system (1). However, we know that just because an equilibrium state occurs does not necessarily mean that it is a stable state.

We shall demonstrate in the remainder of this section that

1 Cases 1 and 3, that is, rotation about the smallest or largest moment of inertia axes, are stable (but not asymptotically stable) equilibrium conditions.
2 Case 2, that is, rotation about the intermediate moment of inertia axis corresponds to an unstable equilibrium state. Thus, the system will naturally diverge from this equilibrium state if it is started in it at the initial time.

To analyze the stability properties of the three equilibrium states we first define the state perturbation variables

$$
\begin{aligned}
\omega_1(t) &= \omega_1^* + \delta\omega_1(t) \\
\omega_2(t) &= \omega_2^* + \delta\omega_2(t) \\
\omega_3(t) &= \omega_3^* + \delta\omega_3(t)
\end{aligned}
\tag{6}
$$

In all three cases ω_1^*, ω_2^*, and ω_3^* are constant. Hence,

$$
\begin{aligned}
\dot{\omega}_1(t) &= \frac{d}{dt}\delta\omega_1(t) \\
\dot{\omega}_2(t) &= \frac{d}{dt}\delta\omega_2(t) \qquad \text{or} \quad \dot{\omega}(t) = \frac{d}{dt}\delta\omega(t) \\
\dot{\omega}_3(t) &= \frac{d}{dt}\delta\omega_3(t)
\end{aligned}
\tag{7}
$$

with

$$
\omega(t) \triangleq \begin{bmatrix} \omega_1(t) \\ \omega_2(t) \\ \omega_3(t) \end{bmatrix} \qquad \omega^* \triangleq \begin{bmatrix} \omega_1^* \\ \omega_2^* \\ \omega_3^* \end{bmatrix} \qquad \delta\omega(t) \triangleq \begin{bmatrix} \delta\omega_1(t) \\ \delta\omega_2(t) \\ \delta\omega_3(t) \end{bmatrix}
\tag{8}
$$

We note that (1) can be written in the vector form

$$
\dot{\omega}(t) = \mathbf{f}(\omega(t))
\tag{9}
$$

where

$$
\mathbf{f}(\omega(t)) = \begin{bmatrix} a_1\,\omega_2(t)\omega_3(t) \\ a_2\,\omega_1(t)\omega_3(t) \\ a_3\,\omega_1(t)\omega_2(t) \end{bmatrix}
\tag{10}
$$

and the scalars a_1, a_2, a_3 are defined [see also inequality (2)] by

$$a_1 = \frac{I_2 - I_3}{I_1} < 0$$

$$a_2 = \frac{I_3 - I_1}{I_2} > 0 \tag{11}$$

$$a_3 = \frac{I_1 - I_2}{I_3} < 0$$

The results of the linearization analysis about any equilibrium state vector ω^* will yield an LTI vector differential equation

$$\frac{d}{dt}\delta\omega(t) = \mathbf{A}_* \, \delta\omega(t) \tag{12}$$

where $\mathbf{A}_*$ is the Jacobian matrix evaluated at the equilibrium state ω^*, that is,

$$\mathbf{A}_* \triangleq \left. \frac{\partial \mathbf{f}(\omega(t))}{\partial \omega(t)} \right|_{\omega(t)=\omega^*} \tag{13}$$

For our problem, since $\mathbf{f}(\omega(t))$ is given by Eq. (10), we readily compute the Jacobian matrix

$$\frac{\partial \mathbf{f}(\omega(t))}{\partial \omega(t)} = \begin{bmatrix} 0 & a_1\,\omega_3(t) & a_1\,\omega_2(t) \\ a_2\,\omega_3(t) & 0 & a_2\,\omega_1(t) \\ a_3\,\omega_2(t) & a_3\,\omega_1(t) & 0 \end{bmatrix} \tag{14}$$

Hence for any equilibrium state ω^* [see Eq. (8)] we have

$$\mathbf{A}_* = \begin{bmatrix} 0 & a_1\,\omega_3^* & a_1\,\omega_2^* \\ a_2\,\omega_3^* & 0 & a_2\,\omega_1^* \\ a_3\,\omega_2^* & a_3\,\omega_1^* & 0 \end{bmatrix} \tag{15}$$

All we have to do is evaluate the eigenvalues of $\mathbf{A}_*$ for each equilibrium state to deduce the stability properties at equilibrium.

CASE 1:   The equilibrium state vector ω^* is characterized by

$$\omega^* = \begin{bmatrix} \omega_1^* \\ 0 \\ 0 \end{bmatrix} \qquad \omega_1^* \neq 0 \tag{16}$$

$$\omega_1(t) = \omega_1^* + \delta\omega_1(t)$$

$$\omega_2(t) = 0 + \delta\omega_2(t) \tag{17}$$

$$\omega_3(t) = 0 + \delta\omega_3(t)$$

The state perturbation vector $\delta\omega(t)$ still satisfies

$$\frac{d}{dt}\delta\omega(t) = \mathbf{A}_* \, \delta\omega(t) \tag{18}$$

The $\mathbf{A}_*$ matrix for this equilibrium state is obtained by substituting (16) into (15) to obtain

$$\mathbf{A}_* = \begin{bmatrix} 0 & 0 & 0 \\ 0 & 0 & a_2\,\omega_1^* \\ 0 & a_3\,\omega_1^* & 0 \end{bmatrix} \tag{19}$$

Let us compute the characteristic polynomial $p(\lambda)$ of $\mathbf{A}_*$:

$$p(\lambda) = \det \begin{bmatrix} \lambda & 0 & 0 \\ 0 & \lambda & -a_2\,\omega_1^* \\ 0 & -a_3\,\omega_1^* & \lambda \end{bmatrix} = \lambda^3 - \lambda a_2\,a_3\,\omega_1^{*2} \tag{20}$$

Hence, the three eigenvalues of $\mathbf{A}_*$ given by (19) are at

$$\lambda_1 = 0$$

$$\lambda_2 = j\omega_1^* \sqrt{-a_2\,a_3} \tag{21}$$

$$\lambda_3 = -j\omega_1^* \sqrt{-a_2\,a_3}$$

because, in view of (11), $-a_2\,a_3$ is a positive scalar. These are nonrepeated eigenvalues on the $j\omega$ axis; hence the system (18) is *stable* (but not asymptotically stable). We therefore conclude that rotation about the smallest moment of inertia is stable, in the sense that small perturbations about this equilibrium state will not grow without bound.

CASE 2: The equilibrium state vector ω^* is characterized by

$$\omega^* = \begin{bmatrix} 0 \\ \omega_2^* \\ 0 \end{bmatrix} \qquad \omega_2^* \neq 0 \tag{22}$$

$$\omega_1(t) = 0 + \delta\omega_1(t)$$

$$\omega_2(t) = \omega_2^* + \delta\omega_2(t) \tag{23}$$

$$\omega_3(t) = 0 + \delta\omega_3(t)$$

The state perturbation vector $\delta\omega(t)$ still satisfies the linearized equation

$$\frac{d}{dt}\delta\omega(t) = \mathbf{A}_* \, \delta\omega(t) \tag{24}$$

The $\mathbf{A}_*$ matrix for this equilibrium state is obtained by substituting (22) into (15) to obtain

$$
\mathbf{A}_* = \begin{bmatrix} 0 & 0 & a_1\,\omega_2^* \\ 0 & 0 & 0 \\ a_3\,\omega_2^* & 0 & 0 \end{bmatrix}
\tag{25}
$$

Its characteristic polynomial is

$$
p(\lambda) = \det \begin{bmatrix} \lambda & 0 & -a_1\,\omega_2^* \\ 0 & \lambda & 0 \\ -a_3\,\omega_2^* & 0 & \lambda \end{bmatrix} = \lambda^3 - \lambda a_1\,a_3\,\omega_2^{*2}
\tag{26}
$$

In this case, $a_1\,a_3$ is a positive number [see Eq. (11)], and hence the three eigenvalues of $\mathbf{A}_*$ given by (25) are all real and

$$
\begin{aligned}
\lambda_1 &= 0 \\
\lambda_2 &= -\omega_2^* \sqrt{a_1\,a_3} \\
\lambda_3 &= +\omega_2^* \sqrt{a_1\,a_3}
\end{aligned}
\tag{27}
$$

Now we note that one of the eigenvalues, λ_3, is positive. Hence, the linearized system (24) is unstable. This means that any small perturbation about this equilibrium will contain a component proportional to

$$
\exp(\omega_2^* \sqrt{a_1\,a_3}\,)t
\tag{28}
$$

Thus, small initial perturbations will grow exponentially, and after a while the linearization will become invalid and we must employ the nonlinear dynamics (1) to study what happens in detail. Nevertheless, our analysis has indicated that pure rotation about the intermediate moment of inertia is not possible.

CASE 3:   The equilibrium state vector $\boldsymbol{\omega}^*$ is

$$
\boldsymbol{\omega}^* = \begin{bmatrix} 0 \\ 0 \\ \omega_3^* \end{bmatrix} \qquad \omega_3^* \neq 0
\tag{29}
$$

and

$$
\begin{aligned}
\omega_1(t) &= 0 + \delta\omega_1(t) \\
\omega_2(t) &= 0 + \delta\omega_2(t) \\
\omega_3(t) &= \omega_3^* + \delta\omega_3(t)
\end{aligned}
\tag{30}
$$

The state perturbation vector $\delta\omega(t)$ satisfies the linearized equation

$$\frac{d}{dt}\,\delta\omega(t) = \mathbf{A}_* \,\delta\omega(t) \tag{31}$$

In this case, $\mathbf{A}_*$ is found by substituting (29) into (15) to obtain

$$\mathbf{A}_* = \begin{bmatrix} 0 & a_1\,\omega_3^* & 0 \\ a_2\,\omega_3^* & 0 & 0 \\ 0 & 0 & 0 \end{bmatrix} \tag{32}$$

From (11) we see that $a_1 a_2$ is a negative number. Hence the eigenvalues of $\mathbf{A}_*$ are given by

$$\lambda_1 = 0$$
$$\lambda_2 = j\omega_3^*\sqrt{-a_1 a_2} \tag{33}$$
$$\lambda_3 = -j\omega_3^*\sqrt{-a_1 a_2}$$

None of the eigenvalues has a positive real part. Hence, the system (31) is stable. Thus, as in case 1, we may conclude that rotation about the largest moment of inertia axis corresponds to a stable equilibrium condition.

7.18 COMPENSATION USING STATE-VARIABLE FEEDBACK

Often the natural response of a linear or linearized system, in the absence of any corrective inputs, possesses certain undesirable characteristics due to initial conditions. Of course, if the unforced system is unstable, the divergence of the state from its desired equilibrium value constitutes such an undesirable response. However, even if the system is stable, its response to initial conditions may also possess undesirable characteristics. In such problems it may happen that

   1 An initial deviation of the actual state of a system from a desirable value takes a long time to be nulled out.
   2 The time history of the state variables may show undesirable oscillations.

To illustrate this issue of desirable vs. undesirable response in more concrete terms let us consider the vertical motion of an aircraft. Most commercial aircraft under FAA Air Traffic Control regulation are supposed to maintain a constant altitude assigned to them; if the pilot wishes to change altitude, he must request permission to do so from the Air Traffic Control center.

   However, upward or downward wind gusts often carry the aircraft some distance from its desired altitude. The pilot or autopilot then attempts to null out

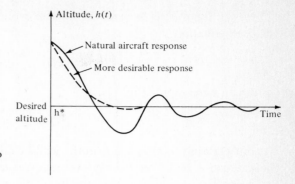

FIGURE 7.18.1
Desirable and undesirable corrections to
an initial altitude error for aircraft.

this altitude error. For certain aircraft, this altitude error may be naturally corrected, i.e., without pilot interference, in the manner shown in Fig. 7.18.1. Eventually, the aircraft returns to its desired altitude (hence this corresponds to an asymptotically stable equilibrium condition). However, the natural aircraft response shown in Fig. 7.18.1 may be undesirable from the passengers' viewpoint, since people (and their stomachs) do not like to be moving up and down all the time. Thus, one may argue that the dashed curve in Fig. 7.18.1 represents a more desirable response. Such response can be attained by either a good pilot or a well-designed autopilot simply by properly adjusting the angle of the elevators on the aircraft as a function of time.

The above example illustrates that an undesirable system response can be often corrected by the judicious choice of the input to the system. The key question is then: Are there any general methods by which one characterizes undesirable and desirable responses, and are there methods of adjusting the system inputs so that a desirable response is attained? The answer is yes. We shall explain the key ideas in the remainder of this section, noting that the basic conceptual techniques are completely analogous to those discussed for systems described by difference equations (see Sec. 4.9).

7.18.1   Naturally Linear Systems

Let us first consider an LTI system whose n-dimensional state vector $\mathbf{x}(t)$ and m-dimensional input vector $\mathbf{u}(t)$ are related by the vector differential equation

$$\dot{\mathbf{x}}(t) = \mathbf{A}\mathbf{x}(t) + \mathbf{B}\mathbf{u}(t) \tag{1}$$

Let us suppose that the zero state, $\mathbf{x}(t) = \mathbf{0}$, is the desired state of the system. Let us further suppose that at the initial time $t = 0$, the initial state

$$\mathbf{x}(0) = \boldsymbol{\xi} \neq \mathbf{0} \tag{2}$$

provides us with the deviations of the state variables from their desired zero values.

If the input is set to zero, the natural evolution of the state is governed by the unforced equation

$$\dot{\mathbf{x}}(t) = \mathbf{A}\mathbf{x}(t) \qquad \mathbf{x}(0) = \boldsymbol{\xi} \tag{3}$$

and it can be computed by

$$\mathbf{x}(t) = e^{\mathbf{A}t}\boldsymbol{\xi} \tag{4}$$

We can certainly evaluate $\mathbf{x}(t)$ numerically using Eq. (4) and look for undesirable characteristics. However, we can express the solution of (3) using the eigenvalues of the $\mathbf{A}$ matrix and in so doing obtain a better feel for what may constitute an undesirable response.

We know that the solution of (3) can be written

$$\mathbf{x}(t) = \sum_{i=1}^{n} \beta_i e^{\lambda_i t} \mathbf{v}_i \tag{5}$$

where the λ_i are the eigenvalues of $\mathbf{A}$, $\mathbf{v}_i$ the corresponding eigenvectors, and the scalars β_i depend on the initial state $\boldsymbol{\xi}$, which is expressed as a linear combination of the eigenvectors

$$\boldsymbol{\xi} = \sum_{i=1}^{n} \beta_i \mathbf{v}_i \tag{6}$$

We remind the reader that the time character of $\mathbf{x}(t)$ is characterized by the exponential time functions $\exp(\lambda_i t)$ and hence by the eigenvalues. Hence, it is reasonable to suspect that the eigenvalues of the $\mathbf{A}$ matrix contain information that can be used to evaluate undesirable characteristics of the time evolution of $\mathbf{x}(t)$.

What constitutes an undesirable response of course depends on the physical nature of the system (1) and the specific application at hand. However, independent of the physical problem, there are certain common undesirable characteristics that can be related to the location of the eigenvalues of the $\mathbf{A}$ matrix in the complex plane. A typical, but by no means exhaustive list, is given below.

Instability   The state vector $\mathbf{x}(t)$ increases without bound,

$$\lim_{t \to \infty} \|\mathbf{x}(t)\| \to \infty \tag{7}$$

so that $\mathbf{x}(t)$ never approaches its desired value $\mathbf{0}$. The clue to this undesirable response is that one or more of the eigenvalues of the $\mathbf{A}$ matrix are in the right half of the complex plane.

Stability but not asymptotic stability   In this case all the eigenvalues of the **A** matrix have nonpositive real parts, but one or more eigenvalues of the **A** matrix are on the imaginary axis of the complex plane. In this case, the state remains bounded, i.e.,

$$\lim_{t \to \infty} \|\mathbf{x}(t)\| \leq M \tag{8}$$

but does not tend to the zero state, i.e.,

$$\lim_{t \to \infty} \|\mathbf{x}(t)\| \neq 0 \tag{9}$$

In such problems the components of the state vector may have constant and/or undamped sinusoidal characteristics.† Now let us suppose that all the eigenvalues of the **A** matrix have negative real parts, so that indeed

$$\lim_{t \to \infty} \|\mathbf{x}(t)\| = 0 \tag{10}$$

and at least the state $\mathbf{x}(t)$ eventually approaches its desired zero value. However, this property alone may not be enough to make $\mathbf{x}(t)$ automatically desirable. In this case, one usually focuses attention on the *dominant* eigenvalues of the **A** matrix, i.e., those closest to the imaginary axis, because, as we have already discussed, the dominant eigenvalues contribute most heavily to the transient response. On the basis of the dominant eigenvalues one can predict certain other responses that may be undesirable.

Stable, lightly damped oscillations   This occurs when one has a dominant pair of complex eigenvalues close to the imaginary axis. Suppose that this dominant pair is described by

$$\begin{aligned} \lambda_1 &= -\sigma_1 + j\omega_1 & \sigma_1 &> 0 \\ \lambda_2 &= -\sigma_1 - j\omega_1 & \omega_1 &> 0 \end{aligned} \tag{11}$$

Then the components of the state vector will have time functions of the form

$$ae^{-\sigma_1 t}\cos(\omega_1 t + \phi) \tag{12}$$

where a and ϕ are constants. Thus, the smaller the value of σ_1 the longer the sinusoidal time functions may persist, and this may constitute an undesirable response (as was the case with the altitude profiles discussed at the beginning of this section).

---

† This was the case with the stable equilibria (cases 1 and 3) of the space vehicle analyzed in Sec. 7.17.

Slow speed of response   Another undesirable characteristic occurs when $\|\mathbf{x}(t)\|$ does not approach zero fast enough. Once more the speed of response is governed by the dominant eigenvalues and in particular by the magnitude of the real part. Thus, the closer the eigenvalues are to the origin or to the imaginary axis the slower the response.

Thus, gross characteristics of the time response can be obtained by visual examination of the location of the eigenvalues of the **A** matrix in the complex plane. In general, the farther away the eigenvalues in the left half of the complex plane the faster the speed of response; also, for complex eigenvalues the smaller the ratio of their imaginary to real parts the less oscillatory the response.

Now let us return to our original problem and see what we can do to improve an unsatisfactory natural response. Intuitively, we would expect to use the input $\mathbf{u}(t)$ for this purpose. Let us suppose that we can measure all the state variables of our system. Let us also assume that we generate the input $\mathbf{u}(t)$ to the system (1) by multiplying the state vector $\mathbf{x}(t)$ by some $m \times n$ *constant* matrix **G**. Thus, $\mathbf{u}(t)$ is generated by means of the relation

$$\mathbf{u}(t) = \mathbf{G}\mathbf{x}(t) \tag{13}$$

Our system is still described by Eq. (1). Substituting Eq. (13) into (1), we obtain the vector differential equation

$$\dot{\mathbf{x}}(t) = \mathbf{A}\mathbf{x}(t) + \mathbf{B}\mathbf{G}\mathbf{x}(t) \tag{14}$$

Factoring out $\mathbf{x}(t)$ in (14), we obtain

$$\dot{\mathbf{x}}(t) = (\mathbf{A} + \mathbf{B}\mathbf{G})\mathbf{x}(t) \tag{15}$$

We shall refer to the unforced system $\dot{\mathbf{x}}(t) = \mathbf{A}\mathbf{x}(t)$ as the *open-loop system* and call **A** the *open-loop-system matrix*. It is common practice to refer to the system $\dot{\mathbf{x}}(t) = (\mathbf{A} + \mathbf{B}\mathbf{G})\mathbf{x}(t)$ as the *closed-loop system* and the matrix $\mathbf{A} + \mathbf{B}\mathbf{G}$ as the *closed-loop-system matrix*; the matrix **G** is called the *state-feedback-gain matrix*.

We now note that the use of feedback, i.e., the generation of $\mathbf{u}(t)$ from $\mathbf{x}(t)$ according to Eq. (13), has resulted in a new system defined by Eq. (15). For the same initial state $\mathbf{x}(0) = \boldsymbol{\xi}$ the closed-loop system will behave differently from the open-loop system, because:

Open-loop state: $\qquad\qquad\qquad \mathbf{x}(t) = e^{\mathbf{A}t}\boldsymbol{\xi} \tag{16}$

Closed-loop state: $\qquad\qquad\qquad \mathbf{x}(t) = e^{(\mathbf{A}+\mathbf{B}\mathbf{G})t}\boldsymbol{\xi} \tag{17}$

Hence, the time nature of the closed-loop state depends on the original matrices **A** and **B** and the particular state-feedback-gain matrix **G** we have selected.

Now let us examine how to analyze the behavior of the closed-loop system when we have selected a particular state-feedback-gain matrix **G**. *Since the closed-loop system (15) is an LTI system, we know that its time response will be governed by the eigenvalues of the closed-loop-system matrix* **A** + **BG**.

Thus, if we let $\mu_1, \mu_2, \ldots, \mu_n$ denote the eigenvalues of **A** + **BG**, and if we let $\mathbf{w}_1, \mathbf{w}_2, \ldots, \mathbf{w}_n$ denote the corresponding eigenvectors, then (assuming that the eigenvalues are distinct) we can write the time evolution of the closed-loop state $\mathbf{x}(t)$ as

$$\mathbf{x}(t) = \sum_{i=1}^{n} \gamma_i e^{\mu_i t} \mathbf{w}_i \tag{18}$$

where

$$\xi = \sum_{i=1}^{n} \gamma_i \mathbf{w}_i \tag{19}$$

It therefore follows that given our original system (which fixes **A** and **B**), once we select a **G** matrix, we can look at the location of the eigenvalues $\mu_1, \mu_2, \ldots, \mu_n$ of the matrix **A** + **BG** and, at least qualitatively analyze the response of the closed-loop system.

This is perfectly all right, but it still does not solve our design problem, which consists of curing the undesirable response of the open-loop system. However, we have gained some insight since we have demonstrated that through feedback we can change our system from

$$\dot{\mathbf{x}}(t) = \mathbf{A}\mathbf{x}(t) \tag{20}$$

to

$$\dot{\mathbf{x}}(t) = (\mathbf{A} + \mathbf{BG})\mathbf{x}(t) \tag{21}$$

and this feedback has *not* changed the linearity or time invariance of the overall system.

Let us then return to the design problem. Suppose that the examination of the eigenvalues $\lambda_1, \lambda_2, \ldots, \lambda_n$ of **A** has led us to dislike the open-loop system. On the other hand, we know at least qualitatively the relationship between eigenvalues and goodness of response. Hence, we can attempt to pose the following question: Suppose that we select the eigenvalues $\mu_1, \mu_2, \ldots, \mu_n$ of the closed-loop-system matrix **A** + **BG** so that they lie in a desirable set of positions in the complex plane. Can we find a state-feedback-gain matrix **G** such that the eigenvalues of **A** + **BG** are indeed the desired ones $\mu_1, \mu_2, \ldots, \mu_n$? The answer to this question is, in general, yes. We shall outline the answer to this question in Theorem 1 below, but first it is convenient to present a definition.

Definition 1 *Let* **A** *be an arbitrary* $n \times n$ *real matrix. Let* **B** *be an arbitrary* $n \times m$ *matrix. Then the pair* $[\mathbf{A}, \mathbf{B}]$ *is said to be a* controllable *pair if and only if the rank of the* $n \times (nm)$ *matrix* **H** *defined by*

$$\mathbf{H} = [\mathbf{B} \; \vdots \; \mathbf{AB} \; \vdots \; \mathbf{A}^2\mathbf{B} \; \vdots \; \cdots \; \vdots \; \mathbf{A}^{n-1}\mathbf{B}] \tag{22}$$

is equal to n. In other words, there are n linearly independent column vectors (out of the nm) in the **H** *matrix.*

It is important to explain how the **H** matrix in Eq. (22) is constructed. Its first m columns are the columns of the $n \times m$ matrix **B**. Its next m columns are the columns of the $n \times m$ matrix **AB**. Its next m columns are the columns of the $n \times m$ matrix **A**^{2}**B**, and so on.

EXAMPLE 1 Suppose that

$$\mathbf{A} = \begin{bmatrix} 0 & 1 & -1 \\ 1 & 0 & 1 \\ 0 & 1 & -1 \end{bmatrix} \qquad \mathbf{B} = \begin{bmatrix} 1 & 0 \\ 0 & 1 \\ 0 & 0 \end{bmatrix}$$

Since **A** is 3×3 $(n = 3)$ and **B** is 3×2 $(m = 2)$, **H** is the 3×6 matrix

$$\mathbf{H} = [\mathbf{B} \; \vdots \; \mathbf{AB} \; \vdots \; \mathbf{A}^2\mathbf{B}]$$

But

$$\mathbf{AB} = \begin{bmatrix} 0 & 1 \\ 1 & 0 \\ 0 & 1 \end{bmatrix} \qquad \text{and} \qquad \mathbf{A}^2\mathbf{B} = \begin{bmatrix} 1 & -1 \\ 0 & 2 \\ 1 & -1 \end{bmatrix}$$

Thus **H** is the 3×6 matrix

$$\mathbf{H} = \begin{bmatrix} 1 & 0 & \vdots & 0 & 1 & \vdots & 1 & -1 \\ 0 & 1 & \vdots & 1 & 0 & \vdots & 0 & 2 \\ 0 & 0 & \vdots & 0 & 1 & \vdots & 1 & -1 \end{bmatrix}$$

We note that the first, second, and fourth columns of **H** are linearly independent. Hence, according to Definition 1, $[\mathbf{A}, \mathbf{B}]$ is a controllable pair.

EXAMPLE 2 Suppose that

$$\mathbf{A} = \begin{bmatrix} 0 & 1 \\ 0 & 1 \end{bmatrix} \qquad \mathbf{B} = \begin{bmatrix} 1 \\ 0 \end{bmatrix}$$

Thus $n = 2$ and $m = 1$. The matrix $\mathbf{H}$ is

$$\mathbf{H} = \begin{bmatrix} \mathbf{B} & \vdots & \mathbf{AB} \end{bmatrix} = \begin{bmatrix} 1 & \vdots & 0 \\ 0 & \vdots & 0 \end{bmatrix}$$

The columns of $\mathbf{H}$ are *not* linearly independent; hence $[\mathbf{A}, \mathbf{B}]$ is *not* a controllable pair.

We are now ready to state the precise theorem of interest.

Theorem 1 *Consider the system* (1) *and suppose that $\mathbf{A}$ is any $n \times n$ real matrix and that $\mathbf{B}$ is any $n \times m$ real matrix. Furthermore, suppose that $[\mathbf{A}, \mathbf{B}]$ is a controllable pair. Then there exists (at least one) $m \times n$ real state-feedback-gain matrix $\mathbf{G}$ such that the eigenvalues of $\mathbf{A} + \mathbf{B}\mathbf{G}$ are exactly equal to prespecified values $\mu_1, \mu_2, \ldots, \mu_n$, provided that any complex μ_i come in complex conjugate pairs.*

We shall not prove this theorem in this book.

This theorem indicates that we can in general construct feedback systems with prespecified eigenvalues, which presumably we pick so that the system has a satisfactory response.

EXAMPLE 3 To illustrate these ideas let us consider a second-order system described by

$$\underbrace{\begin{bmatrix} \dot{x}_1(t) \\ \dot{x}_2(t) \end{bmatrix}}_{\dot{\mathbf{x}}(t)} = \underbrace{\begin{bmatrix} 0 & 1 \\ -1 & 0 \end{bmatrix}}_{\mathbf{A}} \underbrace{\begin{bmatrix} x_1(t) \\ x_2(t) \end{bmatrix}}_{\mathbf{x}(t)} + \underbrace{\begin{bmatrix} 0 \\ 1 \end{bmatrix}}_{\mathbf{B}} \underbrace{u(t)}_{u(t)} \tag{23}$$

To use state-variable feedback we want

$$u(t) = -\mathbf{G}\begin{bmatrix} x_1(t) \\ x_2(t) \end{bmatrix} \tag{24}$$

Hence, $\mathbf{G}$ must be a 1×2 matrix (row vector)

$$u(t) = -[g_1 \quad g_2]\begin{bmatrix} x_1(t) \\ x_2(t) \end{bmatrix} = -[g_1 x_1(t) + g_2 x_2(t)] \tag{25}$$

The eigenvalues of the **A** matrix are at

$$\lambda_1 = j \qquad \lambda_2 = -j \tag{26}$$

Suppose they are undesirable, and hence feedback is needed.

For any values of g_1 and g_2 the closed-loop system is

$$\begin{bmatrix} \dot{x}_1(t) \\ \dot{x}_2(t) \end{bmatrix} = \begin{bmatrix} 0 & 1 \\ -1 & 0 \end{bmatrix} \begin{bmatrix} x_1(t) \\ x_2(t) \end{bmatrix} - \begin{bmatrix} 0 \\ 1 \end{bmatrix} \begin{bmatrix} g_1 & g_2 \end{bmatrix} \begin{bmatrix} x_1(t) \\ x_2(t) \end{bmatrix} \tag{27}$$

or

$$\begin{bmatrix} \dot{x}_1(t) \\ \dot{x}_2(t) \end{bmatrix} = \begin{bmatrix} 0 & 1 \\ -1 & 0 \end{bmatrix} \begin{bmatrix} x_1(t) \\ x_2(t) \end{bmatrix} - \begin{bmatrix} 0 & 0 \\ g_1 & g_2 \end{bmatrix} \begin{bmatrix} x_1(t) \\ x_2(t) \end{bmatrix} \tag{28}$$

or

$$\underbrace{\begin{bmatrix} \dot{x}_1(t) \\ \dot{x}_2(t) \end{bmatrix}}_{\dot{\mathbf{x}}(t)} = \underbrace{\begin{bmatrix} 0 & 1 \\ -1 - g_1 & -g_2 \end{bmatrix}}_{\mathbf{A} - \mathbf{BG}} \underbrace{\begin{bmatrix} x_1(t) \\ x_2(t) \end{bmatrix}}_{\mathbf{x}(t)} \tag{29}$$

Now suppose that we want the eigenvalues of the $\mathbf{A} - \mathbf{BG}$ matrix to be at

$$\mu_1 = -4 \qquad \mu_2 = -6 \tag{30}$$

Since $[\mathbf{A}, \mathbf{B}]$ can be shown to be controllable, Theorem 1 guarantees that we can find g_1 and g_2. The characteristic polynomial $p(\mu)$ of $\mathbf{A} - \mathbf{BG}$ is

$$p(\mu) = \det \begin{bmatrix} \mu & -1 \\ 1 + g_1 & \mu + g_2 \end{bmatrix} = \mu(\mu + g_2) + 1 + g_1$$
$$= \mu^2 + g_2 \mu + 1 + g_1 \tag{31}$$

Since we want to have

$$p(\mu) = (\mu - \mu_1)(\mu - \mu_2) = (\mu + 4)(\mu + 6) = \mu^2 + 10\mu + 24 \tag{32}$$

we must select

$$g_1 = 23 \qquad g_2 = 10 \tag{33}$$

★ **7.18.2   Extension to Tracking Systems**

The notion of using feedback to improve the response characteristics of a dynamical LTI system is not limited to the case where the desirable state is the zero state. It can also be used to null out, in a satisfactory manner, deviations of the actual state of the system $\mathbf{x}(t)$ from a time-varying desired profile $\mathbf{x}^*(t)$.

To be more precise let us consider once more an LTI system

$$\dot{\mathbf{x}}(t) = \mathbf{A}\mathbf{x}(t) + \mathbf{B}\mathbf{u}(t) \tag{34}$$

Let us suppose that if the system (34) starts at a specific initial state

$$\mathbf{x}^*(0) = \boldsymbol{\xi}^* \tag{35}$$

we have found a specific input time function denoted by $\mathbf{u}^*(t)$, $0 \le t < \infty$, such that the resultant state trajectory $\mathbf{x}^*(t)$, given by

$$\mathbf{x}^*(t) = e^{\mathbf{A}t}\left[\boldsymbol{\xi}^* + \int_0^t e^{-\mathbf{A}\tau}\mathbf{B}\mathbf{u}^*(\tau)d\tau \right] \tag{36}$$

represents a desirable state evolution.

Let us now suppose that, due to some disturbances, the initial state is no longer equal to $\boldsymbol{\xi}^*$ and instead we have

$$\mathbf{x}(0) = \boldsymbol{\xi} \ne \boldsymbol{\xi}^* \tag{37}$$

If, under these circumstances, we continue to apply $\mathbf{u}^*(t)$, the state will now be the solution of

$$\dot{\mathbf{x}}(t) = \mathbf{A}\mathbf{x}(t) + \mathbf{B}\mathbf{u}^*(t) \qquad \mathbf{x}(0) = \boldsymbol{\xi} \tag{38}$$

and hence

$$\mathbf{x}(t) = e^{\mathbf{A}t}\left[\boldsymbol{\xi} + \int_0^t e^{-\mathbf{A}\tau}\mathbf{B}\mathbf{u}^*(\tau)\,d\tau \right] \ne \mathbf{x}^*(t) \tag{39}$$

Let $\boldsymbol{\delta}\mathbf{x}(t)$ denote the difference between the actual state $\mathbf{x}(t)$ and the desired state $\mathbf{x}^*(t)$, that is,

$$\boldsymbol{\delta}\mathbf{x}(t) \triangleq \mathbf{x}(t) - \mathbf{x}^*(t) \tag{40}$$

Clearly

$$\boldsymbol{\delta}\mathbf{x}(t) = e^{\mathbf{A}t}(\boldsymbol{\xi} - \boldsymbol{\xi}^*) \tag{41}$$

Thus, it is the **A** matrix that governs the time evolution of $\boldsymbol{\delta}\mathbf{x}(t)$. There can certainly exist cases where $\boldsymbol{\delta}\mathbf{x}(t)$ behaves unsatisfactorily [ideally we want $\boldsymbol{\delta}\mathbf{x}(t)$ to go to zero], and we seek methods of improving its response.

There is an alternate way of seeing how this problem reduces to that considered in Sec. 7.18.1. From Eq. (40) we find

$$\frac{d}{dt}\delta\mathbf{x}(t) = \dot{\mathbf{x}}(t) - \dot{\mathbf{x}}^*(t) \tag{42}$$

But $\dot{\mathbf{x}}(t)$ is given by Eq. (38). Similarly, since $\mathbf{u}^*(t)$ generates $\mathbf{x}^*(t)$, we must have

$$\dot{\mathbf{x}}^*(t) = \mathbf{A}\mathbf{x}^*(t) + \mathbf{B}\mathbf{u}^*(t) \tag{43}$$

Thus, by substituting Eqs. (43) and (38) into Eq. (42) we obtain the dynamics of $\delta\mathbf{x}(t)$

$$\frac{d}{dt}\delta\mathbf{x}(t) = \mathbf{A}\delta\mathbf{x}(t) \qquad \delta\mathbf{x}(0) = \boldsymbol{\xi} - \boldsymbol{\xi}^* \tag{44}$$

Since our desired value for $\delta\mathbf{x}(t)$ is zero, we are facing the same problem as in Sec. 7.18.1. If Eq. (44) yields an unsatisfactory performance, we seek ways of curing it.

The only way that we can affect $\delta\mathbf{x}(t)$ is to change the actual input of the system $\mathbf{u}(t)$ from its preset value $\mathbf{u}^*(t)$. Let us define the control correction vector $\delta\mathbf{u}(t)$ by

$$\mathbf{u}(t) \triangleq \mathbf{u}^*(t) + \delta\mathbf{u}(t) \tag{45}$$

If we do that, the state $\mathbf{x}(t)$ will now evolve according to

$$\dot{\mathbf{x}}(t) = \mathbf{A}\mathbf{x}(t) + \mathbf{B}\mathbf{u}^*(t) + \mathbf{B}\delta\mathbf{u}(t) \qquad \mathbf{x}(0) = \boldsymbol{\xi} \tag{46}$$

which is obtained by substituting Eq. (45) into Eq. (34).

Now we can recompute the dynamics of $\delta\mathbf{x}(t)$ by substituting Eqs. (46) and (43) into Eq. (42) to obtain

$$\frac{d}{dt}\delta\mathbf{x}(t) = \mathbf{A}\delta\mathbf{x}(t) + \mathbf{B}\delta\mathbf{u}(t) \tag{47}$$

Hence, we can use state perturbation feedback, i.e.,

$$\delta\mathbf{u}(t) = \mathbf{G}\delta\mathbf{x}(t) \tag{48}$$

to change the dynamics of $\delta\mathbf{x}(t)$ to satisfy the equation

$$\frac{d}{dt}\delta\mathbf{x}(t) = (\mathbf{A} + \mathbf{B}\mathbf{G})\delta\mathbf{x}(t) \tag{49}$$

We can use then Theorem 1 to deduce that we can find a $\mathbf{G}$ matrix such that the eigenvalues $\mu_1, \mu_2, \ldots, \mu_n$ of $\mathbf{A} + \mathbf{B}\mathbf{G}$ are at prespecified locations.

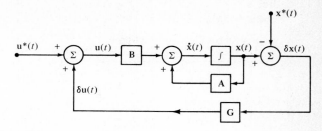

FIGURE 7.18.2
Structure of the feedback systems that nulls out $\delta\mathbf{x}(t)$ when $\mathbf{x}^*(t) = \mathbf{A}\mathbf{x}^*(t) + \mathbf{B}\mathbf{u}^*(t)$, $\mathbf{x}^*(0) = \boldsymbol{\xi}^*$.

The structure of the overall system that makes the actual state $\mathbf{x}(t)$ stay close to the desired state $\mathbf{x}^*(t)$ is shown in Fig. 7.18.2.

7.18.3   Extension to Nonlinear Systems

We saw in Sec. 7.15 how the linearization of a nonlinear system

$$\dot{\mathbf{x}}(t) = \mathbf{f}(\mathbf{x}(t), \mathbf{u}(t)) \tag{50}$$

about a constant equilibrium state $\mathbf{x}^*$ and constant equilibrium control $\mathbf{u}^*$, defined by the relation

$$\mathbf{0} = \mathbf{f}(\mathbf{x}^*, \mathbf{u}^*) \tag{51}$$

led to an LTI system

$$\frac{d}{dt}\delta\mathbf{x}(t) = \mathbf{A}_* \, \delta\mathbf{x}(t) + \mathbf{B}_* \, \delta\mathbf{u}(t) \tag{52}$$

where

$$\mathbf{x}(t) \overset{\triangle}{=} \mathbf{x}^* + \delta\mathbf{x}(t) \qquad \mathbf{u}(t) \overset{\triangle}{=} \mathbf{u}^* + \delta\mathbf{u}(t) \tag{53}$$

$$\mathbf{A}_* = \left.\frac{\partial \mathbf{f}}{\partial \mathbf{x}(t)}\right|_{\mathbf{x}(t)=\mathbf{x}^*,\mathbf{u}(t)=\mathbf{u}^*} \qquad \mathbf{B}_* = \left.\frac{\partial \mathbf{f}}{\partial \mathbf{u}(t)}\right|_{\mathbf{x}(t)=\mathbf{x}^*,\mathbf{u}(t)=\mathbf{u}^*} \tag{54}$$

Once more in these problems we are interested in bringing $\delta\mathbf{x}(t)$ to zero, so that the actual system state $\mathbf{x}(t)$ will approach the desired equilibrium state $\mathbf{x}^*$.

If we set $\delta\mathbf{u}(t) = \mathbf{0}$, so that $\mathbf{u}(t) = \mathbf{u}^*$, then the actual input to the system is the equilibrium constant input $\mathbf{u}^*$. Under these circumstances, the nature of $\delta\mathbf{x}(t)$ is governed by

$$\frac{d}{dt}\delta\mathbf{x}(t) = \mathbf{A}_* \, \delta\mathbf{x}(t) \tag{55}$$

and undesirable characteristics can be deduced by examining the eigenvalues of $\mathbf{A}_*$.

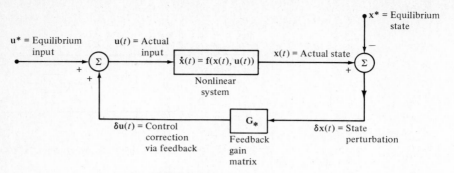

FIGURE 7.18.3
Feedback control of a nonlinear system about an equilibrium.

Once more we can employ state perturbation feedback, i.e.,

$$\delta \mathbf{u}(t) = \mathbf{G}_* \, \delta \mathbf{x}(t) \tag{56}$$

to change the dynamics of $\delta \mathbf{x}(t)$ from that of (55) into

$$\frac{d}{dt} \delta \mathbf{x}(t) = (\mathbf{A}_* + \mathbf{B}_* \, \mathbf{G}_*) \delta \mathbf{x}(t) \tag{57}$$

Once more we can use Theorem 1 to see whether we can find $\mathbf{G}_*$ such that the eigenvalues of $\mathbf{A}_* + \mathbf{B}_* \, \mathbf{G}_*$ are at prespecified locations. Figure 7.18.3 shows the structure of this system.

7.19   CONCLUDING REMARKS

In this chapter we considered systems analysis and design tools that can be used to understand the behavior of continuous-time systems described by vector differential equations. The conceptual tools are completely parallel to those studied for discrete-time dynamical systems (see Chap. 4).

Such analysis and design techniques are used extensively in practical problems. The analysis of open-loop systems and their properties is used to decide whether the behavior of the natural system is satisfactory. If it is not, feedback is used to change the system dynamics; in this manner, the closed-loop system can be given more desirable characteristics.

Two additional topics are needed to complete our treatment of continuous-time dynamical systems. In Chap. 8 we shall bring under a common roof the state-variable approach and the input-output approach and in Chap. 9 we shall present techniques for obtaining numerical solutions, using digital computers, to vector differential equations.

REFERENCES

1 PIELOU, E. C.: "An Introduction to Mathematical Ecology," Wiley, New York, 1969.
2 LOTKA, A. J.: "Elements of Physical Biology," Williams & Wilkins, Baltimore, 1925.
3 ODUM, E. P.: "Fundamentals of Ecology," 2d ed., Saunders, Philadelphia, 1959.
4 OGATA, K.: "State Space Analysis of Control Systems," Prentice-Hall, Englewood Cliffs, N. J. 1967.
5 BROCKETT, R. W.: "Finite Dimensional Linear Systems," Wiley, New York, 1970.
6 CODDINGTON, E. A., and N. LEVINSON: "Theory of Ordinary Differential Equations," McGraw-Hill, New York, 19,55.
7 DeRUSSO, P. M., K. J. ROY, and C. M. CLOSE: "State Variables for Engineers," Wiley, New York, 1965.
8 SCHULTZ, D. G., and J. L. MELSA: "State Functions and Linear Control Systems," McGraw-Hill, New York, 1967.
9 BELLMAN, R.: "Stability Theory of Differential Equations," McGraw-Hill, New York, 1953.
10 ZADEH, L. A., and C. A. DESOER: "Linear System Theory: The State Space Approach," McGraw-Hill, New York, 1963.
11 ROSENBROCK, H. H.: "State Space and Multivariable Theory," Wiley, New York, 1970.
12 DESOER, C. A.: "Notes for a Second Course on Linear Systems," Van Nostrand Reinhold, New York, 1970.

EXERCISES

Section 7.3

7.3.1   In this exercise you set up the dynamical equations of an interconnected set of masses and springs. The basic unit of such a system is defined in Fig. Ex. 7.3.1a. We

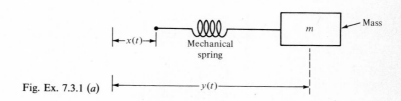

Fig. Ex. 7.3.1 (*a*)

let $x(t)$ be the position of one end of a mechanical spring, and we let $y(t)$ denote the position of its other end; $y(t)$ coincides with the position of a mass m. The mechanical spring is an energy-storage element; it produces a force when contracted or expanded. A nonlinear spring produces a force F which depends on its displacement

$$F_s = \text{function of } y(t) \text{ and } x(t)$$

A linear mechanical spring produces a force

$$F_s(t) = K(y(t) - x(t)) \qquad K \ge 0$$

where K is a characteristic constant of the spring (related to its stiffness). The force produced by the spring influences the motion of the mass. In general, the mass motion is subject to friction. We assume that the friction force F_f is proportional to the velocity $v(t)$ of the mass.

$$F_f(t) = bv(t) \qquad b \ge 0$$
$$v(t) = \dot{y}(t)$$

where b is the friction coefficient. Under these assumptions the motion of the mass is characterized by the second-order LTI equation

$$m\ddot{y}(t) + b\dot{y}(t) + Ky(t) = Kx(t)$$

We now proceed with the formulation of a specific problem; Fig. Ex. 7.3.1b shows the interconnection of mass-spring systems. Let us define the velocity variables

$$x_5(t) = \dot{x}_1(t) \qquad x_6(t) = \dot{x}_2(t) \qquad x_7(t) = \dot{x}_3(t) \qquad x_8(t) = \dot{x}_4(t)$$

Let each mass be characterized by

$$m_1 = 10 \qquad m_2 = 2 \qquad m_3 = 20 \qquad m_4 = 5$$

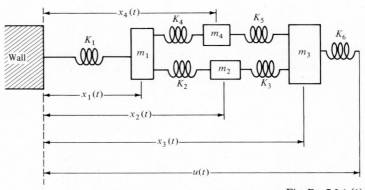

Fig. Ex. 7.3.1 (b)

Let each spring constant be characterized by

$$K_1 = 1 \qquad K_2 = 0.1 \qquad K_3 = 0.2 \qquad K_4 = 2 \qquad K_5 = 1 \qquad K_6 = 0.6$$

Let the friction coefficient associated with each mass be

$$b_1 = 0.5 \qquad b_2 = 0.1 \qquad b_3 = 0.4 \qquad b_4 = 2$$

Let $u(t)$ denote the position of the last spring. We assume that it is controlled from an outside cause [hence $u(t)$ is the input to the system]. Let $\mathbf{x}(t)$ be the 8-dimensional column vector which describes the state of the system (positions and velocities of each mass).

$$\mathbf{x}(t) = \begin{bmatrix} x_1(t) \\ x_2(t) \\ \vdots \\ x_8(t) \end{bmatrix}$$

Write the system equations in the form

$$\dot{\mathbf{x}}(t) = \mathbf{A}\mathbf{x}(t) + \mathbf{B}u(t)$$

Give numerical values for the 8×8 matrix $\mathbf{A}$ and the column vector $\mathbf{B}$.

Section 7.6

7.6.1   Using infinite-series expansions, prove that

$$e^{\mathbf{A}t}e^{\mathbf{A}\tau} = e^{\mathbf{A}(t+\tau)} = e^{\mathbf{A}\tau}e^{\mathbf{A}t}$$

for all real values of t and τ. Use this result to show that

$$(e^{\mathbf{A}t})^{-1} = e^{-\mathbf{A}t}$$

7.6.2   Using infinite-series expansions, prove that

$$\frac{d}{dt}e^{\mathbf{A}t} = \mathbf{A}e^{\mathbf{A}t} = e^{\mathbf{A}t}\mathbf{A} \qquad \text{and} \qquad \frac{d}{dt}e^{\mathbf{A}(t-t_0)} = \mathbf{A}e^{\mathbf{A}(t-t_0)} = e^{\mathbf{A}(t-t_0)}\mathbf{A}$$

7.6.3   Using infinite-series expansions compute $e^{\mathbf{A}t}$ where $\mathbf{A}$ is:

(a) $\mathbf{A} = \begin{bmatrix} a & 0 \\ 0 & b \end{bmatrix}$ (b) $\mathbf{A} = \begin{bmatrix} 0 & 1 \\ 0 & 0 \end{bmatrix}$

(c) $\mathbf{A} = \begin{bmatrix} 0 & 1 \\ -1 & 0 \end{bmatrix}$ (d) $\mathbf{A} = \begin{bmatrix} \alpha & \omega \\ -\omega & \alpha \end{bmatrix}$

7.6.4   (a) Prove that $e^{\mathbf{A}+\mathbf{B}} = e^{\mathbf{A}}e^{\mathbf{B}}$ if and only if $\mathbf{A}$ and $\mathbf{B}$ commute, that is, $\mathbf{A}\mathbf{B} = \mathbf{B}\mathbf{A}$.

(b) Prove that if $\mathbf{A}$ is block diagonal, i.e.,

$$\mathbf{A} = \begin{bmatrix} \mathbf{A}_1 & \mathbf{0} & \cdots & \mathbf{0} \\ \mathbf{0} & \mathbf{A}_2 & \cdots & \cdots \\ \vdots & \vdots & \vdots & \vdots \\ \mathbf{0} & \cdots & \cdots & \mathbf{A}_n \end{bmatrix} \quad \text{then} \quad e^{\mathbf{A}t} = \begin{bmatrix} e^{\mathbf{A}_1 t} & \mathbf{0} & \cdots & \mathbf{0} \\ \mathbf{0} & e^{\mathbf{A}_2 t} & \cdots & \cdots \\ \vdots & \vdots & \vdots & \vdots \\ \mathbf{0} & \cdots & \cdots & e^{\mathbf{A}_n t} \end{bmatrix}$$

7.6.5 (Computer exercise) Write a digital-computer subroutine that computes $e^{\mathbf{A}t}$ for any given $n \times n$ matrix $\mathbf{A}$ and scalar t. Use $n = 15$ as the maximum dimension for $\mathbf{A}$. Use the following truncated series to compute $e^{\mathbf{A}t}$

$$e^{\mathbf{A}t} = \sum_{k=0}^{M} \frac{(\mathbf{A}t)^k}{k!} \qquad 0! \triangleq 1$$

In order to get a feel for the dependence of the accuracy of computation upon the value of M, carry out the following experiment. Let $\mathbf{A}$ be the 3×3 matrix

$$\mathbf{A} = \begin{bmatrix} 0 & 1 & 0 \\ 0 & 0 & 1 \\ -6 & -11 & -6 \end{bmatrix}$$

Print out the value of $e^{\mathbf{A}}$ for $M = 5, 10, 15, 20, 25, 30, 35, 40, 45, 50$. Devise an automated stopping rule of the determination of M, given an accuracy requirement on the norm of $e^{\mathbf{A}t}$.

7.6.6 (Computer exercise) Often it is desired to compute $e^{\mathbf{A}t}$ for a given matrix $\mathbf{A}$ and several values of t. Suppose that we are interested in its value for $t = 0, \Delta, 2\Delta, \ldots, k\Delta, \ldots$, where Δ is a positive number. To do this computation we can utilize the following property of the matrix exponential

$$e^{\mathbf{A}(t+\tau)} = e^{\mathbf{A}t} e^{\mathbf{A}\tau}$$

Hence

$$e^{\mathbf{A}2\Delta} = (e^{\mathbf{A}\Delta})^2 \qquad \text{and} \qquad e^{\mathbf{A}3\Delta} = (e^{\mathbf{A}\Delta})^3$$

Write a continuation of the digital-computer subroutine of Exercise 7.6.5 which can be used to print out the elements of $e^{\mathbf{A}t}$ as a function of time. As a test case, consider the matrix

$$\mathbf{A} = \begin{bmatrix} 0 & 1 & 0 \\ 0 & 0 & 1 \\ -1 & -1 & -1 \end{bmatrix}$$

Use $M = 30$, $\Delta = 0.1$, and print the elements of $e^{\mathbf{A}t}$ for $0 \leq t \leq 5.0$ every 0.1 s.

Section 7.7

7.7.1 Find $e^{\mathbf{A}t}$ and $\int_0^t e^{\mathbf{A}\tau} d\tau$ for

$$\mathbf{A} = \begin{bmatrix} 0 & 1 & 2 \\ 0 & 0 & 2 \\ 0 & 0 & 0 \end{bmatrix}$$

7.7.2   Consider a second-order LTI system described by the vector differential equation

$$\dot{\mathbf{x}}(t) = \mathbf{A}\mathbf{x}(t) \qquad \begin{array}{c} \mathbf{x}(t) \in R_2 \\[4pt] \mathbf{A} = 2 \times 2 \text{ constant matrix} \end{array}$$

We have conducted the following tests upon the system.

Test 1:   When $\mathbf{x}(0) = \begin{bmatrix} 2 \\ 0 \end{bmatrix}$ $\mathbf{x}(t) = \begin{bmatrix} 2 \\ 0 \end{bmatrix}$ for all $t \geq 0$

Test 2:   When $\mathbf{x}(0) = \begin{bmatrix} 3 \\ 3 \end{bmatrix}$ $\mathbf{x}(t) = 3\begin{bmatrix} e^t \\ e^t \end{bmatrix}$ for all $t \geq 0$

Determine the numerical values of the elements of the matrix $\mathbf{A}$.

7.7.3   The following facts are known about the linear system $\dot{\mathbf{x}}(t) = \mathbf{A}\mathbf{x}(t)$.

If $\mathbf{x}(0) = \begin{bmatrix} 1 \\ -1 \end{bmatrix}$ then $\mathbf{x}(t) = \begin{bmatrix} e^{-t} \\ -e^{-t} \end{bmatrix}$

If $\mathbf{x}(0) = \begin{bmatrix} 2 \\ 1 \end{bmatrix}$ then $\mathbf{x}(t) = \begin{bmatrix} 5e^{-t} - 3e^{-2t} \\ -5e^{-t} + 6e^{-2t} \end{bmatrix}$

Find $e^{\mathbf{A}t}$ and hence $\mathbf{A}$.

7.7.4   Consider the illustrated RLC circuit with a variable resistance.

(*a*)   Write the state equation of this system using x_1 for the inductor current and x_2 for the capacitor voltage.

(*b*)   Let the initial state be

$$\mathbf{x}(0) = \begin{bmatrix} 1 \\ -1 \end{bmatrix}$$

Find $\mathbf{x}(t)$ for $t \geq 0$ as a function of R.

(*c*)   Find R such that

$$\mathbf{x}(t) = e^{\lambda_1 t} \begin{bmatrix} 1 \\ -1 \end{bmatrix}$$

for some λ_1. What is the relation of $[\begin{smallmatrix} 1 \\ -1 \end{smallmatrix}]$ to λ_1?

(*d*)   For this value of R find another initial state ξ such that $\mathbf{x}(t) = e^{\lambda_2 t}\xi$.

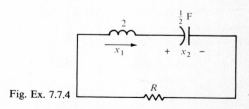

Fig. Ex. 7.7.4

7.7.5 Consider the time-varying system

$$\dot{\mathbf{x}}(t) = \mathbf{A}(t)\mathbf{x}(t) \qquad \mathbf{x}(0) = \begin{bmatrix} 1 \\ 1 \end{bmatrix}$$

where $\mathbf{A}(t) = \begin{bmatrix} \sigma_1(t) & \omega(t) \\ \omega(t) & \sigma_2(t) \end{bmatrix}$

$$\sigma_1(t) = \begin{cases} 2 & t < 2 \\ 1 & t \geq 2 \end{cases} \qquad \sigma_2(t) = \begin{cases} 1 & t < 2 \\ 1 & t \geq 2 \end{cases} \qquad \omega(t) = \begin{cases} 0 & t < 2 \\ 3 & t \geq 2 \end{cases}$$

Find $\mathbf{x}(t)$ as a function of t.

7.7.6 Let $\mathbf{X}(t)$ be a time-varying $n \times n$ matrix. Suppose that it satisfies the matrix differential equation

$$\dot{\mathbf{X}}(t) = \mathbf{A}\mathbf{X}(t) \qquad \mathbf{X}(0) = \mathbf{X}_0 \tag{1}$$

where $\mathbf{A}$ is a constant $n \times n$ matrix. Equation (1) of course can be viewed as shorthand for the set of n^2 scalar equations

$$\dot{x}_{ij}(t) = \sum_{k=1}^{n} a_{ik} x_{kj}(t) \qquad i, j = 1, 2, \ldots, n$$

Prove that the solution to (1) exists, is unique, and is given by

$$\mathbf{X}(t) = e^{\mathbf{A}t}\mathbf{X}_0$$

7.7.7 Determine the solution to the matrix differential equations

$$\dot{\mathbf{X}}(t) = \mathbf{X}(t)\mathbf{B} \qquad \mathbf{X}(0) = \mathbf{X}_0$$

$$\dot{\mathbf{X}}(t) = \mathbf{A}\mathbf{X}(t) + \mathbf{X}(t)\mathbf{B} \qquad \mathbf{X}(0) = \mathbf{X}_0$$

7.7.8 Consider two systems described by the differential equations

$$\dot{\mathbf{x}}(t) = \mathbf{A}\mathbf{x}(t) \qquad \mathbf{x}(0) = \xi \tag{1}$$

$$\dot{\mathbf{y}}(t) = -\mathbf{A}'\mathbf{y}(t) \qquad \mathbf{y}(0) = \eta \tag{2}$$

System (2) is said to be the adjoint system to (1). Prove that

$$\langle \mathbf{x}(t), \mathbf{y}(t) \rangle = \langle \xi, \eta \rangle \qquad \text{for all } t$$

What is the geometric significance of this?

7.7.9 Consider the vector differential equation

$$\dot{\mathbf{x}}(t) = \mathbf{A}\mathbf{x}(t) \qquad \mathbf{x}(0) = \xi$$

Suppose that the $n \times n$ matrix $\mathbf{A}$ is skew-symmetric, that is, $\mathbf{A}' = -\mathbf{A}$, and n is an even number.

(a) Let $\|\mathbf{x}(t)\|$ denote the euclidean norm of $\mathbf{x}(t)$. Prove that

$$\|\mathbf{x}(t)\| = \|\boldsymbol{\xi}\| \qquad \text{for all } t$$

(b) Consider an arbitrary network whose elements are LTI capacitors and inductors. There are no independent voltage or current sources in the network, there are no capacitor loops or inductor cut sets, and there are an equal number of capacitors and inductors. Analyze the network carefully and see if you can use the results of part (a) to deduce general properties of such networks. (This is the hard part.)

7.7.10 Consider the network shown. Let $v_1(t)$ and $v_2(t)$ denote the capacitor voltages (state variables).

(a) Write the vector differential equations for this network.
(b) Determine $v_1(t)$ and $v_2(t)$ when $v_1(0) = v_2(0) = 1$.

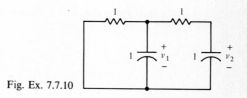

Fig. Ex. 7.7.10

7.7.11 Analyze and solve using vector differential equations the standard LTI *RLC* network shown.

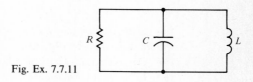

Fig. Ex. 7.7.11

Section 7.9

7.9.1 Consider a third-order system with a single input $u(t)$, a single output $y(t)$, and a 3-dimensional state vector $\mathbf{x}(t)$. Suppose that the state $\mathbf{x}(t)$ satisfies the LTI vector differential equation

$$\dot{\mathbf{x}}(t) = \mathbf{A}\mathbf{x}(t) + \mathbf{b}u(t) \tag{1}$$

where
$$\mathbf{A} = \begin{bmatrix} 0 & 1 & 0 \\ 0 & 0 & 1 \\ -6 & -11 & -6 \end{bmatrix} \qquad \mathbf{b} = \begin{bmatrix} 0 \\ 0 \\ 1 \end{bmatrix} \tag{2}$$

Suppose that the output $y(t)$ is related to the state by

$$y(t) = \mathbf{c}'\mathbf{x}(t) \tag{3}$$

where
$$\mathbf{c}' = \begin{bmatrix} 1 & 1 & 1 \end{bmatrix} \tag{4}$$

Hint: The eigenvalues of $\mathbf{A}$ are -1, -2, and -3.

(a)   Eliminate the state variables x_1, x_2, and x_3 and obtain a single third-order differential equation relating $y(t)$ and its time derivatives to the input $u(t)$ and its derivatives.

(b)   Compute the exponential matrix $e^{\mathbf{A}t}$ analytically.

(c)   Use the results of part (b) to compute $\mathbf{x}(t)$ for all $t \geq 0$ and $y(t)$ for all $t \geq 0$ given that at $t = 0$ the initial state is given by

$$\mathbf{x}(0) = \begin{bmatrix} 1 \\ -1 \\ 0 \end{bmatrix} \tag{5}$$

and
$$u(t) = 0 \qquad \text{for all } t \tag{6}$$

(d)   For $t \geq 0$, compute $\mathbf{x}(t)$ and $y(t)$ when the system is initially at rest, that is, $\mathbf{x}(0) = \mathbf{0}$, and when the input is a unit step, i.e.,

$$u(t) = \begin{cases} 0 & \text{for } t < 0 \\ 1 & \text{for } t \geq 0 \end{cases} \tag{7}$$

(e)   Solve $y(t)$ for the same conditions as in part (d) but compute it using the differential equation obtained in part (a). Verify that the answers are identical.

(f)   Compute $\mathbf{x}(t)$ for $t \geq 0$ when $\mathbf{x}(0)$ is given by (5) and $u(t)$ by (7).

(g)   From part (a) show that the input-to-output system function is

$$H(s) = \frac{s^2 + s + 1}{s^3 + 6s^2 + 11s + 6}$$

Show that

$$H(s) = \mathbf{c}'(s\mathbf{I} - \mathbf{A})^{-1}\mathbf{b}$$

where $\mathbf{A}$, $\mathbf{b}$, and $\mathbf{c}'$ are given by Eqs. (2) and (4). Convince yourself that the poles of $H(s)$ are precisely the eigenvalues of $\mathbf{A}$. Can you find an analogous relation for the zeros of $H(s)$?

7.9.2   Consider an LTI system described by

$$\dot{x}(t) = Ax(t) + bu(t) \qquad x(0) = 0$$

$$y(t) = c'x(t)$$

(a)   Suppose that $u(t)$ is a *narrow pulse* defined by

$$u(t) = \begin{cases} 0 & \text{for} \quad t < 0 \\ \dfrac{1}{\Delta} & \text{for} \quad 0 \le t \le \Delta \\ 0 & \text{for} \quad t > \Delta \end{cases}$$

Suppose also that **A** is nonsingular. Determine $y(t)$ as $\Delta \to 0$ for all $t > \Delta$. (Evaluate all integrals.) You have evaluated the *impulse response* of an LTI system described in state-variable form.

(b)   Now suppose that $u(t)$ is a unit step, i.e.,

$$u(t) = \begin{cases} 0 & \text{for } t > 0 \\ 1 & \text{for } t \ge 0 \end{cases}$$

Compute $y(t)$ in this case; we call it the *step response*.

(c)   Convince yourself that the step response is the integral of the impulse response.

(d)   Now suppose that $u(t)$ is a unit ramp, i.e.,

$$u(t) = \begin{cases} 0 & \text{for } t < 0 \\ t & \text{for } t \ge 0 \end{cases}$$

Evaluate the ramp response $y(t)$. Show that the *ramp response* is the integral of the step response.

(e)   Generalize these ideas for the class of inputs

$$u(t) = \begin{cases} 0 & t < 0 \\ \dfrac{1}{k!} t^k & t \ge 0, k = 2, 3, \ldots \end{cases}$$

7.9.3   Consider two LTI dynamical systems S_1 and S_2 described as follows:

$$\left. \begin{aligned} \dot{x}_1(t) &= A_1 x_1(t) + b_1 u_1(t) \\ y_1(t) &= c_1' x_1(t) \end{aligned} \right\} S_1 \qquad \left. \begin{aligned} \dot{x}_2(t) &= A_2 x_2(t) + b_2 u_2(t) \\ y_2(t) &= c_2' x_2(t) \end{aligned} \right\} S_2$$

There are two ways of connecting these two systems in series: $y_1(t)$ being the input to S_2 or $y_2(t)$ being the input to S_1; more precisely

Connection 1: $u_1(t) = u(t)$ $u_2(t) = y_1(t)$ $y_2(t) = y(t)$

Connection 2: $u_2(t) = u(t)$ $u_1(t) = y_2(t)$ $y_1(t) = y(t)$

Deduce precise conditions under which for the same input $u(t)$ we observe the same output $y(t)$ for all $t \ge 0$ for both interconnections.

7.9.4   A convex set S in the n-dimensional euclidean space R_n is a set such that for any two vectors $\mathbf{x}$ and $\mathbf{y}$ in S and any scalar β, $0 \leq \beta \leq 1$, the vector

$$\mathbf{z} = \beta\mathbf{x} + (1 - \beta)\mathbf{y}$$

is also a vector in the set S.

(*a*)   Now consider the LTI system

$$\dot{\mathbf{x}}(t) = \mathbf{A}\mathbf{x}(t) + \mathbf{B}\mathbf{u}(t) \qquad \mathbf{x}(0) = \mathbf{0}$$

Suppose that the input vector $\mathbf{u}(t)$ belongs to some convex set U for each value of time $0 \leq t \leq T$. Now for each input $\mathbf{u}(t)$, $0 \leq t \leq T$, we have a state $\mathbf{x}_u(T)$ [we use u as a subscript to stress that the value of the state at time T depends on the choice of $\mathbf{u}(t)$, $0 \leq t \leq T$]. Prove that the vectors $\mathbf{x}_u(T)$ form a convex set.

(*b*)   Is the same true if $\mathbf{x}(0) = \xi$?

Section 7.10

7.10.1   Consider an arbitrary LTI network containing positive R, L, and C's. Suppose that there are no dependent or independent sources in the network. Let $\mathbf{x}(t)$ denote the state vector (inductor currents and capacitor voltages) for this network. Then we know that the state vector satisfies a vector differential equation

$$\dot{\mathbf{x}}(t) = \mathbf{A}\mathbf{x}(t) \tag{1}$$

We also know that the energy $E(t)$ stored in the network is given by

$$E(t) = \tfrac{1}{2}\langle \mathbf{x}(t), \mathbf{Q}\mathbf{x}(t)\rangle \geq 0 \tag{2}$$

where $\mathbf{Q}$ is a positive definite symmetric matrix (what is it?) so that the energy is always positive unless $\mathbf{x}(t) = \mathbf{0}$. Physically, in such a network, we expect that energy will be dissipated in the resistors. Hence we would expect that the rate of change of stored energy would be negative, i.e.,

$$E(t) < 0 \tag{3}$$

Prove that $\mathbf{A}$ must be a stable matrix, i.e., the eigenvalues of $\mathbf{A}$ must have negative real parts. (You need to be clever to do this.)

7.10.2   Consider the continuous-time system

$$\dot{\mathbf{x}}(t) = \mathbf{A}\mathbf{x}(t) \qquad \mathbf{x}(0) = \xi \tag{1}$$

Suppose that we observe the state of the system every T s. Define

$$\mathbf{x}(k) \triangleq \mathbf{x}(kT) \qquad k = 0, 1, 2, \ldots \tag{2}$$

(a) Show that $\mathbf{x}(k)$ satisfies a difference equation of the form

$$\mathbf{x}(k + 1) = \boldsymbol{\Phi}\mathbf{x}(k) \tag{3}$$

and determine the relation between $\boldsymbol{\Phi}$ and $\mathbf{A}$.

(b) Suppose all the (assumed distinct) eigenvalues of $\mathbf{A}$ have negative real parts. Prove that the discrete-time system (3) is asymptotically stable.

Section 7.15

7.15.1 (Hydraulic reservoir problem) You are given the following mathematical models for the flow of water in a reservoir system:

Flow in valves:
$$F = C_p \sqrt{\Delta P_{\text{pipe}}} \tag{1}$$

where F = flow, ft³/min
ΔP_{pipe} = pressure drop from one end of pipe to other, lb/in²
C_p = const

Flow in valves:

$$F = C_v r \sqrt{\Delta P_{\text{valve}}} \tag{2}$$

where r = valve position, $0 \le r \le 1$. Note that for a line with a pipe *and* a valve

$$F_{\text{pipe}} = F_{\text{valve}} \quad \text{and} \quad \Delta P_{\text{pipe}} + \Delta P_{\text{valve}} = \Delta P_{\text{total}}$$

Tanks:

$$\frac{d}{dt}(V_L) = \sum F_i \tag{3}$$

where V_L = volume of liquid, ft³
F_i = flows into tank

A reservoir system consists of a large hilltop reservoir feeding two smaller town reservoirs. The hilltop reservoir has the following normal operating-point values:

Liquid volume = 10^9 ft³ = V_R.
Liquid surface area = 10^7 ft² = A_R.
The surface area changes 2 percent for each 1 percent change in V_L near the operating point.
Rainfall = $F_6 = 10^4$ ft³/min (a constant, since all the rain falling into the basin flows into the reservoir).
Evaporation is proportional to surface area (you can figure out F_5).

The normal average water use by the towns is $F_1 = 10^3$ ft³/min and $F_2 = 2 \times 10^3$ ft³/min. Both towns have the same type of town reservoir. In normal operation these reservoirs have 10^4 ft³ liquid volume and 10^2 ft² surface area. These reservoirs

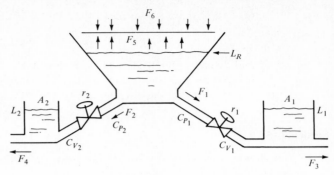

Fig. Ex. 7.15.1

have vertical walls. Neglect evaporation for the town reservoirs. The valves V_1 and V_2 which are half open at the normal operating point, are described by

$$C_{V_1} = C_{V_2} = 10^2 \qquad \text{refer to Eq. (2)}$$

The pipes are described by

$$C_{p1} = C_{p2} = 10^2 \qquad \text{refer to Eq. (1)}$$

The density of water is 62.4 lb/ft³.

(a) What is the difference in elevation between L_1 and L_R and between L_2 and L_R during the normal operating condition?

(b) Formulate the state-variable equations for this system, using L_1, L_2, and L_R as the state variables. Consider F_3, F_4, F_5, F_6, r_1, and r_2 to be inputs.

(c) Find the linearized state-variable system about the normal operating point. What are the eigenvalues of this system? Is it stable?

Section 7.16

7.16.1 A circuit employing a nonlinear resistor is shown in Fig. Ex. 7.16.1a

(a) Write the state-variable representation of the behavior of the circuit choosing as state variables

$$x_1 = i \qquad x_2 = v_R$$

The behavior of the nonlinear resistor is shown in Fig. Ex. 7.16.1b.

(b) The circuit has three equilibrium points:

$$x_1 = 0.0378 \qquad x_2 = 6.22$$

$$x_1 = 0.05 \qquad x_2 = 5.0$$

$$x_1 = 0.0622 \qquad x_2 = 3.78$$

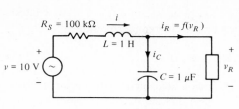

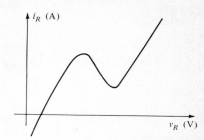

Fig. Ex. 7.16.1 (*a*)   Fig. Ex. 7.16.1 (*b*)

Construct linearized models of the circuit in the vicinity of each of the equilibrium points.

(*c*)   Which equilibrium point or points correspond to stable behavior?

7.16.2   Consider the network shown. Except for the nonlinear conductance, all elements are linear and time-invariant. The nonlinear conductance is defined by the following current-voltage characteristics

$$i_G = v_G^3 - 5v_G^2 + 4v_G = v_G(v_G - 1)(v_G - 4)$$

(*a*)   Using cut-set analysis, determine the differential equations satisfied by the network state variables, $v(t)$ and $i(t)$. Clearly define your tree branches, links, cut sets, cut-set matrices, etc.

(*b*)   Determine *all* equilibrium conditions for this network (there is more than one).

(*c*)   Examine the stability of each equilibrium condition found in part (*b*). State your reasoning *clearly* and exhibit your calculations.

(*d*)   Discuss the response of the network when the initial conditions are

$$v(0) = 0.1\text{V} \qquad i(0) = 0\text{A}$$

Detailed equations and exact numerical plots are not required. However, you should be able to *sketch* the behavior of $v(t)$ and $i(t)$ in the vicinity of the equilibrium, as well as the qualitative behavior away from the equilibrium point.

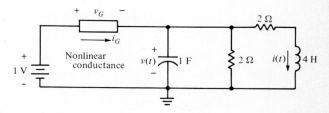

Fig. Ex. 7.16.2

Section 7.18

7.18.1 This exercise introduces a very important concept in system theory, system controllability. Consider an LTI system described by the differential equation

$$\dot{\mathbf{x}}(t) = \mathbf{A}\mathbf{x}(t) + \mathbf{B}\mathbf{u}(t) \tag{1}$$

Suppose the pair $[\mathbf{A}, \mathbf{B}]$ is controllable. Then show that there exists an input vector $\mathbf{u}(t)$, $0 \leq t \leq T$, with T finite and arbitrary, such that the state of the system (1) can be transferred from any initial state $\mathbf{x}(0) = \boldsymbol{\xi}$ to any terminal state $\mathbf{x}(T) = \boldsymbol{\theta}$. *Hint:* First consider the terminal state $\boldsymbol{\theta} = \mathbf{0}$. Set up the equation for the control. Then use the Cayley-Hamilton theorem to write

$$e^{-\mathbf{A}t} = \sum_{i=0}^{n-1} \beta_i(t) \mathbf{A}^i$$

under the assumption that the eigenvalues of $\mathbf{A}$ are distinct.

7.18.2 A notion related to that of controllability is observability. Consider the LTI system

$$\begin{aligned} \dot{\mathbf{x}}(t) &= \mathbf{A}\mathbf{x}(t) + \mathbf{B}\mathbf{u}(t) \\ \mathbf{y}(t) &= \mathbf{C}\mathbf{x}(t) \end{aligned} \tag{1}$$

where $\mathbf{A}$ is an $n \times n$ matrix and $\mathbf{C}$ is an $r \times n$ matrix. We say that the pair $[\mathbf{A}, \mathbf{C}]$ is observable if

$$\text{Rank } [\mathbf{C}' \mid \mathbf{A}'\mathbf{C}' \mid \mathbf{A}'^2\mathbf{C}' \mid \cdots \mid \mathbf{A}'^{n-1}\mathbf{C}'] = n$$

The system (1) is said to be observable if knowledge of $\mathbf{y}(t)$ and $\mathbf{u}(t)$, $0 \leq t \leq T$, where T is finite, can be used to determine uniquely the initial state $\mathbf{x}(0)$ of the system. Prove that the system is observable if and only if the pair $[\mathbf{A}, \mathbf{C}]$ is observable. *Hint:* Set $\mathbf{u}(t) = \mathbf{0}$. Assume that the knowledge of $\mathbf{y}(t)$ over a finite interval of time implies that we can calculate $\dot{\mathbf{y}}(t)$, $\ddot{\mathbf{y}}(t)$, ... exactly. Look for n linearly independent equations in n unknowns. Once more the Cayley-Hamilton theorem will come in handy.

7.18.3 Consider the system

$$\dot{\mathbf{x}}(t) = \begin{bmatrix} 0 & 1 & 0 \\ 0 & 0 & 1 \\ 0 & 0 & 0 \end{bmatrix} \mathbf{x}(t) + \begin{bmatrix} 0 & 0 \\ 1 & 0 \\ 0 & 1 \end{bmatrix} \mathbf{u}(t)$$

(*a*) Verify that the system is controllable.

(*b*) Suppose that we want to use state-variable feedback so that the eigenvalues of the closed-loop system are all at -1. Find a 2×3 constant matrix $\mathbf{G}$ such that the feedback $\mathbf{u}(t) = \mathbf{G}\mathbf{x}(t)$ will accomplish this requirement.

(*c*) Is the matrix $\mathbf{G}$ unique?

7.18.4   Consider the uncontrollable system

$$\dot{x}(t) = \begin{bmatrix} 0 & 1 & 0 \\ 0 & 0 & 1 \\ 0 & 0 & 0 \end{bmatrix} x(t) + \begin{bmatrix} 0 \\ 1 \\ 0 \end{bmatrix} u(t)$$

Is it possible to find a 1×3 matrix $\mathbf{G}$ such that the feedback $u(t) = \mathbf{G}x(t)$ will result in having all the eigenvalues of the closed-loop matrix at $\lambda = -1$?

7.18.5   A broom is to be balanced in a vertical plane by attaching one end to a model railroad engine running on a straight track, as shown. Control is accomplished by changing the linear velocity of the engine on the track. It is assumed that the mass m of the broom is concentrated at a distance of L ft from the pivot point on the engine and that the broom is free to rotate only about the pivot point in a plane parallel to the axis of the engine. Friction effects and the mass of the engine are neglected, and the linear velocity of the engine can be changed instantaneously in proportion to the voltage applied to the dc motor that drives the engine. For small angular velocities and small angular deviations of the position of the broom relative to the vertical, the angular motion of the broom is governed by the differential equation

$$L \frac{d^2\theta}{dt^2} - g\theta = \frac{dv}{dt}$$

where g is the acceleration of gravity and v is the linear velocity of the engine.

(*a*)   The state-variable representation of the system is

$$\dot{x} = \mathbf{A}x + \mathbf{b}u$$

If $x_1 = \theta$ and $x_2 = d\theta/dt$ are the states of the system, find A, **b**, and u.

(*b*)   If $L = 4$ ft and $g = 32$ ft/s², find the eigenvalues of A. If the engine is running at constant velocity, is the system stable?

(*c*)   The control u is made a function of the state by using sensors which measure θ and $d\theta/dt$. The outputs of the sensors are passed through adjustable-gain amplifiers whose outputs are added to produce $u(t)$. Thus

Fig. Ex. 7.18.5

$$u = \mathbf{k}'\mathbf{x} \qquad \text{where} \qquad \mathbf{k} = \begin{bmatrix} k_1 \\ k_2 \end{bmatrix}$$

If the eigenvalues of the controlled closed-loop system are to be located at $-0.5 \pm j1.5$, what are the required values of the feedback gains k_1 and k_2?

(*d*) Show that the eigenvalues of the controlled system can be located anywhere in the complex plane if state feedback is used subject to the restriction that complex eigenvalues must be a conjugate pair.

7.18.6 (Observers) Since the state of a dynamical system summarizes all the information concerning its past behavior, it is frequently important to determine the state of a system when that state cannot be measured directly. Suppose that the system is described by the usual linear state equations

$$S_1: \quad \begin{aligned} \dot{\mathbf{x}} &= \mathbf{Ax} \qquad \mathbf{x}(0) = \mathbf{x}_0 \\ y &= \mathbf{c}'\mathbf{x} \end{aligned} \qquad \boxed{S_1} \to y$$

where $\mathbf{x}(t) \in R^n$ is the state, the measured output y is a scalar, and $\mathbf{A}$ and $\mathbf{c}$ are constant. The goal is to design a second system S_2 which estimates the state of S_1 by looking at the measured output y

The output y of S_1 serves as the input to S_2 and the state $\mathbf{z}(t)$ of S_2 serves as an estimate of $\mathbf{x}(t)$. The system S_2 is called an *observer*. Suppose the observer is also an LTI system

$$S_2: \qquad \mathbf{z}(t) = \mathbf{Fz}(t) + \mathbf{g}y(t) \qquad \mathbf{z}(0) = \mathbf{z}_0$$

(*a*) By considering the error $\mathbf{e}(t) \triangleq \mathbf{x}(t) - \mathbf{z}(t)$, find a value for the $\mathbf{F}$ matrix (in terms of $\mathbf{A}$, $\mathbf{c}$, and $\mathbf{g}$) so that the error satisfies a homogeneous linear equation of the form

$$\dot{\mathbf{e}} = \mathbf{Me} \qquad \mathbf{e}(0) = \mathbf{e}_0$$

(*1*) What is $\mathbf{M}$ (in terms of $\mathbf{A}$, $\mathbf{G}$, and $\mathbf{c}$)?

(*2*) Is there any $\mathbf{z}_0$ so that the error is zero for all time? Why will the error always be zero under the conditions you specify?

Now suppose

$$\mathbf{A} = \begin{bmatrix} -1 & 2 \\ 0 & -3 \end{bmatrix} \qquad \mathbf{c} = \begin{bmatrix} 1 \\ 0 \end{bmatrix} \qquad \mathbf{c}' = \begin{bmatrix} 1 & 0 \end{bmatrix} \qquad \mathbf{g} = \begin{bmatrix} g_1 \\ g_2 \end{bmatrix}$$

(*b*) What is the characteristic polynomial of $\mathbf{M}$, the error system matrix (in terms of g_1, g_2, and constants)?

(c) Suppose that the desired values of the error eigenvalues are μ_1 and μ_2. What values should g_1 and g_2 have (in terms of μ_1, μ_2, and constants)?

(d) Since the initial state $\mathbf{x}_0$ is unknown, it is impossible to ensure that the error will always be zero. One realistic design criterion is that the error should become small before system S_1 "does much," in other words, the error should be small before time $t = T$, where T is the shortest time constant associated with S_1.

(1) What are the time constants of S_1 (for the **A** matrix given above)?

(2) How should g_1 and g_2 be chosen so that the error is no larger than 1 percent of its initial value ($\mathbf{e}_0 = \mathbf{x}_0 - \mathbf{z}_0$) by the time t gets to $T = $ shortest time constant of S_1?

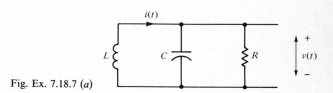

Fig. Ex. 7.18.7 (a)

7.18.7 Consider the LTI circuit shown:

(a) Write down the state equations in the form $\dot{\mathbf{x}}(t) = \mathbf{A}\mathbf{x}(t)$. (The use of cut-set analysis is optional.)

(b) Assume that $v(t)$ can be measured but $i(t)$ cannot; however, we wish to estimate $i(t)$ from observations of $v(t)$ alone. Assuming you have a differentiator, how can $i(t)$ be determined (values of L, C, and R are known)? Briefly explain the practical disadvantages of using differentiators.

(c) An alternate method is to build a circuit called an observer whose input is the observation $v(t)$ and whose output is an estimate $\hat{i}(t)$ of $i(t)$. Such a circuit is illustrated in Fig. Ex. 7.18.7b, where $e(t)$ is a dependent source and $e(t) = v(t)$. We want $\hat{i}(t)$ to approach $i(t)$ as quickly as possible. In particular, we would like

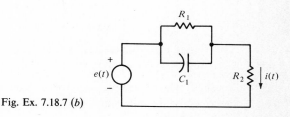

Fig. Ex. 7.18.7 (b)

$$\frac{d}{dt}(\hat{i} - i) = -\frac{1}{\tau}(\hat{i} - i) \tag{1}$$

Calculate values of $R_1, R_2,$ and C_1 in terms of $R, L, C,$ and τ such that (1) is satisfied.

(d)   Determine the behavior of the observer as $\tau \to 0$ and compare with part (b).

(e)   Set $R = 1\,\Omega,\quad C = \frac{1}{16}F,\quad L = \frac{1}{3}H,\quad i(0) = 3\text{ A};\quad v(0) = 4\text{ V},\quad v_{c_1}(0) = 0$ V. Solve for $i(t), v(t), \hat{i}(t),$ and calculate

$$\frac{\hat{i}(t) - i(t)}{i(t)} \qquad \text{for } t \geq 0$$

(f)   Comment on appropriate values of τ.

7.18.8   Consider the following dynamic population exchange between two species

$$x_1(t) = \text{population of prey}$$

$$x_2(t) = \text{population of predators}$$

(see Sec. 7.4). Suppose that changes in population are described by the differential equations

$$\dot{x}_1(t) = x_1(t) - 0.1x_1(t)x_2(t) \qquad \dot{x}_2(t) = x_2(t) - 2.5\frac{x_2^2(t)}{x_1(t)}$$

(a)   Is there a positive equilibrium population condition?

(b)   Is this equilibrium stable?

7.18.9   Consider the following prey-predator population model involving two species

$$x_1(t) = x_1(t) - 0.10x_1(t)x_2(t)$$

$$x_2(t) = -0.5x_2(t) + 0.02x_1(t)x_2(t)$$

(a)   Is there a positive equilibrium population condition?

(b)   Is this equilibrium stable?

(c)   Suppose that we wish to stabilize the population at the equilibrium that you have found in part (a) by externally changing $x_1(t)$ and $x_2(t)$ according to the model

$$\dot{x}_1(t) = x_1(t) - 0.10x_1(t)x_2(t) + u_1(t)$$

$$\dot{x}_2(t) = -0.5x_2(t) + 0.02x_1(t)x_2(t) + u_2(t)$$

Using perturbation-feedback ideas, determine ways of selecting $u_1(t)$ and $u_2(t)$ such that the equilibrium is locally asymptotically stable.

7.18.10   Consider two populations $x_1(t)$ and $x_2(t)$ that compete for the same resources described by the differential equations

$$\dot{x}_1(t) = a_1 x_1(t) - b_{11} x_1^2(t)$$

$$\dot{x}_2(t) = a_2 x_2(t) - b_{21} x_1(t)x_2(t) - b_{22} x_2^2(t)$$

where all the constants are positive.

(*a*) Determine conditions on the constants such that there is a unique positive population condition.

(*b*) Examine the local stability of that equilibrium condition.

7.18.11 Consider a third-order LTI system described by the equations

$$\dot{\mathbf{x}}(t) = \mathbf{A}\mathbf{x}(t) + \mathbf{b}u(t) \tag{1}$$

$$\mathbf{x}(t) \in R_3 \qquad u(t) \in R_1 \tag{2}$$

Assume that

$$\mathbf{A} = \begin{bmatrix} 0 & 1 & 0 \\ 0 & 0 & 1 \\ 0 & 0 & 0 \end{bmatrix} \qquad \mathbf{b} = \begin{bmatrix} 0 \\ 0 \\ 1 \end{bmatrix} \tag{3}$$

(*a*) An engineer claims that he can stabilize the system using output feedback, i.e.,

$$u(t) = -Kx_1(t) \tag{4}$$

Show that this is not possible. Construct the locus of the eigenvalues of the closed-loop system as K changes in the region $0 \le K \le \infty$, and comment on what happens.

(*b*) Suppose that you are allowed to use full state-variable feedback, i.e.,

$$u(t) = -[g_1 x_1(t) + g_2 x_2(t) + g_3 x_3(t)] \tag{5}$$

Determine the values of g_1, g_2, g_3 such that the eigenvalues of the closed-loop system matrix are at -1, -2, and -3.

7.18.12 The following conditions will be useful in the sequel. Consider a general second-order LTI system

$$\frac{d}{dt}\begin{bmatrix} x_1(t) \\ x_2(t) \end{bmatrix} = \begin{bmatrix} a_{11} & a_{12} \\ a_{21} & a_{22} \end{bmatrix}\begin{bmatrix} x_1(t) \\ x_2(t) \end{bmatrix}$$

(*a*) Derive necessary and sufficient conditions on a_{ij} for the system to be asymptotically stable.

This problem illustrates the stability considerations in the design of continuous-flow stirred-tank reactors (CSTR). A well-stirred region uniform in composition C and temperature T is the conceptual starting point for the simplest reactor model. By noting that the accumulation of a component of interest is the algebraic sum of inflow, outflow, and loss by chemical reaction, one develops the mass balance

$$V\frac{dC}{dt} = q(C_0 - C) - VR \tag{1}$$

where V = constant reaction volume
   q = flow rate in and out
   C_0 = feed concentration of component of interest
   R = reaction rate per unit volume

Because the reaction rate is ordinarily a function of C and T, another relationship is needed to account for the temperature variation with time. This is the energy balance

$$VC_P\frac{dT}{dt} = qC_P(T_0 - T) + \Delta H\ VR - U(T - T_w) \tag{2}$$

in which, except for the last, the terms parallel those in the prior equation. The exceptional term arises from the need to allow for heat transfer across the boundary of the system by means other than the material flow. Accordingly,

   C_P = heat capacity per unit volume of flowing materials and
       contents of reactor
   T_0 = feed temperature
   ΔH = molar heat of reaction, taken positive for exothermic
       systems
   U = overall heat-transfer coefficient, in units which include area
       of transfer surface
   T_w = ambient, coolant, or heating-fluid temperature

The parallelism between the two conservation equations can be made even more obvious by defining a reduced temperature $\eta = C_P T/\Delta H$. Using this substitution in Eq. (2) gives

$$V\frac{d\eta}{dt} = q(\eta_0 - \eta) + VR - \frac{U}{C_p}(\eta - \eta_w) \tag{3}$$

which together with (1) and appropriate initial conditions determines the concentration and temperature in the reactor at all later times. Since provision has been made for flow into and out of the system, the model is identified as a CSTR. The rate of reaction time $R(C, T)$ is given by

$$R(C, T) = K_0 C \exp(-Q/T) \tag{4}$$

The following parameters are for a first-order exothermic reaction, where C is the concentration of a reactant being destroyed:

$$Q = 10^4\ \text{K} \qquad C_0 = 1\ \text{g mol/l} \qquad T_0 = T_w = 350\ \text{K}$$

$$\frac{\Delta H}{C_p} = 200\ \text{K/(g mole)(1)} \qquad \frac{U}{VC_p} = 1\ \text{min}^{-1} \qquad \frac{q}{V} = 1\ \text{min}^{-1} \tag{5}$$

$$K_0 = e^{25}\ \text{min}^{-1}$$

(b) Reduce the problem of finding the steady-state or equilibrium solutions to (1) and (3) to that of solving a single transcendental equation in one variable. For the parameters given in Eq. (5) verify that $T_s = 354, 400$, and 441 °K satisfy the equilibrium condition, and find the corresponding values of C_s, the steady-state concentrations.

(c) Derive the linearized equations of the unforced system about an equilibrium η_s, C_s. Let $x_1(t) = \eta(t) - \eta_s$ and $x_2(t) = C(t) - C_s$. (Do not substitute numerical values yet.)

(d) Using the parameter values in Eq. (5), find conditions in terms of C_s and η_s which are necessary and sufficient for the linearized system to be asymptotically stable [use part (a)]. Determine which of the equilibrium states found in part (b) are asymptotically stable.

(e) The system can be controlled by changing the rate of flow of the cooling fluid, and this will alter the effective heat-transfer coefficient U. Therefore consider $u(t) = [\tilde{U}(t) - U]/C_p V$ as the control input, where $\tilde{U}(t)$ is the effective heat-transfer coefficient, which can be varied with time. Now derive the linearized dynamic equations about the equilibrium pair C_s, η_s, U, to give an equation of the form

$$\dot{\mathbf{x}}(t) = \mathbf{A}\mathbf{x}(t) + \mathbf{b}u(t) \tag{6}$$

(f) A state feedback control is now used to modify the system response. Let

$$u(t) = K_1 x_1(t) + K_2 x_2(t) \tag{7}$$

For the steady state with $T_s = 400$ K evaluate the matrices $\mathbf{A}$ and $\mathbf{b}$ in Eq. (6), and then derive the region in $K_1 K_2$ space for which the closed-loop system is asymptotically stable. Also find values of K_1 and K_2 such that the eigenvalues of the closed-loop system are at -3 and -4 min^{-1}.

8
THE RELATION OF THE STATE-VARIABLE AND INPUT-OUTPUT DESCRIPTION FOR LINEAR TIME-INVARIANT DYNAMICAL SYSTEMS

SUMMARY

This chapter interrelates the two most common methods of representing systems, especially those characterized by a single input and a single output. The traditional method consists of relating the output(s) and the input(s) by means of high-order differential equations with constant coefficients; this method is useful in deducing the input-output *system function* (through the use of complex exponential functions) and the input-output *transfer function* (through the use of Laplace transform methods). The more modern method is to use the internal state variables of the system to provide the link between the input and output signal(s). Since both approaches are alternate ways of characterizing the same system, there are strong interrelations. The purpose of this chapter is to bring these similarities into focus and to present methods by which one can go back and forth between the input-output description and the state-variable description.

FIGURE 8.1.1
The input-output representation of a system.

8.1   THE INPUT-OUTPUT APPROACH

Since we have so far discussed the state-variable method for the mathematical description of dynamical systems almost exclusively, in this section we shall review the basic ideas behind the so-called input-output approach. We assume that the reader has encountered this method in a course on introductory circuit theory (or the·equivalent) and in a first course dealing with ordinary differential equations. For this reason, we shall not elaborate upon this method extensively, simply presenting the basic ideas and definitions to the extent needed to establish the correspondence between the input-output method and the state-variable method.

Throughout this chapter we shall discuss almost exclusively the case of LTI single-input—single-output (SISO) systems.

The scalar time function $y(t)$ will be used to denote the system *output* as a function of time. The scalar time function $u(t)$ will be used to denote the *system input*. From a conceptual point of view the input-output approach is strictly a cause-and-effect approach. The values of the input time function $u(t)$ represent the cause, and the values of the output $y(t)$ represent the effect. The system is viewed as a black box, i.e., as a system whose internal structure and interconnections are of no particular interest. For example, a hi-fi amplifier is often viewed as a black box; all one is interested in is the fidelity of the output signal with respect to the input signal.

Figure 8.1.1 shows the usual way of visualizing this cause-and-effect relationship.

8.1.1   Input-Output Description

If we consider a lumped dynamical system, the most general mathematical description is of the form

$$f\left(y^{(n)}(t), y^{(n-1)}(t), \ldots, y^{(1)}(t), y(t), u^{(m)}(t), \ldots, u^{(1)}(t), u(t), t\right) = 0 \qquad (1)$$

where $y^{(k)}(t)$ is a shorthand notation for the kth time derivative of the output $y(t)$, that is,

$$y^{(k)}(t) \triangleq \frac{d^k y(t)}{dt^k} \triangleq D^k y(t) \qquad (2)$$

with

$$y^{(0)}(t) \triangleq y(t) \triangleq D^0 y(t) \tag{3}$$

Similarly, $u^{(m)}(t)$ denotes the mth time derivative of the input time function, i.e.,

$$u^{(m)}(t) \triangleq \frac{d^m u(t)}{dt^m} \triangleq D^m u(t) \tag{4}$$

with

$$u^{(0)}(t) \triangleq u(t) \triangleq D^0 u(t) \tag{5}$$

A system described by the scalar equation (1) is said to be a nonlinear time-varying SISO system.

Definition 1   *A linear time-invariant, single-input–single-output system is described by the following scalar differential equation relating the input $u(t)$ and the output $y(t)$*

$$y^{(n)}(t) + \alpha_{n-1} y^{(n-1)}(t) + \cdots + \alpha_1 y^{(1)}(t) + \alpha_0 y(t)$$
$$= \beta_m u^{(m)}(t) + \beta_{m-1} u^{(m-1)}(t) + \cdots + \beta_1 u^{(1)}(t) + \beta_0 u(t) \tag{6}$$

An alternate way of writing Eq. (6) is by utilizing the D operator, defined by (2), as follows†

$$\{D^n + \alpha_{n-1} D^{n-1} + \cdots + \alpha_1 D + \alpha_0\} y(t)$$
$$= \{\beta_m D^m + \beta_{m-1} D^{m-1} + \cdots + \beta_1 D + \beta_0\} u(t) \tag{7}$$

In Eqs. (6) and (7)

$$\alpha_0, \alpha_1, \alpha_2, \ldots, \alpha_{n-1} = \text{real constant scalars} \tag{8}$$

$$\beta_0, \beta_1, \beta_2, \ldots, \beta_m = \text{real constant scalars} \tag{9}$$

In physical systems the degrees m and n of the highest derivatives of the input and output, respectively, are constrained by the inequality

$$m \le n \tag{10}$$

Accurate modeling, especially at high frequencies, often yields

---

† Thus $\{D^3 + 4D - 3\} y(t) = \{5 D^2 - 4D\} u(t)$ means

$$\frac{d^3 y(t)}{dt^3} + 4\frac{dy(t)}{dt} - 3y(t) = 5\frac{d^2 u(t)}{dt^2} - 4\frac{du(t)}{dt}$$

$$m \leq n - 1 \tag{11}$$

The scalar n is called the *order* of the system.

8.1.2 System-Function or Transfer-Function Representation

Such representations are valid only for LTI systems. We shall discuss later how the system function or transfer function is obtained.

Definition 2 *An LTI SISO system described by the input-output differential equation (6) or (7) is said to have the system or transfer function H(s), where s is a complex scalar, defined by*

$$H(s) \triangleq \frac{B(s)}{A(s)} \triangleq \frac{\beta_m s^m + \beta_{m-1} s^{m-1} + \cdots + \beta_1 s + \beta_0}{s^n + \alpha_{n-1} s^{n-1} + \cdots + \alpha_1 s + \alpha_0} \tag{12}$$

The complex polynomial B(s)

$$B(s) = \beta_m s^m + \cdots + \beta_1 s + \beta_0 \tag{13}$$

is called the numerator *polynomial of H(s). The complex polynomial A(s)*

$$A(s) = s^n + \alpha_{n-1} s^{n-1} + \cdots + \alpha_1 s + \alpha_0 \tag{14}$$

is called the denominator *polynomial of H(s).*

Definition 3 *The (real or complex-conjugate) roots $z_1, z_2, \ldots, z_m$ of the numerator polynomial B(s) are called the* zeros *of the system or transfer function H(s). Clearly*

$$B(s) = \beta_m(s - z_1)(s - z_2) \cdots (s - z_m) = \beta_m \prod_{k=1}^{m} (s - z_k) \tag{15}$$

The (real or complex-conjugate) roots $s_1, s_2, \ldots, s_n$ of the denominator polynomial A(s) are called the poles *of the system or transfer function H(s). Clearly,*

$$A(s) = (s - s_1)(s - s_2) \cdots (s - s_n) = \prod_{k=1}^{n} (s - s_k) \tag{16}$$

When this factored form of the numerator and denominator polynomials is used, $H(s)$ of Eq. (12) can also be written

$$H(s) = \beta_m \frac{(s - z_1)(s - z_2) \cdots (s - z_m)}{(s - s_1)(s - s_2) \cdots (s - s_n)} \tag{17}$$

8.1.3   Partial-Fraction Expansion

If

$$m \leq n - 1 \tag{18}$$

i.e., if the system has at least one less zero than the number of poles, and if the poles of $H(s)$ are distinct, i.e.,

$$s_i \neq s_k \qquad \text{for all } i, k = 1, 2, \ldots, n \tag{19}$$

then it is possible to express $H(s)$ in the form

$$H(s) = \frac{r_1}{s - s_1} + \frac{r_2}{s - s_2} + \cdots + \frac{r_n}{s - s_n} = \sum_{k=1}^{n} \frac{r_k}{s - s_k} \tag{20}$$

where the r_k's are (real or complex) scalars. The scalar r_k is called the *residue* of $H(s)$ at the pole s_k. The residues are computed according to the formula

$$r_k \triangleq H(s)(s - s_k)\big|_{s=s_k} = \frac{(s_k - z_1)(s_k - z_2) \cdots (s_k - z_m)}{(s_k - s_1) \cdots (s_k - s_{k-1})(s_k - s_{k+1}) \cdots (s_k - s_n)} \tag{21}$$

8.1.4   The Derivation of the System Function Using Complex Exponential Signals

In this section we summarize the key ideas behind the derivation of the system function $H(s)$ using complex excitations.

Let us consider the LTI SISO system as described by the input-output differential equation (6) or (7). Let us suppose that the poles (natural frequencies) $s_1, s_2, \ldots, s_n$ are distinct. Let us suppose that the input $u(t)$ is suddenly applied complex exponential excitation function at time $t = 0$, that is,

$$u(t) = \hat{u}e^{st} \qquad t \geq 0 \tag{22}$$

where $\hat{u}$ is a complex scalar and s is also a complex scalar.

We know† that the structure of the solution of the differential equation (6) or (7) can be expressed as a sum of exponential time functions; i.e., it takes the form, provided $s \neq s_i$ for all $i = 1, 2, \ldots, n$,

$$y(t) = \hat{y}e^{st} + \gamma_1 e^{s_1 t} + \gamma_2 e^{s_2 t} + \cdots + \gamma_n e^{s_n t} \tag{23}$$

where s is the complex exponent of the input excitation [see Eq. (22)] and the s_i's are the system natural frequencies (poles). The scalars $\hat{y}$, γ_1, γ_2, $\ldots$, γ_n are in general

---

† See, for example, "Basic Concepts," chap. 8.

complex numbers that depend upon the system parameters, the system initial conditions, and the parameters u and s of the input $u(t)$.

Now let us suppose that the complex input frequency s is *dominant* over the natural frequencies s_i of the system; more precisely, we shall assume that

$$\text{Re}(s) > \text{Re}(s_i) \qquad \text{for all } i = 1, 2, \ldots, n \tag{24}$$

Under this assumption, as $t \to \infty$, the largest and dominant contribution to the output time function $y(t)$ of (23) is due to the term $\hat{y}e^{st}$. Thus, if we describe the limiting behavior of $y(t)$ as $t \to \infty$, by the so-called *steady-state output* $y_{ss}(t)$, we have

$$y_{ss}(t) = \hat{y}e^{st} \tag{25}$$

Using this analysis, we define the system function $H(s)$ as

$$y_{ss}(t) = H(s)u(t) \tag{26}$$

But one can show that

$$\hat{y} = \frac{\beta_m s^m + \cdots + \beta_1 s + \beta_0}{s^n + \alpha_{n-1} s^{n-1} + \cdots + \alpha_1 s + \alpha_0} \hat{u} \tag{27}$$

Hence, from Eqs. (22), (25), and (26) we obtain

$$\hat{y}e^{st} = H(s)\hat{u}e^{st} \tag{28}$$

so that $H(s)$ turns out to be given by Eq. (12).

8.1.5 The Derivation of the Transfer Function $H(s)$ Using the Laplace Transform Method†

The use of the Laplace transform method assumes that the system is initially (at time $t = 0$) at rest; i.e., all initial conditions are zero. Under this assumption, the transfer function $H(s)$ of the system is *defined* to be the ratio of the Laplace transform of the output to the Laplace transform of the input.

Let

$$\hat{y}(s) = \mathcal{L}\{y(t)\} \tag{29}$$

denote the Laplace transform of the output $y(t)$, and let

$$\hat{u}(s) = \mathcal{L}\{u(t)\} \tag{30}$$

denote the Laplace transform of the input $u(t)$.

---

† This material can be omitted by the reader who is not familiar with Laplace transforms with no loss in continuity. Appendix C contains some basic definitions dealing with Laplace transforms.

We consider the input-output differential equation of our system, i.e.,

$$\{D^n + \alpha_{n-1} D^{n-1} + \cdots + \alpha_1 D + \alpha_0\}y(t) = \{\beta_m D^m + \cdots + \beta_1 D + \beta_0\}u(t) \quad (31)$$

By taking the Laplace transform of both sides of (31), by utilizing the linearity property of the Laplace transform (see Appendix C), and by using the fact that the system is initially at rest, we know that

$$\mathcal{L}\{D^k y(t)\} = s^k \hat{y}(s) \quad (32)$$

$$\mathcal{L}\{D^j u(t)\} = s^j \hat{u}(s) \quad (33)$$

so that

$$\mathcal{L}\left\{\{D^n + \alpha_{n-1} D^{n-1} + \cdots + \alpha_1 D + \alpha_0\}y(t)\right\}$$

$$= (s^n + \alpha_{n-1} s^{n-1} + \cdots + \alpha_1 s + \alpha_0)\hat{y}(s) \quad (34)$$

and

$$\mathcal{L}\left\{\{\beta_m D^m + \cdots + \beta_0\}u(t)\right\} = (\beta_m s^m + \cdots + \beta_1 s + \beta_0)\hat{u}(s) \quad (35)$$

Therefore,

$$(s^n + \alpha_{n-1} s^{n-1} + \cdots + \alpha_1 s + \alpha_0)\hat{y}(s) = (\beta_m s^m + \cdots + \beta_1 s + \beta_0)\hat{u}(s) \quad (36)$$

By definition, we have that the transfer function $H(s)$ is given by

$$H(s) \triangleq \frac{\hat{y}(s)}{\hat{u}(s)} = \frac{\beta_m s^m + \cdots + \beta_1 s + \beta_0}{s^n + \alpha_{n-1} s^{n-1} + \cdots + \alpha_1 s + \alpha_0} \quad (37)$$

8.1.6 Remarks

This method of input-output description of LTI systems and networks by means of their system function has occupied a central position in network and system analysis. The fact that this method is in tune with the use of the Laplace transform to solve differential equations has made the input-output description extremely popular.

The more recent importance of nonlinear networks can be viewed as one of the causes of introducing the *state-variable* method into the analysis of networks. For systems described by nonlinear differential equations, the neat properties of complex exponentials (and Laplace transforms) disappear. For this reason, one cannot carry over the elegant results associated with the system-functions approach to nonlinear networks and systems.

We have seen in Chaps. 6 and 7 that the state-variable method of analyzing networks and systems can be used for both linear and nonlinear networks. Thus, the state-variable method, *viewed as an analysis technique*, does not rely either on the linearity of the network elements or on the neat properties of complex exponential functions. In this respect, the state-variable method of analysis is more general than the input-output approach, and more powerful too!

Several questions arise, however, when one wishes to analyze the *same* SISO LTI network or system by the input-output and the state-variable approach. We have briefly reviewed above what we mean by the input-output description. So let us consider the same LTI system with a single input $u(t)$ and single output $y(t)$ and with $\mathbf{x}(t)$ its n-dimensional state vector. We have seen in Chap. 7 that the state-variable method of analysis leads to the pair of equations

$$\dot{\mathbf{x}}(t) = \mathbf{A}\mathbf{x}(t) + \mathbf{b}u(t) \tag{38}$$

$$y(t) = \mathbf{c}'\mathbf{x}(t) + du(t) \tag{39}$$

where $\mathbf{A} = n \times n$ real constant matrix
  $\mathbf{b}$ = real constant column n-vector
  $\mathbf{c}$ = real constant column n-vector
  d = real constant scalar

Since we have assumed that we are analyzing the *same* system using two distinct methods, the signals $u(t)$ and $y(t)$ in Eqs. (6), (38), and (39) must turn out identical. (It would be highly embarrassing if the analysis of the same system by two different methods resulted in different solutions. We assure the reader that this is not the case.) If we assume that both analysis methods yield the same answers, then it is only natural to inquire about *the relationships that exist between the two methods when applied to LTI SISO systems*. This is precisely the purpose of this chapter. We want to make the interrelationships crystal clear so that the student will feel at ease with both methods and alternate between them at will.

The interrelationships we develop are more than qualitative and theoretical ones. They are also quantitative because questions of the following type must be answered.

1 We know that in the input-output approach, the response is governed by the poles (natural frequencies) of the system function; on the other hand, we know that in the state-variable approach, the response is governed by the eigenvalues of the **A** matrix. Is there a relation between the poles and eigenvalues? As we shall see, they are identical.

2 How can one go from a system-function to a state-variable representation? Is this transformation from input-output to state variables unique?

3 How can one go from a state-variable to an input-output representation? Is this transformation unique?

Some thought should convince the reader that if we can answer all these questions, we can utilize either approach with confidence. For example, we can couple the ease with which we can write network equations using impedances to the computational advantages offered by the state-variable approach.

8.2 COMMON ASSUMPTIONS

Let us suppose that we are dealing with a LTI SISO system or network with a single input $u(t)$ and a single output $y(t)$. Let us suppose that we analyze the same system with both the input-output and the state-variable methods. Each results in a different mathematical relationship, as summarized in Table 8.1.

We derived the system function from the input-output differential equation under certain assumptions which are summarized in the left column of Table 8.2. We shall make the same or equivalent assumptions about the state-variable representation; these assumptions are summarized in the right column of Table 8.2. We have seen the implications of these assumptions for the input-output method. To review our results, we found that the steady-state response $y_{ss}(t)$ was given by

$$y_{ss}(t) = \frac{B(s)}{A(s)} \hat{u}e^{st} \triangleq H(s)\hat{u}e^{st} \triangleq \hat{y}e^{st} \tag{1}$$

where

$$B(s) \triangleq \beta_m s^m + \beta_{m-1}s^{m-1} + \cdots + \beta_1 s + \beta_0 \tag{2}$$

$$A(s) \triangleq s^n + \alpha_{n-1}s^{n-1} + \cdots + \alpha_1 s + \alpha_0 \tag{3}$$

$$H(s) \triangleq \text{system function} \tag{4}$$

$$s_1, s_2, \ldots, s_n = \text{poles of } H(s) \quad \text{that is} \quad A(s_i) = 0 \tag{5}$$

Table 8.1 MATHEMATICAL EQUATIONS

Input-output relations:
$\{D^n + \alpha_{n-1}D^{n-1} + \cdots + \alpha_1 D + \alpha_0\}y(t) = \{\beta_m D^m + \beta_{m-1}D^{m-1} + \cdots + \beta_1 D + \beta_0\}u(t)$

State-variable relations:
$\dot{\mathbf{x}}(t) = \mathbf{A}\mathbf{x}(t) + \mathbf{b}u(t)$ $y(t) = \mathbf{c}'\mathbf{x}(t) + du(t)$

Table 8.2 COMMON ASSUMPTIONS

No.	Input-output method	State-variable method
1	System initially (at $t = 0$) at rest	System initially (at $t = 0$) at rest, i.e., $\mathbf{x}(0) = \mathbf{0}$
2	Input is a complex exponential, that is, $u(t) = \hat{u}e^{st}$, where $\hat{u}$ = complex scalar and s = complex driving frequency	Input is a complex exponential, that is, $u(t) = \hat{u}e^{st}$, where $\hat{u}$ = complex scalar and s = complex driving frequency
3	The driving frequency s is dominant over the natural frequencies (poles) $s_1, s_2, \ldots, s_n$ of the system; more precisely $\mathrm{Re}(s) > \mathrm{Re}(s_i)$; $i = 1, 2, \ldots, n$, where the s_i are the roots of the polynomial $q(\lambda) = \lambda^n + \alpha_{n-1}\lambda^{n-1} + \cdots + \alpha_1\lambda + \alpha_0$	The driving frequency s is dominant over the eigenvalues $\lambda_1, \lambda_2, \ldots, \lambda_n$ of the matrix $\mathbf{A}$; more precisely, $\mathrm{Re}(s) > \mathrm{Re}(\lambda_i)$; $i = 1, 2, \ldots, n$, where the λ_i are the roots of the characteristic polynomial of $\mathbf{A}$, $p(\lambda) = \det(\lambda\mathbf{I} - \mathbf{A})$
4	Attention is focused on the steady-state $y_{ss}(t)$ part of the output $y(t)$	Attention is focused on the steady-state $y_{ss}(t)$ part of the output $y(t)$
5	The system is stable	The system is stable†

† This is a simplifying assumption that can easily be removed.

In the next section we shall use the assumptions stated in the right column of Table 8.2 and investigate their implications in the state-variable method of representation.

8.3 FROM STATE VARIABLES TO SYSTEM FUNCTIONS

Consider an LTI network or system whose state vector $\mathbf{x}(t)$, scalar input $u(t)$, and scalar output $y(t)$ are related by:

(State equation): $$\dot{\mathbf{x}}(t) = \mathbf{A}\mathbf{x}(t) + \mathbf{b}u(t) \qquad \mathbf{x}(0) = \mathbf{x}_0 \tag{1}$$

(Output equation): $$y(t) = \mathbf{c}'\mathbf{x}(t) + du(t) \tag{2}$$

We let $\lambda_1, \lambda_2, \ldots, \lambda_n$ denote the eigenvalues of $\mathbf{A}$, that is, the roots of the characteristic polynomial

$$p(\lambda) = \det(\lambda\mathbf{I} - \mathbf{A}) \tag{3}$$

Now let us assume that the system is initially (at $t = 0$) at rest, i.e.,

$$\mathbf{x}(0) = \mathbf{0} \tag{4}$$

and that $u(t)$ is a complex exponential

$$u(t) = \hat{u}e^{st} \qquad \text{for all } t \geq 0 \tag{5}$$

Under these assumptions, Eqs. (1) and (2) reduce to

$$\dot{\mathbf{x}}(t) = \mathbf{A}\mathbf{x}(t) + \mathbf{b}\hat{u}e^{st} \qquad \mathbf{x}(0) = \mathbf{0} \tag{6}$$

$$y(t) = \mathbf{c}'\mathbf{x}(t) + d\hat{u}e^{st} \tag{7}$$

Clearly, if we can solve the vector differential equation (6), we can substitute $\mathbf{x}(t)$ into Eq. (7) and deduce $y(t)$.

8.3.1   Solution of the Vector Differential Equation

The complete solution of Eq. (6) is (see Chap. 7)

$$\mathbf{x}(t) = e^{\mathbf{A}t} \int_0^t e^{-\mathbf{A}\tau} \mathbf{b}\hat{u}e^{s\tau} d\tau \tag{8}$$

Since $\mathbf{b}$ is a constant vector and $\hat{u}$ is a constant complex scalar, Eq. (8) can be written

$$\mathbf{x}(t) = e^{\mathbf{A}t} \int_0^t e^{s\tau}e^{-\mathbf{A}\tau} d\tau \, \mathbf{b}\hat{u} \tag{9}$$

The next step combines the two exponentials in the integrand. Since $e^{s\tau}$ is a scalar and $e^{-\mathbf{A}\tau}$ is an $n \times n$ matrix, we must use a trick

$$e^{s\tau}e^{-\mathbf{A}\tau} = e^{s\tau}\mathbf{I}e^{-\mathbf{A}\tau} = e^{s\mathbf{I}\tau}e^{-\mathbf{A}\tau} \tag{10}$$

To justify Eq. (10) we note that

$$e^{s\tau}\mathbf{I} = \begin{bmatrix} e^{s\tau} & 0 & \cdots & 0 \\ 0 & e^{s\tau} & \cdots & 0 \\ \multicolumn{4}{c}{\dotfill} \\ 0 & 0 & \cdots & e^{s\tau} \end{bmatrix} = e^{s\mathbf{I}\tau} \tag{11}$$

We next claim that

$$e^{s\mathbf{I}\tau}e^{-\mathbf{A}\tau} = e^{(s\mathbf{I}-\mathbf{A})\tau} \tag{12}$$

To justify Eq. (12) we recall (see Sec. 7.6) that $e^{\mathbf{A}}e^{\mathbf{B}} = e^{\mathbf{A}+\mathbf{B}}$ if $\mathbf{AB} = \mathbf{BA}$, that is, if $\mathbf{A}$ and $\mathbf{B}$ commute. Since the identity matrix $\mathbf{I}$ commutes with any matrix, we deduce that Eq. (12) is correct.

Hence, the solution $\mathbf{x}(t)$ is

$$\mathbf{x}(t) = e^{\mathbf{A}t}\left(\int_0^t e^{(s\mathbf{I}-\mathbf{A})\tau} d\tau\right)\mathbf{b}\hat{u} \tag{13}$$

We must now evaluate the integral. Let us recall that if $\mathbf{B}$ is a nonsingular matrix, then

$$\int_0^t e^{\mathbf{B}\tau} d\tau = e^{\mathbf{B}\tau}\Big|_0^t \mathbf{B}^{-1} = (e^{\mathbf{B}t} - \mathbf{I})\mathbf{B}^{-1} \tag{14}$$

To apply Eq. (14) to the evaluation of the integral in Eq. (13) we must show that the matrix $s\mathbf{I} - \mathbf{A}$ is nonsingular. To do this let us recall the definition of the eigenvalues $\lambda_1, \lambda_2, \ldots, \lambda_n$ which are the roots of the characteristic polynomial

$$p(\lambda) = \det(\lambda\mathbf{I} - \mathbf{A}) \tag{15}$$

Thus

$$p(\lambda_i) = \det(\lambda_i\mathbf{I} - \mathbf{A}) = 0 \qquad i = 1, 2, \ldots, n \tag{16}$$

This means that the matrix $\lambda_i\mathbf{I} - \mathbf{A}$ is singular.

Now we make the assumption that s is dominant over the eigenvalues of $\mathbf{A}$, that is,

$$\text{Re}(s) > \text{Re}(\lambda_i) \qquad i = 1, 2, \ldots, n \tag{17}$$

In particular, Eq. (17) implies that s is not equal to any of the eigenvalues, i.e.,

$$s \neq \lambda_i \qquad i = 1, 2, \ldots, n \tag{18}$$

This means that

$$\det(s\mathbf{I} - \mathbf{A}) \neq 0 \tag{19}$$

and, hence, that the matrix $(s\mathbf{I} - \mathbf{A})$ is nonsingular.

Since we have shown that $(s\mathbf{I} - \mathbf{A})^{-1}$ exists, we can evaluate the integral in Eq. (13) with the help of Eq. (14) to obtain

$$\mathbf{x}(t) = e^{\mathbf{A}t}(e^{(s\mathbf{I}-\mathbf{A})t} - \mathbf{I})(s\mathbf{I} - \mathbf{A})^{-1}\mathbf{b}\hat{u} \tag{20}$$

Now we split the matrix $e^{(s\mathbf{I}-\mathbf{A})t}$ into

$$e^{(s\mathbf{I}-\mathbf{A})t} = e^{s\mathbf{I}t}e^{-\mathbf{A}t} = e^{st}\mathbf{I}e^{-\mathbf{A}t} = e^{st}e^{-\mathbf{A}t} \tag{21}$$

by an argument opposite to that used to derive Eq. (12) from Eq. (10). Substitution of Eq. (21) into Eq. (20) yields

$$\mathbf{x}(t) = e^{\mathbf{A}t}(e^{st}e^{-\mathbf{A}t} - \mathbf{I})(s\mathbf{I} - \mathbf{A})^{-1}\mathbf{b}\hat{u} \tag{22}$$

$$= (e^{st}e^{\mathbf{A}t}e^{-\mathbf{A}t} - e^{\mathbf{A}t})(s\mathbf{I} - \mathbf{A})^{-1}\mathbf{b}\hat{u} \tag{23}$$

$$= (s\mathbf{I} - \mathbf{A})^{-1}\mathbf{b}\hat{u}e^{st} - e^{\mathbf{A}t}(s\mathbf{I} - \mathbf{A})^{-1}\mathbf{b}\hat{u} \tag{24}$$

8.3.2 The Output Equation

Let us now substitute Eq. (24) into the output equation (7) to obtain

$$y(t) = \mathbf{c}'(s\mathbf{I} - \mathbf{A})^{-1}\mathbf{b}\hat{u}e^{st} - \mathbf{c}'e^{\mathbf{A}t}(s\mathbf{I} - \mathbf{A})^{-1}\mathbf{b}\hat{u} + d\hat{u}e^{st} \qquad (25)$$

We now use the fact that our assumption that s is dominant over the eigenvalues of $\mathbf{A}$ [see Eq. (17)] to deduce that

$$\text{As } t \to \infty \qquad |e^{st}| \gg \|e^{\mathbf{A}t}\| \qquad (26)$$

Equation (26) is even more easily justified if we assume that the system is stable, i.e., $\text{Re}(\lambda_i) < 0$, which means that

$$\text{As } t \to \infty \qquad \|e^{\mathbf{A}t}\| \to 0 \qquad (27)$$

In this case if we define the *steady-state output* $y_{ss}(t)$ (just as we did in Sec. 8.1) by

$$y_{ss}(t) \triangleq \mathbf{c}'(\mathbf{I} - \mathbf{A})^{-1}\mathbf{b}\hat{u}e^{st} + d\hat{u}e^{st} \qquad (28)$$

then

$$\text{As } t \to \infty \qquad y(t) \approx y_{ss}(t) \qquad (29)$$

8.3.3 Conclusions

What conclusion can we draw at this point? We have shown that the steady-state output $y_{ss}(t)$ has the form

$$y_{ss}(t) = \hat{y}e^{st} \qquad (30)$$

where $\hat{y}$ is a complex scalar given by

$$\hat{y} \triangleq [\mathbf{c}'(s\mathbf{I} - \mathbf{A})^{-1}\mathbf{b} + d]\hat{u} \qquad (31)$$

and that the time variation of the steady-state output is identical to that of the input (e^{st}) under the assumption that s is dominant over the eigenvalues of $\mathbf{A}$.

8.3.4 The System Function in Terms of State-Variable Quantities

Let us now recall that given an input $u(t) = \hat{u}e^{st}$ resulting in the steady-state output $y_{ss}(t) = \hat{y}e^{st}$, the system function $H(s)$ satisfies the relation (see Sec. 8.1)

$$\hat{y} = H(s)\hat{u} \qquad (32)$$

If we compare Eqs. (31) and (32), we immediately deduce that the system function is

$$H(s) = \mathbf{c}'(s\mathbf{I} - \mathbf{A})^{-1}\mathbf{b} + d \tag{33}$$

where the $n \times n$ matrix $\mathbf{A}$, the column vectors $\mathbf{b}$ and $\mathbf{c}$, and the scalar d are quantities in the state-variable representation

$$\dot{\mathbf{x}}(t) = \mathbf{A}\mathbf{x}(t) + \mathbf{b}u(t) \qquad y(t) = \mathbf{c}'\mathbf{x}(t) + du(t) \tag{34}$$

Thus, we have demonstrated that we can go from the state-variable representation to the system function $H(s)$.

8.3.5 Poles and Eigenvalues: One and the Same

We continue our examination of the interrelationships between the input-output and state-variable representations by examining Eq. (33) a bit more closely.

First, let us recall that the derivation of the system function $H(s)$ from the input-output point of view resulted in a more specific form for $H(s)$, namely, that of the ratio of two polynomials

$$H(s) = \frac{B(s)}{A(s)} = \frac{\beta_m s^m + \beta_{m-1} s^{m-1} + \cdots + \beta_1 s + \beta_0}{s^n + \alpha_{n-1} s^{n-1} + \cdots + \alpha_1 s + \alpha_0} \qquad m \le n \tag{35}$$

The *poles* of $H(s)$ were defined to be the roots $s_1, s_2, \ldots, s_n$ of the denominator polynomial

$$A(s) = s^n + \alpha_{n-1} s^{n-1} + \cdots + \alpha_1 s + \alpha_0 \tag{36}$$

Now let us examine the expression for $H(s)$ given by Eq. (33) derived from the state-variable representation. In particular, let us consider the $n \times n$ matrix $(s\mathbf{I} - \mathbf{A})^{-1}$ which appears in that expression.

The general form of the $n \times n$ matrix $s\mathbf{I} - \mathbf{A}$ is

$$s\mathbf{I} - \mathbf{A} = \begin{bmatrix} s - a_{11} & -a_{12} & \cdots & -a_{1n} \\ -a_{21} & s - a_{22} & \cdots & -a_{2n} \\ \cdots\cdots\cdots\cdots\cdots\cdots\cdots\cdots\cdots\cdots\cdots \\ -a_{n1} & -a_{n2} & \cdots & s - a_{nn} \end{bmatrix} \tag{37}$$

where the a_{ij}'s are the real scalar elements of $\mathbf{A}$. Now, the inverse matrix $(s\mathbf{I} - \mathbf{A})^{-1}$ can be written (see Appendix A, Sec. A.7)

$$(s\mathbf{I} - \mathbf{A})^{-1} = \frac{1}{\det(s\mathbf{I} - \mathbf{A})} \mathbf{Q}(s) \tag{38}$$

where $\mathbf{Q}(s)$ is an $n \times n$ matrix and $\det(s\mathbf{I} - \mathbf{A})$ is, of course, a scalar quantity. We claim that each element $q_{ij}(s)$ of the $n \times n$ matrix $\mathbf{Q}(s)$ is a simple polynomial in s whose degree is at most $n - 1$. Why? The reason is that $q_{ij}(s)$ is obtained by eliminating the jth row and ith column of the matrix $s\mathbf{I} - \mathbf{A}$ and computing the determinant of the resultant $(n - 1) \times (n - 1)$ matrix. Since [see Eq. (37)] each element of $s\mathbf{I} - \mathbf{A}$ is a polynomial of at most first degree in s, we conclude that the elements of $\mathbf{Q}(s)$ are simple polynomials in s whose degree is at most $n - 1$.

If we now substitute Eq. (38) into Eq. (33), we can see that

$$H(s) = \frac{1}{\det(s\mathbf{I} - \mathbf{A})} \mathbf{c}'\mathbf{Q}(s)\mathbf{b} + d \tag{39}$$

But $\mathbf{c}'$ is a $1 \times n$ matrix (row vector), $\mathbf{Q}(s)$ is an $n \times n$ matrix, and $\mathbf{b}$ is an $n \times 1$ matrix (column vector). It follows that

$$\mathbf{c}'\mathbf{Q}(s)\mathbf{b} = \text{scalar} \tag{40}$$

In point of fact, since the elements of $\mathbf{c}$ and $\mathbf{b}$ are independent of s, we deduce, in view of our discussion on the form of the elements of $\mathbf{Q}(s)$, that

$$\mathbf{c}'\mathbf{Q}(s)\mathbf{b} = \text{polynomial in } s \text{ of at most } n - 1 \text{ degree} \tag{41}$$

Now, since d is a real scalar, Eq. (39) can also be written

$$H(s) = \frac{\mathbf{c}'\mathbf{Q}(s)\mathbf{b} + d\det(s\mathbf{I} - \mathbf{A})}{\det(s\mathbf{I} - \mathbf{A})} \tag{42}$$

which implies that the system function is the ratio of two polynomials in s. If we compare Eqs. (42) and (35), we can make the identification

$$\boxed{B(s) = \mathbf{c}'\mathbf{Q}(s)\mathbf{b} + d\det(s\mathbf{I} - \mathbf{A})} \tag{43}$$

$$\tag{44}$$

$$\boxed{A(s) = \det(s\mathbf{I} - \mathbf{A})}$$

Hence, the poles of $H(s)$ must be the roots of the polynomial $\det(s\mathbf{I} - \mathbf{A})$. But $\det(s\mathbf{I} - \mathbf{A})$ is precisely the characteristic polynomial of the matrix $\mathbf{A}$, and its roots are the eigenvalues of $\mathbf{A}$. Hence, *the eigenvalues of $\mathbf{A}$ and the poles of the system function $H(s)$ are identical.*

In a similar manner, we can deduce that the zeros of the system function $H(s)$

are simply the roots of

$$\mathbf{c}'\mathbf{Q}(s)\mathbf{b} + d\det(s\mathbf{I} - \mathbf{A}) \tag{45}$$

Note that if the number of zeros (m) is the same as the number of poles (n), then $d \neq 0$. If $m < n$, then $d = 0$ because $\det(s\mathbf{I} - \mathbf{A})$ is an nth-degree polynomial.

Remarks In this section, we derived the system function of a LTI SISO system from its state-variable representation. This enables us to conclude that the poles of the system function are identical to the eigenvalues of the matrix $\mathbf{A}$ that appears in the state-vector differential equation $\dot{\mathbf{x}}(t) = \mathbf{A}\mathbf{x}(t) + \mathbf{b}u(t)$. With this identification in mind, if we reexamine the common assumptions of Table 8.2, we conclude that they are indeed the same.

8.3.6 Derivation of the Transfer Function Using Laplace Transforms[†]

It is a very straightforward matter to obtain the transfer function of a system starting from its state-variable description using Laplace transforms. The reader should review Appendix C at this point since we are going to use the notion of Laplace transforms of vector-valued time functions.

Once more we start with the state-variable description of our LTI SISO system; we assume that it starts at rest so that $\mathbf{x}(0) = \mathbf{0}$

$$\dot{\mathbf{x}}(t) = \mathbf{A}\mathbf{x}(t) + \mathbf{b}u(t) \qquad \mathbf{x}(0) = \mathbf{0} \tag{46}$$

$$y(t) = \mathbf{c}'\mathbf{x}(t) + du(t) \tag{47}$$

Let

$$\hat{\mathbf{x}}(s) \triangleq \mathcal{L}\{\mathbf{x}(t)\} \qquad \hat{u}(s) \triangleq \mathcal{L}\{u(t)\} \qquad \hat{y}(s) = \mathcal{L}\{y(t)\} \tag{48}$$

By taking Laplace transforms of both sides of (46) we obtain

$$s\hat{\mathbf{x}}(s) = \mathbf{A}\hat{\mathbf{x}}(s) + \mathbf{b}\hat{u}(s) \tag{49}$$

Equation (49) implies that

$$(s\mathbf{I} - \mathbf{A})\hat{\mathbf{x}}(s) = \mathbf{b}\hat{u}(s) \tag{50}$$

hence,

$$\boxed{\hat{\mathbf{x}}(s) = (s\mathbf{I} - \mathbf{A})^{-1}\mathbf{b}\hat{u}(s)} \tag{51}$$

[†] Again the reader not familiar with Laplace transforms can omit this section with no loss in continuity.

Next, we take Laplace transforms of both sides of Eq. (47) to deduce

$$\hat{y}(s) = \mathbf{c}'\hat{\mathbf{x}}(s) + d\hat{u}(s) \qquad (52)$$

The next step is to eliminate the vector $\hat{\mathbf{x}}(s)$ between Eqs. (51) and (52) to obtain

$$\hat{y}(s) = \mathbf{c}'(s\mathbf{I} - \mathbf{A})^{-1}\mathbf{b}\hat{u}(s) + d\hat{u}(s) \qquad (53)$$

By definition, the transfer function $H(s)$ is

$$H(s) = \frac{\hat{y}(s)}{\hat{u}(s)} \qquad (54)$$

and therefore it follows that

$$H(s) = \mathbf{c}'(s\mathbf{I} - \mathbf{A})^{-1}\mathbf{b} + d \qquad (55)$$

We now note that Eqs. (33) and (55) are identical in structure. The only difference is the interpretation of the scalar s. In the *system function* (33), s was the complex driving frequency of the exponential excitation; in the *transfer function* (55), s is the complex Laplace transform variable. (This is why we make the distinction between system function and transfer function.)

Since Eqs. (33) and (55) are identical in structure, the development of Eqs. (35) to (45) is identical for the case of transfer functions also. Thus we conclude that

1 The poles of the transfer function $H(s)$ are identical to the eigenvalues of the **A** matrix.

2 The zeros of the transfer function $H(s)$ are the roots of

$$\mathbf{c}'\mathbf{Q}(s)\mathbf{b} + d\det(s\mathbf{I} - \mathbf{A}) \qquad (56)$$

8.4 ALTERNATE STATE-VARIABLE REPRESENTATIONS AND THEIR IN-PUT-OUTPUT INVARIANCE

We are now confident that starting from a state-variable representation, we can find the system function. To increase our level of confidence and to illustrate the fact that *many state-variable representations can result in the same input-output system function* (so that the transformation from system function to state variables is not one to one) we shall briefly examine the implications of some changes of variables.

Let us start with a system or network whose state-variable representation is

$$\dot{\mathbf{x}}(t) = \mathbf{A}\mathbf{x}(t) + \mathbf{b}u(t) \tag{1}$$

$$y(t) = \mathbf{c}'\mathbf{x}(t) + du(t) \tag{2}$$

where $\mathbf{x}(t)$ is a column n-vector. Let $H(s)$ be the system function

$$H(s) = \mathbf{c}'(s\mathbf{I} - \mathbf{A})^{-1}\mathbf{b} + d \tag{3}$$

Let $\lambda_1, \lambda_2, \ldots, \lambda_n$ be the eigenvalues of $\mathbf{A}$. Suppose that $\mathbf{M}$ is a matrix which is similar to $\mathbf{A}$. We recall (see Appendix A, Secs. A.8 to A.10) that

1 If $\mathbf{M}$ and $\mathbf{A}$ are similar $n \times n$ matrices, they both have identical eigenvalues.

2 If $\mathbf{M}$ and $\mathbf{A}$ are similar matrices, there exists a nonsingular $n \times n$ matrix, say $\mathbf{P}$, such that

$$\mathbf{M} = \mathbf{P}^{-1}\mathbf{A}\mathbf{P} \tag{4}$$

Let us now define a new column n-vector $\mathbf{z}(t)$ by

$$\mathbf{z}(t) \triangleq \mathbf{P}^{-1}\mathbf{x}(t) \quad \text{or} \quad \mathbf{x}(t) = \mathbf{P}\mathbf{z}(t) \tag{5}$$

Hence, since $\mathbf{P}$ is a constant matrix,

$$\dot{\mathbf{z}}(t) = \mathbf{P}^{-1}\dot{\mathbf{x}}(t) \tag{6}$$

Next, we premultiply both sides of Eq. (1) by $\mathbf{P}^{-1}$ to obtain

$$\mathbf{P}^{-1}\dot{\mathbf{x}}(t) = \mathbf{P}^{-1}\mathbf{A}\mathbf{x}(t) + \mathbf{P}^{-1}\mathbf{b}u(t) \tag{7}$$

Substitution of Eqs. (5) and (6) into Eq. (7) yields

$$\dot{\mathbf{z}}(t) = \mathbf{P}^{-1}\mathbf{A}\mathbf{P}\mathbf{z}(t) + \mathbf{P}^{-1}\mathbf{b}u(t) \tag{8}$$

Substitution of Eq. (5) into the output equation (2) yields

$$y(t) = \mathbf{c}'\mathbf{P}\mathbf{z}(t) + du(t) \tag{9}$$

Use of Eq. (4) results in the new set of equations

$$\dot{\mathbf{z}}(t) = \mathbf{M}\mathbf{z}(t) + \mathbf{P}^{-1}\mathbf{b}u(t) \tag{10}$$

$$y(t) = \mathbf{c}'\mathbf{P}\mathbf{z}(t) + du(t) \tag{11}$$

Let us see what we have done up to now. If we compare Eqs. (1) and (2) with Eqs. (10) and (11), we see that by leaving the input $u(t)$ and output $y(t)$ unchanged, we have generated by a change of variables a new state vector $\mathbf{z}(t)$ for the same system. Since there is an infinite number of matrices which are similar to $\mathbf{A}$, the reader should be convinced that *we can start from one set of state equations and generate (by changing variables) an infinite number of alternate state-variable representations while still retaining the original input* [$u(t)$] *and output* [$y(t)$] *variables.*

To remove any doubts that may still linger, let us write the system function $G(s)$ of the system described by the state representation (10) and (11)

$$G(s) = \mathbf{c}'\mathbf{P}(s\mathbf{I} - \mathbf{M})^{-1}\mathbf{P}^{-1}\mathbf{b} + d \tag{12}$$

If everything we have said is true, we should be able to prove that

$$G(s) = H(s) \tag{13}$$

because this internal change of state variables should not affect the input-output properties of our system.

To prove that Eq. (13) holds, we substitute Eq. (4) into Eq. (12) to obtain

$$G(s) = \mathbf{c}'\mathbf{P}(s\mathbf{I} - \mathbf{P}^{-1}\mathbf{A}\mathbf{P})^{-1}\mathbf{P}^{-1}\mathbf{b} + d \tag{14}$$

If we write the identity matrix $\mathbf{I}$ as

$$\mathbf{I} = \mathbf{P}^{-1}\mathbf{P} \tag{15}$$

then

$$G(s) = \mathbf{c}'\mathbf{P}(s\mathbf{P}^{-1}\mathbf{P} - \mathbf{P}^{-1}\mathbf{A}\mathbf{P})^{-1}\mathbf{P}^{-1}\mathbf{b} + d \tag{16}$$

$$= \mathbf{c}'\mathbf{P}(\mathbf{P}^{-1}s\mathbf{I}\mathbf{P} - \mathbf{P}^{-1}\mathbf{A}\mathbf{P})^{-1}\mathbf{P}^{-1}\mathbf{b} + d \tag{17}$$

$$= \mathbf{c}'\mathbf{P}[\mathbf{P}^{-1}(s\mathbf{I} - \mathbf{A})\mathbf{P}]^{-1}\mathbf{P}^{-1}\mathbf{b} + d \tag{18}$$

$$= \mathbf{c}'\mathbf{P}(\mathbf{P}^{-1})(s\mathbf{I} - \mathbf{A})^{-1}(\mathbf{P}^{-1})^{-1}\mathbf{P}^{-1}\mathbf{b} + d \tag{19}^\dagger$$

$$= \mathbf{c}'(s\mathbf{I} - \mathbf{A})^{-1}\mathbf{b} + d \tag{20}$$

If we now compare Eqs. (20) and (3), we deduce that indeed $G(s) = H(s)$, as promised.

8.5 PHILOSOPHICAL COMMENTS: SYSTEM FUNCTIONS, STATE VARIABLES, AND SIMULATION

We have now sufficient quantitative results to discuss in an intelligent manner the differences between the input-output and the state-variable approaches. The input-output approach considers systems or networks as black boxes and simply relates the effect (output) $y(t)$ to the cause (input) $u(t)$. The input-output approach does not

---

† In the derivation of Eq. (19) from Eq. (18) we have used the fact that if $\mathbf{A}$, $\mathbf{B}$, $\mathbf{C}$ are nonsingular matrices, then $(\mathbf{A}\mathbf{B}\mathbf{C})^{-1} = \mathbf{C}^{-1}\mathbf{B}^{-1}\mathbf{A}^{-1}$.

take into account what is inside the black box; it may be an airplane, a network, an electromechanical system, etc. As such, the emphasis is on the physical nature of the input and output signals and not in their interrelation through the physical energy-storage elements which are inside the black box.

The state-variable approach is based on what is inside the black box. If it is a network, the natural state variables are the capacitor voltages and inductor currents (or charges and flux linkages), quantities that are intimately related to the energy-storage elements. If we deal with a mechanical system (composed of interconnected masses, springs, and dashpots), the natural state variables are positions and velocities; once more these are quantities related to the potential and kinetic energies stored in the springs and masses, respectively. From the state-variable approach we can get from the inside of the black box and deduce, as we have demonstrated, its input-output behavior say in terms of the system function. Thus, the transformation from a given state-variable representation to the input-output representation is one to one.

On the other hand, we have shown in Sec. 8.4 that different state-variable representations yield the same input-output behavior. Thus the transformation from the input-output representation to a state-variable representation is not one to one. From a physical point of view this makes sense; an electric network and a mechanical system can obey the same input-output differential equation; however, their energy state variables are different physical quantities.

If we consider an LTI system initially at rest, we can view the input-output approach as yielding an abstract linear transformation which maps *input functions of time* into *output functions of time*. We can think of each state-variable description of the same system as corresponding to the specifications of a coordinate system (set of bases) for this abstract input-output linear transformation.

This type of analogy has more than pure intellectual value. It is of extreme use in system analysis and design. To illustrate this point let us consider a typical example. In aircraft design one often relates aerodynamic forces due to aileron and/or elevator deflections to, say, angle (yaw, pitch) deflections of the aircraft. Such relations often take the form of a system function relating, say, an input (elevator deflection) to an output (pitch angle). The engineer is interested in finding the response of the aircraft. However, it may be expensive, impractical, or even dangerous to carry out these experiments in flight. So he must start with the black-box system-function description and do something with it on the ground. What can he do? He can attempt to *simulate* the black box by another kind of physical system, which is certainly not the aircraft. He may attempt to construct a network whose input-output behavior is identical to that of the aircraft, or he may wish to carry out an analog-computer simulation, which is in essence a continuous-time information-processing system constructed by interconnecting integrators, summers, and amplifiers.

What is the essence of such simulations? The engineer starts with the input-output mathematical description of one black box (the airplane). He wants to construct another black box (network, analog-computer information-processing system) with the same input-output behavior. However, by virtue of the fact that he must construct the second black box he must start from the inside; hence, he must start with some state-variable representations for his simulated black box which lead to the given system function.

For this reason it is important to be able to find ways of going from system-function representations to state-variable representations (keeping in mind the nonunique aspects). Analog-computer information-processing systems offer an excellent vehicle; this approach is further enhanced by virtue of the fact that analog-computer information-processing systems can themselves be simulated on general-purpose digital computers† or special-purpose digital machines, e.g., digital differential analyzers (DDA). From a simulation point of view, the accuracies of digital and analog computers (say compared with those of networks) make them ideal for engineering studies.

For this reason we consider in the next section certain standard ways of obtaining a state-variable representation from an input-output system function and explaining them using analog-computer information-processing systems.

8.6   FROM SYSTEM FUNCTIONS TO STATE-VARIABLE REPRESENTA-TIONS

In this section we shall discuss certain somewhat standard ways of constructing state-variable representations from input-output relations.

8.6.1   The Basic Objective

Throughout this section we shall deal with SISO LTI systems described by their system function

$$\frac{\beta_m s^m + \beta_{m-1} s^{m-1} + \cdots + \beta_1 s + \beta_0}{s^n + \alpha_{n-1} s^{n-1} + \cdots + \alpha_1 s + \alpha_0} \qquad m \leq n \qquad (1)$$

We remind the reader that this implies that the input $u(t)$ and the output $y(t)$ are related by the nth-order differential equation

$$\{D^n + \alpha_{n-1} D^{n-1} + \cdots + \alpha_1 D + \alpha_0\} y(t)$$
$$= \{\beta_m D^m + \beta_{m-1} D^{m-1} + \cdots + \beta_1 D + \beta_0\} u(t) \qquad (2)$$

---

† Which simply replace continuous integration by numerical integration.

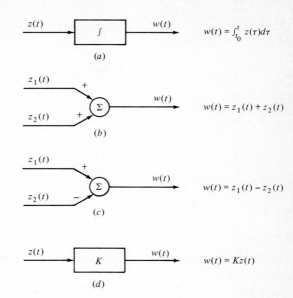

FIGURE 8.6.1
Ideal elements available on an analog computer: (*a*) ideal integrator; (*b*) ideal adder; (*c*) ideal subtractor; (*d*) ideal gain.

where $D(.)$ is the differential operator, i.e.,

$$D^k z(t) \triangleq \frac{d^k z(t)}{dt^k} \qquad (3)$$

Our basic objective is to find a state-variable representation for such systems. In other words, we wish to represent such systems by a pair of equations of the form

$$\dot{\mathbf{x}}(t) = \mathbf{A}\mathbf{x}(t) + \mathbf{b}u(t) \qquad (4)$$

$$y(t) = \mathbf{c}'\mathbf{x}(t) + du(t) \qquad (5)$$

We seek ways of specifying the elements of the matrix $\mathbf{A}$, of the column vectors $\mathbf{b}$ and $\mathbf{c}$, and of the scalar d in terms of the given values of the real scalars $\alpha_0, \alpha_1, \alpha_2, \ldots, \alpha_{n-1}$ and $\beta_0, \beta_1, \beta_2, \ldots, \beta_m$ which appear in the input-output description (1) or (2).

8.6.2 Analog-Computer Interpretations

To visualize the implications of the mathematical equations, we shall use analog-computer information-processing systems. Such a procedure not only enhances physical intuition but is of value in generating realistic simulation schemes for practical use.

To be more specific we shall assume that we have available the information-processing elements of Fig. 8.6.1 consisting of integrators, adders, subtracters, and

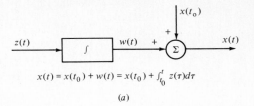

$$x(t) = x(t_0) + w(t) = x(t_0) + \int_{t_0}^{t} z(\tau)d\tau$$

(a)

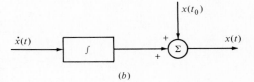

FIGURE 8.6.2
Integration with initial conditions.

(b)

constant-gain amplifiers. These elements perform the mathematical operations indicated in Fig. 8.6.1. Such elements can be built with extreme accuracy and are available on an analog computer. Furthermore, the indicated mathematical operations can be executed with extreme accuracy by a digital computer.

A key operation which can be performed is that of definite integration. To stress this, consider the interconnection of an integrator and a summer shown in Fig. 8.6.2a. In this case, the relation between $x(t)$ and $z(t)$ is the memory relation

$$x(t) = x(t_0) + \int_{t_0}^{t} z(\tau) \, d\tau \tag{6}$$

What is the relation between $x(t)$ and $z(t)$? If we differentiate both sides of Eq. (6), we obtain [since $x(t_0)$ is a constant]

$$\dot{x}(t) = \frac{d}{dt} \int_{t_0}^{t} z(\tau) \, d\tau = z(t) \tag{7}$$

Hence, as indicated in Fig. 8.6.2b, if we introduce at $t \geq t_0$ the time derivative $\dot{x}(t)$ of a signal $x(t)$ into the integrator, and if we add the initial value $x(t_0)$ of $x(t)$, we recover the signal $x(t)$.

The main value of this operation can be explained by considering a scalar differential equation, say,

$$\dot{x}(t) = -4x(t) + 3u(t) \qquad x(0) = 5 \tag{8}$$

The simulation of this differential equation is shown in Fig. 8.6.3. In other words, one starts with signals $x(t)$ and $u(t)$ and forms the right-hand side of Eq. (8) to generate $\dot{x}(t)$. The initial condition $x(0) = 5$ is added to the output of the integrator.

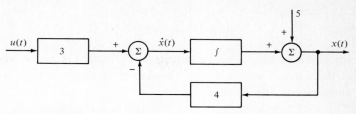

FIGURE 8.6.3
Analog-computer simulation of $\dot{x}(t) = -4x(t) + 3u(t)$, $x(0) = 5$.

These basic building blocks can be used to simulate the differential equation satisfied by the state vector $\mathbf{x}(t)$

$$\dot{\mathbf{x}}(t) = \mathbf{A}\mathbf{x}(t) + \mathbf{b}u(t) \tag{9}$$

Suppose that at the initial time $t = 0$ we have the initial condition

$$\mathbf{x}(0) = \boldsymbol{\xi} \tag{10}$$

Let us write Eqs. (9) and (10) component by component $(i = 1, 2, \ldots, n)$

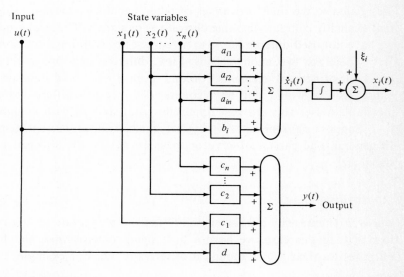

FIGURE 8.6.4
Analog-computer simulation of $x_i(t) = \sum_{j=1}^{n} a_{ij} x_j(t) + b_i u(t)$, $y(t) = \sum_{i=1}^{n} c_i x_i(t) + du(t)$.

$$\dot{x}_i(t) = \sum_{j=1}^{n} a_{ij} x_j(t) + b_i u(t) \qquad x_i(0) = \xi_i \tag{11}$$

where the a_{ij} are the real elements of the matrix $\mathbf{A}$ and the b_i are the real elements of the column vector $\mathbf{b}$. The simulation of this equation, utilizing n integrators is illustrated in Fig. 8.6.4.

The output equation can also be simulated since

$$y(t) = \sum_{i=1}^{n} c_i x_i(t) + du(t) \tag{12}$$

where $c_1, c_2, \ldots, c_n$ are the elements of the column vector $\mathbf{c}$. This is also shown in Fig. 8.6.4.

In this manner, we can use the basic elements of Fig. 8.6.1 to simulate a given state-variable representation. The dimension of the state vector $\mathbf{x}(t)$ is equal to the number of integrators that must be used. The elements a_{ij} of the $n \times n$ matrix $\mathbf{A}$, the elements b_i of the column n-vector $\mathbf{b}$, the elements c_i of the column n-vector $\mathbf{c}$, and the scalar d appear as amplifier gains.

8.6.3 Differentiators Not Allowed

Although certain signals (inputs to integrators) in analog-computer representations are related to the time derivatives of other signals, we are not using any elements that explicitly compute the time derivatives of signals. The reader may wonder why we have not used differentiators to give ourselves additional freedom.

There are two main reasons. First, differentiators are extremely difficult to build out of network elements.† Second, even if one could build an ideal differentiator, its use would not be recommended because differentiators amplify the inevitable noise. This is a very important point, and we wish to elaborate upon it.

Suppose $s(t)$ represents a physical signal (voltage, current, temperature, etc.). In general, one cannot observe or measure $s(t)$ exactly. Instead one measures a signal $z(t)$

$$z(t) = s(t) + n(t) \tag{13}$$

where $n(t)$ represents a noise signal. Often the magnitude of $n(t)$ is much smaller than $s(t)$; in this case, we have a high *signal-to-noise ratio*, and for all practical purposes we treat $z(t)$ as though it were really $s(t)$. For example, suppose $s(t)$ is a

---

† If one could build a real differentiator, one would be able to generate a *true* impulse (infinite amplitude in zero time) by driving it with a step. Try your luck building it with physical network elements.

sine wave

$$s(t) = \sin t \tag{14}$$

and that the noise is a high-frequency small-amplitude sinusoid

$$n(t) = 10^{-3} \sin 10^3 t \tag{15}$$

Then

$$z(t) = \sin t + 10^{-3} \sin 10^3 t \approx s(t) \tag{16}$$

because the signal-to-noise ratio is 1,000.

Now suppose that we have ideal differentiators. If we introduce the noisy signal $z(t)$ into this device, its output would be

$$\dot{z}(t) = \dot{s}(t) + \dot{n}(t) = \cos t + \cos 10^3 t \tag{17}$$

The signal-to-noise ratio of $\dot{z}(t)$ is now unity and, so we cannot by any stretch of the imagination claim that $\dot{z}(t) \approx \dot{s}(t)$. The troubles compound if we introduce $\dot{z}(t)$ into another differentiator to obtain

$$\ddot{z}(t) = \ddot{s}(t) + \ddot{n}(t) = -\sin t - 10^3 \sin 10^3 t \tag{18}$$

and the signal-to-noise ratio has been further reduced to 1:1,000. Now we have lost our signal completely.

For these two reasons ideal or approximate differentiating networks are not used. On the other hand, integrators do not suffer from such shortcoming as they *improve* the signal-to-noise ratio. Thus if we introduce $z(t)$ into an integrator, we obtain, at its output, a signal-to-noise ratio of 10^6. Hence, our use of integrators does not suffer from either practical or theoretical shortcomings.

8.6.4   State-Variable Representation for System Functions Containing Only Poles

We start our development of a state-variable representation of a system by considering a simple example. Suppose we are given a system with the system (or transfer) function

$$H(s) = \frac{1}{s^3 + 3s^2 + 5s + 6} \tag{19}$$

Note that this system has three poles and no zeros. For such a system, the output $y(t)$ is related to the input $u(t)$ by the LTI third-order differential equation

$$\{D^3 + 3D^2 + 5D + 6\}y(t) = u(t) \tag{20}$$

For such systems it is easy to define the state variables by means of the output $y(t)$ and its time derivatives. Toward this goal let us define

$$x_1(t) \triangleq y(t) \tag{21}$$

$$x_2(t) \triangleq Dy(t) \tag{22}$$

$$x_3(t) \triangleq D^2 y(t) \tag{23}$$

and examine whether $x_1(t)$, $x_2(t)$, $x_3(t)$ qualify as state variables. By differentiating Eqs. (21) to (23) we obtain

$$\dot{x}_1(t) = Dy(t) = x_2(t) \tag{24}$$

$$\dot{x}_2(t) = D^2 y(t) = x_3(t) \tag{25}$$

$$\dot{x}_3(t) = D^3 y(t) \tag{26}$$

But from Eq. (20)

$$D^3 y(t) = -3 D^2 y(t) - 5 Dy(t) - 6y(t) + u(t) \tag{27}$$

Substituting Eqs. (21), (24), (25) into (27), we obtain

$$D^3 y(t) = -3 x_3(t) - 5 x_2(t) - 6 x_1(t) + u(t) \tag{28}$$

From Eqs. (26) and (28) we conclude that

$$\dot{x}_3(t) = -6 x_1(t) - 5 x_2(t) - 3 x_3(t) + u(t) \tag{29}$$

Thus, the three variables $x_1(t)$, $x_2(t)$, $x_3(t)$ satisfy the first-order differential equations

$$\begin{aligned}
\dot{x}_1(t) &= x_2(t) \\
\dot{x}_2(t) &= x_3(t) \\
\dot{x}_3(t) &= -6 x_1(t) - 5 x_2(t) - 3 x_3(t) + u(t)
\end{aligned} \tag{30}$$

This set of equations can be written in vector form

$$\frac{d}{dt} \begin{bmatrix} x_1(t) \\ x_2(t) \\ x_3(t) \end{bmatrix} = \begin{bmatrix} 0 & 1 & 0 \\ 0 & 0 & 1 \\ -6 & -5 & -3 \end{bmatrix} \begin{bmatrix} x_1(t) \\ x_2(t) \\ x_3(t) \end{bmatrix} + \begin{bmatrix} 0 \\ 0 \\ 1 \end{bmatrix} u(t) \tag{31}$$

$$\underbrace{\qquad}_{\dot{\mathbf{x}}(t)} \qquad \underbrace{\qquad}_{\mathbf{A}} \qquad \underbrace{\qquad}_{\mathbf{x}(t)} \quad \underbrace{\quad}_{\mathbf{b}} \; \underbrace{}_{u(t)}$$

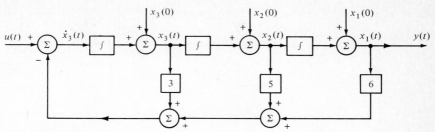

FIGURE 8.6.5
Analog-computer representation of the system function $H(s) = 1/(s^3 + 3s^2 + 5s + 6)$.
The state variables are the signals $x_1(t)$, $x_2(t)$ and $x_3(t)$.

Also, Eq. (21) can be written in vector form

$$y(t) = \underbrace{[1 \quad 0 \quad 0]}_{\mathbf{c}'} \underbrace{\begin{bmatrix} x_1(t) \\ x_2(t) \\ x_3(t) \end{bmatrix}}_{\mathbf{x}(t)} + \underbrace{0\, u(t)}_{d\, u(t)} \tag{32}$$

Clearly, Eqs. (31) and (32) are of the form

$$\dot{\mathbf{x}}(t) = \mathbf{A}\mathbf{x}(t) + \mathbf{b}u(t)$$
$$y(t) = \mathbf{c}'\mathbf{x}(t) + du(t) \tag{33}$$

Therefore, we conclude that $x_1(t)$, $x_2(t)$, $x_3(t)$ qualify as state variables. The analog-computer representation for this system is shown in Fig. 8.6.5; the state variables are at the integrator outputs.

The basic ideas of this example can readily be extended to higher-order systems. We shall leave it to the reader to verify the following lemma.

Lemma 1   *Given the system function*

$$H(s) = \frac{1}{s^n + \alpha_{n-1} s^{n-1} + \cdots + \alpha_1 s + \alpha_0} \tag{34}$$

which implies the input-output differential equation

$$\{D^n + \alpha_{n-1} D^{n-1} + \cdots + \alpha_1 D + \alpha_0\} y(t) = u(t) \tag{35}$$

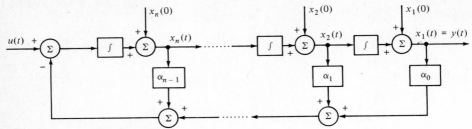

FIGURE 8.6.6
Analog-computer representation of the system function $H(s) = 1/(s^n + \alpha_{n-1}s^{n-1} + \cdots + \alpha_1 s + \alpha_0)$. The state variables are the signals $x_1(t), x_2(t), \ldots, x_n(t)$. The input is $u(t)$ and the output $y(t)$.

then the column n-vector $\mathbf{x}(t)$

$$\mathbf{x}(t) = \begin{bmatrix} x_1(t) \\ x_2(t) \\ \vdots \\ x_n(t) \end{bmatrix} = \begin{bmatrix} y(t) \\ Dy(t) \\ \vdots \\ D^{n-1}y(t) \end{bmatrix} \tag{36}$$

qualifies as a state vector. Furthermore, in state-variable form, the system is described by

$$\frac{d}{dt}\underbrace{\begin{bmatrix} x_1(t) \\ x_2(t) \\ \vdots \\ x_n(t) \end{bmatrix}}_{\dot{\mathbf{x}}(t)} = \underbrace{\begin{bmatrix} 0 & 1 & 0 & \cdots & 0 \\ 0 & 0 & 1 & \cdots & 0 \\ \cdots\cdots\cdots\cdots\cdots\cdots\cdots\cdots\cdots \\ -\alpha_0 & -\alpha_1 & -\alpha_2 & \cdots & -\alpha_{n-1} \end{bmatrix}}_{\mathbf{A}} \underbrace{\begin{bmatrix} x_1(t) \\ x_2(t) \\ \vdots \\ x_n(t) \end{bmatrix}}_{\mathbf{x}(t)} + \underbrace{\begin{bmatrix} 0 \\ 0 \\ \vdots \\ 1 \end{bmatrix}}_{\mathbf{b}} \underbrace{u(t)}_{u(t)} \tag{37}$$

$$y(t) = \underbrace{[1 \quad 0 \quad \cdots \quad 0]}_{\mathbf{c}'} \underbrace{\begin{bmatrix} x_1(t) \\ x_2(t) \\ \vdots \\ x_n(t) \end{bmatrix}}_{\mathbf{x}(t)} + \underbrace{0 \; u(t)}_{d \; u(t)} \tag{38}$$

The analog-computer representation of this system is illustrated in Fig. 8.6.6. Note that the constants $\alpha_0, \alpha_1, \ldots, \alpha_{n-1}$ appear as gains in the feedback channels.

8.6.5   The Standard Controllable State Representation

We now consider a system function with both zeros and poles. In this section we present a method for system functions which contain at least one more pole than zeros. As before, we shall illustrate the main ideas by an example and then present the extension to arbitrary system functions.

Let us consider a system with the system function†

$$H(s) = \frac{10s^2 + 4s + 7}{s^3 + 3s^2 + 5s + 6} \tag{39}$$

which implies the input-output differential equation

$$\{D^3 + 3D^2 + 5D + 6\}y(t) = \{10D^2 + 4D + 7\}u(t) \tag{40}$$

We seek a state-variable representation for this system. To find this representation we shall exploit the superposition principle enjoyed by linear systems. Let us consider the system

$$\{D^3 + 3D^2 + 5D + 6\}y_1(t) = u(t) \tag{41}$$

and assume that it is initially at rest. For any given input $u(t)$ the output is $y_1(t)$. If we now apply the input $\beta_0 u(t)$, linearity implies that the output is $\beta_0 y_1(t)$. In fact, all the state variables will be multiplied by β_0 since we have shown that the zero-state response obeys the principle of superposition.

Now suppose that instead of $u(t)$ we apply its derivative $Du(t)$. In this case, the output of the system is $Dy_1(t)$. To see this, differentiate both sides of Eq. (41).

$$D\{D^3 + 3D^2 + 5D + 6\}y_1(t) = Du(t) \tag{42}$$

or

$$\{D^4 + 3D^3 + 5D^2 + 6D\}y_1(t) = Du(t) \tag{43}$$

or

$$\{D^3 + 3D^2 + 5D + 6\}Dy_1(t) = Du(t) \tag{44}$$

Again, due to linearity, the output will be $4Dy_1(t)$ when the input is $4Du(t)$. Similarly, when the input is $10D^2u(t)$, the output will be $10D^2y_1(t)$. Finally, the superposition principle also implies that when the input is

---

† Note that the system function (39) and the system function (19) have the same denominator polynomial. There is a reason for this, as we shall see later.

$$\{10D^2 + 4D + 7\}u(t) \tag{45}$$

[which is the right-hand side of Eq. (40)], the output is going to be

$$y(t) = \{10D^2 + 4D + 7\}y_1(t) \tag{46}$$

Now let us recall that we know the state-variable representation for the system (41). By the results of Sec. 8.6.4 the state variables $x_1(t)$, $x_2(t)$, $x_3(t)$ are defined in terms of $y_1(t)$ by

$$x_1(t) = y_1(t)$$
$$x_2(t) = Dy_1(t) \tag{47}$$
$$x_3(t) = D^2 y_1(t)$$

If we substitute Eq. (47) into Eq. (46), we can readily see that the true output $y(t)$ can be expressed as

$$y(t) = 7x_1(t) + 4x_2(t) + 10x_3(t) \tag{48}$$

We therefore conclude that the system (39) [or (40)] can be represented in the state-variable form

$$\frac{d}{dt}\underbrace{\begin{bmatrix} x_1(t) \\ x_2(t) \\ x_3(t) \end{bmatrix}}_{\dot{\mathbf{x}}(t)} = \underbrace{\begin{bmatrix} 0 & 1 & 0 \\ 0 & 0 & 1 \\ -6 & -5 & -3 \end{bmatrix}}_{\mathbf{A}} \underbrace{\begin{bmatrix} x_1(t) \\ x_2(t) \\ x_3(t) \end{bmatrix}}_{\mathbf{x}(t)} + \underbrace{\begin{bmatrix} 0 \\ 0 \\ 1 \end{bmatrix}}_{\mathbf{b}} \underbrace{u(t)}_{u(t)} \tag{49}$$

$$y(t) = \underbrace{[7 \quad 4 \quad 10]}_{\mathbf{c}'} \underbrace{\begin{bmatrix} x_1(t) \\ x_2(t) \\ x_3(t) \end{bmatrix}}_{\mathbf{x}(t)} + \underbrace{0\,u(t)}_{d\,u(t)} \tag{50}$$

If we compare Eqs. (49) and (50) with Eqs. (37) and (38), we deduce that the net effect of having zeros in the system function is reflected in the change of the row vector $\mathbf{c}'$ in the state-variable representation; the matrix $\mathbf{A}$ and the column vector $\mathbf{b}$ are the same.

Additional insight can be gained by finding the analog-computer representation of Eqs. (49) and (50). This is done in Fig. 8.6.7. If we compare Figs. 8.6.5 and 8.6.7, we deduce that the feedback channels remain identical (since the $\mathbf{A}$ matrix remains the same). If we have zeros, the output $y(t)$ is a linear combination of more than one state variable (since $\mathbf{c}'$ has changed).

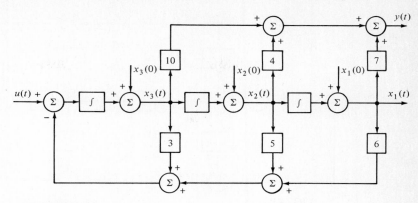

FIGURE 8.6.7
Analog-computer representation of system function $H(s) = |10s^2 + 4s + 7$ $/(s^3 + 3s^2 + 5s + 6)$; the state variables are $x_1(t)$, $x_2(t)$, and $x_3(t)$. The input is $u(t)$, and the output is $y(t)$. Note that the coefficients of the numerator polynomial appear as gains in the feedforward channels and that the coefficients of the denominator polynomial appear as gains in the feedback channels.

The basic ideas of this example can be readily extended. Once more we leave it to the reader to verify the following lemma.

Lemma 2   *Given the system function*

$$H(s) = \frac{\beta_{n-1} s^{n-1} + \cdots + \beta_1 s + \beta_0}{s^n + \alpha_{n-1} s^{n-1} + \cdots + \alpha_1 s + \alpha_0} \tag{51}$$

which implies the input-output differential equation

$$\{D^n + \alpha_{n-1} D^{n-1} + \cdots + \alpha_1 D + \alpha_0\} y(t) = \{\beta_{n-1} D^{n-1} + \cdots + \beta_1 D + \beta_0\} u(t) \tag{52}$$

then a valid state-variable (controllable) representation is

$$\frac{d}{dt} \underbrace{\begin{bmatrix} x_1(t) \\ x_2(t) \\ \vdots \\ x_n(t) \end{bmatrix}}_{\dot{\mathbf{x}}(t)} = \underbrace{\begin{bmatrix} 0 & 1 & 0 & \cdots & 0 \\ 0 & 0 & 1 & \cdots & 0 \\ \vdots & \vdots & \vdots & \vdots & \vdots \\ -\alpha_0 & -\alpha_1 & -\alpha_2 & \cdots & -\alpha_{n-1} \end{bmatrix}}_{\mathbf{A}} \underbrace{\begin{bmatrix} x_1(t) \\ x_2(t) \\ \vdots \\ x_n(t) \end{bmatrix}}_{\mathbf{x}(t)} + \underbrace{\begin{bmatrix} 0 \\ 0 \\ \vdots \\ 1 \end{bmatrix}}_{\mathbf{b}} \underbrace{u(t)}_{u(t)} \tag{53}$$

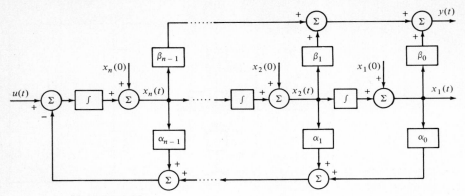

FIGURE 8.6.8
Analog-computer representation of the system function $H(s) = (\beta_{n-1}s^{n-1} + \cdots + \beta_1 s + \beta_0)/(s^n + \alpha_{n-1}s^{n-1} + \cdots + \alpha_1 s + \alpha_0)$. The state variables are $x_1(t)$, $x_2(t)$, ..., $x_n(t)$. The input is $u(t)$, and the output is $y(t)$. Note that the coefficients of the numerator polynomial appear as gains in the feedforward channels, while the roots of the denominator polynomial appear as gains in the feedback channels.

$$y(t) = \underbrace{[\beta_0 \quad \beta_1 \quad \cdots \quad \beta_{n-1}]}_{c'} \underbrace{\begin{bmatrix} x_1(t) \\ x_2(t) \\ \vdots \\ x_n(t) \end{bmatrix}}_{\mathbf{x}(t)} + \underbrace{0 \, u(t)}_{d \, u(t)} \tag{54}$$

The analog-computer representation of Eqs. (53) and (54) is illustrated in Fig. 8.6.8.

8.6.6   The Standard Observable Representation

We now present another popular technique for deducing a state-variable representation for system function containing both poles and zeros. We saw in Sec. 8.6.5 that the effect of zeros is to create the output via feedforward channels from the state variables. In the representation below we shall see that the effect of the zeros is exhibited by feeding the input $u(t)$ to the inputs of the integrators.

We shall illustrate this technique by considering the same example as before. Thus, we shall assume that we are given the system function

$$H(s) = \frac{10s^2 + 4s + 7}{s^3 + 3s^2 + 5s + 6} \tag{55}$$

or, equivalently, the input-output differential equation

$$\{D^3 + 3D^2 + 5D + 6\}y(t) = \{10D^2 + 4D + 7\}u(t) \tag{56}$$

Let us define the variables $x_1(t)$, $x_2(t)$, $x_3(t)$ by the equations

$$x_1(t) = y(t) - b_0 u(t) \tag{57}$$

$$x_2(t) = Dy(t) - b_0 Du(t) - b_1 u(t) \tag{58}$$

$$x_3(t) = D^2 y(t) - b_0 D^2 u(t) - b_1 Du(t) - b_2 u(t) \tag{59}$$

where b_0, b_1, b_2 are as yet undetermined constants. Our objective is to examine whether the $x_i(t)$ qualify as state variables. To be sure, these defining equations have been pulled out of a hat; the reader will soon see why they work.

Let us differentiate both sides of Eq. (57) to obtain

$$\dot{x}_1(t) = Dy(t) - b_0 Du(t) \tag{60}$$

and substitute it in Eq. (58). In so doing we see that

$$\dot{x}_1(t) = x_2(t) + b_1 u(t) \tag{61}$$

Clearly the right-hand side of Eq. (61) does not contain any derivatives of the input, as required in a state-variable representation.

Next, we differentiate both sides of Eq. (58) and substitute the resultant equation into Eq. (59) to obtain

$$\dot{x}_2(t) = x_3(t) + b_2 u(t) \tag{62}$$

Next we differentiate both sides of Eq. (59) to obtain

$$\dot{x}_3(t) = D^3 y(t) - b_0 D^3 u(t) - b_1 D^2 u(t) - b_2 Du(t) \tag{63}$$

But the input-output differential equation (56) implies that

$$D^3 y(t) = -3D^2 y(t) - 5Dy(t) - 6y(t) + 10D^2 u(t) + 4Du(t) + 7u(t) \tag{64}$$

We substitute Eqs. (57) to (59) in Eq. (64) to find

$$D^3 y(t) = -3[x_3(t) + b_0 D^2 u(t) + b_1 Du(t) + b_2 u(t)] - 5[x_2(t) + b_0 Du(t) + b_1 u(t)]$$
$$- 6[x_1(t) + b_0 u(t)] + 10D^2 u(t) + 4Du(t) + 7u(t) \tag{65}$$

We now substitute Eq. (65) into Eq. (63) and collect terms to obtain

$$\dot{x}_3(t) = -6x_1(t) - 5x_2(t) - 3x_3(t) - b_0 D^3 u(t) - (3b_0 + b_1 - 10)D^2 u(t)$$
$$- (5b_0 + 3b_1 + b_2 - 4)Du(t) - (6b_0 + 5b_1 + 3b_2 - 7)u(t) \tag{66}$$

In order for $x_3(t)$ to qualify as a state variable the right-hand side of its differential equation (66) must *not* contain any terms involving input derivatives. Since b_0, b_1, and b_2 have not been specified, we can now choose them so that the coefficients of the terms involving $D^3 u(t)$, $D^2 u(t)$, and $Du(t)$ are zero. In so doing we obtain the equations

$$b_0 = 0$$

$$3b_0 + b_1 - 10 = 0 \tag{67}$$

$$5b_0 + 3b_1 + b_2 - 4 = 0$$

which are readily solved by successive substitution to obtain

$$b_0 = 0 \qquad b_1 = 10 \qquad b_2 = -26 \tag{68}$$

We substitute Eq. (68) into Eqs. (61), (62), and (66), and we conclude that the $x_i(t)$ satisfy the differential equations

$$\dot{x}_1(t) = x_2(t) + 10u(t)$$

$$\dot{x}_2(t) = x_3(t) - 26u(t) \tag{69}$$

$$\dot{x}_3(t) = -6x_1(t) - 5x_2(t) - 3x_3(t) + 35u(t)$$

while the output $y(t)$ is related to the state variables, in view of Eqs. (57) and (68), by

$$y(t) = x_1(t) \tag{70}$$

We can now conclude that the $x_i(t)$ qualify as state variables. In vector form, Eqs. (69) and (70) are

$$\frac{d}{dt} \underbrace{\begin{bmatrix} x_1(t) \\ x_2(t) \\ x_3(t) \end{bmatrix}}_{\dot{\mathbf{x}}(t)} = \underbrace{\begin{bmatrix} 0 & 1 & 0 \\ 0 & 0 & 1 \\ -6 & -5 & -3 \end{bmatrix}}_{\mathbf{A}} \underbrace{\begin{bmatrix} x_1(t) \\ x_2(t) \\ x_3(t) \end{bmatrix}}_{\mathbf{x}(t)} + \underbrace{\begin{bmatrix} 10 \\ -26 \\ 35 \end{bmatrix}}_{\mathbf{b}} \underbrace{u(t)}_{u(t)} \tag{71}$$

$$y(t) = \underbrace{\begin{bmatrix} 1 & 0 & 0 \end{bmatrix}}_{\mathbf{c}'} \underbrace{\begin{bmatrix} x_1(t) \\ x_2(t) \\ x_3(t) \end{bmatrix}}_{\mathbf{x}(t)} + \underbrace{0 \ u(t)}_{d \ u(t)} \tag{72}$$

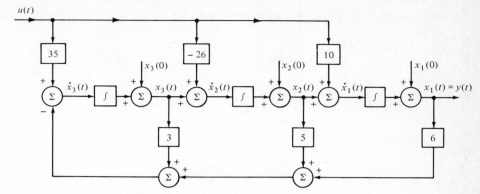

FIGURE 8.6.9
Alternate analog-computer representation of the system function $H(s) = (10s^2 + 4s + 7)/(s^3 + 3s^2 + 6)$. The state variables are $x_1(t)$, $x_2(t)$, and $x_3(t)$. The input is $u(t)$, and the output is $y(t)$.

If we compare Eqs. (71) and (72) with Eqs. (49) and (50), we see that the **A** matrix is still the same. The difference of the controllable and observable state representations is manifested in the change of the vectors **b** and **c**.

Additional insight is provided by comparing the analog-computer representations. Figure 8.6.9 illustrates the analog-computer diagram for systems (69) and (70) [or (71) and (72)]. The effect of the zeros is exhibited by feedforwarding the input $u(t)$ to the integrator inputs. If we compare Figs. 8.6.5, 8.6.7, and 8.6.9, we observe that the feedback channels are identical; this is due to the fact that the denominator polynomial of $H(s)$ or, equivalently, the matrix **A** in the state-variable representation are the same.

It is possible to generalize the techniques presented in this example and deduce the following lemma:

Lemma 3 *Given the system function*

$$H(s) = \frac{\beta_n s^n + \beta_{n-1} s^{n-1} + \cdots + \beta_1 s + \beta_0}{s^n + \alpha_{n-1} s^{n-1} + \cdots + \alpha_1 s + \alpha_0} \tag{73}$$

or, equivalently, the input-output differential equation

$$\{D^n + \alpha_{n-1} D^{n-1} + \cdots + \alpha_1 D + \alpha_0\} y(t) = \{\beta_n D^n + \cdots + \beta_1 D + \beta_0\} u(t) \tag{74}$$

then a valid state-variable (observable) representation is

$$\frac{d}{dt}\underbrace{\begin{bmatrix} x_1(t) \\ x_2(t) \\ \vdots \\ x_n(t) \end{bmatrix}}_{\dot{\mathbf{x}}(t)} = \underbrace{\begin{bmatrix} 0 & 1 & 0 & \cdots & 0 \\ 0 & 0 & 1 & \cdots & 0 \\ \vdots & \vdots & \vdots & \vdots & \vdots \\ -\alpha_0 & -\alpha_1 & -\alpha_2 & \cdots & -\alpha_{n-1} \end{bmatrix}}_{\mathbf{A}} \cdot \underbrace{\begin{bmatrix} x_1(t) \\ x_2(t) \\ \vdots \\ x_n(t) \end{bmatrix}}_{\mathbf{x}(t)} + \underbrace{\begin{bmatrix} b_1 \\ b_2 \\ \vdots \\ b_n \end{bmatrix}}_{\mathbf{b}} \underbrace{u(t)}_{u(t)} \tag{75}$$

$$y(t) = \underbrace{[1 \quad 0 \quad 0 \quad \cdots \quad 0]}_{\mathbf{c}'} \underbrace{\begin{bmatrix} x_1(t) \\ x_2(t) \\ \vdots \\ x_n(t) \end{bmatrix}}_{\mathbf{x}(t)} + \underbrace{\beta_n u(t)}_{d\, u(t)} \tag{76}$$

where the elements $b_1, b_2, \ldots, b_n$ of the column n-vector $\mathbf{b}$ are easily computed (by successive substitution) by the equations

$$b_0 = \beta_n$$
$$b_1 = \beta_{n-1} - b_0 \alpha_{n-1}$$
$$b_2 = \beta_{n-2} - b_0 \alpha_{n-2} - b_1 \alpha_{n-1}$$
$$b_3 = \beta_{n-3} - b_0 \alpha_{n-3} - b_1 \alpha_{n-2} - b_2 \alpha_{n-1} \tag{77}$$
$$\vdots$$
$$b_n = \beta_0 - \sum_{i=0}^{n-1} \alpha_i b_i$$

The analog-computer representation of Eqs. (75) and (76) is illustrated in Fig. 8.6.10.

8.6.7 Concluding Remarks

We have demonstrated ways of generating state-variable representations from the system-function representations from the system-function description. A system function with n poles is represented by n state variables; this in turn requires n integrators for the analog-computer representation (independent of the number of zeros as long as the number of zeros does not exceed the number of poles).

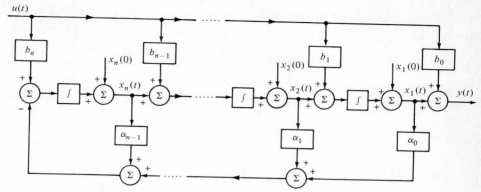

FIGURE 8.6.10
Alternate analog-computer representation of the system function $H(s) = (\beta_n s^n + \cdots + \beta_1 s + \beta_0)/(s^n + \alpha_{n-1} s^{n-1} + \cdots + \alpha_1 s + \alpha_0)$. The state variables are $x_1(t)$, $x_2(t)$, ..., $x_n(t)$. The input is $u(t)$, and the output is $y(t)$. The gains $b_0, b_1, \ldots, b_n$ are computed from Eq. (77).

It should be clear that the mapping of system functions into state-variable representations is not unique. To see this, recall that in Secs. 8.6.5 and 8.6.6 we derived two distinct state-variable representations for the same system function. In fact one can find an infinite number of state-variable representations for the same system functions. To prove this, let us suppose that we have found one state-variable representation of the form

$$\dot{\mathbf{x}}(t) = \mathbf{A}\mathbf{x}(t) + \mathbf{b}u(t)$$
$$y(t) = \mathbf{c}'\mathbf{x}(t) + du(t) \tag{78}$$

for a given system function $H(s)$. Let $\mathbf{P}$ be an arbitrary nonsingular matrix. Let $\mathbf{z}(t)$ be a column n-vector defined by

$$\mathbf{z}(t) = \mathbf{P}\mathbf{x}(t) \tag{79}$$

Then it is easy to show that $\mathbf{z}(t)$ satisfies the equations

$$\dot{\mathbf{z}}(t) = \hat{\mathbf{A}}\mathbf{z}(t) + \hat{\mathbf{b}}u(t)$$
$$y(t) = \hat{\mathbf{c}}'\mathbf{z}(t) + du(t) \tag{80}$$

where

$$\hat{\mathbf{A}} = \mathbf{P}\mathbf{A}\mathbf{P}^{-1} \qquad \hat{\mathbf{b}} = \mathbf{P}\mathbf{b} \qquad \hat{\mathbf{c}}' = \mathbf{c}'\mathbf{P}^{-1} \tag{81}$$

From Eq. (80) we conclude that $\mathbf{z}(t)$ also qualifies as a state vector for the given system function $H(s)$.

The nonuniqueness of the state-variable representations, as we have mentioned, can easily be explained on physical grounds. Two or more physically different dynamical systems or networks may exhibit identical input-output characteristics when viewed as black boxes. The existence of more than one state-variable representation for this black box is the consequence of the fact that the state variables correspond to distinct physical variables for physically different dynamical systems.

8.7   DISCUSSION

This concludes our examination of the interrelationships between the input-output and the state-variable descriptions of LTI dynamical systems. We hope that the reader has understood both the similarities and the differences between these two methods of modeling.

At this point the reader may ask: *which method should I use?* Unfortunately, one cannot give a clear-cut answer, because it depends on the application at hand.†
For some applications the state-varaible approach is preferable; for others, the system-function approach is intimately suited to the use of Laplace transforms; for network analysis it is convenient since it utilizes the notion of impedance. On the other hand, the state-variable approach is more natural since it can be extended to multiinput-multioutput linear time-varying and nonlinear networks and systems. For such systems the notions of system function, poles, and zeros are not available, and state variables offer the only vehicle for rigorous mathematical modeling.

REFERENCES

The references cited in Chap. 7 are also appropriate for the material in this chapter.

EXERCISES

Section 8.3

8.3.1   Consider a SISO system with the state-variable description

$$\dot{\mathbf{x}}(t) = \mathbf{A}\mathbf{x}(t) + \mathbf{b}u(t) \qquad y(t) = \mathbf{c}'\mathbf{x}(t)$$

---

† This is similar to the question: *Should I use binary or decimal arithmetic?* Clearly, the answer depends on the application.

Deduce the input-output system function $H(s)$ for:

(a) $A = \begin{bmatrix} 0 & 1 & 0 \\ 1 & 0 & 1 \\ 2 & 1 & 0 \end{bmatrix}$ $b = \begin{bmatrix} 1 \\ -1 \\ 0 \end{bmatrix}$ $c = \begin{bmatrix} 2 \\ 1 \\ 3 \end{bmatrix}$

(b) $A = \begin{bmatrix} 0 & 2 \\ -2 & 0 \end{bmatrix}$ $b = \begin{bmatrix} 1 \\ -1 \end{bmatrix}$ $c = \begin{bmatrix} 1 \\ 0 \end{bmatrix}$

(c) $\dot{A} = \begin{bmatrix} 0 & 1 & 0 \\ 0 & 0 & 1 \\ 0 & 0 & 0 \end{bmatrix}$ $b = \begin{bmatrix} 1 \\ 2 \\ 3 \end{bmatrix}$ $c = \begin{bmatrix} 1 \\ 0 \\ 0 \end{bmatrix}$

8.3.2   Consider the two interconnected systems

$$\left. \begin{array}{l} \dot{x}_1 = A_1 x_1 + b_1 u(t) \\ y_1(t) = c_1' x_1(t) \end{array} \right\} S_1 \qquad \left. \begin{array}{l} \dot{x}_2 = A_2 x_2 + b_2 y_1(t) \\ y(t) = c_2' x_2(t) \end{array} \right\} S_2$$

(a)   Determine the system function $H(s)$ from $u(t)$ to $y(t)$.
(b)   Prove that $H(s) = H_1(s) H_2(s)$, where $H_1(s)$ is the system function of the system S_1 and $H_2(s)$ is the system function of S_2.

8.3.3   Consider the two interconnected systems

$$\left. \begin{array}{l} \dot{x}_1 = A_1 x_1 + b_1 u(t) \\ y_1(t) = c_1' x_1(t) \end{array} \right\} S_1 \qquad \left. \begin{array}{l} \dot{x}_2 = A_2 x_2 + b_2 u(t) \\ y_2(t) = c_2' x_2(t) \end{array} \right\} S_2$$

Let $y(t) = y_1(t) + y_2(t)$.

(a)   Determine the system function $H(s)$ from $u(t)$ to $y(t)$.
(b)   Show that $H(s) = H_1(s) + H_2(s)$, where $H_1(s)$ and $H_2(s)$ are the system functions for the systems S_1 and S_2, respectively.

Section 8.6.6

8.6.1   Find the standard controllable and standard observable state representations for the following system functions. In each case draw the corresponding analog-computer representation

(a) $H(s) = \dfrac{s^4}{s^4 + 1}$ (b) $H(s) = \dfrac{s^3 + s^2 + s + 1}{s^3 + 2s^2 + 3s + 4}$

(c) $H(s) = \dfrac{s^2 + 1}{s^3 + 2s^2}$ (d) $H(s) = \dfrac{5s^3 + 9s^2 - 3}{s^3 - 6s^3 + 8s^2 + 4}$

8.6.2   Consider a system function of the form

$$H(s) = \frac{1}{\displaystyle\prod_{i=1}^{n} (s - s_i)}$$

where the poles are real and distinct.

(a) Express $H(s)$ in the form

$$H(s) = \sum_{i=1}^{n} \frac{r_i}{s - s_i}$$

The r_i's are called the residues (see Sec. 8.1).

(b) Determine a suitable set of state variables $x_1(t)$, $x_2(t)$, ..., $x_n(t)$ satisfying the equations

$$\dot{x}(t) = \mathbf{S}x(t) + \mathbf{b}u(t)$$

$$y(t) = \sum_{i=1}^{n} x_i(t)$$

where $\mathbf{S}$ is the diagonal $n \times n$ matrix

$$\mathbf{S} = \begin{bmatrix} s_1 & 0 & \cdots & 0 \\ 0 & s_2 & \cdots & 0 \\ \cdots\cdots\cdots\cdots\cdots \\ 0 & 0 & \cdots & s_n \end{bmatrix}$$

Determine the relation of the elements of the column n-vector $\mathbf{b}$ to the residues r_i of the system function. Draw the analog-computer representation.

(c) Repeat parts (a) and (b) for the system function

$$H(s) = \frac{\prod\limits_{j=1}^{m} (s - \mu_j)}{\prod\limits_{i-1}^{n} (s - s_i)} \qquad m < n$$

What happens if one of the zeros, say μ_1, is equal to one of the poles, say s_1? Draw the analog-computer diagram and explain in terms of the physical interconnections.

8.6.3 Consider a system described by the system function

$$H(s) = \frac{s + 1}{(s + 1)(s + 2)}$$

Note that one of the zeros cancels one of the poles, so that from the input-output point of view this system looks as if it had the system function

$$H'(s) = \frac{1}{s + 2}$$

The purpose of this exercise is to investigate the implications of pole-zero cancellation from a state-variable point of view.

(a) Find the standard controllable state representation for $H(s)$. Draw the analog-computer diagram. What is the implication of the pole-zero cancellation in terms of the analog-computer diagram? Now form the so-called *observability matrix*.

$$\begin{bmatrix} \uparrow & \vdots & \uparrow \\ \mathbf{c} & \vdots & \mathbf{A'c} \\ \downarrow & \vdots & \downarrow \end{bmatrix}$$

i.e., the matrix whose first column vector is $\mathbf{c}$ and its second $\mathbf{A'c}$, and prove that it is singular.

(b) Find the standard observable state representation for $H(s)$. Determine the analog-computer representation and deduce the implications of the pole-zero cancellation. Form the so-called *controllability* matrix

$$\begin{bmatrix} \uparrow & \vdots & \uparrow \\ \mathbf{b} & \vdots & \mathbf{Ab} \\ \downarrow & \vdots & \downarrow \end{bmatrix}$$

i.e., the matrix whose first column vector is $\mathbf{b}$ and second $\mathbf{Ab}$, and prove that it is singular.

(c) Now consider the system function

$$H(s) = \frac{s - \mu_1}{(s - s_1)(s - s_2)}$$

Show that in the standard controllable representation the observability matrix

$$\begin{bmatrix} \uparrow & \vdots & \uparrow \\ \mathbf{c} & \vdots & \mathbf{A'c} \\ \downarrow & \vdots & \downarrow \end{bmatrix}$$

is singular if and only if $\mu_1 = s_1$ or $\mu_1 = s_2$. Show that in the standard observable representation the controllability matrix

$$\begin{bmatrix} \uparrow & \vdots & \uparrow \\ \mathbf{b} & \vdots & \mathbf{Ab} \\ \downarrow & \vdots & \downarrow \end{bmatrix}$$

is singular if and only if $\mu_1 = s_1$ or $\mu_1 = s_2$.

8.6.4 Suppose that $\dot{\mathbf{x}} = \mathbf{Ax} + \mathbf{b}u(t)$; $y(t) = x_1(t)$ is the standard observable representation for a system function $H(s)$. Prove that if the $n \times n$ controllability matrix

$$\begin{bmatrix} \uparrow & \vdots & \uparrow & \vdots & \uparrow & \vdots & & \vdots & \uparrow \\ \mathbf{b} & \vdots & \mathbf{Ab} & \vdots & \mathbf{A^2b} & \vdots & \cdots & \vdots & \mathbf{A^{n-1}b} \\ \downarrow & \vdots & \downarrow & \vdots & \downarrow & \vdots & & \vdots & \downarrow \end{bmatrix}$$

is singular, then at least one of the zeros of $H(s)$ is the same as one of the poles of $H(s)$.

8.6.5 Suppose that $\dot{\mathbf{x}} = \mathbf{Ax} + \mathbf{b}u(t)$, $y(t) = \mathbf{c'x}(t)$ is the standard controllable representation for a system function $H(s)$. Prove that if the $n \times n$ observability matrix

$$\begin{bmatrix} \uparrow & \vdots & \uparrow & \vdots & \uparrow & \vdots & & \vdots & \uparrow \\ \mathbf{c} & \vdots & \mathbf{A'c} & \vdots & \mathbf{A'^2c} & \vdots & \cdots & \vdots & \mathbf{A'^{n-1}c} \\ \downarrow & \vdots & \downarrow & \vdots & \downarrow & \vdots & & \vdots & \downarrow \end{bmatrix}$$

is singular, then at least one of the zeros of $H(s)$ is the same as one of the poles of $H(s)$.

8.6.6   Another easy state-variable representation for systems with both poles and zeros is the following. If the system transfer function is

$$H(s) = \frac{\beta_m s^m + \cdots + \beta_1 s + \beta_0}{s^n + \alpha_{n-1} s^{n-1} + \cdots + \alpha_1 s + \alpha_0} \qquad m \le n - 1$$

then it also has the state-variable representation

$$\dot{\mathbf{x}}(t) = \mathbf{A}\mathbf{x}(t) + \mathbf{b}u(t) \qquad y(t) = \mathbf{c}'\mathbf{x}(t)$$

where

$$\mathbf{A} = \begin{bmatrix} 0 & 0 & \cdots & 0 & -\alpha_0 \\ 1 & 0 & \cdots & 0 & -\alpha_1 \\ 0 & 1 & \cdots & 0 & -\alpha_2 \\ \multicolumn{5}{c}{\dotfill} \\ 0 & 0 & \cdots & 1 & -\alpha_{n-1} \end{bmatrix} \qquad \mathbf{b} = \begin{bmatrix} \beta_0 \\ \cdot \\ \cdot \\ \cdot \\ \beta_m \\ 0 \\ \cdot \\ \cdot \\ 0 \end{bmatrix}$$

$$\mathbf{c}' = \begin{bmatrix} 0 & 0 & \cdots & 0 & 1 \end{bmatrix}$$

Devise a procedure, first by an example and then in general, that verifies the above representation. Draw an analog-computer diagram.

NUMERICAL INTEGRATION METHODS

9

SUMMARY

This chapter provides a brief introduction to different methods for obtaining numerical solutions to systems of differential equations using a digital computer. These numerical methods are equally applicable to the solution of linear and nonlinear differential equations. However, because of their special structure and properties linear differential equations can also be solved by special techniques, as we have remarked in Chap. 7. On the other hand, when we deal with nonlinear differential equations, such special solution methods are not available and one must rely exclusively on numerical solution methods to calculate the solutions of nonlinear differential equations.

9.1 INTRODUCTION

In this chapter, we shall be using the digital computer to calculate solutions of vector differential equations which are either linear or nonlinear. As we have remarked in the preceding three chapters, such equations arise when one considers dynamic networks and other physical systems whose variables are most naturally defined as continuous or piecewise continuous functions of time.

From the start we stress that numerical integration techniques will *never* be able to give us *exactly* the time function we seek, because the numerical description of a time function will require an infinite set of values of the function for each value of time, even if we are concerned with the numerical description of a scalar time function over a finite interval of time. No present or future computer is capable of generating an infinite table in a finite interval of time.

Because of such considerations, we must accept from the start that any numerical integration method will provide us with only an *approximation* to the true desired solution. We must be content with describing a continuous time function by a finite set of values of the function defined at specified discrete (and finite in number) values of time.

However, the engineer does have control over the degree of approximation, simply by increasing the number of time points over which the value of the function is described. Nevertheless, it should be clear that such an increase in the degree of accuracy can be attained only with an increase in computation time. Thus, in any given problem a decision has to be made as to the appropriate trade-off between solution accuracy and computer time.

However, the above sources of error are not the only ones encountered in numerical solution techniques; additional approximations must be made even in the evaluation of the value of a time function at discrete instants of time. These errors strongly depend upon the type of algorithm utilized; many algorithms are available for doing essentially the same task, but they have individual characteristics pertaining to

1 Programming complexity
2 Real-time computational requirements
3 Accuracy characteristics

In this chapter we discuss some popular algorithms and present a qualitative discussion of their distinct accuracy-complexity characteristics.

We start our discussion with numerical methods for evaluating definite integrals of known time-varying vector-valued functions. Such a discussion is necessary because the numerical evaluation of definite time integrals is required as a subroutine when we tackle the more complex problem of numerical integration of vector differential equations.

9.2 NUMERICAL EVALUATION OF DEFINITE INTEGRALS

Although the problem of numerical evaluation of definite integrals is a universal problem in science and engineering, it is important to motivate our development by considering precisely where in our development so far we have encountered this problem. We have seen in Chaps. 6 to 8 that a great variety of networks and systems are described by LTI vector differential equations of the form

$$\dot{\mathbf{x}}(t) = \mathbf{A}\mathbf{x}(t) + \mathbf{B}\mathbf{u}(t) \qquad \mathbf{x}(t_0) = \xi \tag{1}$$

where $\mathbf{x}(t) \in R_n$ = state vector
$\mathbf{u}(t) \in R_m$ = input vector
$\mathbf{A}$ = constant $n \times n$ matrix
$\mathbf{B}$ = constant $n \times m$ matrix

We saw in Chap. 7 that the solution of Eq. (1) is given by

$$\mathbf{x}(t) = e^{\mathbf{A}(t-t_0)}\xi + e^{\mathbf{A}t}\int_{t_0}^{t} e^{-\mathbf{A}\tau}\mathbf{B}\mathbf{u}(\tau)\,d\tau \tag{2}$$

In order to compute the solution $\mathbf{x}(t)$ for any given value of t, it becomes necessary to calculate the integral

$$\int_{t_0}^{t} \mathbf{g}(\tau)\,d\tau \triangleq \int_{t_0}^{t} e^{-\mathbf{A}\tau}\mathbf{B}\mathbf{u}(\tau)\,d\tau \tag{3}$$

where we define the n-dimensional vector $\mathbf{g}(\tau)$ by

$$\mathbf{g}(\tau) \triangleq e^{-\mathbf{A}\tau}\mathbf{B}\mathbf{u}(\tau) \tag{4}$$

For certain types of input functions $\mathbf{u}(\tau)$, for example, constant and exponential, the integral (3) can be evaluated analytically. On the other hand, for arbitrary time functions $\mathbf{u}(\tau)$, it may not be possible to evaluate the integral in closed form, and hence one must resort to numerical means of calculating the definite vector-valued integral in (3) for any given values of t_0 and t.

These remarks point out that even for linear vector differential equations one may have to resort to numerical integral evaluation to calculate the time evolution of the state. In the remainder of this section we shall outline certain numerical methods that can be used in conjunction with digital computers.

9.2.1 Splitting Up the Time Interval

Let us suppose that we wish to evaluate the definite vector-valued integral

$$\mathbf{y}(t_0, t_n) = \int_{t_0}^{t_n} \mathbf{x}(\tau)\,d\tau \tag{5}$$

where both $\mathbf{x}(\tau)$ and $\mathbf{y}$ are n-dimensional vectors. We denote the value of the integral by the vector $\mathbf{y}(t_0, t_n)$ to stress its dependence upon the end limits t_0 and t_n.

The first step is to discretize the time interval

$$t_0 \leq t \leq t_n \tag{6}$$

into subintervals. This requires the selection of certain values of time

$$t_1, t_2, \ldots, t_{n-1} \tag{7}$$

such that

$$t_0 < t_1 < t_2 < \cdots < t_{n-1} < t_n \tag{8}$$

In this manner, the finite time interval (6) has been subdivided into n subintervals. These subintervals need not be of the same length; however, in practice they are often selected to be of the same size (this leads to a slightly simpler digital-computer subroutine).

In most cases of practical importance, the components $x_i(\tau)$, $i = 1, 2, \ldots, n$, of the vector $\mathbf{x}(\tau)$ are piecewise continuous functions of time. In this case, in order to preserve accuracy it is important to select a subset of the times t_i to coincide with the times at which the time functions $x_i(\tau)$ are discontinuous. In this manner, one guarantees that all components $x_i(\tau)$ of $\mathbf{x}(\tau)$ are continuous-time functions in the time interval

$$t_k < \tau < t_{k+1} \qquad k = 0, 1, \ldots, n - 1$$

Once this basic time discretization has been accomplished, the integral (5) can be decomposed *exactly* into the sum

$$\mathbf{y}(t_0, t_n) = \int_{t_0}^{t_n} \mathbf{x}(\tau)\, d\tau = \sum_{k=0}^{n-1} \int_{t_k}^{t_{k+1}} \mathbf{x}(\tau)\, d\tau \tag{9}$$

Hence, it suffices to calculate the individual integrals

$$\int_{t_k}^{t_{k+1}} \mathbf{x}(\tau)\, d\tau \tag{10}$$

where, by construction, $\mathbf{x}(\tau)$ is assumed to be continuous for all τ,

$$t_k < \tau < t_{k+1} \tag{11}$$

We shall next present three popular methods for evaluating the integral (10), the Euler method, the trapezoidal method, and Simpson's rule.

9.2.2 The Euler Method

In the Euler method we use the approximation

$$\int_{t_k}^{t_{k+1}} \mathbf{x}(\tau)\, d\tau \approx (t_{k+1} - t_k)\mathbf{x}(t_k) \tag{12}$$

If the vector $\mathbf{x}(\tau)$ has a discontinuity at the left limit t_k, in the sense that one or more of the elements $x_i(\tau)$ has a discontinuity at $\tau = t_k$, then the vector $\mathbf{x}(t_k)$ in (12) may not be well defined.

In this case one should use the value $\mathbf{x}(t_{k^+})$ instead of $\mathbf{x}(t_k)$ in (12); this is formally defined as the limit from the right, i.e.,

$$\mathbf{x}(t_{k^+}) = \lim_{\epsilon \to 0} \mathbf{x}(t_k + \epsilon) \qquad \epsilon > 0 \tag{13}$$

The basic idea of the Euler method is to view the vector $\mathbf{x}(\tau)$ as being constant over the integration time interval $t_k \leq \tau \leq t_{k+1}$.

9.2.3   The Trapezoidal Method

In the trapezoidal method the evaluation of the definite integral is given by the formula

$$\int_{t_k}^{t_{k+1}} \mathbf{x}(\tau)\, d\tau \approx \tfrac{1}{2}(t_{k+1} - t_k)[\mathbf{x}(t_k) + \mathbf{x}(t_{k+1})] \tag{14}$$

If $\mathbf{x}(\tau)$ has a discontinuity at $\tau = t_k$, then in the computation of (14) one should use instead of $\mathbf{x}(t_k)$ the limit from the right

$$\mathbf{x}(t_{k^+}) = \lim_{\epsilon \to 0} \mathbf{x}(t_k + \epsilon) \qquad \epsilon > 0 \tag{15}$$

If $\mathbf{x}(\tau)$ has a discontinuity at $\tau = t_{k+1}$, then in (14) instead of $\mathbf{x}(t_{k+1})$ one should use the limit from the left, i.e.,

$$\mathbf{x}(t_{k+1^-}) = \lim_{\epsilon \to 0} \mathbf{x}(t_{k+1} - \epsilon) \qquad \epsilon > 0 \tag{16}$$

Basically, in the trapezoidal method one approximates $\mathbf{x}(\tau)$, $t_k \leq \tau \leq t_{k+1}$, by its average value (defined by its end-point values)

$$\mathbf{x}(\tau) \approx \tfrac{1}{2}[\mathbf{x}(t_k) + \mathbf{x}(t_{k+1})] \tag{17}$$

Another way of viewing the trapezoidal method is to view it as giving no integration error provided all the components of $\mathbf{x}(\tau)$ are linear time functions. To see this suppose that

$$x_i(\tau) = x_i(t_k) + a_i(\tau - t_k) \qquad i = 1, 2, \ldots, n \tag{18}$$

Let $\mathbf{a}$ denote the n-dimensional column vector whose components are the slopes a_i. In this case,

$$\mathbf{x}(\tau) = \mathbf{x}(t_k) + \mathbf{a}(\tau - t_k) \tag{19}$$

and we can evaluate analytically the definite integral

$$\int_{t_k}^{t_{k+1}} \mathbf{x}(\tau)\, d\tau = \int_{t_k}^{t_{k+1}} [\mathbf{x}(t_k) + \mathbf{a}(\tau - t_k)]\, d\tau \tag{20}$$
$$= \mathbf{x}(t_k)(t_{k+1} - t_k) + \tfrac{1}{2}\mathbf{a}(t_{k+1} - t_k)^2$$

On the other hand, by evaluating (19) at $\tau = t_{k+1}$ we have

$$\mathbf{x}(t_{k+1}) = \mathbf{x}(t_k) + \mathbf{a}(t_{k+1} - t_k) \tag{21}$$

Hence,

$$\mathbf{a} = [\mathbf{x}(t_{k+1}) - \mathbf{x}(t_k)]\frac{1}{t_{k+1} - t_k} \tag{22}$$

Substituting (22) into (20), we obtain

$$\int_{t_k}^{t_{k+1}} \mathbf{x}(\tau)\, d\tau = \tfrac{1}{2}(t_{k+1} - t_k)[\mathbf{x}(t_{k+1}) + \mathbf{x}(t_k)] \tag{23}$$

which is precisely the right-hand side of (14).

9.2.4   Simpson's Rule

In Simpson's rule, we find the middle point of the time interval $t_k \leq \tau \leq t_{k+1}$, denoted by $\hat{\tau}_k$ and given by

$$\boxed{\hat{\tau}_k = \tfrac{1}{2}(t_k + t_{k+1})} \tag{24}$$

Then, the definite integral is approximated as follows:

$$\boxed{\int_{t_k}^{t_{k+1}} \mathbf{x}(\tau)\, d\tau \approx \tfrac{1}{6}(t_{k+1} - t_k)[\mathbf{x}(t_k) + 4\mathbf{x}(\hat{\tau}_k) + \mathbf{x}(t_{k+1})]} \tag{25}$$

By the very way of selecting the t_k to coincide at least with all the discontinuities of $\mathbf{x}(\tau)$, we are assured that $\hat{\tau}_k$ is not a point of discontinuity of $\mathbf{x}(\tau)$; discontinuities at t_k and/or t_{k+1} are handled exactly as in the trapezoidal method.

The way to understand Simpson's rule is to prove that it yields no integration error when all the elements $x_i(\tau)$ of the vector $\mathbf{x}(\tau)$ are quadratic (parabolas). So let us suppose that the elements $x_i(.)$ have the time structure

$$x_i(\tau) = x_i(t_k) + a_i(\tau - t_k) + b_i(\tau - t_k)^2 \qquad i = 1, 2, \ldots, n \tag{26}$$

Let $\mathbf{a}$ be the n-dimensional vector whose elements are the a_i's, and let $\mathbf{b}$ be the n-dimensional vector whose elements are the b_i's. In this case, (26) can be written in the vector form

$$\mathbf{x}(\tau) = \mathbf{x}(t_k) + \mathbf{a}(\tau - t_k) + \mathbf{b}(\tau - t_k)^2 \tag{27}$$

For the $\mathbf{x}(\tau)$ of (27) we can calculate the definite integral analytically

$$\int_{t_k}^{t_{k+1}} \mathbf{x}(\tau)\, d\tau = \int_{t_k}^{t_{k+1}} [\mathbf{x}(t_k) + \mathbf{a}(\tau - t_k) + \mathbf{b}(\tau - t_k)^2]\, d\tau$$
$$= (t_{k+1} - t_k)\mathbf{x}(t_k) + \tfrac{1}{2}(t_{k+1} - t_k)^2 \mathbf{a} + \tfrac{1}{3}(t_{k+1} - t_k)^3 \mathbf{b} \tag{28}$$

The next step is to determine $\mathbf{a}$ and $\mathbf{b}$ in terms of $\mathbf{x}(\hat{\tau}_k)$ and $\mathbf{x}(t_{k+1})$, where $\hat{\tau}_k$ is given by (24). From Eq. (27) we obtain

$$\mathbf{x}(\hat{\tau}_k) = \mathbf{x}(t_k) + \tfrac{1}{2}\mathbf{a}(t_{k+1} - t_k) + \tfrac{1}{4}\mathbf{b}(t_{k+1} - t_k)^2$$
$$\mathbf{x}(t_{k+1}) = \mathbf{x}(t_k) + \mathbf{a}(t_{k+1} - t_k) + \mathbf{b}(t_{k+1} - t_k)^2 \tag{29}$$

We can now solve (29) for the vectors $\mathbf{a}$ and $\mathbf{b}$ to obtain

$$\mathbf{a} = \frac{1}{t_{k+1} - t_k}[-3\mathbf{x}(t_k) + 4\mathbf{x}(\hat{\tau}_k) - \mathbf{x}(t_{k+1})]$$
$$\mathbf{b} = \frac{2}{(t_{k+1} - t_k)^2}[\mathbf{x}(t_k) - 2\mathbf{x}(\hat{\tau}_k) + \mathbf{x}(t_{k+1})] \tag{30}$$

Substituting (30) into (28), we obtain after some algebra

$$\int_{t_k}^{t_{k+1}} \mathbf{x}(\tau)\, d\tau = \tfrac{1}{6}(t_{k+1} - t_k)[\mathbf{x}(t_k) + 4\mathbf{x}(\hat{\tau}_k) + \mathbf{x}(t_{k+1})] \tag{31}$$

which is the right-hand side of Eq. (25).

9.2.5 More Complex Algorithms

The above approach can be extended to the construction of more complex numerical integration algorithms simply by finding the algorithm that yields no integration error when the elements of $\mathbf{x}(\tau)$ are polynomials of a specified degree in $\tau - t_k$.[†] For example, after Simpson's rule, the next least complicated algorithm is derived by approximating the elements of $\mathbf{x}(\tau)$ as cubic polynomials in $\tau - t_k$, the next as quartic polynomials in $\tau - t_k$, and so on.

[†] For the scalar version, see "Basic Concepts," chap. 9. The generalization to the vector case is completely straightforward. After all, vector integration is defined component by component.

⋆ 9.2.6 Accuracy of Numerical Integration

Since different integration algorithms are obtained by making polynomial approximations to the time-varying vector $\mathbf{x}(\tau)$, it is obvious that the errors will decrease as the approximating polynomial increases in degree. On the other hand, this does not clearly focus the effects of the basic integration step size

$$h = t_{k+1} - t_k \tag{32}$$

which, for the sake of concreteness, we shall assume to be the same for all $k = 0, 1, 2, \ldots, N - 1$.

Intuitively, we expect that the smaller the value of h the better the approximation of $\mathbf{x}(\tau)$ by a constant (Euler's method), by a linear time function (trapezoidal method), or by a parabolic time function (Simpson's rule).

The key to understanding integration errors is to compute the error introduced in the value of $\mathbf{x}(\tau)$ by a finite polynomial approximation. Taylor's theorem (see Appendix B) allows us to compute the error, as we truncate a Taylor series at an arbitrary point.

So let us consider the problem of integrating the vector-valued time function $\mathbf{g}(\tau)$. As we have remarked, the various integration algorithms are exact provided that $\mathbf{g}(\tau)$ is an n-order polynomial vector with respect to τ.

Let us suppose that we expand $\mathbf{g}(\tau)$ about t_k in a truncated Taylor series, and we let $\mathbf{g}_a(\tau)$ denote this truncated polynomial of, say, degree m

$$\mathbf{g}_a(\tau) = \mathbf{g}(t_k) + \mathbf{g}^{(1)}(t_k)(\tau - t_k) + \frac{1}{2!}\mathbf{g}^{(2)}(t_k)(\tau - t_k)^2$$
$$+ \cdots + \tfrac{1}{m!}\mathbf{g}^{(m)}(t_k)(\tau - t_k)^m \tag{33}$$

where $\mathbf{g}^{(j)}(t_k)$ denotes the jth time derivative of $\mathbf{g}(t)$ evaluated at $\tau = t_k$.

Now according to out discussion we have

Euler method Characterized by $m = 0$

Trapezoidal method Characterized by $m = 1$

Simpson's rule Characterized by $m = 2$

Hence, depending on the integration method we select (characterized by m), we have means for computing *exactly* the integral (vector-valued)

$$\mathbf{y}_m \overset{\triangle}{=} \int_{t_k}^{t_{k+1}} \mathbf{g}_a(\tau)\, d\tau \tag{34}$$

when $\mathbf{g}_a(\tau)$ is given by Eq. (33). Let

$$\mathbf{y} \overset{\triangle}{=} \int_{t_k}^{t_{k+1}} \mathbf{g}(\tau)\, d\tau \tag{35}$$

denote the *exact* value (vector-valued) of the desired integral.

The integration-error vector, denoted by $\boldsymbol{\epsilon}_m$, is then simply defined by

$$\boldsymbol{\epsilon}_m \triangleq \mathbf{y} - \mathbf{y}_m = \int_{t_k}^{t_{k+1}} [\mathbf{g}(\tau) - \mathbf{g}_a(\tau)] \, d\tau \qquad (36)$$

and we can use $\|\boldsymbol{\epsilon}_m\|$ as a scalar measure of the integration error.

However, from Taylor's theorem (see Appendix B) we know that there exists a value of τ, denoted by $\hat{\tau}$, such that

$$t_k \leq \hat{\tau} \leq \tau \leq t_{k+1} \qquad (37)$$

so that

$$\mathbf{g}(\tau) - \mathbf{g}_a(\tau) = \frac{1}{(m+1)!} \mathbf{g}^{(m+1)}(\hat{\tau})(\tau - t_k)^{m+1} \qquad (38)$$

It therefore follows that

$$\boldsymbol{\epsilon}_m = \int_{t_k}^{t_{k+1}} \frac{1}{(m+1)!} \mathbf{g}^{m+1}(\hat{\tau})(\tau - t_k)^{m+1} \, d\tau \qquad (39)$$

Hence,

$$\|\boldsymbol{\epsilon}_m\| \leq \int_{t_k}^{t_{k+1}} \frac{1}{(m+1)!} \|\mathbf{g}^{(m+1)}(\hat{\tau})\| (\tau - t_k)^{m+1} \, d\tau \qquad (40)$$

Define $\|\mathbf{g}_{\max}^{(m+1)}\|$ to denote the maximum value of $\|\mathbf{g}^{(m+1)}(\hat{\tau})\|$ when $\hat{\tau}$ is restricted in the time interval from t_k to t_{k+1}. Clearly,

$$\|\mathbf{g}^{(m+1)}(\hat{\tau})\| \leq \|\mathbf{g}_{\max}^{(m+1)}\| \qquad (41)$$

From Eqs. (40) and (41) we deduce that

$$\begin{aligned}
\|\boldsymbol{\epsilon}_m\| &\leq \frac{\|\mathbf{g}_{\max}^{(m+1)}\|}{(m+1)!} \int_{t_k}^{t_{k+1}} (\tau - t_k)^{m+1} \, d\tau \\
&= \frac{\|\mathbf{g}_{\max}^{(m+1)}\|}{(m+2)!} h^{m+2}
\end{aligned} \qquad (42)$$

where h is given by Eq. (32).

From the above considerations we conclude that for any integration algorithm characterized by m the integration error

1 Is proportional to the largest value of the norm of the $(m+1)$st time derivative

2 Is proportional to the time interval of integration h raised to the $(m+2)$d power

In particular we have for the three methods

Euler method $(m = 0)$ The integration error depends on the maximum size $\|\mathbf{g}_{\text{max}}^{(1)}\|$ of the first derivative and is proportional to h^2. Thus if the function we are integrating is characterized by, for example,

$$\|\mathbf{g}_{\text{max}}^{(1)}\| \leq 10$$

and we are interested in an integration accuracy

$$\|\boldsymbol{\epsilon}_0\| \leq 10^{-5}$$

we must select h to be

$$h^2 < 10^{-6} \quad \text{or} \quad h < 10^{-3}$$

Trapezoidal method $(m = 1)$ The integration error depends upon the maximum size $\|\mathbf{g}_{\text{max}}^{(2)}\|$ of the second derivative and is proportional to h^3. Thus, if the function we are integrating is characterized, say, by

$$\|\mathbf{g}_{\text{max}}^{(2)}\| \leq 100$$

and we are interested in an integration accuracy

$$\|\boldsymbol{\epsilon}_1\| \leq 10^{-4}$$

we must select h such that

$$h^3 < 10^{-6} \quad \text{or} \quad h < 10^{-2}$$

Simpson's rule $(m = 2)$ The integration error depends upon the maximum size $\|\mathbf{g}_{\text{max}}^{(3)}\|$ of the third derivative and is proportional to h^4. Thus, if the function we are integrating is characterized by

$$\|\mathbf{g}_{\text{max}}^{(3)}\| \leq 100$$

and we are interested in an integration accuracy

$$\|\boldsymbol{\epsilon}_2\| \leq 10^{-2}$$

we should select h so that

$$h^4 < 10^{-4} \quad \text{or} \quad h < 10^{-1}$$

9.3 INTRODUCTION TO THE NUMERICAL SOLUTION OF VECTOR DIF-FERENTIAL EQUATIONS

In the remainder of this chapter we shall discuss techniques for the numerical solution of vector differential equations. Such equations describe the relation of the state vector $\mathbf{x}(t)$ to the input vector $\mathbf{u}(t)$ in nonlinear dynamical systems. We have already seen how they arise in electric networks which contain nonlinear resistors, capacitors, inductors, and dependent sources; and in ecological and aerospace systems.

In general, nonlinear vector differential equations do not admit analytical closed-form solutions. Graphical integration becomes awkward when the dimension of the state vector is greater than 2. In such cases one must resort to numerical solution techniques. Fortunately, the digital computer can indeed carry out numerical integration extremely rapidly so that one can study nonlinear systems without too much dogwork. Indeed, special subroutines minimize the required programming effort, e.g., in SHARE.

In the remainder of this chapter we shall focus our attention on certain standard solution techniques. The material is the vector counterpart of that of the scalar case which the reader has probably been exposed to (see, for example, "Basic Concepts," chap. 9).

As usual, the transition from the scalar to the vector case is easy from the conceptual point of view. The use of vector notation contributes to notational clarity.

9.3.1 The Main Problem

If we consider a nonlinear network or dynamical system with state vector $\mathbf{x}(t)$ and input vector $\mathbf{u}(t)$, they are related by means of the vector differential equation

$$\dot{\mathbf{x}}(t) = \mathbf{f}(\mathbf{x}(t), \mathbf{u}(t)) \qquad \mathbf{x}(t_0) = \boldsymbol{\xi} \tag{1}$$

We shall assume (see Chap. 7) that this equation admits a unique solution in the interval of interest.

If the input vector $\mathbf{u}(t) = \boldsymbol{\alpha} = \text{const}$ for all $t \geq t_0$, then (1) reduces to

$$\dot{\mathbf{x}}(t) = \mathbf{f}(\mathbf{x}(t), \boldsymbol{\alpha}) \qquad \mathbf{x}(t_0) = \boldsymbol{\xi} \tag{2}$$

and if we define

$$\mathbf{f}_1(\mathbf{x}(t)) \triangleq \mathbf{f}(\mathbf{x}(t), \boldsymbol{\alpha}) \tag{3}$$

then we deal with the *time-invariant* equation

$$\dot{\mathbf{x}}(t) = \mathbf{f}_1(\mathbf{x}(t)) \qquad \mathbf{x}(t_0) = \boldsymbol{\xi} \tag{4}$$

If the input $\mathbf{u}(t)$ is specified to be some time function, say, $\mathbf{u}(t) = \boldsymbol{\beta}(t)$ for all $t \geq t_0$, then by defining

$$\mathbf{f}_2(\mathbf{x}(t), t) \triangleq \mathbf{f}(\mathbf{x}(t), \boldsymbol{\beta}(t)) \tag{5}$$

we can readily see that we deal with a *time-varying* equation of the form

$$\dot{\mathbf{x}}(t) = \mathbf{f}_2(\mathbf{x}(t), t) \qquad \mathbf{x}(t_0) = \boldsymbol{\xi} \tag{6}$$

For these reasons we henceforth suppress the input dependence in the equation and write

$$\dot{\mathbf{x}}(t) = \mathbf{f}(\mathbf{x}(t), t) \qquad \mathbf{x}(t_0) = \boldsymbol{\xi} \tag{7}$$

or

$$\dot{\mathbf{x}}(t) = \mathbf{f}(\mathbf{x}(t)) \qquad \mathbf{x}(t_0) = \boldsymbol{\xi} \tag{8}$$

so that the notation will not become unduly cumbersome.

9.3.2   The Equivalent Integral Equation

The solution $\mathbf{x}(t)$ at time t of Eq. (7) is given *exactly* by the integral equation

$$\mathbf{x}(t) = \boldsymbol{\xi} + \int_{t_0}^{t} \mathbf{f}(\mathbf{x}(\tau), \tau)\, d\tau \tag{9}$$

This is an implicit equation because $\mathbf{x}(.)$ appears on both sides of the equation.

9.3.3   The Discretization of Time

The numerical solution of differential or integral equations cannot be efficiently obtained for all values of $t \geq t_0$. One must compromise and accept as a solution a technique which generates the value of a solution vector only at discrete instants of time denoted by

$$t_0, t_1, t_2, \ldots, t_k, t_{k+1}, \ldots \tag{10}$$

These time increments need not necessarily be the same. However, for ease of programming, they are usually selected equidistant. In this case the scalar

$$h = t_{k+1} - t_k = \text{const} \qquad \text{for all } k = 0, 1, 2, \ldots \tag{11}$$

is called the *integration step size*. Clearly

$$t_{k+1} = t_k + h \qquad \text{for all } k = 0, 1, 2, \ldots \tag{12}$$

The selection of the step size h is up to the engineer. It directly governs the accuracy of the numerical solution to be obtained.

For any given time discretization we have

$$\mathbf{x}(t_1) = \mathbf{x}(t_0) + \int_{t_0}^{t_1} \mathbf{f}(\mathbf{x}(\tau), \tau) \, d\tau$$

$$\mathbf{x}(t_2) = \mathbf{x}(t_1) + \int_{t_1}^{t_2} \mathbf{f}(\mathbf{x}(\tau), \tau) \, d\tau$$

$$\vdots \tag{13}$$

$$\mathbf{x}(t_{k+1}) = \mathbf{x}(t_k) + \int_{t_k}^{t_{k+1}} \mathbf{f}(\mathbf{x}(\tau), \tau) \, d\tau$$

$$\vdots$$

We seek to evaluate the sequence of vectors

$$\mathbf{x}(t_1), \mathbf{x}(t_2), \ldots \mathbf{x}(t_k), \ldots \tag{14}$$

and *accept* this discrete set as a solution, in lieu of the continuous-time solution vector $\mathbf{x}(t)$.

From Eq. (13) we can see that to determine $\mathbf{x}(t_{k+1})$, given $\mathbf{x}(t_k)$, we need to evaluate the integral

$$\int_{t_k}^{t_{k+1}} \mathbf{f}(\mathbf{x}(\tau), \tau) \, d\tau \tag{15}$$

We stress that the integrand is *not* known because $\mathbf{x}(\tau)$, $t_k \leq \tau \leq t_{k+1}$, is not known [if it were we would let $\tau = t_{k+1}$ to find $\mathbf{x}(\tau) = \mathbf{x}(t_{k+1})$] . Hence, one must resign oneself to the fact that any numerical solution will be subject to some error. In other words, *we shall not be able to evaluate the sequence of solution vectors* (14) *exactly.* What will be generated is the sequence of vectors

$$\hat{\mathbf{x}}(t_1), \hat{\mathbf{x}}(t_2), \ldots, \hat{\mathbf{x}}(t_k), \ldots \tag{16}$$

such that (it is hoped)

$$\hat{\mathbf{x}}(t_k) \approx \mathbf{x}(t_k) \qquad \text{for all } k = 1, 2, \ldots \tag{17}$$

i.e., $\|\mathbf{x}(t_k) - \hat{\mathbf{x}}(t_k)\|$ is "small."

9.3.4 General Approaches

As in the scalar case, there are two main methods for finding numerical solutions to vector differential equations: *open and closed methods.* Roughly speaking, *open methods* approximate the value of the integral (15) by approximating the integrand $\mathbf{f}(\mathbf{x}(\tau), \tau)$, $t_k \leq \tau \leq t_{k+1}$, by explicit equations that involve the previous value $\mathbf{x}(t_k)$.

Closed methods, however, are in a sense *trial-and-error methods*, iterative in nature, that take into account the fact that the integral equation (9) involves the sought for solution function $\mathbf{x}(.)$ on both sides of the equation. We shall examine each technique in the remainder of this chapter.

The study of numerical methods for solving nonlinear differential equations is a scientific discipline with a long history. Needless to say, the general availability of digital computers has contributed to a renewed interest in this important area. Many special numerical methods have been described in the literature. They differ in the amount of programming complexity required, as well as the amount of actual computer time needed to solve a given equation for a given accuracy. Some methods always utilize a fixed step size h; others are adaptive in the sense that the algorithms contain provisions for changing the value of h so as to guarantee a particular prespecified accuracy and yet minimize computation time. It is beyond the scope of this book to delve deeply into such techniques, but it is important to realize that there is extensive knowledge of this class of problem from both a theoretical and an applied viewpoint.

9.4   OPEN METHODS FOR NUMERICAL SOLUTION OF DIFFERENTIAL EQUATIONS

We seek to evaluate approximately the sequence of vectors

$$\mathbf{x}(t_{k+1}) = \mathbf{x}(t_k) + \int_{t_k}^{t_{k+1}} \mathbf{f}(\mathbf{x}(\tau), \tau) \, d\tau \tag{1}$$

under the assumptions that

1 $\mathbf{x}(t_0) = \boldsymbol{\xi}$ is known.
2   The analytical dependence of $\mathbf{f}(.,.)$ upon $\mathbf{x}(\tau)$ and τ is available in the sense that if we let $\mathbf{x}(\tau) = \boldsymbol{\alpha}$ and $\tau = \beta$, we can compute $\mathbf{f}(\boldsymbol{\alpha}, \beta)$ at once.
3   The times $t_0, t_1, \ldots$ have been specified.

The open methods for the scalar case can be developed† by a mixture of analytic and graphical arguments. Unfortunately, geometrical concepts loose their appeal in multidimensional euclidean spaces and we must rely on analytic tools.

9.4.1   The Euler Method

The simplest (and most inaccurate) method is the Euler method, which states that

$$\boxed{\mathbf{x}(t_{k+1}) \approx \mathbf{x}(t_k) + (t_{k+1} - t_k)\mathbf{f}(\mathbf{x}(t_k), t_k) = \mathbf{x}(t_k) + h\mathbf{f}(\mathbf{x}(t_k), t_k)} \tag{2}$$

† See "Basic Concepts," chap. 9.

The Euler method assumes that

$$\mathbf{f}(\mathbf{x}(\tau), \tau) = \text{const} = \mathbf{f}(\mathbf{x}(t_k), t_k) \tag{3}$$

for all τ in the interval $t_k \leq \tau \leq t_{k+1}$. Under this assumption, i.e., that the integrand is a constant vector, the integral in (1) is evaluated analytically.

9.4.2 The Heun Method

The Heun method states that

$$\mathbf{x}(t_{k+1}) \approx \mathbf{x}(t_k) + \frac{h}{2}(\mathbf{f}_k{}^A + \mathbf{f}_k{}^B) \tag{4}$$

where the vectors $\mathbf{f}_k{}^A$ and $\mathbf{f}_k{}^B$ are defined as follows:

$$\mathbf{f}_k{}^A \triangleq \mathbf{f}(\mathbf{x}(t_k), t_k) \tag{5}$$

$$\mathbf{f}_k{}^B \triangleq \mathbf{f}(\mathbf{x}(t_k) + h\mathbf{f}_k{}^A, t_k + h) \tag{6}$$

To explain the Heun method we write Eq. (4) in the form

$$\frac{\mathbf{x}(t_{k+1}) - \mathbf{x}(t_k)}{h} = \tfrac{1}{2}(\mathbf{f}_k^A + \mathbf{f}_k^B) \tag{7}$$

Thus the left-hand side is approximately the derivative $\dot{\mathbf{x}}(t)$ for $t_k \leq t \leq t_{k+1}$. Now from (5) we see that $\mathbf{f}_k{}^A$ is the derivative vector at t_k and $\mathbf{f}_k{}^B$ is the derivative vector at t_{k+1} *if one used the Euler method*. These two derivatives are then averaged to obtain a better estimate of the actual derivative vector.

There is another way of explaining the Heun method. Let us recall (see Sec. 9.2) that the trapezoidal integration method for evaluating the integral of a known function is

$$\int_{t_0}^{T} \mathbf{g}(\tau) \, d\tau = \tfrac{1}{2}(T - t_0)[\mathbf{g}(T) + \mathbf{g}(t_0)] \tag{8}$$

If we try to use the trapezoidal method to evaluate the integral in the exact expression

$$\mathbf{x}(t_{k+1}) = \mathbf{x}(t_k) + \int_{t_k}^{t_{k+1}} \mathbf{f}(\mathbf{x}(\tau), \tau) \, d\tau \tag{9}$$

we obtain

$$\int_{t_k}^{t_{k+1}} \mathbf{f}(\mathbf{x}(\tau), \tau) \approx \tfrac{1}{2}(t_{k+1} - t_k)[\mathbf{f}(\mathbf{x}(t_{k+1}), t_{k+1}) + \mathbf{f}(\mathbf{x}(t_k), t_k)] \tag{10}$$

We note that $f(x(t_k), t_k) = f_k^A$ is known. However, $f(x(t_{k+1}), t_{k+1})$ is *not* known because $x(t_{k+1})$ is *not* known; however, the Euler method states that

$$x(t_{k+1}) \approx x(t_k) + hf_k^A \tag{11}$$

and this approximate value is substituted in (10) to obtain the approximation

$$f(x(t_{k+1}), t_{k+1}) \approx f(x(t_k) + hf_k^A, t_k + h) \triangleq f_k^B \tag{12}$$

In this manner, one can deduce that *the Heun method utilizes trapezoidal integration and an estimate provided by the Euler method to calculate at an intermediate step the approximate value of* $x(t_{k+1})$.

9.4.3   The Fourth-Order Runge-Kutta (FORK) Method

A very popular solution method is the FORK method, which states that

$$x(t_{k+1}) \approx x(t_k) + \frac{h}{6}(f_k^A + 2f_k^B + 2f_k^C + f_k^D) \tag{13}$$

where

$$f_k^A = f(x(t_k), t_k)$$

$$f_k^B = f(x(t_k) + \frac{h}{2}f_k^A, t_k + \frac{h}{2})$$

$$f_k^C = f(x(t_k) + \frac{h}{2}f_k^B, t_k + \frac{h}{2}) \tag{14}$$

$$f_k^D = f(x(t_k) + hf_k^C, t_k + h)$$

The FORK method attempts to obtain a good estimate of the true derivative by attempting to evaluate an estimate of the derivative vector at the middle $t_k + h/2$ of the integration interval $t_k \leq \tau \leq t_{k+1}$.

9.4.4   Illustration of Integration Methods for the Equation $\dot{x}(t) = Ax(t)$

It is instructive to consider the LTI vector differential equation

$$\dot{x}(t) = Ax(t) \tag{15}$$

and calculate the approximate solutions using each one of the three methods. To make the correspondence clear, we note that (15) is a special case of

$$\dot{x}(t) = f(x(t), t) \tag{16}$$

with

$$f(x(t), t) = Ax(t) \tag{17}$$

The analytical solution   Given the value of $\mathbf{x}(t_k)$, and letting $h = t_{k+1} - t_k$, we know that the *exact* solution of (15) is given by

$$\mathbf{x}(t_{k+1}) = e^{\mathbf{A}h}\mathbf{x}(t_k) = \left(\mathbf{I} + h\mathbf{A} + \frac{h^2}{2!}\mathbf{A}^2 + \frac{h^3}{3!}\mathbf{A}^3 + \cdots\right)\mathbf{x}(t_k) \qquad (18)$$

We shall demonstrate below that the three methods for numerical solution simply truncate the infinite series for the matrix exponential at different points.

The Euler method   The Euler method states [see Eq. (2)] in general that

$$\mathbf{x}(t_{k+1}) \approx \mathbf{x}(t_k) + h\mathbf{f}(\mathbf{x}(t_k), t_k) \qquad (19)$$

In view of (17) we have

$$\mathbf{f}(\mathbf{x}(t_k), t_k) = \mathbf{A}\mathbf{x}(t_k) \qquad (20)$$

Hence, from (19) and (20) we deduce that

$$\mathbf{x}(t_{k+1}) = \mathbf{x}(t_k) + h\mathbf{A}\mathbf{x}(t_k) = (\mathbf{I} + h\mathbf{A})\mathbf{x}(t_k) \qquad (21)$$

We therefore conclude that for LTI equations the Euler method simply approximates the matrix exponential by its first two terms, i.e.,

$$e^{\mathbf{A}h} \approx \mathbf{I} + h\mathbf{A} \qquad (22)$$

and its accuracy is of the order h^2, because this is the next dominant term in the infinite series for sufficiently small h.

The Heun method   The Heun method states [see Eqs. (4) to (6)] that

$$\mathbf{x}(t_{k+1}) = \mathbf{x}(t_k) + \tfrac{h}{2}\left(\mathbf{f}_k{}^A + \mathbf{f}_k{}^B\right)$$
$$\mathbf{f}_k{}^A = \mathbf{f}(\mathbf{x}(t_k), t_k) \qquad (23)$$
$$\mathbf{f}_k{}^B = \mathbf{f}(\mathbf{x}(t_k) + h\mathbf{f}_k{}^A, t_k + h)$$

In view of (17) we find that

$$\mathbf{f}_k{}^A = \mathbf{A}\mathbf{x}(t_k) \qquad (24)$$
$$\mathbf{f}_k{}^B = \mathbf{A}[\mathbf{x}(t_k) + h\mathbf{A}\mathbf{x}(t_k)] = (\mathbf{A} + h\mathbf{A}^2)\mathbf{x}(t_k) \qquad (25)$$

Hence, we readily conclude that

$$\mathbf{x}(t_{k+1}) \approx \mathbf{x}(t_k) + \frac{h}{2}[\mathbf{A}\mathbf{x}(t_k) + (\mathbf{A} + h\mathbf{A}^2)\mathbf{x}(t_k)] \qquad (26)$$

which simplifies to

$$x(t_{k+1}) \approx \left(I + hA + \frac{h^2}{2!}A^2\right)x(t_k) \tag{27}$$

On the basis of the above development we conclude that, for LTI differential equations, the Heun method approximates the matrix exponential by its first three terms, i.e.,

$$e^{Ah} \approx I + hA + \frac{h^2}{2!}A^2 \tag{28}$$

and its accuracy is of the order h^3.

The FORK method   The FORK method states that

$$x(t_{k+1}) = x(t_k) + \frac{h}{6}(\mathbf{f}_k^A + 2\mathbf{f}_k^B + 2\mathbf{f}_k^C + \mathbf{f}_k^D) \tag{29}$$

From Eqs. (14) and (17) we find that

$$\mathbf{f}_k^A = Ax(t_k) \tag{30}$$

$$\mathbf{f}_k^B = A\left[x(t_k) + \frac{h}{2}\mathbf{f}_k^A\right] = \left(A + \frac{h}{2}A^2\right)x(t_k) \tag{31}$$

$$\mathbf{f}_k^C = A\left[x(t_k) + \frac{h}{2}\mathbf{f}_k^B\right] = \left(A + \frac{h}{2}A^2 + \frac{h^2}{4}A^3\right)x(t_k) \tag{32}$$

$$\mathbf{f}_k^D = A[x(t_k) + h\mathbf{f}_k^C] = \left(A + hA^2 + \frac{h^2}{2}A^3 + \frac{h^3}{4}A^4\right)x(t_k) \tag{33}$$

Substituting Eqs. (30) to (33) into Eq. (29), we obtain after some algebra

$$x(t_{k+1}) \approx \left(I + hA + \frac{h^2}{2!}A^2 + \frac{h^3}{3!}A^3 + \frac{h^4}{4!}A^4\right)x(t_k) \tag{34}$$

We therefore can conclude that for LTI differential equations the FORK method approximates the matrix exponential by its first five terms, i.e.,

$$e^{Ah} \approx I + hA + \frac{h^2}{2!}A^2 + \frac{h^3}{3!}A^3 + \frac{h^4}{4!}A^4 \tag{35}$$

and its integration accuracy is of the order of h^5.

9.5 NUMERICAL COMPARISON OF INTEGRATION METHODS FOR THE VAN DER POL EQUATION†

In this section a specific numerical example is given to illustrate the effects of changing the step size h upon the accurary of the three basic numerical algorithms:

1 The Euler method
2 The Heun method
3 The FORK method

The differential equation we use is a celebrated second-order nonlinear differential equation known as the *Van der Pol equation*. In the days of vacuum tubes, this equation was useful in analyzing circuits that acted as nonlinear oscillators.

The general form of the Van der Pol equation is

$$\frac{d^2 y(t)}{dt^2} - p[1 - y^2(t)]\frac{dy(t)}{dt} + y(t) = 0 \tag{1}$$

where p is a constant parameter. In our example, we select

$$p = 1 \tag{2}$$

so that we deal with the equation

$$\frac{d^2 y(t)}{dt^2} - [1 - y^2(t)]\frac{dy(t)}{dt} + y(t) = 0 \tag{3}$$

To transform Eq. (3) into state-variable form we define

$$x_1(t) \triangleq y(t) \qquad x_2(t) \triangleq \frac{dy(t)}{dt} \tag{4}$$

so that the state-variable description of (3) is

$$\dot{x}_1(t) = x_2(t)$$
$$\dot{x}_2(t) = -x_1(t) + [1 - x_1^2(t)]x_2(t) \tag{5}$$

The initial conditions are

$$x_1(0) = 0 \qquad x_2(0) = 0 \tag{6}$$

The reader can readily verify that although (6) represents an equilibrium state of (5), it is an unstable equilibrium state because the linearized perturbation equations are

---

† We are grateful to Mr. A.H. Sarris for the results of this section.

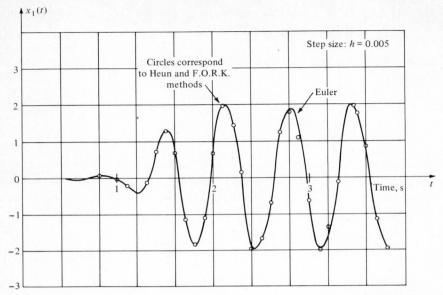

FIGURE 9.5.1

$$\frac{d}{dt}\begin{bmatrix} \delta x_1(t) \\ \delta x_2(t) \end{bmatrix} = \begin{bmatrix} 0 & 1 \\ -1 & 1 \end{bmatrix}\begin{bmatrix} \delta x_1(t) \\ \delta x_2(t) \end{bmatrix} \qquad (7)$$

$$\underbrace{\frac{d}{dt}\,\delta\mathbf{x}(t)}_{} \qquad \underbrace{\mathbf{A}}_{} \qquad \underbrace{\delta\mathbf{x}(t)}_{}$$

and the eigenvalues of the **A** matrix are at

$$\lambda = \tfrac{1}{2} \pm j\frac{\sqrt{3}}{2} \qquad (8)$$

so that they are in the right half of the complex plane.

Due to the instability of the initial state, small roundoff computer errors will accumulate, and the solution of (5) will diverge from (0,0). This type of instability, in turn, will have significant effects upon the numerical integration accuracy as a function of the specific integration routine employed and the integration step size h used. Figure 9.5.1 shows the numerical solution when the step size was selected to be small ($h = 0.005$). For such a small step size, all three methods gave approximately the same answer. Hence, for all practical purposes, we can accept the time function $x_1(t)$ of Fig. 9.5.1 as the "correct" solution of the Van der Pol equation.

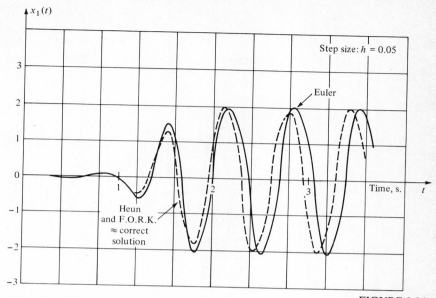

FIGURE 9.5.2

Figure 9.5.2 shows the effects of increasing the step size by a factor of 10 ($h = 0.05$). Such a decrease in step size has not affected the numerical accuracy of the Heun and FORK methods significantly (in the sense that their solution in Fig. 9.5.2 is almost identical to the "correct" solution of Fig. 9.5.1). On the other hand, this has started to affect the accuracy of the Euler method (the solid curve in Fig. 9.5.2). Note that as time goes on, the integration errors of the Euler method keep piling up, and the integration accuracy degrades more and more.

Figure 9.5.3 shows the effects of further increasing the step size by another factor of 10 ($h = 0.5$). In this case, the Euler method diverges rapidly, and the numerical solution bears no resemblance to the correct solution. At this step size we start seeing errors accumulating in the Heun method (dashed curve); on the other hand, the FORK method still gives us excellent accuracy (the FORK solution in Fig. 9.5.3 is almost identical to the "correct" solution of Fig. 9.5.1).

If we increase the integration step size still further, all methods eventually diverge. Figure 9.5.4 illustrates this point with an integration step size $h = 1.0$. The FORK tries to hang on until about 1.7, but after that time it loses accuracy and diverges.

We hope that this specific numerical comparison has helped the reader to appreciate the idiosyncrasies of each numerical integration method and the impor-

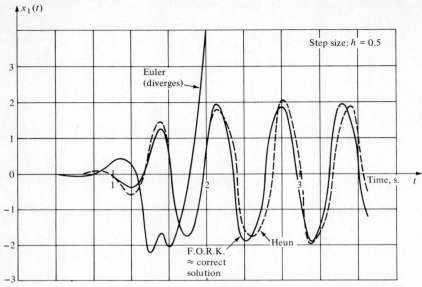

FIGURE 9.5.3

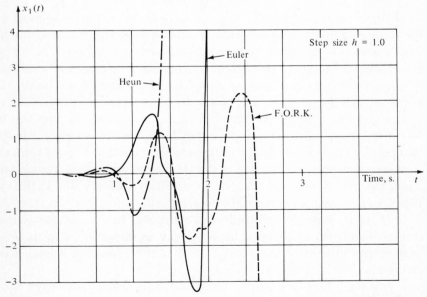

FIGURE 9.5.4

tance of the step size h upon integration accuracy and/or divergence of the numerical method.

★ 9.6 CLOSED METHODS FOR NUMERICAL SOLUTION OF DIFFEREN-TIAL EQUATIONS

We have seen that the open-solution methods attempt to evaluate the integral in the expression

$$\mathbf{x}(t_{k+1}) = \mathbf{x}(t_k) + \int_{t_k}^{t_{k+1}} \mathbf{f}(\mathbf{x}(t), t)\, dt \tag{1}$$

by explicit formulas which can be written down once the value of $\mathbf{x}(t_k)$ has been obtained. The accuracy is improved by trying to obtain a better approximation of the derivative $\dot{\mathbf{x}}(t)$, $t_k \leq t \leq t_{k+1}$, as is done in the Heun and Runge-Kutta methods. At any rate, the problem of evaluating $\mathbf{x}(t_{k+1})$ may involve preliminary calculations, but it does not involve any trial-and-error iterative approaches.

Closed-solution methods use iteration to solve Eq. (1). To keep the notation as simple as possible, we shall illustrate the ideas involved by considering time-invariant differential equations of the form

$$\dot{\mathbf{x}}(t) = \mathbf{f}(\mathbf{x}(t)) \qquad \mathbf{x}(t_0) = \boldsymbol{\xi} \tag{2}$$

so that we deal with the evaluation of the sequence

$$\mathbf{x}(t_{k+1}) = \mathbf{x}(t_k) + \int_{t_k}^{t_{k+1}} \mathbf{f}(\mathbf{x}(t))\, dt \tag{3}$$

9.6.1 Why Iteration?

Let us examine Eq. (3) carefully. Abstractly, it can be viewed as an equation of the form

$$\mathbf{x}(t_{k+1}) = \mathbf{x}(t_k) + \mathbf{g}(\mathbf{x}(t_{k+1}), \mathbf{x}(t_k)) \tag{4}$$

where $\mathbf{g}(.)$ represents the value of the integral in (3). If we think of $\mathbf{x}(t_k)$, $\mathbf{f}(.)$, t_{k+1} and t_k as known and $\mathbf{x}(t_{k+1})$ as the unknown, Eq. (4) or (3) is of the form

$$\mathbf{y} = \mathbf{m}(\mathbf{y}) \qquad \text{where } \mathbf{y} = \mathbf{x}(t_{k+1}) \tag{5}$$

Thus, the unknown $\mathbf{y}$ appears on both sides of the equation. We know that we can attempt to solve such equation using Picard's or Newton's method or some other iterative technique. This is the basic idea of closed methods, namely, to use some iterative algorithm in Eq. (3) which converges to the true solution $\mathbf{x}(t_{k+1})$.

9.6.2 Some Difficulties

If we want to solve Eq. (5) by, say, Picard's method, we must know the function $\mathbf{m}(.)$ analytically. In our problem, this implies that we must know the function $\mathbf{g}(.)$ analytically in Eq. (4), i.e.,

$$\mathbf{g}(\mathbf{x}(t_{k+1}), \mathbf{x}(t_k)) = \int_{t_k}^{t_{k+1}} \mathbf{f}(\mathbf{x}(t)) \, dt \tag{6}$$

In general, we cannot evaluate the integral in closed form; thus $\mathbf{g}(.)$ is *not* known analytically. Hence, it is not obvious how one can apply the Picard algorithm.

9.6.3 Use of Numerical Integration Methods

To eliminate this problem of lack of analyticity one must compromise and accept the fact that one cannot obtain an exact formula for $\mathbf{g}(.)$ in Eq. (4); instead, one can use a numerical integration algorithm to obtain an approximate equation.

In the remainder of this section we shall use the *trapezoidal integration method* to obtain an analytical expression. In other words, we let

$$\int_{t_k}^{t_{k+1}} \mathbf{f}(\mathbf{x}(t)) \, dt \approx \underbrace{\tfrac{1}{2}(t_{k+1} - t_k)}_{h} [\underbrace{\mathbf{f}(\mathbf{x}(t_{k+1}))}_{\text{Unknown}} + \underbrace{\mathbf{f}(\mathbf{x}(t_k))}_{\text{Known}}] \tag{7}$$

Thus instead of the exact expression (3) we have the approximate relation

$$\mathbf{x}(t_{k+1}) \approx \mathbf{x}(t_k) + \tfrac{1}{2}h\mathbf{f}(\mathbf{x}(t_k)) + \tfrac{1}{2}h\mathbf{f}(\mathbf{x}(t_{k+1})) \tag{8}$$

where the step size h is defined, as usual, by

$$h \triangleq t_{k+1} - t_k \tag{9}$$

We remind the reader that *the error associated with the trapezoidal integration method is of the order h^3*; thus, we have some idea, for any given value of h, of the errors involved in replacing the approximate relation (7) by the equality

$$\boxed{\mathbf{x}(t_{k+1}) = \mathbf{x}(t_k) + \tfrac{1}{2}h\mathbf{f}(\mathbf{x}(t_k)) + \tfrac{1}{2}\mathbf{f}(\mathbf{x}(t_{k+1}))} \tag{10}$$

We have now obtained in (10) an analytical expression; it is an implicit algebraic equation in $\mathbf{x}(t_{k+1})$; note that the vector $\mathbf{x}(t_k) + \frac{1}{2}h\mathbf{f}(\mathbf{x}(t_k))$ is viewed as a known constant.

We can now apply the Picard algorithm to Eq. (10) to construct a sequence of vectors

$$\mathbf{x}^0(t_{k+1}), \, \mathbf{x}^1(t_{k+1}), \, \ldots, \, \mathbf{x}^j(t_{k+1}), \, \ldots \tag{11}$$

If this sequence of vectors converges, then

$$\lim_{j \to \infty} \mathbf{x}^j(t_{k+1}) = \mathbf{x}(t_{k+1}) \tag{12}$$

is indeed the solution of Eq. (10). The Picard algorithm is

$$\boxed{\mathbf{x}^{j+1}(t_k) = \mathbf{x}(t_k) + \tfrac{1}{2}h\mathbf{f}(\mathbf{x}(t_k)) + \tfrac{1}{2}h\mathbf{f}(\mathbf{x}^j(t_{k+1}))} \tag{13}$$

for $j = 0, 1, 2, \ldots$.

9.6.4 The Initial Guess

We can start the Picard algorithm by making the initial guess $\mathbf{x}^0(t_{k+1})$ using any one of the open numerical solution methods examined in Sec. 9.5; for example, we can use the Euler method to obtain the initial guess

$$\boxed{\mathbf{x}^0(t_{k+1}) = \mathbf{x}(t_k) + h\mathbf{f}(\mathbf{x}(t_k))} \tag{14}$$

Then we set $j = 1$ in Eq. (13) to obtain $\mathbf{x}^1(t_{k+1})$ and so on.

9.6.5 Convergence

We have seen (Chap. 3) that iterative algorithms may or may not converge. For the Picard algorithm we do have sufficient conditions to guarantee its convergence. Let us recall (Sec. 3.5) that the convergence of the Picard algorithm is guaranteed if for every vector $\mathbf{y}$ and $\mathbf{z}$ in some region of convergence Ω the condition

$$\|\mathbf{x}(t_k) + \tfrac{1}{2}h\mathbf{f}(\mathbf{x}(t_k)) + \tfrac{1}{2}h\mathbf{f}(\mathbf{y}) - \mathbf{x}(t_k) - \tfrac{1}{2}h\mathbf{f}(\mathbf{x}(t_k)) - \tfrac{1}{2}h\mathbf{f}(\mathbf{z})\|$$

$$\le L\|\mathbf{y} - \mathbf{z}\| \qquad 0 \le L < 1 \tag{15}$$

holds. The inequality (15) reduces to

$$\|\mathbf{f}(\mathbf{y}) - \mathbf{f}(\mathbf{z})\| \le \frac{2L}{h}\|\mathbf{y} - \mathbf{z}\| \qquad 0 \le L < 1 \tag{16}$$

The next question is: How can we guarantee that (16) indeed holds?

To answer this very basic question we must digress for a moment. What we are trying to do is evaluate, by numerical means, the solution to a vector differential equation of the form

$$\dot{\mathbf{x}}(t) = \mathbf{f}(\mathbf{x}(t)) \qquad \mathbf{x}(t_0) = \boldsymbol{\xi} \tag{17}$$

We are implicitly making the assumption that (17) admits a unique solution. In Sec. 7.14 we proved that Eq. (17) *admits a unique solution in some subset* Ω *of* R_n *if there exists a positive constant* M, $0 \leq M < \infty$, *called Lipschitz constant of the differential equation* (17) *such that*

$$\|\mathbf{f}(\mathbf{y}) - \mathbf{f}(\mathbf{z})\| \leq M\|\mathbf{y} - \mathbf{z}\| \qquad \text{for all } \mathbf{y}, \mathbf{z} \in \Omega \tag{18}$$

If we assume that (17) admits a unique solution in Ω, then such a constant M exists. Let us therefore suppose henceforth that we know M. In this case, by comparing (16) and (18) we deduce the relation

$$\frac{2L}{h} = M \tag{19}$$

or

$$\frac{hM}{2} = L \tag{20}$$

Since $L < 1$ guarantees the convergence of the Picard algorithm, it follows that *if the integration step size h is chosen to be*

$$0 < h < \frac{2}{M} \tag{21}$$

then the Picard algorithm will converge to the solution of (10). Thus, the larger the value of the Lipschitz constant M the smaller we must select the integration step size h. This makes physical sense since the magnitude of M is a measure of how big the norm of the Jacobian matrix $\|\partial \mathbf{f}/\partial \mathbf{x}\|$ can become. If $\mathbf{f}(.)$ has large partial derivatives, one must select a small integration step to obtain accurate answers.

9.6.6   When Do We Stop?

In any practical situation we must stop the Picard algorithm at some finite number $j = J$ once we are convinced that we are close enough to the solution. We know that if we let $j \to \infty$, we shall still be in error since the trapezoidal integration method has an error of the order of h^3. Our initial guess, obtained by the Euler method, has an initial error of the order of h, that is,

$$\|\mathbf{x}^0(t_{k+1}) - \mathbf{x}(t_{k+1})\| \approx h \tag{22}$$

The convergent Picard algorithm reduces the initial error as a function of the Lipschitz constant L. If we set the accuracy at h^3 (the inherent error of the trapezoidal integration method), then the step $j = J$ at which we stop the Picard algorithm is defined by the relation

$$\|\mathbf{x}^J(t_{k+1}) - \mathbf{x}(t_{k+1})\| < h^3 \tag{23}$$

But

$$\|\mathbf{x}^J(t_{k+1}) - \mathbf{x}(t_{k+1})\| \leq L^J \|\mathbf{x}^0(t_{k+1}) - \mathbf{x}(t_{k+1})\| \tag{24}$$

$$\leq L^J h \tag{25}$$

Hence, we must have

$$L^J \approx h^2 \tag{26}$$

or

$$J \approx 2 \frac{\log h}{\log L} \tag{27}$$

To illustrate these ideas let us suppose that the Lipschitz constant of our differential equation is $M = 200$. The step size h must be chosen [see Eq. (21)] so that

$$h < \frac{2}{M} = \frac{2}{200} = 10^{-2}$$

To play it safe let us select

$$h = \tfrac{1}{2} \times 10^{-2}$$

Then [see Eq. (20)]

$$L = \frac{hM}{2} = \tfrac{1}{2} \times 10^{-2} \times 100 = \tfrac{1}{2}$$

Then we stop the Picard algorithm after

$$J = 2 \frac{\log h}{\log L} = 2 \frac{-2 - \log 2}{-\log 2} \approx 3.86 = 4 \text{ steps}$$

9.6.7 Summary

To determine the solution of a differential equation $\dot{\mathbf{x}}(t) = \mathbf{f}(\mathbf{x}(t))$, $\mathbf{x}(t_0) = \xi$, using the described method proceed as follows:

1 Evaluate the Lipschitz constant M of the differential equation.
2 Select step size h such that

$$0 < h < \frac{2}{M} \tag{28}$$

3   Calculate a number J by

$$J = 2\frac{\log h}{\log(hM/2)} \tag{29}$$

4   Let $t_1 = t_0 + h$, $t_2 = t_1 + h$, $\ldots$, $t_{k+1} = t_k + h$, $\ldots$.
5   Set $k = 0$. Compute

$$\mathbf{x}^0(t_{k+1}) = \mathbf{x}(t_k) + h\mathbf{f}(\mathbf{x}(t_k)) \tag{30}$$

Apply the Picard algorithm for

$$0 \le j \le J \tag{31}$$

$$\mathbf{x}^{j+1}(t_{k+1}) = \mathbf{x}(t_k) + \tfrac{1}{2}h\mathbf{f}(\mathbf{x}(t_k)) + \tfrac{1}{2}h\mathbf{f}(\mathbf{x}^j(t_{k+1})) \tag{32}$$

Set

$$\mathbf{x}(t_{k+1}) = \mathbf{x}^{J+1}(t_{k+1}) \tag{33}$$

6   Set $k = k + 1$ and go to step 5.

REFERENCES

1   MOURSUND, D. G., and C. S. DURIS: "Elementary Theory and Application of Numerical Analysis," McGraw-Hill, New York, 1967.
2   RALSTON, A.: "A First Course in Numerical Analysis," McGraw-Hill, New York, 1965.
3   TODD, J.: "Survey of Numerical Analysis," McGraw-Hill, New York, 1962.
4   MCCALLA, T. R.: "Introduction to Numerical Methods and FORTRAN Programming," Wiley, New York, 1967.
5   KETTER, R. L., and S. P. PRAWEL, JR.: "Modern Methods of Engineering Computation," McGraw-Hill, New York, 1969.
6   HAMMING, R. W.: "Numerical Methods for Scientists and Engineers," McGraw-Hill, New York, 1962.

EXERCISES

Section 9.4

9.4.1   (Computer exercise)   Consider the differential equations satisfied by a pendulum

$$\ddot{\theta}(t) + \frac{g}{l}\sin\theta(t) = 0$$

where $\theta(t)$ = angular displacement from vertical
g = acceleration of gravity = 32 ft/s²
l = length of pendulum = 1 ft

Start from the initial conditions $\theta(0) = 60°$ and $\theta(0) = 0$. Compare the solutions you obtain using the Euler, Heun, and FORK method for step sizes $h = 1$, $h = 0.1$, $h = 0.01$, and $h = 0.001$.

9.4.2   Consider a harmonic oscillator defined by

$$\dot{x}_1(t) = x_2(t) \qquad \dot{x}_2(t) = -x_1(t) \tag{1}$$

(a)   State the equations of the Euler method for the numerical integration of (1). Prove that there does not exist a value of a step size h that will yield a *stable* calculation. *Hint*: The system (1) is stable; consider the stability of the discrete system.

(b)   Write the equations of the Heun algorithm. Is there a value of h that can guarantee the stability of the numerical integration algorithm?

9.4.3   (Computer exercise)   Consider the motion of the asymmetrical space body analyzed in Sec. 7.15. For specific values of moments of inertia $I_1 = 1$, $I_2 = 2$, $I_3 = 3$ we obtain the nonlinear equations of motion

$$\dot{\omega}_1(t) = -\omega_2(t)\omega_3(t)$$

$$\dot{\omega}_2(t) = \omega_1(t)\omega_3(t)$$

$$\dot{\omega}_3(t) = -\tfrac{1}{3}\omega_1(t)\omega_2(t)$$

Consider the initial condition

$$\omega_1(0) = 0.1 \qquad \omega_2(0) = 1.0 \qquad \omega_3(0) = 0$$

Compare the numerical solutions obtained through the use of the Euler, Heun and FORK methods for step sizes $h = 1$, $h = 0.1$, and $h = 0.01$.

9.4.4   (Computer exercise)   Many ecological systems describe the interactions between populations that depend upon each other for survival. A simple example can be provided by the interactions of two species whose populations at each instant of time are denoted by $x_1(t)$ and $x_2(t)$. Suppose that the population dynamics are related by the nonlinear differential equations

$$\dot{x}_1(t) = -2x_1(t) + x_1(t)x_2(t)$$

$$\dot{x}_2(t) = -x_1(t) + x_1(t)x_2(t) \tag{1}$$

(a)   Determine the equilibrium states of Eq. (1).
(b)   Test the stability properties of each equilibrium state [meaningful only for $x_1(t) \geq 0$, $x_2(t) \geq 0$].
(c)   Using a Heun method with a step size of $h = 10^{-2}$, analyze the changes

in population for the following initial condition:

(1) x_1 (0) = 0.5, $x_2(0)$ = 0.5
(2) x_1 (0) = 2.0, $x_2(0)$ = 0.5
(3) x_1 (0) = 1.0, $x_2(0)$ = 3.0
(4) x_1 (0) = 1.25, $x_2(0)$ = 2.25

Section 9.6

9.6.1   Consider the LTI differential equation $\dot{x}(t) = Ax(t)$. Determine exactly how the closed integration method described in Sec. 9.6 works out for this case. Demonstrate that at each step of the Picard algorithm one simply adds one more term to the truncated series approximation for the matrix exponential e^{Ah}.

9.6.2   Use the approach outlined in Sec. 9.6 to derive another closed-form method but use Simpson's rule to obtain an analytical expression for the integral of Eq. (6).

9.6.3   (Computer homework) Consider the Van der Pol equation discussed in Sec. 9.5. Apply the closed integration method described in Sec. 9.6 to obtain the solution using the step sizes $h = 1$, $h = 0.1$, and $h = 0.01$. Do not forget to put an appropriate stopping rule to the Picard method; you can always do this by requiring the integration accuracy to be better than 10^{-3}.

APPENDIX A
VECTORS, MATRICES, AND LINEAR ALGEBRA

A.1 MOTIVATION

This appendix gives a brief introduction to vectors and matrices and defines some of their operations. This minimal background should be sufficient for our use of linear-algebra techniques for the analysis and design of networks and systems. Reading this appendix should give the student a good understanding of the operational and geometric interpretation of some of the definitions and operations. Since linear-algebra tools are used throughout this text, the student should review the appendix thoroughly.

A.2 COLUMN VECTORS: DEFINITIONS AND ARITHMETIC OPERATIONS

Consider n numbers $x_1, x_2, \ldots, x_n$, which may be real or complex. These numbers can be arranged so as to define a new object $\mathbf{x}$, called a *column n-vector*,

$$\mathbf{x} = \begin{bmatrix} x_1 \\ x_2 \\ \vdots \\ x_n \end{bmatrix} \qquad \mathbf{x} \in R_n \qquad (1)$$

A lowercase boldface letter will always denote a column vector in this text, for example, $\mathbf{x}$, $\mathbf{a}$, $\boldsymbol{\delta}$.

The numbers $x_1, x_2, \ldots, x_n$ form the *components* of the column vector $\mathbf{x}$; the number x_i is the ith component of $\mathbf{x}$. For example,

$$\mathbf{x} = \begin{bmatrix} 2 \\ -3 \\ 4 \end{bmatrix} \tag{2}$$

is a column 3-vector, and -3 is its second component.

With every column n-vector $\mathbf{x}$ we can associate its *transpose* vector, denoted by $\mathbf{x}'$ and defined by

$$\mathbf{x}' = [x_1 \quad x_2 \quad \cdots \quad x_n] \tag{3}$$

This is often called a *row n-vector*.

If every component $x_1, x_2, \ldots, x_n$ of the column vector $\mathbf{x}$ is a real number, then $\mathbf{x}$ is called a real column vector. Otherwise, it is called a complex-valued column vector.

The column vectors $\mathbf{x}$ and $\mathbf{y}$ are said to be equal (written $\mathbf{x} = \mathbf{y}$) if and only if

1 They have the same number of components, say, n.
2 For each i, $i = 1, 2, \ldots, n$, the equality

$$x_i = y_i$$

holds.

We shall now proceed with the definition of the operations of addition of two column vectors and multiplication of a column vector by a scalar so that we develop a vector algebra.

A.2.1 Addition of Column Vectors

Consider the column vectors

$$\mathbf{x} = \begin{bmatrix} x_1 \\ x_2 \\ \vdots \\ x_n \end{bmatrix} \qquad \mathbf{y} = \begin{bmatrix} y_1 \\ y_2 \\ \vdots \\ y_n \end{bmatrix} \qquad \mathbf{z} = \begin{bmatrix} x_1 + y_1 \\ x_2 + y_2 \\ \vdots \\ x_n + y_n \end{bmatrix} \tag{4}$$

We define the sum of the column vectors $\mathbf{x}$ and $\mathbf{y}$ to be the column vector $\mathbf{z}$

$$\mathbf{x} + \mathbf{y} = \mathbf{z} \tag{5}$$

In other words, to add two column vectors one adds their respective components and uses the sum as the components of another vector. The following examples illustrate this point.

EXAMPLE 1   Consider the real column 3-vectors

$$\mathbf{x} = \begin{bmatrix} 1 \\ 2 \\ -3 \end{bmatrix} \qquad \text{and} \qquad \mathbf{y} = \begin{bmatrix} -3 \\ 5 \\ 0 \end{bmatrix}$$

Then

$$\mathbf{x} + \mathbf{y} = \begin{bmatrix} 1 - 3 \\ 2 + 5 \\ -3 + 0 \end{bmatrix} = \begin{bmatrix} -2 \\ 7 \\ -3 \end{bmatrix}$$

EXAMPLE 2 Consider the two column 2-vectors **x** and **y** whose components are complex numbers

$$\mathbf{x} = \begin{bmatrix} 2 + j3 \\ -1 - j2 \end{bmatrix} \quad \text{and} \quad \mathbf{y} = \begin{bmatrix} -1 + j2 \\ 5 + j2 \end{bmatrix}$$

Then

$$\mathbf{x} + \mathbf{y} = \begin{bmatrix} 1 + j5 \\ 4 \end{bmatrix}$$

The operation of column-vector addition has the following important properties:

1 The set of all column *n*-vectors is *closed* under addition. In other words, the sum of two column *n*-vectors is also a column *n*-vector.

2 Vector addition is *commutative*, i.e.,

$$\mathbf{x} + \mathbf{y} = \mathbf{y} + \mathbf{x} \tag{6}$$

3 Vector addition is *associative*, i.e.,

$$(\mathbf{x} + \mathbf{y}) + \mathbf{z} = \mathbf{x} + (\mathbf{y} + \mathbf{z}) \tag{7}$$

4 There exists a column 0-*vector*, denoted **0** and defined by

$$\mathbf{0} = \begin{bmatrix} 0 \\ 0 \\ \vdots \\ 0 \end{bmatrix} \tag{8}$$

such that for all column *n*-vectors **x**

$$\mathbf{x} + \mathbf{0} = \mathbf{x} \tag{9}$$

5 If + is the ordinary addition operation, there is a unique column vector $-\mathbf{x}$ associated with every column vector **x** such that

$$\mathbf{x} + (-\mathbf{x}) = \mathbf{0} \tag{10}$$

where

$$\mathbf{x} = \begin{bmatrix} x_1 \\ x_2 \\ \vdots \\ x_n \end{bmatrix} \quad -\mathbf{x} = \begin{bmatrix} -x_1 \\ -x_2 \\ \vdots \\ -x_n \end{bmatrix} \tag{11}$$

A.2.2 Multiplication of a Column Vector by a Scalar

We now turn our attention to the multiplication of a column vector by a scalar (real or complex-valued).

Let β be a scalar and $\mathbf{x}$ be a column n-vector. Then the product $\beta\mathbf{x}$ is defined by

$$\beta\mathbf{x} = \beta\begin{bmatrix} x_1 \\ x_2 \\ \vdots \\ x_n \end{bmatrix} = \begin{bmatrix} \beta x_1 \\ \beta x_2 \\ \vdots \\ \beta x_n \end{bmatrix} = \mathbf{x}\beta \tag{12}$$

Thus, the product of a scalar with a column n-vector is a column n-vector.

EXAMPLE 3 The multiplication is illustrated by the following specific examples.

$$3\begin{bmatrix} 5 \\ -1 \\ 4 \end{bmatrix} = \begin{bmatrix} 15 \\ -3 \\ 12 \end{bmatrix}$$

$$(3 + j2)\begin{bmatrix} 2 \\ 3 - j2 \\ 1 + j1 \end{bmatrix} = \begin{bmatrix} 6 + j4 \\ 13 \\ 1 + j5 \end{bmatrix}$$

The operation of column-vector multiplication by a constant has the following properties:

1 The set of all column n-vectors is *closed* under multiplication; i.e., if β is a scalar and $\mathbf{x}$ is a column n-vector, then $\beta\mathbf{x}$ is also a column n-vector.

2 The operation is distributive in the sense, for α and β scalars,

$$(\alpha + \beta)\mathbf{x} = \alpha\mathbf{x} + \beta\mathbf{x} \tag{13}$$

$$\alpha(\mathbf{x} + \mathbf{y}) = \alpha\mathbf{x} + \alpha\mathbf{y} \tag{14}$$

3 The operation is associative, i.e.,

$$\alpha(\beta\mathbf{x}) = (\alpha\beta)\mathbf{x} \tag{15}$$

4 There is an identity; i.e.,

$$1\mathbf{x} = \mathbf{x} \tag{16}$$

The above properties can be directly verified from the given definition (12).

We defined the operation of multiplication by Eq. (12). This is often called premultiplication of a vector by a constant. We define postmultiplication in the same way, i.e.,

$$\mathbf{x}b = \begin{bmatrix} x_1 \\ x_2 \\ \vdots \\ x_n \end{bmatrix} b = \begin{bmatrix} x_1 b \\ x_2 b \\ \vdots \\ x_n b \end{bmatrix} = \begin{bmatrix} bx_1 \\ bx_2 \\ \vdots \\ bx_n \end{bmatrix} = b\begin{bmatrix} x_1 \\ x_2 \\ \vdots \\ x_n \end{bmatrix} = b\mathbf{x} \tag{17}$$

Thus, it makes no difference whether we pre- or postmultiply a vector with a scalar; the resulting vector is the same. Thus,

$$\alpha \mathbf{x} = \mathbf{x} \alpha \tag{18}$$

A.3 LINEAR DEPENDENCE AND INDEPENDENCE

Consider a set of vectors $\mathbf{x}_1, \mathbf{x}_2, \ldots, \mathbf{x}_m$ in R_n. Let us construct a vector $\mathbf{x}$ in R_n in the following way:

$$\mathbf{x} = c_1 \mathbf{x}_1 + c_2 \mathbf{x}_2 + \cdots + c_m \mathbf{x}_m = \sum_{i=1}^{m} c_i \mathbf{x}_i \tag{1}$$

where $c_1, c_2, \ldots, c_m$ are scalars. We can see that since each vector $c_i \mathbf{x}_i$ is a vector in R_n, so is the vector $\mathbf{x}$. If $\mathbf{x}$ is a vector defined by Eq. (1), we say that the vector $\mathbf{x}$ is a *linear combination* of the vectors $\mathbf{x}_i$, $i = 1, 2, \ldots, m$.

A set of vectors $\mathbf{x}_1, \mathbf{x}_2, \ldots, \mathbf{x}_m$ is said to be *linearly dependent* if

$$c_1 \mathbf{x}_1 + c_2 \mathbf{x}_2 + \cdots + c_m \mathbf{x}_m = \mathbf{0} \tag{2}$$

and there is at least one scalar $c_i \neq 0$. If, on the other hand, Eq. (2) implies that all the c_i are zero, i.e., Eq. (2) holds if and only if

$$c_i = 0 \qquad \text{for all } i = 1, 2, \ldots, m \tag{3}$$

then we say that the set of vectors $\mathbf{x}_1, \mathbf{x}_2, \ldots, \mathbf{x}_m$ is *linearly independent*.

EXAMPLE 1   Consider the three column 2-vectors

$$\mathbf{x}_1 = \begin{bmatrix} 1 \\ 2 \end{bmatrix} \qquad \mathbf{x}_2 = \begin{bmatrix} 0 \\ 1 \end{bmatrix} \qquad \mathbf{x}_3 = \begin{bmatrix} 2 \\ 4 \end{bmatrix} \tag{4}$$

We want to find whether they are linearly dependent. To do so we form the sum $c_1 \mathbf{x}_1 + c_2 \mathbf{x}_2 + c_3 \mathbf{x}_3 = \mathbf{0}$, which yields

$$\begin{bmatrix} c_1 \\ 2c_1 \end{bmatrix} + \begin{bmatrix} 0 \\ c_2 \end{bmatrix} + \begin{bmatrix} 2c_3 \\ 4c_3 \end{bmatrix} = \begin{bmatrix} 0 \\ 0 \end{bmatrix} = \begin{bmatrix} c_1 + 2c_3 \\ 2c_1 + c_2 + 4c_3 \end{bmatrix}$$

We want to see whether we can find nonzero values of c_1 or c_2 or c_3 such that the equality holds. This can be done by choosing $c_2 = 0$, $c_1 = 2$, and $c_3 = -1$. Thus, according to our definition, the three column vectors of Eq. (4) are linearly dependent.

There is a close relation between the number n of the components of each vector $\mathbf{x}_i$ and the number m of the set of vectors $\mathbf{x}_1, \mathbf{x}_2, \ldots, \mathbf{x}_m$ whose linear dependence or independence we are considering. We investigate this relation later. For the time being we wish to examine a very useful set of vectors which is linearly independent. This set is composed of the n column n-vectors $\mathbf{e}_1, \mathbf{e}_2, \ldots, \mathbf{e}_n$, called the *natural basis vectors*, defined by

$$\mathbf{e}_1 = \begin{bmatrix} 1 \\ 0 \\ \vdots \\ 0 \end{bmatrix} \qquad \mathbf{e}_2 = \begin{bmatrix} 0 \\ 1 \\ \vdots \\ 0 \end{bmatrix}, \ldots, \mathbf{e}_n = \begin{bmatrix} 0 \\ 0 \\ \vdots \\ 1 \end{bmatrix} \tag{5}$$

Note that the number n of vectors in our set is the same as the number of components of each vector $\mathbf{e}_i$. To show that this set is a linearly independent set we observe that

$$\sum_{i=1}^{n} c_i \mathbf{e}_i = \begin{bmatrix} c_1 \\ c_2 \\ \vdots \\ c_n \end{bmatrix} = \begin{bmatrix} 0 \\ 0 \\ \vdots \\ 0 \end{bmatrix} \tag{6}$$

implies that $c_1 = c_2 = \cdots = c_n = 0$. Thus, the vectors $\mathbf{e}_1, \mathbf{e}_2, \ldots, \mathbf{e}_n$ are linearly independent.

A.4   GEOMETRIC INTERPRETATION

We shall now examine some elementary geometric properties of vectors. To do this we observe that when we deal with a column vector

$$\mathbf{x} = \begin{bmatrix} x_1 \\ x_2 \\ \vdots \\ x_n \end{bmatrix} \tag{1}$$

where $x_1, x_2, \ldots, x_n$ are real numbers, we are arranging n numbers in some particular order. We can see that this technique of arranging numbers is similar to the concept of an *ordered n-tuple* $(x_1, x_2, \ldots, x_n)$.

If each scalar x_i is an element of the set of real numbers, which we usually denote by R, then

$$x_i \in R, \qquad x_2 \in R, \ldots, \qquad x_n \in R \tag{2}$$

We can now define the product set

$$R_n \triangleq R \times R \times R \cdots \times R \tag{3}$$

as the set of all ordered n-tuples $(x_1, x_2, \ldots, x_n)$. *We called this set R_n the n-dimensional euclidean space.* A particular example is for $n = 2$, that is, the space R_2, which is the familiar euclidean plane illustrated in Fig. A.4.1.

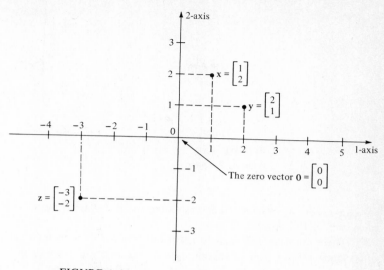

FIGURE A.4.1
The space R_2, that is, the euclidean plane, and several vectors in R_2.

Now let us consider column vectors with two components. For example, let us consider the vectors

$$\mathbf{x} = \begin{bmatrix} 1 \\ 2 \end{bmatrix} \qquad \mathbf{y} = \begin{bmatrix} 2 \\ 1 \end{bmatrix} \qquad \mathbf{z} = \begin{bmatrix} -3 \\ -2 \end{bmatrix}$$

Is it possible to give some geometrical significance to these vectors? This is a matter of convenience, intuition,[†] and above all convention. It is reasonable to agree to represent vectors with two components as *points* in the euclidean plane R_2. To do this we use the first component of each vector to measure a distance along the 1 axis and the second component to measure a distance along the 2 axis. In this manner, we can visualize $\mathbf{x}$, $\mathbf{y}$, and $\mathbf{z}$ as illustrated in Fig. A.4.1. Thus, observe that each vector in R_2 has a distinct identity as a point in the plane. In point of fact although $\mathbf{x}$ and $\mathbf{y}$ have the same numbers (1 and 2) as components, $\mathbf{x}$ and $\mathbf{y}$ represent distinct points.

This geometric representation is not limited to vectors with two components. If we consider vectors with three components, e.g.,

$$\mathbf{x} = \begin{bmatrix} 2 \\ 1 \\ 1 \end{bmatrix} \qquad \mathbf{y} = \begin{bmatrix} 1 \\ 2 \\ 1 \end{bmatrix}$$

---

† It is highly unlikely that geometry is popular on a planet where all men are blind.

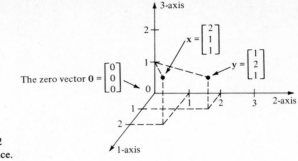

FIGURE A.4.2
Point in R_3 space.

we can agree to represent them as points in the 3-dimensional euclidean space R_3, as illustrated in Fig. A.4.2.

If we are daring enough to carry our physical and geometrical intuition into higher dimensions, we can think of column n-vectors as points in the n-dimensional euclidean space R_n. As long as we follow reasonable intuition and logic, we can generalize and enhance our geometric intuition to spaces of arbitrary dimension.

Another popular geometric representation of vectors consists of

1 Locating a point as above
2 Drawing from the origin **0** to the point an arrow pointing in the direction from the origin to the point

This representation is illustrated in Fig. A.4.3 for the vectors

$$\mathbf{x} = \begin{bmatrix} 1 \\ 2 \end{bmatrix} \qquad \mathbf{y} = \begin{bmatrix} 3 \\ 1 \end{bmatrix}$$

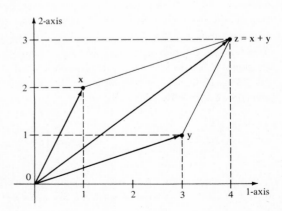

FIGURE A.4.3
The directed-arrow representation of vectors and the parallelogram method of addition.

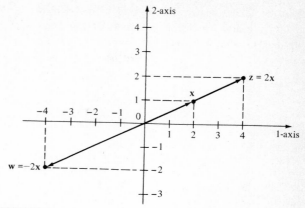

FIGURE A.4.4
Graphical interpretation of multiplication of a vector by a scalar.

Indeed if we let the vector $\mathbf{z}$ be the sum of the vectors $\mathbf{x}$ and $\mathbf{y}$

$$\mathbf{z} = \mathbf{x} + \mathbf{y} = \begin{bmatrix} 1 \\ 2 \end{bmatrix} + \begin{bmatrix} 3 \\ 1 \end{bmatrix} = \begin{bmatrix} 4 \\ 3 \end{bmatrix}$$

we see in Fig. A.4.3 the familiar rule for adding directed-arrow vectors. Of course, one can give a similar interpretation to the operation of vector multiplication by a constant. This concept is illustrated in Fig. A.4.4, where we represent the vectors

$$\mathbf{x} = \begin{bmatrix} 2 \\ 1 \end{bmatrix} \qquad \mathbf{z} = 2\mathbf{x} = \begin{bmatrix} 4 \\ 2 \end{bmatrix} \qquad \mathbf{w} = -2\mathbf{x} = \begin{bmatrix} -4 \\ -2 \end{bmatrix}$$

Multiplication of a vector by a positive real number changes the magnitude but not the direction of the vector.

A.5 MATRICES AND THEIR BASIC ARITHMETIC OPERATIONS

A matrix is defined by a set of scalars a_{ij}, where $i = 1, 2, \ldots, n$ and $j = 1, 2, \ldots, m$, arranged in the following way:

$$\mathbf{A} = \begin{bmatrix} a_{11} & a_{12} & \cdots & a_{1m} \\ a_{21} & a_{22} & \cdots & a_{2m} \\ \vdots & \vdots & \vdots & \vdots \\ a_{n1} & a_{n2} & \cdots & a_{nm} \end{bmatrix} \tag{1}$$

Such a matrix is said to have n rows and m columns. We then say that $\mathbf{A}$ is an $n \times m$ matrix.

Thus an $n \times m$ matrix has a total of *nm elements* a_{ij}. We denote matrices by uppercase boldface letters, for example, $\mathbf{A}$, $\mathbf{B}$, $\mathbf{\Phi}$. The elements a_{ij} of a matrix $\mathbf{A}$ may be real or complex; we then say that $\mathbf{A}$ is a real or complex matrix. Thus, the matrix

$$\mathbf{A} = \begin{bmatrix} 1 & 2 & 0 \\ -1 & 0 & 3 \\ 5 & 10 & -4 \\ 2 & 2 & 2 \end{bmatrix} \tag{2}$$

is a real 4×3 matrix while the matrix

$$\mathbf{B} = \begin{bmatrix} 1+j2 & 2-j3 & 5 \\ j9 & -j7 & 0 \end{bmatrix} \tag{3}$$

is a complex 2×3 matrix.

Two matrices $\mathbf{A}$ and $\mathbf{B}$ are equal, written $\mathbf{A} = \mathbf{B}$, if and only if

1   They both have the same number of rows (say n) and the same number of columns (say m).
2   For each $i, i = 1, 2, \ldots, n$, and each $j, j = 1, 2, \ldots, m$,

$$a_{ij} = b_{ij} \tag{4}$$

Associated with every matrix $\mathbf{A}$ we have its *transpose matrix* $\mathbf{A}'$. If $\mathbf{A}$ is the matrix given by Eq. (1), then its transpose $\mathbf{A}'$ is given by

$$\mathbf{A}' = \begin{bmatrix} a_{11} & a_{21} & \cdots & a_{n1} \\ a_{12} & a_{22} & \cdots & a_{n2} \\ \vdots & \vdots & \vdots & \vdots \\ a_{1m} & a_{2m} & \cdots & a_{nm} \end{bmatrix} \tag{5}$$

Thus, to form $\mathbf{A}'$ from $\mathbf{A}$ one uses the first column of $\mathbf{A}$ as the first row of $\mathbf{A}'$, the second column of $\mathbf{A}$ as the second row of $\mathbf{A}'$ and so on. Thus, if $\mathbf{A}$ is an $n \times m$ matrix, $\mathbf{A}'$ is an $m \times n$ matrix. For example,

$$\mathbf{A} = \begin{bmatrix} 1 & 2 & 3 \\ 4 & 5 & 6 \end{bmatrix} \qquad \mathbf{A}' = \begin{bmatrix} 1 & 4 \\ 2 & 5 \\ 3 & 6 \end{bmatrix}$$

It should be clear from the above construction that for any matrix $\mathbf{A}$

$$(\mathbf{A}')' = \mathbf{A} \tag{6}$$

i.e., the transpose of the transpose of $\mathbf{A}$ is $\mathbf{A}$ itself.

We now proceed with the definition of three matrix operations, addition of matrices, multiplication of a matrix by a scalar, and multiplication of two matrices.

A.5.1 Matrix Addition

Let $\mathbf{A}$ and $\mathbf{B}$ be $n \times m$ matrices with respective elements a_{ij} and b_{ij}, $i = 1, 2, \ldots, n$ and $j = 1, 2, \ldots, m$. Then the sum of $\mathbf{A}$ and $\mathbf{B}$ is an $n \times m$ matrix, say $\mathbf{C}$, with elements c_{ij} such that

$$c_{ij} = a_{ij} + b_{ij} \tag{7}$$

In other words,

$$
\mathbf{A} + \mathbf{B} =
\begin{bmatrix}
a_{11} & a_{12} & \cdots & a_{1m} \\
a_{21} & a_{22} & \cdots & a_{2m} \\
\vdots & \vdots & \vdots & \vdots \\
a_{n1} & a_{n2} & \cdots & a_{nm}
\end{bmatrix}
+
\begin{bmatrix}
b_{11} & b_{12} & \cdots & b_{1m} \\
b_{21} & b_{22} & \cdots & b_{2m} \\
\vdots & \vdots & \vdots & \vdots \\
b_{n1} & b_{n2} & \cdots & b_{nm}
\end{bmatrix}
$$

$$
\begin{bmatrix}
a_{11} + b_{11} & a_{12} + b_{12} & \cdots & a_{1m} + b_{1m} \\
a_{21} + b_{21} & a_{22} + b_{22} & \cdots & a_{2m} + b_{2m} \\
\vdots & \vdots & \vdots & \vdots \\
a_{n1} + b_{n1} & a_{n2} + b_{n2} & \cdots & a_{nm} + b_{nm}
\end{bmatrix}
= \mathbf{C} \tag{8}
$$

For example,

$$
\begin{bmatrix}
1 & 2 & 3 \\
4 & -5 & 6
\end{bmatrix}
+
\begin{bmatrix}
3 & 0 & -1 \\
1 & 4 & -3
\end{bmatrix}
=
\begin{bmatrix}
4 & 2 & 2 \\
5 & -1 & 3
\end{bmatrix}
\tag{9}
$$

Thus to add two matrices, add their corresponding elements to obtain the corresponding element in the matrix of the sum.

The operation of matrix addition has the following important properties:

1 The set of all $n \times m$ matrices is *closed* under addition, i.e., the sum of two $n \times m$ matrices is an $n \times m$ matrix. Note that one cannot add, say, a 2×3 matrix to a 3×2 matrix.

2 Matrix addition is *commutative*, i.e.,

$$\mathbf{A} + \mathbf{B} = \mathbf{B} + \mathbf{A} \tag{10}$$

3 Matrix addition is *associative*, i.e.,

$$(\mathbf{A} + \mathbf{B}) + \mathbf{C} = \mathbf{A} + (\mathbf{B} + \mathbf{C}) \tag{11}$$

4 There is a 0-matrix, denoted by $\mathbf{0}$ and defined as a matrix having all its elements equal to zero, such that

$$\mathbf{A} + \mathbf{0} = \mathbf{A} \tag{12}$$

5 If $+$ is the ordinary sum, then there is a unique matrix $-\mathbf{A}$ associated with $\mathbf{A}$ such that

$$\mathbf{A} + (-\mathbf{A}) = \mathbf{0} \tag{13}$$

If a_{ij} are the elements of **A** and if α_{ij} are the elements of $(-\mathbf{A})$ then

$$\alpha_{ij} = -a_{ij} \tag{14}$$

for all $i = 1, 2, \ldots, n$ and $j = 1, 2, \ldots, m$.

A.5.2 Multiplication of a Matrix by a Scalar

Let us consider an $n \times m$ matrix **A** with elements a_{ij}. Let β be a scalar. Then the multiplication of the matrix **A** by the scalar β yields an $n \times m$ matrix, say **B**, whose elements b_{ij} are given by

$$b_{ij} = \beta a_{ij} \tag{15}$$

In other words,

$$\mathbf{B} = \beta\mathbf{A} = \mathbf{A}\beta = \beta\begin{bmatrix} a_{11} & a_{12} & \cdots & a_{1m} \\ a_{21} & a_{22} & \cdots & a_{2m} \\ \vdots & \vdots & \vdots & \vdots \\ a_{n1} & a_{n2} & \cdots & a_{nm} \end{bmatrix} = \begin{bmatrix} \beta a_{11} & \beta a_{12} & \cdots & \beta a_{1m} \\ \beta a_{21} & \beta a_{22} & \cdots & \beta a_{2m} \\ \vdots & \vdots & \vdots & \vdots \\ \beta a_{n1} & \beta a_{n2} & \cdots & \beta a_{nm} \end{bmatrix} \tag{16}$$

This operation has the following properties:

1 The set of all $n \times m$ matrices is *closed* under multiplication by a scalar; i.e., if **A** is an $n \times m$ matrix and β is a scalar, then $\beta\mathbf{A} = \mathbf{A}\beta$ is an $n \times m$ matrix.

2 The operation is *distributive*; i.e., for scalars α and β and matrices **A** and **B**

$$(\alpha + \beta)\mathbf{A} = \alpha\mathbf{A} + \beta\mathbf{A} \tag{17}$$

$$\alpha(\mathbf{A} + \mathbf{B}) = \alpha\mathbf{A} + \alpha\mathbf{B} \tag{18}$$

3 The operation is *associative*; i.e.,

$$\alpha(\beta\mathbf{A}) = (\alpha\beta)\mathbf{A} \tag{19}$$

4 There is an *identity* 1; that is,

$$1\mathbf{A} = \mathbf{A} \tag{20}$$

These properties are immediate consequences of Eq. (16).

EXAMPLE 1

$$2\begin{bmatrix} 1 & 2 & 3 \\ 4 & 5 & 6 \end{bmatrix} = \begin{bmatrix} 2 & 4 & 6 \\ 8 & 10 & 12 \end{bmatrix}$$

A.5.3 Multiplication of Two Matrices

In multiplying one matrix by another certain ground rules must be observed if the product of two matrices is to make sense.

Let **A** be an $n \times m$ matrix with elements a_{ij}. Let **B** be an $m \times q$ matrix with elements

b_{jk}. Thus, the three indices i, j, and k take the values

$$i = 1, 2, \ldots, n \qquad j = 1, 2, \ldots, m \qquad k = 1, 2, \ldots, q \qquad (21)$$

We define the product of **A** with **B** to be an $n \times q$ matrix **C** with elements c_{ik} given by

$$c_{ik} = \sum_{j=1}^{m} a_{ij} b_{jk} \qquad (22)$$

for each i and each k. In matrix notation, the product is written

$$\mathbf{C} = \mathbf{AB} \qquad (23)$$

It is very important to realize that in order for the product **AB** to make sense *the number of columns of* **A** *(which is m) must be equal to the number of rows of* **B** *(which is m). Then the product matrix* **C** = **AB** *has the same number (n) of rows as the first matrix* **A** *and the same number of columns (q) as the second matrix* **B**. To emphasize this point we have

If **A** is an $n \times m$ matrix, and if **B** is an $p \times q$ matrix,

AB makes sense if and only if $m = p$ $\qquad (24)$

BA makes sense if and only if $q = n$

Let us illustrate the meaning of Eq. (22) by the following example. Let **A** be a 2×2 matrix and let **B** be a 2×3 matrix:

$$\mathbf{A} = \begin{bmatrix} a_{11} & a_{12} \\ a_{21} & a_{22} \end{bmatrix} \qquad \mathbf{B} = \begin{bmatrix} b_{11} & b_{12} & b_{13} \\ b_{21} & b_{22} & b_{23} \end{bmatrix} \qquad (25)$$

Since the number of columns (two) of **A** is the same as the number of rows (two) of **B**, the product **AB** is defined and is a 2×3 matrix:

$$\begin{bmatrix} a_{11} & a_{12} \\ a_{21} & a_{22} \end{bmatrix} \begin{bmatrix} b_{11} & b_{12} & b_{13} \\ b_{21} & b_{22} & b_{23} \end{bmatrix} = \begin{bmatrix} a_{11} b_{11} + a_{12} b_{12} & a_{11} b_{12} + a_{12} b_{22} & a_{11} b_{13} + a_{12} b_{23} \\ a_{21} b_{11} + a_{22} b_{21} & a_{21} b_{12} + a_{22} b_{22} & a_{21} b_{13} + a_{22} b_{23} \end{bmatrix} \qquad (26)$$

$$\underbrace{}_{A} \quad \underbrace{}_{B} \qquad \underbrace{}_{C}$$

Thus, for example, the element c_{12} of **C** is

$$c_{12} = a_{11} b_{12} + a_{12} b_{22} = \sum_{j=1}^{2} a_{1j} b_{j2} \qquad (27)$$

while the element c_{21} is

$$c_{21} = a_{21} b_{11} + a_{22} b_{21} = \sum_{j=1}^{2} a_{2j} b_{j1} \qquad (28)$$

Under our rules the product **BA** is not defined.

EXAMPLE 2

$$\begin{bmatrix} 1 & 2 & 3 \\ 4 & 5 & 6 \end{bmatrix} \begin{bmatrix} 1 & -1 \\ 0 & 2 \\ 2 & 3 \end{bmatrix} = \begin{bmatrix} 7 & 12 \\ 16 & 24 \end{bmatrix}$$

If a matrix A has the same number, say n, of rows and vectors, we say that A is a *square* $n \times n$ matrix. If both A and B are square $n \times n$ matrices, then the two products

$$C = AB \qquad D = BA \tag{29}$$

are well defined and both C and D are square $n \times n$ matrices. The question that arises is: Are AB and BA related? In general,

$$AB \neq BA \tag{30}$$

and this fact is often stated as: A and B do not commute. To see this we note that according to the definition (22),

$$c_{ik} = \sum_{j=1}^{n} a_{ij} b_{jk} \tag{31}$$

$$d_{ik} = \sum_{j=1}^{n} b_{ij} a_{jk} \tag{32}$$

and, in general, $c_{ik} \neq d_{ik}$.

EXAMPLE 3 To illustrate that matrices do not commute, we consider the 2×2 matrices

$$A = \begin{bmatrix} 1 & 2 \\ 3 & 4 \end{bmatrix} \qquad B = \begin{bmatrix} 4 & 5 \\ 6 & 7 \end{bmatrix}$$

Then

$$AB = \begin{bmatrix} 16 & 19 \\ 36 & 43 \end{bmatrix} \qquad BA = \begin{bmatrix} 19 & 28 \\ 27 & 40 \end{bmatrix}$$

so that $AB \neq BA$.

The operation of matrix multiplication has the following properties:

1 Matrix multiplication is *associative*; i.e.,

$$A(BC) = (AB)C \tag{33}$$

2 Matrix multiplication is *distributive*; i.e.,

$$A(B + C) = AB + AC \tag{34}$$

$$(A + B)C = AC + BC \tag{35}$$

3 If A is an $n \times m$ matrix, then there exists an $n \times n$ *identity* matrix I_n such that

$$I_n A = A \tag{36}$$

and an $m \times m$ identity matrix $\mathbf{I}_m$ such that

$$\mathbf{A}\mathbf{I}_m = \mathbf{A} \tag{37}$$

In general the $q \times q$ identity matrix $\mathbf{I}_q$ is given by

$$\mathbf{I}_q = \begin{bmatrix} 1 & 0 & 0 & \cdots & 0 \\ 0 & 1 & 0 & \cdots & 0 \\ 0 & 0 & 1 & \cdots & 0 \\ \vdots & \vdots & \vdots & \vdots & \vdots \\ 0 & 0 & 0 & \cdots & 1 \end{bmatrix} \tag{38}$$

i.e., a matrix with 1's along the main diagonal and zeros everywhere else.

A.5.4 Multiplication of Vectors and Matrices

In Sec. A.2 we defined a column n-vector, say $\mathbf{x}$, by

$$\mathbf{x} = \begin{bmatrix} x_1 \\ x_2 \\ \vdots \\ x_n \end{bmatrix} \tag{39}$$

A column n-vector can also be thought as an $n \times 1$ matrix, i.e., a matrix with n rows and one column. On the other hand, the transpose vector $\mathbf{x}'$ is given by the row vector

$$\mathbf{x}' = [x_1 \quad x_2 \quad \cdots \quad x_n] \tag{40}$$

Clearly, the vector (40) can be interpreted as a matrix with one row and n columns; that is, $\mathbf{x}'$ is a $1 \times n$ matrix.

Since both column and row vectors can be interpreted as matrices, we can follow the rules for matrix multiplication to obtain additional results when we deal with vectors.

First let us consider the problem of postmultiplying an $n \times m$ matrix $\mathbf{A}$ with a column m-vector $\mathbf{x}$. Since $\mathbf{x}$ is interpreted as an $m \times 1$ matrix, it follows that the product $\mathbf{A}\mathbf{x}$ should be an $n \times 1$ matrix [because $(n \times m) \times (m \times 1) = n \times 1$], which in turn is simply an n-vector. Thus, if we let

$$\mathbf{A}\mathbf{x} = \mathbf{y} \tag{41}$$

or, more fully,

$$\underbrace{\begin{bmatrix} a_{11} & a_{12} & \cdots & a_{1m} \\ a_{21} & a_{22} & \cdots & a_{2m} \\ \vdots & \vdots & \vdots & \vdots \\ a_{n1} & a_{n2} & \cdots & a_{nm} \end{bmatrix}}_{\mathbf{A}} \underbrace{\begin{bmatrix} x_1 \\ x_2 \\ \vdots \\ x_m \end{bmatrix}}_{\mathbf{x}} = \underbrace{\begin{bmatrix} y_1 \\ y_2 \\ \vdots \\ y_m \end{bmatrix}}_{\mathbf{y}} \tag{42}$$

then the components y_i of the vector $\mathbf{y}$ are given, in view of Eq. (22), by

$$y_i = \sum_{j=1}^{m} a_{ij} x_j \qquad i = 1, 2, \ldots, n \tag{43}$$

As a general example suppose $\mathbf{A}$ is a 3×2 matrix and that $\mathbf{x}$ is a column 2-vector. Then

$$\underbrace{\begin{bmatrix} a_{11} & a_{12} \\ a_{21} & a_{22} \\ a_{31} & a_{32} \end{bmatrix}}_{\mathbf{A}} \underbrace{\begin{bmatrix} x_1 \\ x_2 \end{bmatrix}}_{\mathbf{x}} = \underbrace{\begin{bmatrix} a_{11} x_1 + a_{12} x_2 \\ a_{21} x_1 + a_{22} x_2 \\ a_{31} x_1 + a_{32} x_2 \end{bmatrix}}_{\mathbf{y}} \tag{44}$$

EXAMPLE 4

$$\begin{bmatrix} 0 & 1 & 2 & 3 \\ 0 & 0 & 4 & 1 \end{bmatrix} \begin{bmatrix} 4 \\ 3 \\ 0 \\ 1 \end{bmatrix} = \begin{bmatrix} 6 \\ 1 \end{bmatrix}$$

Now we consider the multiplication of a row vector with a matrix. Suppose that $\mathbf{z}$ is a row n-vector (or a $1 \times n$ matrix)

$$\mathbf{z} = [z_1 \quad z_2 \quad \cdots \quad z_n]$$

Let $\mathbf{A}$ be an $n \times m$ matrix. Then the product $\mathbf{zA}$ will be a $1 \times m$ matrix [since $(1 \times n) \times (n \times m) = 1 \times m$] or, equivalently, a row vector, say $\mathbf{w}$, with m elements. Thus,

$$\mathbf{zA} = \mathbf{w} \tag{45}$$

or, more fully,

$$\underbrace{\begin{bmatrix} z_1 & z_2 & \cdots & z_n \end{bmatrix}}_{\mathbf{z}} \underbrace{\begin{bmatrix} a_{11} & a_{12} & \cdots & a_{1m} \\ a_{21} & a_{22} & \cdots & a_{2m} \\ \vdots & \vdots & \vdots & \vdots \\ a_{n1} & a_{n2} & \cdots & a_{nm} \end{bmatrix}}_{\mathbf{A}} = \underbrace{\begin{bmatrix} w_1 & w_2 & \cdots & w_m \end{bmatrix}}_{\mathbf{w}} \tag{46}$$

where

$$w_j = \sum_{i=1}^{n} a_{ij} z_i \qquad j = 1, 2, \ldots, m \tag{47}$$

EXAMPLE 5

$$[1 \quad 2 \quad -1] \begin{bmatrix} 1 & -2 \\ 2 & 4 \\ 0 & 3 \end{bmatrix} = [5 \quad 3]$$

We conclude this section with the results obtained when one multiplies a row vector by a column vector and vice versa.

Suppose that $\mathbf{x}$ is a row vector with n components, i.e., a $1 \times n$ matrix, and that $\mathbf{y}$ is a column n-vector, i.e., an $n \times 1$ matrix. Then the product

$$\begin{bmatrix} x_1 & x_2 & \cdots & x_n \end{bmatrix} \begin{bmatrix} y_1 \\ y_2 \\ \vdots \\ y_n \end{bmatrix} = \sum_{i=1}^{n} x_i y_i \tag{48}$$

is a 1×1 matrix, i.e., a *scalar*. Indeed note that if $\mathbf{x}$ and $\mathbf{y}$ are both column n-vectors,

$$\mathbf{x} = \begin{bmatrix} x_1 \\ x_2 \\ \vdots \\ x_n \end{bmatrix} \qquad \mathbf{y} = \begin{bmatrix} y_1 \\ y_2 \\ \vdots \\ y_n \end{bmatrix} \tag{49}$$

then Eq. (48) takes the form

$$\mathbf{x}'\mathbf{y} = \sum_{i=1}^{n} x_i y_i \tag{50}$$

because $\mathbf{x}'$ is a row vector.

Finally, the product

$$\begin{bmatrix} y_1 \\ y_2 \\ \vdots \\ y_n \end{bmatrix} \begin{bmatrix} x_1 & x_2 & \cdots & x_n \end{bmatrix} = \begin{bmatrix} y_1 x_1 & y_1 x_2 & \cdots & y_1 x_n \\ y_2 x_1 & y_2 x_2 & \cdots & y_2 x_n \\ \vdots & \vdots & \vdots & \vdots \\ y_n x_1 & y_n x_2 & \cdots & y_n x_n \end{bmatrix} \tag{51}$$

yields an $n \times n$ matrix. If $\mathbf{x}$ and $\mathbf{y}$ are the column vectors of Eq. (49), the left-hand side of Eq. (51) can also be written

$$\mathbf{y}\mathbf{x}' \tag{52}$$

which is, of course, equal to the $n \times n$ matrix of Eq. (51). Often, if $\mathbf{x}$ and $\mathbf{y}$ are both n-column vectors, $\mathbf{x}'\mathbf{y}$ is called their *inner* (or scalar) product while the $n \times n$ matrix $\mathbf{x}\mathbf{y}'$ is called their *outer* product.

A.6 GENERAL DEFINITIONS REGARDING SQUARE MATRICES

Let us suppose that $\mathbf{A}$ is an $n \times n$ matrix; we have called these matrices *square* as they have the same number n of rows and columns. Recall that the transpose matrix $\mathbf{A}'$ of $\mathbf{A}$ is a matrix obtained by interchanging the rows and columns.

Definition 1   *If*

$$A = A' \tag{1}$$

then **A** *is called a symmetric matrix.*

Definition 2   *If*

$$A = -A' \tag{2}$$

then **A** *is called a skew-symmetric matrix.*

For example, the matrices

$$\begin{bmatrix} 1 & 2 & 3 \\ 2 & 5 & -4 \\ 3 & -4 & 0 \end{bmatrix} \quad \begin{bmatrix} 2+j5 & 3+j \\ 3+j & -6+j2 \end{bmatrix} \quad \begin{bmatrix} 0 & 1 \\ 1 & 0 \end{bmatrix}$$

are symmetric. The matrices

$$\begin{bmatrix} 0 & 1 \\ -1 & 0 \end{bmatrix} \quad \begin{bmatrix} 0 & 3+j5 \\ -3-j5 & 0 \end{bmatrix}$$

are skew-symmetric.

Definition 3   *If* **A** *is a matrix with elements* a_{ij} *which are complex and* **A*** *is the matrix with elements* a_{ij}^{*} *(where * denotes complex conjugate), and if*

$$A = A^{*\prime} \tag{3}$$

(read as complex-conjugate transpose), then we say that **A** *is a hermitian matrix.*

For example, the matrices

$$\begin{bmatrix} 1 & 2 & 3 \\ 2 & 2 & 4 \\ 3 & 4 & 5 \end{bmatrix} \quad \begin{bmatrix} 5 & 3+j2 \\ 3-j2 & 7 \end{bmatrix}$$

are hermitian.

Given a square matrix, its main diagonal is defined by the elements $a_{11}, a_{22}, \ldots, a_{nn}$

$$A = \begin{bmatrix} a_{11} & a_{12} & \cdots & a_{1n} \\ a_{21} & a_{22} & \cdots & a_{2n} \\ \vdots & \vdots & \ddots & \vdots \\ a_{n1} & a_{n2} & \cdots & a_{nn} \end{bmatrix} \begin{array}{l} \\ \\ \\ \text{Main diagonal} \end{array} \tag{4}$$

A square matrix **D** whose elements not on the main diagonal equal zero is called a diagonal matrix. Thus, the general form for **D** is

$$D = \begin{bmatrix} d_{11} & 0 & \cdots & 0 \\ 0 & d_{22} & \cdots & 0 \\ \vdots & \vdots & \vdots & \vdots \\ 0 & 0 & \cdots & d_{nn} \end{bmatrix} \tag{5}$$

The identity matrix

$$I = \begin{bmatrix} 1 & 0 & \cdots & 0 \\ 0 & 1 & \cdots & 0 \\ \vdots & \vdots & \vdots & \vdots \\ 0 & 0 & \cdots & 1 \end{bmatrix} \tag{6}$$

is a special diagonal matrix.

The *determinant*[†] of a square matrix A is a scalar denoted by

$$\det A \tag{7}$$

The *trace* of a square matrix A, denoted by tr A, is also a scalar and is equal to the sum of the elements of A along the main diagonal, i.e.,

$$\text{tr } A = a_{11} + a_{22} + \cdots + a_{nn} = \sum_{i=1}^{n} a_{ii} \tag{8}$$

We leave it to the reader to prove that the following relations hold (provided that each operation makes sense):

$$\det(A + B) \neq \det A + \det B \tag{9}$$
$$\det AB = \det A \det B \tag{10}$$
$$\det AB = \det BA \tag{11}$$
$$\det A = \det A' \tag{12}$$
$$\text{tr}(A + B) = \text{tr } A + \text{tr } B \tag{13}$$
$$\text{tr } AB \neq \text{tr } A \text{ tr } B \tag{14}$$
$$\text{tr } AB = \text{tr } BA \tag{15}$$

EXAMPLE 1 Suppose that

$$A = \begin{bmatrix} 1 & 0 & 3 \\ 2 & 1 & 2 \\ 1 & -1 & 4 \end{bmatrix}$$

Then

$$\det A = -3 \qquad \text{tr } A = 1 + 1 + 4 = 6$$

---

† We assume that the student knows how to evaluate determinants.

A.7 THE INVERSE OF A MATRIX

Let **A** be an $n \times n$ matrix. Suppose that

$$\det \mathbf{A} \neq 0 \tag{1}$$

Then **A** is called a *nonsingular* matrix; otherwise, it is called *singular*.

If **A** is a nonsingular $n \times n$ matrix, then there exists a *unique* $n \times n$ nonsingular matrix, denoted by $\mathbf{A}^{-1}$ and called the *inverse* of the **A** matrix. It has the property

$$\mathbf{A}^{-1}\mathbf{A} = \mathbf{A}\mathbf{A}^{-1} = \mathbf{I} \tag{2}$$

where **I** is the $n \times n$ identity matrix.

We remark that

$$(\mathbf{A}^{-1})^{-1} = \mathbf{A} \tag{3}$$

Singular matrices do not have inverses.

We now present a general formula for evaluating the inverse of an $n \times n$ matrix. Let **A** be an $n \times n$ matrix with elements a_{ij} $(i,j = 1, 2, \ldots, n)$. Consider the element a_{ij} which is at the intersection of the ith row and of the jth column. Then the *minor* m_{ij} is defined as the determinant of the $(n-1) \times (n-1)$ matrix obtained from **A** by deleting the ith row and the jth column. The *cofactor* β_{ij} of the element a_{ij} is simply related to the minor by

$$\beta_{ij} = (-1)^{i+j} m_{ij} \tag{4}$$

For example, suppose that

$$\mathbf{A} = \begin{bmatrix} 1 & 3 & 2 \\ 1 & 4 & 3 \\ 1 & 3 & 4 \end{bmatrix}$$

Then, for $i = 3, j = 2$,

$$m_{32} = \det \begin{bmatrix} 1 & 3 & 2 \\ 1 & 4 & 3 \\ 1 & 3 & 4 \end{bmatrix} = \det \begin{bmatrix} 1 & 2 \\ 1 & 3 \end{bmatrix} = 1$$

$$\beta_{32} = (-1)^{3+2} m_{32} = (-1)(1) = -1$$

The inverse matrix $\mathbf{A}^{-1}$ can be computed from all the cofactors by the formula

$$\mathbf{A}^{-1} = \frac{1}{\det \mathbf{A}} \begin{bmatrix} \beta_{11} & \beta_{21} & \cdots & \beta_{n1} \\ \beta_{12} & \beta_{22} & \cdots & \beta_{n2} \\ \vdots & \vdots & \vdots & \vdots \\ \beta_{1n} & \beta_{2n} & \cdots & \beta_{nn} \end{bmatrix} \tag{5}$$

A.7.1   Special Case of a 2×2 matrix

Suppose that

$$\mathbf{A} = \begin{bmatrix} a_{11} & a_{12} \\ a_{21} & a_{22} \end{bmatrix} \qquad \det \mathbf{A} \neq 0 \tag{6}$$

Then

$$\mathbf{A}^{-1} = \begin{bmatrix} \dfrac{a_{22}}{\det \mathbf{A}} & -\dfrac{a_{12}}{\det \mathbf{A}} \\ -\dfrac{a_{21}}{\det \mathbf{A}} & \dfrac{a_{11}}{\det \mathbf{A}} \end{bmatrix} = \dfrac{1}{\det \mathbf{A}} \begin{bmatrix} a_{22} & -a_{12} \\ -a_{21} & a_{11} \end{bmatrix} \tag{7}$$

A.7.2   Some Facts about Inverses

One can verify by several means, in particular using the construction method above, that the following properties are true for nonsingular matrices:

$$(\mathbf{A}')^{-1} = (\mathbf{A}^{-1})'$$

$$(\mathbf{AB})^{-1} = \mathbf{B}^{-1}\mathbf{A}^{-1} \tag{8}$$

$$\mathbf{Ax} = \mathbf{0} \text{ for some } \mathbf{x} \neq \mathbf{0} \text{ implies that } \mathbf{A} \text{ is singular} \tag{9}$$

$$\text{and hence } \det \mathbf{A} = 0 \tag{10}$$

Note that

$$\det \mathbf{AA}^{-1} = \det \mathbf{I} = 1 = \det \mathbf{A} \det \mathbf{A}^{-1} \tag{11}$$

and so

$$\det \mathbf{A}^{-1} = \frac{1}{\det \mathbf{A}} \tag{12}$$

A.8   SIMILAR MATRICES

Let $\mathbf{A}$ and $\mathbf{B}$ be $n \times n$ matrices. If there exists an $n \times n$ nonsingular matrix $\mathbf{P}$ such that

$$\mathbf{B} = \mathbf{P}^{-1}\mathbf{AP} \tag{1}$$

or

$$\mathbf{PB} = \mathbf{AP} \tag{2}$$

or

$$\mathbf{PBP}^{-1} = \mathbf{A} \tag{3}$$

then we say that $\mathbf{A}$ and $\mathbf{B}$ are similar matrices.

We shall prove certain properties of similar matrices after we introduce the notion of eigenvalues and eigenvectors. For the time being, we can show that if $\mathbf{A}$ and $\mathbf{B}$ are similar matrices, then

$$\det \mathbf{B} = \det \mathbf{A} \tag{4}$$

To prove Eq. (4) we note

$$\det \mathbf{B} = \det \mathbf{P}^{-1}\mathbf{AP} = \det \mathbf{P}^{-1}\det \mathbf{A} \det \mathbf{P} = \det \mathbf{A}$$

Also

$$\operatorname{tr} \mathbf{B} = \operatorname{tr} \mathbf{A} \tag{5}$$

because

$$\operatorname{tr} \mathbf{B} = \operatorname{tr} \mathbf{P}^{-1}\mathbf{AP} = \operatorname{tr} \mathbf{APP}^{-1} = \operatorname{tr} \mathbf{AI} = \operatorname{tr} \mathbf{A} \tag{6}$$

A.9 EIGENVALUES AND EIGENVECTORS

In this section we shall discuss one of the most important concepts of square matrices, their eigenvalues and eigenvectors. We have discussed that a vector can be used to define a direction in the euclidean n-dimensional space. Geometrically, an eigenvector of a matrix $\mathbf{A}$ is a vector, say ξ, such that the effect of operating upon ξ with $\mathbf{A}$ is to produce a vector $\mathbf{A}\xi$ which is collinear to ξ; that is, it either points in the same or in the opposite direction. The amount λ by which the eigenvector ξ has expanded or shrunk is called an eigenvalue of the matrix $\mathbf{A}$ corresponding to the eigenvector ξ. In general, a matrix $\mathbf{A}$ has many eigenvectors and hence many eigenvalues. The formal definition follows.

Definition 1 *Suppose that* $\mathbf{A}$ *is an* $n \times n$ *matrix. A nonzero n-column vector* ξ *is called an* eigenvector *of* $\mathbf{A}$ *if there is a scalar* λ *such that*

$$\mathbf{A}\xi = \lambda\xi \tag{1}$$

In this case λ *is called an* eigenvalue *of* $\mathbf{A}$ *corresponding to the eigenvector* ξ.

Eigenvalues are also often called *characteristic values* or *characteristic roots*.

Suppose that ξ is an eigenvector of $\mathbf{A}$ and that λ is the corresponding eigenvalue. By definition, we have

$$\mathbf{A}\xi - \lambda\xi = \mathbf{0} \tag{2}$$

which can be written in terms of components as

$$\sum_{j=1}^{n} a_{ij}\xi_j - \lambda\xi_j = 0 \qquad i = 1, 2, \ldots, n \tag{3}$$

This is simply the system of linear equations

$$
\begin{aligned}
(a_{11} - \lambda)\xi_1 + a_{12}\xi_2 + \cdots + a_{1n}\xi_n &= 0 \\
a_{21}\xi_1 + (a_{22} - \lambda)\xi_2 + \cdots + a_{2n}\xi_n &= 0 \\
&\vdots \\
a_{n1}\xi_1 + a_{n2}\xi_2 + \cdots + (a_{nn} - \lambda)\xi_n &= 0
\end{aligned}
\tag{4}
$$

Equation (4) can be written in the matrix form

$$
\underbrace{\begin{bmatrix}
a_{11} - \lambda & a_{12} & \cdots & a_{1n} \\
a_{21} & a_{22} - \lambda & \cdots & a_{2n} \\
\vdots & \vdots & \vdots & \vdots \\
a_{n1} & a_{n2} & \cdots & a_{nn} - \lambda
\end{bmatrix}}_{\mathbf{A} - \lambda\mathbf{I}}
\underbrace{\begin{bmatrix}
\xi_1 \\ \xi_2 \\ \vdots \\ \xi_n
\end{bmatrix}}_{\xi}
=
\underbrace{\begin{bmatrix}
0 \\ 0 \\ \vdots \\ 0
\end{bmatrix}}_{\mathbf{0}}
\tag{5}
$$

Since $\xi_1, \xi_2, \ldots, \xi_n$ are not all zero, this means that the matrix in Eq. (5) must be singular. In fact, this matrix is

$$
\mathbf{A} - \lambda\mathbf{I}
\tag{6}
$$

and since it is singular, its determinant must be zero [see Eq. (10) of Sec. A.7]. Thus,

$$
\det(\mathbf{A} - \lambda\mathbf{I}) = 0 \quad \text{or} \quad \det(\lambda\mathbf{I} - \mathbf{A}) = 0
\tag{7}
$$

Let us recapitulate the development so far. We have shown that if λ is an eigenvalue of the matrix $\mathbf{A}$, then the relation $\det(\mathbf{A} - \lambda\mathbf{I}) = 0$ must hold. This means that λ must be a root of Eq. (7). We remark that the quantity $\det(\mathbf{A} - \lambda\mathbf{I})$ is simply a polynomial in λ.

Let us see what happens in R_2. In this case, the matrix $\mathbf{A} - \lambda\mathbf{I}$ takes the form

$$
\mathbf{A} - \lambda\mathbf{I} = \begin{bmatrix}
a_{11} - \lambda & a_{12} \\
a_{21} & a_{22} - \lambda
\end{bmatrix}
\tag{8}
$$

so that Eq. (7) reduces to

$$
\det(\mathbf{A} - \lambda\mathbf{I}) = (a_{11} - \lambda)(a_{22} - \lambda) - a_{12}a_{21} = \lambda^2 - (a_{11} + a_{22})\lambda + (a_{11}a_{22} - a_{12}a_{21}) = 0
\tag{9}
$$

In this case the eigenvalue λ is a root of the second-order polynomial in Eq. (9).

It can be shown, in general, that if $\mathbf{A}$ is an $n \times n$ matrix, then

$$
p(z) \triangleq \det(\mathbf{A} - z\mathbf{I}) = (-1)^n z^n + c_1 z^{n-1} + \cdots + c_n
\tag{10}
$$

is an nth-degree polynomial called the *characteristic polynomial of the matrix* $\mathbf{A}$. Thus, *the eigenvalue λ is a root of the characteristic polynomial of* $\mathbf{A}$. Since the characteristic polynomial of an $n \times n$ matrix is of the nth degree, it has n roots, thus, there are n eigenvalues of $\mathbf{A}$ (they need not be distinct).

In general, if **A** is a real matrix, i.e., the element of **A** are real scalars, then the coefficients of its characteristic polynomial are also real. On the other hand, we know that the roots of a polynomial with real coefficients are not necessarily all real. This implies that the eigenvalues $\lambda_1, \lambda_2, \ldots, \lambda_n$ of **A** are not necessarily real; if, however, an eigenvalue, say λ_k, is complex,

$$\lambda_k = \sigma_k + j\omega_k$$

then there must exist another eigenvalue, say λ_i, such that

$$\lambda_i = \sigma_k - j\omega_k = \lambda_k^*$$

In short, any complex-valued eigenvalues must come in complex-conjugate pairs.

Theorem 1 *Similar matrices have the same characteristic polynomial and, therefore, the same eigenvalues.*

PROOF   Suppose **A** and **B** are similar matrices. Then

$$\mathbf{B} = \mathbf{P}^{-1}\mathbf{A}\mathbf{P} \tag{11}$$

Let

$$p_A(z) \triangleq \det(\mathbf{A} - z\mathbf{I}) \tag{12}$$

and

$$p_B(z) \triangleq \det(\mathbf{B} - z\mathbf{I}) \tag{13}$$

be their respective characteristic polynomials. But from Eqs. (11) and (13) we have

$$
\begin{aligned}
p_B(z) &= \det(\mathbf{B} - z\mathbf{I}) = \det(\mathbf{P}^{-1}\mathbf{A}\mathbf{P} - z\mathbf{I}) \\
&= \det(\mathbf{P}^{-1}\mathbf{A}\mathbf{P} - z\mathbf{I}\mathbf{P}^{-1}\mathbf{P}) \\
&= \det[\mathbf{P}^{-1}\mathbf{A}\mathbf{P} - \mathbf{P}^{-1}(z\mathbf{I})\mathbf{P}] \\
&= \det[\mathbf{P}^{-1}(\mathbf{A} - z\mathbf{I})\mathbf{P}] \\
&= (\det \mathbf{P}^{-1})[\det(\mathbf{A} - z\mathbf{I})](\det \mathbf{P}) \\
&= \det(\mathbf{A} - z\mathbf{I}) \triangleq p_A(z)
\end{aligned}
\tag{14}
$$

Once we have shown that similar matrices have the same eigenvalues, we can investigate the relationship between their eigenvectors. Suppose that

$$\mathbf{A}\boldsymbol{\xi} = \lambda\boldsymbol{\xi} \tag{15}$$

$$\mathbf{B}\boldsymbol{\eta} = \lambda\boldsymbol{\eta} \tag{16}$$

so that $\boldsymbol{\xi}$ is an eigenvector of **A** and $\boldsymbol{\eta}$ is an eigenvector of **B**, both associated with the same eigenvalue λ. If **A** and **B** are similar, then

$$\mathbf{B}\eta = \mathbf{P}^{-1}\mathbf{AP}\eta = \lambda\eta \tag{17}$$

or

$$\mathbf{A}(\mathbf{P}\eta) = \lambda(\mathbf{P}\eta) \tag{18}$$

so that $\mathbf{P}\eta$ is the eigenvector of **A** associated with λ. From Eqs. (18) and (15) it is obvious that the eigenvectors ξ and η are related by the nonsingular matrix **P** as follows:

$$\xi = \mathbf{P}\eta \qquad \text{or} \qquad \eta = \mathbf{P}^{-1}\xi \tag{19}$$

Computing the eigenvalues of a matrix is not a trivial task since one must find the roots of an nth-degree (characteristic) polynomial. However, there are many computer subroutines that can be used to compute the eigenvalues of a matrix.

The computation of eigenvectors is not as straightforward because an eigenvector of a matrix is defined to within a scalar multiplier. To see this suppose that ξ is an eigenvector of **A**; then, by definition,

$$\mathbf{A}\xi = \lambda\xi \tag{20}$$

Suppose that α is a nonzero scalar. Then,

$$\alpha\mathbf{A}\xi = \alpha\lambda\xi \tag{21}$$

or,

$$\mathbf{A}(\alpha\xi) = \lambda(\alpha\xi) \tag{22}$$

so that, according to our definition, $\alpha\xi$ is *also* an eigenvector. Thus, the eigenvectors define certain invariant directions in R_n. These are straight lines collinear with an eigenvector.

EXAMPLE 1   In this example we illustrate several of the ideas presented in this section. First, consider the two 2×2 real matrices **A** and **B**

$$\mathbf{A} = \begin{bmatrix} 0 & 1 \\ -2 & 3 \end{bmatrix} \qquad \mathbf{B} = \begin{bmatrix} 1 & 3 \\ 0 & 2 \end{bmatrix}$$

First we claim that these matrices are similar. To do that we must show that there is a nonsingular 2×2 matrix **P** such that $\mathbf{B} = \mathbf{P}^{-1}\mathbf{AP}$. We claim that the matrix

$$\mathbf{P} = \begin{bmatrix} 1 & -1 \\ 1 & 1 \end{bmatrix}$$

whose inverse is

$$\mathbf{P}^{-1} = \frac{1}{2}\begin{bmatrix} 1 & 1 \\ -1 & 1 \end{bmatrix}$$

can be used to verify that **A** and **B** are similar. This verification involves standard matrix multiplication, and as such is left to the reader.

Next we compute the characteristic polynomials $p_A(z)$ and $p_B(z)$ of **A** and **B**, respectively.

$$p_A(z) = \det(\mathbf{A} - z\mathbf{I}) = \det \begin{bmatrix} -z & 1 \\ -2 & 3 - z \end{bmatrix} = -z(3 - z) + 2 = z^2 - 3z + 2$$

$$p_B(z) = \det(\mathbf{B} - z\mathbf{I}) = \det \begin{bmatrix} 1 - z & 3 \\ 0 & 2 - z \end{bmatrix} = (1 - z)(2 - z) = z^2 - 3z + 2$$

Thus, as we expected, $p_A(z) = p_B(z)$. Now the roots of the characteristic polynomial

$$p(z) = z^2 - 3z + 2$$

are the eigenvalues of **A** or **B**. Let the eigenvalues be denoted by λ_1 and λ_2. Clearly, the roots of $p(z)$ are at $z = 1$ and $z = 2$, so that

$$\lambda_1 = 1 \qquad \lambda_2 = 2$$

Now let us compute the eigenvectors, say $\boldsymbol{\xi}^1$ and $\boldsymbol{\xi}^2$, of **A**. If $\boldsymbol{\xi}^1$ is associated with λ_1 and $\boldsymbol{\xi}^2$ is associated with λ_2, we have, by definition,

$$A\boldsymbol{\xi}^1 = 1\boldsymbol{\xi}^1 \qquad \text{and} \qquad A\boldsymbol{\xi}^2 = 2\boldsymbol{\xi}^2$$

Let

$$\boldsymbol{\xi}^1 = \begin{bmatrix} \xi_1^{\,1} \\ \xi_2^{\,1} \end{bmatrix} \qquad \boldsymbol{\xi}^2 = \begin{bmatrix} \xi_1^{\,2} \\ \xi_2^{\,2} \end{bmatrix}$$

Then we have

$$\begin{bmatrix} 0 & 1 \\ -2 & 3 \end{bmatrix} \begin{bmatrix} \xi_1^{\,1} \\ \xi_2^{\,1} \end{bmatrix} = \begin{bmatrix} \xi_1^{\,1} \\ \xi_2^{\,1} \end{bmatrix}$$

so that

$$\xi_2^{\,1} = \xi_1^{\,1}$$
$$-2\xi_1^{\,1} + 3\xi_2^{\,1} = \xi_2^{\,1}$$

From these equations we conclude that any vector with equal components is an eigenvector of **A** corresponding to λ_1. Thus, $\boldsymbol{\xi}^1$ has the general form

$$\boldsymbol{\xi}^1 = \alpha \begin{bmatrix} 1 \\ 1 \end{bmatrix} \qquad \alpha = \text{any real nonzero scalar}$$

Now we turn our attention to $\boldsymbol{\xi}^2$; it is defined by

$$\begin{bmatrix} 0 & 1 \\ -2 & 3 \end{bmatrix} \begin{bmatrix} \xi_1^{\,2} \\ \xi_2^{\,2} \end{bmatrix} = 2 \begin{bmatrix} \xi_1^{\,2} \\ \xi_2^{\,2} \end{bmatrix}$$

or

$$\xi_2^2 = 2\xi_1^2$$
$$-2\xi_1^2 + 3\xi_2^2 = 2\xi_2^2$$

Thus, ξ^2 can be any vector of the form

$$\xi^2 = \beta \begin{bmatrix} 1 \\ 2 \end{bmatrix} \qquad \beta = \text{any real nonzero scalar}$$

EXAMPLE 2

$$A = \begin{bmatrix} 1 & 0 & 1 \\ 2 & 2 & 1 \\ 0 & 0 & 3 \end{bmatrix}$$

To compute the eigenvalues of **A** we form the matrix

$$A - \lambda I = \begin{bmatrix} 1 - \lambda & 0 & 1 \\ 2 & 2 - \lambda & 1 \\ 0 & 0 & 3 - \lambda \end{bmatrix}$$

The characteristic polynomial is

$$p(\lambda) = \det(A - \lambda I) = (1 - \lambda)(2 - \lambda)(3 - \lambda)$$

Hence, the eigenvalues of **A** are

$$\lambda_1 = 1 \qquad \lambda_2 = 2 \qquad \lambda_3 = 3$$

Let **x** be the eigenvector associated with λ_1. Let **y** be the eigenvector associated with λ_2. Let **z** be the eigenvector associated with λ_3. To find **x** we have the equality

$$\begin{bmatrix} 1 & 0 & 1 \\ 2 & 2 & 1 \\ 0 & 0 & 3 \end{bmatrix} \begin{bmatrix} x_1 \\ x_2 \\ x_3 \end{bmatrix} = 1 \begin{bmatrix} x_1 \\ x_2 \\ x_3 \end{bmatrix}$$

which yields

$$x_1 + x_3 = x_1$$
$$2x_1 + 2x_2 + x_3 = x_2$$
$$3x_3 = x_3$$

Hence

$$x_3 = 0$$

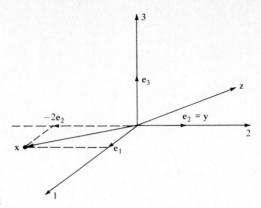

FIGURE A.9.1

If we choose $x_1 = 1$, then $x_2 = -2$. Thus

$$\mathbf{x} = \begin{bmatrix} 1 \\ -2 \\ 0 \end{bmatrix}$$

To find $\mathbf{y}$ we have the equality

$$\begin{bmatrix} 1 & 0 & 1 \\ 2 & 2 & 1 \\ 0 & 0 & 3 \end{bmatrix} \begin{bmatrix} y_1 \\ y_2 \\ y_3 \end{bmatrix} = 2 \begin{bmatrix} y_1 \\ y_2 \\ y_3 \end{bmatrix}$$

which yields

$$y_1 + y_3 = 2y_1$$
$$2y_1 + 2y_2 + y_3 = 2y_2$$
$$3y_3 = 2y_3$$

Hence $y_3 = 0$, $y_1 = 0$, and we can choose $y_2 = 1$. Thus

$$\mathbf{y} = \begin{bmatrix} 0 \\ 1 \\ 0 \end{bmatrix}$$

To find $\mathbf{z}$ we have the equality

$$\begin{bmatrix} 1 & 0 & 1 \\ 2 & 2 & 1 \\ 0 & 0 & 3 \end{bmatrix} \begin{bmatrix} z_1 \\ z_2 \\ z_3 \end{bmatrix} = 3 \begin{bmatrix} z_1 \\ z_2 \\ z_3 \end{bmatrix}$$

which yields

$$z_1 + z_3 = 3z_1$$

$$2z_1 + 2z_2 + z_3 = 3z_2$$

$$3z_3 = 3z_3$$

Hence, we can choose $z_3 = 1$, which implies $z_1 = \frac{1}{2}$, $z_2 = 2$. Therefore

$$\mathbf{z} = \begin{bmatrix} \frac{1}{2} \\ 2 \\ 1 \end{bmatrix}$$

These eigenvectors are illustrated in Fig. A.9.1.

A.10 ADDITIONAL FACTS ABOUT EIGENVALUES, EIGENVECTORS, AND SIMILARITY TRANSFORMATIONS

There is an intimate relation between the eigenvalues and eigenvectors of a matrix and the similarity transformation. The types of result one obtains differ according as the eigenvalues of a matrix are distinct or some (or all) are repeated.

A.10.1 Distinct Eigenvalues

Let $\mathbf{A}$ be an $n \times n$ real matrix. Let $\lambda_1, \lambda_2, \ldots, \lambda_n$ denote the eigenvalues of $\mathbf{A}$, and let $\xi_1, \xi_2, \ldots, \xi_n$ denote its eigenvectors, respectively. We assume that the eigenvalues are distinct; i.e.,

$$\lambda_i \neq \lambda_j \qquad \begin{array}{l} \text{for all } i \neq j \\ i, j = 1, 2, \ldots, n \end{array} \tag{1}$$

Let Λ be the $n \times n$ diagonal matrix which is formed by the eigenvalues of $\mathbf{A}$, that is,

$$\Lambda = \begin{bmatrix} \lambda_1 & 0 & \cdots & 0 \\ 0 & \lambda_2 & \cdots & 0 \\ \vdots & \vdots & \vdots & \vdots \\ 0 & 0 & \cdots & \lambda_n \end{bmatrix} \tag{2}$$

We shall now state some useful theorems whose proofs are left as exercises for the reader.

Theorem 1 *The eigenvalues of Λ are $\lambda_1, \lambda_2, \ldots, \lambda_n$ and the eigenvectors are the natural basis vectors*

$$\mathbf{e}_1 = \begin{bmatrix} 1 \\ 0 \\ \vdots \\ 0 \end{bmatrix}, \quad \mathbf{e}_2 = \begin{bmatrix} 0 \\ 1 \\ \vdots \\ 0 \end{bmatrix}, \ldots, \mathbf{e}_n = \begin{bmatrix} 0 \\ 0 \\ \vdots \\ 1 \end{bmatrix} \tag{3}$$

Theorem 2   *The matrix* **A** *and the diagonal eigenvalue matrix* **Λ** *are similar matrices, i.e., there exists a nonsingular* $n \times n$ *matrix* **P** *such that*

$$\Lambda = \mathbf{P}^{-1}\mathbf{AP} \quad or \quad \mathbf{A} = \mathbf{P}\Lambda\mathbf{P}^{-1} \tag{4}$$

An important question is: How easy is it to find the matrix **P** that accomplishes the similarity transformation (4)? Explicit equations can be obtained for an important special case, related to the structure of the matrix **A**, that arises frequently in system and network analysis.

Definition 1   *The* $n \times n$ *real matrix* **A** *is said to be in* companion form *if it has the structure*

$$\mathbf{A} = \begin{bmatrix} 0 & 1 & 0 & \cdots & 0 \\ 0 & 0 & 1 & \cdots & 0 \\ \vdots & \vdots & \vdots & \vdots & \vdots \\ 0 & 0 & 0 & \cdots & 1 \\ -a_0 & -a_1 & -a_1 & \cdots & -a_{n-1} \end{bmatrix} \tag{5}$$

Theorem 3   *The characteristic polynomial of* **A** *given by Eq.* (5) *is*

$$p(\lambda) = \det(\lambda\mathbf{I} - \mathbf{A}) = \lambda^n + a_{n-1}\lambda^{n-1} + \cdots + a_1\lambda + a_0 \tag{6}$$

Theorem 4   *The matrix* **P** *that accomplishes the similarity transformation* (4) *when* **A** *is in the companion form* (5) *and where* **Λ** *is the diagonal eigenvalue matrix given by* (2) *is the so-called* Vandemonde matrix *given by*

$$\mathbf{P} = \begin{bmatrix} 1 & 1 & \cdots & 1 \\ \lambda_1 & \lambda_2 & \cdots & \lambda_n \\ \lambda_1^2 & \lambda_2^2 & \cdots & \lambda_n^2 \\ \vdots & \vdots & \vdots & \vdots \\ \lambda_1^{n-1} & \lambda_2^{n-1} & \cdots & \lambda_n^{n-1} \end{bmatrix} \tag{7}$$

Theorem 5   *The eigenvectors* $\boldsymbol{\xi}_1, \boldsymbol{\xi}_2, \ldots, \boldsymbol{\xi}_n$ *of a companion-form matrix* **A**, *given by* (5), *are the columns of the Vandemonde matrix,* **P**, *given by* (7), *that is,*

$$\boldsymbol{\xi}_k = \begin{bmatrix} 1 \\ \lambda_k \\ \lambda_k^2 \\ \vdots \\ \lambda_k^{n-1} \end{bmatrix} \quad k = 1, 2, \ldots, n \tag{8}$$

Theorem 6 *If* **A** *has distinct eigenvalues, then its eigenvectors are all linearly independent.*

EXAMPLE 1   Consider the 3×3 matrix in companion form

$$\mathbf{A} = \begin{bmatrix} 0 & 1 & 0 \\ 0 & 0 & 1 \\ 0 & -2 & 3 \end{bmatrix}$$

The characteristic polynomial of **A** is

$$p(\lambda) = \det(\lambda\mathbf{I} - \mathbf{A}) = \lambda^3 - 3\lambda_2 + 2\lambda = \lambda(\lambda - 1)(\lambda - 2)$$

Hence **A** has three distinct eigenvalues

$$\lambda_1 = 0 \qquad \lambda_2 = 1 \qquad \lambda_3 = 2$$

The diagonal eigenvalue matrix Λ is

$$\mathbf{A} = \begin{bmatrix} 0 & 0 & 0 \\ 0 & 1 & 0 \\ 0 & 0 & 2 \end{bmatrix}$$

The matrix **P** such that $\Lambda = \mathbf{P}^{-1}\mathbf{AP}$ is the Vandemonde matrix

$$\mathbf{P} = \begin{bmatrix} 1 & 1 & 1 \\ 0 & 1 & 2 \\ 0 & 1 & 4 \end{bmatrix}$$

The three eigenvectors ξ_1, ξ_2, ξ_3 of **A** are

$$\xi_1 = \begin{bmatrix} 1 \\ 0 \\ 0 \end{bmatrix} \qquad \xi_2 = \begin{bmatrix} 1 \\ 1 \\ 1 \end{bmatrix} \qquad \xi_3 = \begin{bmatrix} 1 \\ 2 \\ 4 \end{bmatrix}$$

A.10.2 Repeated Eigenvalues

Once more let us consider an $n \times n$ real matrix **A**. Suppose that its characteristic polynomial is

$$p(\lambda) = \det(\lambda\mathbf{I} - \mathbf{A}) = \prod_{k=1}^{p} (\lambda - \lambda_k)^{m_k} \tag{9}$$

In this case some of the eigenvalues of **A** are repeated.

Definition 2   *The scalar* m_k *in Eq.* (9) *is called the* multiplicity *of the eigenvalue* λ_k. *Note that*

$$m_1 + m_2 + \cdots + m_p = n \tag{10}$$

Definition 3   *Let* **A** *be an* $n \times n$ *real matrix with eigenvalues* $\lambda_1, \lambda_2, \ldots, \lambda_p$ *of multiplicity* $m_1, m_2, \ldots, m_p$, *respectively. To each eigenvalue* λ_k *of multiplicity* m_k *we associate an* $m_k \times m_k$ *matrix* $\mathbf{J}_k$ *of the form*

$$\mathbf{J}_k = \begin{bmatrix} \lambda_k & 1 & 0 & \cdots & 0 & 0 \\ 0 & \lambda_k & 1 & \cdots & 0 & 0 \\ \vdots & \vdots & \vdots & \vdots & \vdots & \vdots \\ 0 & 0 & 0 & \cdots & \lambda_k & 1 \\ 0 & 0 & 0 & \cdots & 0 & \lambda_k \end{bmatrix} \tag{11}$$

which we call a Jordan block matrix *associated with* λ_k. *The* $n \times n$ *matrix* **J** *given by*

$$\mathbf{J} = \begin{bmatrix} \mathbf{J}_1 & \mathbf{0} & \cdots & \mathbf{0} \\ \mathbf{0} & \mathbf{J}_2 & \cdots & \mathbf{0} \\ \vdots & \vdots & \vdots & \vdots \\ \mathbf{0} & \mathbf{0} & \cdots & \mathbf{J}_p \end{bmatrix} \tag{12}$$

i.e., the matrix formed by placing the Jordan block matrices $\mathbf{J}_k$, *given by* (11), *along its diagonal, is called the* Jordan canonical matrix *associated with* **A**.

Theorem 7   *The eigenvalues of the Jordan canonical matrix are given by*

$$\det(\lambda \mathbf{I} - \mathbf{J}) = \prod_{k=1}^{p} (\lambda - \lambda_k)^{m_k} \tag{13}$$

Theorem 8   *There exists an* $n \times n$ *nonsingular matrix* **P** *such that*

$$\mathbf{J} = \mathbf{P}^{-1}\mathbf{A}\mathbf{P} \quad \text{or} \quad \mathbf{A} = \mathbf{P}\mathbf{J}\mathbf{P}^{-1} \tag{14}$$

Theorem 9   *Suppose that* **A** *is in the companion form* (5). *Suppose that* **A** *has a single eigenvalue* λ *of multiplicity* n (λ *must be real*), *that is,*

$$\det(s\mathbf{I} - \mathbf{A}) = (s - \lambda)^n \tag{15}$$

Then

$$\mathbf{J} = \begin{bmatrix} \lambda & 1 & 0 & \cdots & 0 & 0 \\ 0 & \lambda & 1 & \cdots & 0 & 0 \\ \vdots & \vdots & \vdots & \vdots & \vdots & \vdots \\ 0 & 0 & 0 & \cdots & \lambda & 1 \\ 0 & 0 & 0 & \cdots & 0 & \lambda \end{bmatrix} \tag{16}$$

and the matrix **P** *that accomplishes the similarity transformation* (14) *is given by*

$$\mathbf{P} = \begin{bmatrix} 1 & 0 & 0 & 0 & 0 & 0 & 0 & 0 & \cdots \\ \lambda & 1 & 0 & 0 & 0 & 0 & 0 & 0 & \cdots \\ \lambda^2 & 2\lambda & 1 & 0 & 0 & 0 & 0 & 0 & \cdots \\ \lambda^3 & 3\lambda^2 & 3\lambda & 1 & 0 & 0 & 0 & 0 & \cdots \\ \lambda^4 & 4\lambda^3 & 6\lambda^2 & 4\lambda & 1 & 0 & 0 & 0 & \cdots \\ \lambda^5 & 5\lambda^4 & 10\lambda^3 & 10\lambda^2 & 5\lambda & 1 & 0 & 0 & \cdots \\ \lambda^6 & 6\lambda^5 & 15\lambda^4 & 20\lambda^3 & 15\lambda^2 & 6\lambda & 1 & 0 & \cdots \\ \lambda^7 & 7\lambda^6 & 21\lambda^5 & 35\lambda^4 & 35\lambda^3 & 21\lambda^2 & 7\lambda & 1 & \cdots \\ \vdots & \vdots & \vdots & \vdots & \vdots & \vdots & \vdots & \vdots & \vdots \\ \lambda^{n-1} & \cdot & \cdot & \cdot & \cdot & \cdot & \cdot & \cdot & \end{bmatrix} \tag{17}$$

Remark Careful examination of Eq. (17) yields the following information. The ith row of **P** is constructed by the binomial expansion of

$$(\lambda + 1)^{i-1} \qquad i = 1, 2, \ldots, n$$

An easy way to remember the coefficients of the binomial expansion is by the Pascal triangle

$$
\begin{array}{ccccccccccccc}
 & & & & & & 1 & & & & & & \\
 & & & & & 1 & & 1 & & & & & \\
 & & & & 1 & & 2 & & 1 & & & & \\
 & & & 1 & & 3 & & 3 & & 1 & & & \\
 & & 1 & & 4 & & 6 & & 4 & & 1 & & \\
 & 1 & & 5 & & 10 & & 10 & & 5 & & 1 & \\
1 & & 6 & & 15 & & 20 & & 15 & & 6 & & 1 \\
\end{array}
$$

$$\cdots\cdots\cdots\cdots\cdots\cdots\cdots\cdots\cdots\cdots\cdots$$

in which each element is the sum of the two elements above it, e.g.,

$$
\begin{array}{ccccccc}
1 & & & 4 & 6 & & & 4 \\
 & \searrow & \swarrow & & & \searrow & \swarrow & \\
 & & 5 & & & & 10 &
\end{array}
$$

EXAMPLE 7   Consider the 3×3 matrix **A** in companion form

$$\mathbf{A} = \begin{bmatrix} 0 & 1 & 0 \\ 0 & 0 & 1 \\ 8 & -12 & 6 \end{bmatrix}$$

Its characteristic polynomial is

$$\det(\lambda\mathbf{I} - \mathbf{A}) = \lambda^3 - 6\lambda^2 + 12\lambda - 8 = (\lambda - 2)^3$$

Hence $\mathbf{A}$ has a repeated eigenvalue at $\lambda = 2$ whose multiplicity is 3. The Jordan matrix (16) takes the form

$$\mathbf{J} = \begin{bmatrix} 2 & 1 & 0 \\ 0 & 2 & 1 \\ 0 & 0 & 2 \end{bmatrix}$$

The matrix $\mathbf{P}$ that accomplishes $\mathbf{J} = \mathbf{P}^{-1}\mathbf{A}\mathbf{P}$ is given by [see Eq. (17)]

$$\mathbf{P} = \begin{bmatrix} 1 & 0 & 0 \\ 2 & 1 & 0 \\ 4 & 4 & 1 \end{bmatrix}$$

Theorem 10   *Suppose that $\mathbf{A}$ is an $n \times n$ real matrix in companion form. Let λ_k, $k = 1, 2, \ldots, p$ denote its eigenvalues, and $m_1, m_2, \ldots, m_p$ their multiplicity. Let $\mathbf{J}$ denote the Jordan canonical matrix of $\mathbf{A}$ given by (12). Then the $n \times n$ nonsingular matrix $\mathbf{P}$ that accomplishes the similarity transformation (14) is given by*

$$\mathbf{P} = \begin{bmatrix} \mathbf{P}_1 & \vdots & \mathbf{P}_2 & \vdots & \cdots & \vdots & \mathbf{P}_p \end{bmatrix} \tag{18}$$

where each submatrix $\mathbf{P}_k$, $k = 1, 2, \ldots, p$, is an $n \times m_k$ matrix and

$$\mathbf{P}_k = \begin{bmatrix} 1 & 0 & 0 & 0 & \cdots & 0 \\ \lambda_k & 1 & 0 & 0 & \cdots & 0 \\ \lambda_k^2 & 2\lambda & 1 & 0 & \cdots & 0 \\ \lambda_k^3 & 3\lambda_k^2 & 3\lambda_k & 1 & \cdots & 0 \\ \lambda_k^4 & 4\lambda_k^3 & 6\lambda_k^2 & 4\lambda_k & \cdots & 0 \\ \vdots & \vdots & \vdots & \vdots & \vdots & \vdots \\ \lambda_k^{n-1} & \cdot & \cdot & \cdot & \cdot & 1 \end{bmatrix} \tag{19}$$

where m_k is the multiplicity of the eigenvalue λ_k.

EXAMPLE 2   Consider the 6×6 matrix $\mathbf{A}$ in companion form

$$\mathbf{A} = \begin{bmatrix} 0 & 1 & 0 & 0 & 0 & 0 \\ 0 & 0 & 1 & 0 & 0 & 0 \\ 0 & 0 & 0 & 1 & 0 & 0 \\ 0 & 0 & 0 & 0 & 1 & 0 \\ 0 & 0 & 0 & 0 & 0 & 1 \\ -2 & 3 & 3 & 6 & 0 & 3 \end{bmatrix}$$

Its characteristic polynomial is

$$\det(\lambda\mathbf{I} - \mathbf{A}) = \lambda^6 - 3\lambda^5 - 6\lambda^3 - 3\lambda^2 - 3\lambda + 2 = (\lambda - 1)^3(\lambda - 2)(\lambda + 1)^2$$

Hence the eigenvalues of **A** are specified by

$$\lambda_1 = 1 \text{ with multiplicity } m_1 = 3$$

$$\lambda_2 = 2 \text{ with multiplicity } m_2 = 1$$

$$\lambda_3 = -1 \text{ with multiplicity } m_3 = 2$$

The 6×6 Jordan canonical matrix of **A** is given by

$$
\mathbf{J} = \left[
\begin{array}{ccc:c:cc}
1 & 1 & 0 & 0 & 0 & 0 \\
0 & 1 & 1 & 0 & 0 & 0 \\
0 & 0 & 1 & 0 & 0 & 0 \\ \hdashline
0 & 0 & 0 & 2 & 0 & 0 \\ \hdashline
0 & 0 & 0 & 0 & -1 & 1 \\
0 & 0 & 0 & 0 & 0 & -1
\end{array}
\right]
$$

where the dashed submatrices are the Jordan block matrices

$$
\mathbf{J}_1 = \begin{bmatrix} 1 & 1 & 0 \\ 0 & 1 & 1 \\ 0 & 0 & 1 \end{bmatrix}
\qquad
\mathbf{J}_2 = 2
\qquad
\mathbf{J}_3 = \begin{bmatrix} -1 & 1 \\ 0 & -1 \end{bmatrix}
$$

The 6×6 matrix **P** that accomplishes the similarity transformation $\mathbf{J} = \mathbf{P}^{-1}\mathbf{A}\mathbf{P}$ is

$$
\mathbf{P} = \left[
\begin{array}{ccc:c:cc}
1 & 0 & 0 & 1 & 1 & 0 \\
1 & 1 & 0 & 2 & -1 & 1 \\
1 & 2 & 1 & 4 & 1 & -2 \\
1 & 3 & 3 & 8 & -1 & 3 \\
1 & 4 & 6 & 16 & 1 & -4 \\
1 & 5 & 10 & 32 & -1 & 5
\end{array}
\right]
$$

$$\underbrace{\hphantom{xxxx}}_{\mathbf{P}_1} \quad \underbrace{\hphantom{xx}}_{\mathbf{P}_2} \quad \underbrace{\hphantom{xxxx}}_{\mathbf{P}_3}$$

The reduction of a matrix to either a diagonal form or Jordan canonical form can be used to prove the following theorems.

Theorem 11   *The trace of a matrix is the sum of its eigenvalues. The determinant of a matrix is equal to the product of its eigenvalues.*

Theorem 12   *If **A** is nonsingular with eigenvalues $\lambda_1, \lambda_2, \ldots, \lambda_n$, then the eigenvalues of $\mathbf{A}^{-1}$ are $1/\lambda_1, 1/\lambda_2, \ldots, 1/\lambda_n$.*

Theorem 13   **A** *and* **A**′ *have the same eigenvalues.*

Theorem 14   *If* $\mathbf{A} = \mathbf{A}'$, *then the eigenvalues are real.*

Theorem 15   *If the eigenvalues of* $\mathbf{A}$ *are* $\lambda_1, \lambda_2, \ldots, \lambda_n$, *then the eigenvalues of* $\mathbf{A}^k$ *are* $\lambda_1^k, \lambda_2^k, \ldots, \lambda_n^k$. *Also the eigenvectors of* $\mathbf{A}^k$ *are the same as those of* $\mathbf{A}$.

Theorem 16   (Cayley-Hamilton theorem)   *Let* $p(\lambda) = \det(\lambda \mathbf{I} - \mathbf{A}) = \lambda^n + a_{n-1}\lambda^{n-1} + \cdots + a_1\lambda + a_0$ *be the characteristic polynomial of* $\mathbf{A}$. *Then* $\mathbf{A}^n + a_{n-1}\mathbf{A}^{n-1} + \cdots + a_1\mathbf{A} + a_0\mathbf{I} = \mathbf{0}$.

Theorem 17   *Suppose that* $\mathbf{A}$ *has* n *distinct eigenvalues. Let* $\mathbf{v}_1, \mathbf{v}_2, \ldots, \mathbf{v}_n$ *denote the eigenvectors of* $\mathbf{A}$. *Then the eigenvectors are linearly independent.*

A.11   SCALAR PRODUCTS IN EUCLIDEAN SPACES

First we introduce a notation to denote scalar products in R_n. Thus if $\mathbf{x}$ and $\mathbf{y}$ are two vectors in R_n, we denote their scalar product by

$$\langle \mathbf{x}, \mathbf{y} \rangle \tag{1}$$

The properties of the scalar product are

$$\langle \mathbf{x}, \mathbf{y} \rangle = \langle \mathbf{y}, \mathbf{x} \rangle \tag{2}$$

$$\langle a\mathbf{x}, \mathbf{y} \rangle = a\langle \mathbf{x}, \mathbf{y} \rangle \qquad a \text{ scalar} \tag{3}$$

$$\langle \mathbf{x} + \mathbf{y}, \mathbf{z} \rangle = \langle \mathbf{x}, \mathbf{z} \rangle + \langle \mathbf{y}, \mathbf{z} \rangle \tag{4}$$

$$\langle \mathbf{x}, \mathbf{x} \rangle \geq 0 \tag{5}$$

We can also define the euclidean norm $\|\mathbf{x}\|$ of $\mathbf{x}$ by

$$\|\mathbf{x}\| = \sqrt{\langle \mathbf{x}, \mathbf{x} \rangle} \tag{6}$$

A.11.1   Orthogonality and the Natural Basis

We have a good idea what we mean by orthogonal vectors in R_2 and R_3. In particular, the geometrical feel we have for the *natural basis* $\{\mathbf{e}_1, \mathbf{e}_2, \ldots, \mathbf{e}_n\}$ in R_n is that

1   The natural basis vectors are mutually orthogonal.
2   The magnitude of each natural-basis vector is unity.

We are now in a position to discuss orthogonality in terms of scalar products and magnitudes in terms of norms. In particular, the mutual orthogonality of the natural-basis vectors is expressed by

$$\langle \mathbf{e}_i, \mathbf{e}_j \rangle = 0 \qquad \text{for } i \neq j \tag{7}$$

and the fact that each natural-basis vector has unit length is expressed by

$$\|\mathbf{e}_i\| = \sqrt{\langle \mathbf{e}_i, \mathbf{e}_i \rangle} = 1 \qquad \text{for } i = 1, 2, \dots, n \tag{8}$$

In general, a basis $\{\mathbf{b}_1, \mathbf{b}_2, \dots, \mathbf{b}_n\}$ in R_n is called *orthonormal* if

$$\langle \mathbf{b}_i, \mathbf{b}_j \rangle = 0 \qquad \text{for } i \neq j$$
$$\langle \mathbf{b}_i, \mathbf{b}_i \rangle = 1 \tag{9}$$

The natural basis is of course an orthonormal basis.

A.11.2 The Formula for Scalar Products in R_n

Let us suppose that $\mathbf{x}$ and $\mathbf{y}$ are two vectors in R_n. Let us *choose* the natural basis in R_n so that $\mathbf{x}$ and $\mathbf{y}$ are represented by

$$\mathbf{x} = x_1 \mathbf{e}_1 + x_2 \mathbf{e}_2 + \cdots + x_n \mathbf{e}_n = \sum_{i=1}^{n} x_i \mathbf{e}_i$$
$$\mathbf{y} = y_1 \mathbf{e}_1 + y_2 \mathbf{e}_2 + \cdots + y_n \mathbf{e}_n = \sum_{j=1}^{n} y_j \mathbf{e}_j \tag{10}$$

Now we have a quantitative description of the vectors $\mathbf{x}$ and $\mathbf{y}$ in terms of their coordinates with respect to the *same* natural basis in R_n. We now show that this will enable us to compute their scalar product $\langle \mathbf{x}, \mathbf{y} \rangle$ numerically. This calculation involves the properties (2) to (5) of the scalar product as well as the fact that the natural basis is orthonormal.

$$\langle \mathbf{x}, \mathbf{y} \rangle = \left\langle \sum_{i=1}^{n} x_i \mathbf{e}_i, \sum_{j=1}^{n} y_j \mathbf{e}_j \right\rangle$$
$$= \sum_{i=1}^{n} x_i \left\langle \mathbf{e}_i, \sum_{j=1}^{n} y_j \mathbf{e}_j \right\rangle \tag{11}$$
$$= \sum_{i=1}^{n} \sum_{j=1}^{n} x_i y_j \langle \mathbf{e}_i, \mathbf{e}_j \rangle$$

Now we use Eqs. (7) and (8) to deduce

$$\langle \mathbf{x}, \mathbf{y} \rangle = \sum_{i=1}^{n} x_i y_i \tag{12}$$

which yields the numerical value of the scalar product.

In the special case that $\mathbf{x} = \mathbf{y}$ we obtain

$$\langle \mathbf{x}, \mathbf{x} \rangle = \sum_{i=1}^{n} x_i^2 \tag{13}$$

so that what we have called the euclidean norm is

$$\|\mathbf{x}\| = \sqrt{\langle \mathbf{x}, \mathbf{x} \rangle} = \sqrt{\sum_{i=1}^{n} x_i^2} \tag{14}$$

a formula which is in agreement with our geometrical intuition.

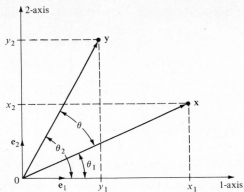

FIGURE A.11.1
Two vectors **x** and **y** in the plane.

A.11.3 Scalar Products and Dot Products

To enhance the reader's geometrical intuition we shall now show that the scalar product just defined is the same as the dot product familiar from physics courses. The discussion will be carried out in the plane (R_2).

Let **x** and **y** be two vectors in R_2. Let us select the natural basis $\{e_1, e_2\}$ and let

$$\mathbf{x} = x_1 \mathbf{e}_1 + x_2 \mathbf{e}_2 \qquad \mathbf{y} = y_1 \mathbf{e}_1 + y_2 \mathbf{e}_2 \tag{15}$$

Then the scalar product is

$$\langle \mathbf{x}, \mathbf{y} \rangle = x_1 y_1 + x_2 y_2 \tag{16}$$

and

$$\|\mathbf{x}\| = \sqrt{x_1^2 + x_2^2} \qquad \|\mathbf{y}\| = \sqrt{y_1^2 + y_2^2} \tag{17}$$

Now let us consider the situation illustrated in Fig. A.11.1, where we have represented **x** and **y** as directed-arrow vectors. Let us recall that the dot product (denoted by $\mathbf{x} \cdot \mathbf{y}$) is

$$\text{Dot product of } \mathbf{x} \text{ and } \mathbf{y} = \mathbf{x} \cdot \mathbf{y} = (\text{length of } \mathbf{x}) \cdot (\text{length of } \mathbf{y}) \cos \theta$$

$$= \sqrt{x_1^2 + x_2^2} \sqrt{y_1^2 + y_2^2} \cos \theta \tag{18}$$

$$= \|\mathbf{x}\| \cdot \|\mathbf{y}\| \cos \theta$$

But

$$\theta = \theta_2 - \theta_1 \tag{19}$$

(see Fig. A.11.1) and

$$\cos \theta = \cos(\theta_2 - \theta_1) = \cos \theta_2 \cos \theta_1 + \sin \theta_2 \sin \theta_1 \tag{20}$$

Clearly, from Fig. A.11.1 we have

$$\sin \theta_1 = \frac{x_2}{\|\mathbf{x}\|} \qquad \cos \theta_1 = \frac{x_1}{\|\mathbf{x}\|}$$
$$\sin \theta_2 = \frac{y_2}{\|\mathbf{y}\|} \qquad \cos \theta_2 = \frac{y_1}{\|\mathbf{y}\|} \tag{21}$$

Hence, the dot product is

$$\mathbf{x} \cdot \mathbf{y} = \|\mathbf{x}\| \cdot \|\mathbf{y}\| \left(\frac{x_2 y_2}{\|\mathbf{x}\| \cdot \|\mathbf{y}\|} + \frac{x_1 y_1}{\|\mathbf{x}\| \cdot \|\mathbf{y}\|} \right) = x_1 y_1 + x_2 y_2 \tag{22}$$

From Eqs. (22) and (16) we deduce that

$$\mathbf{x} \cdot \mathbf{y} = \langle \mathbf{x}, \mathbf{y} \rangle \tag{23}$$

i.e., the scalar product is the same as the dot product.

This correspondence allows us to find the angle between any two high-dimensional vectors $\mathbf{x}$ and $\mathbf{y}$ by the formula

$$\text{Angle between } \mathbf{x} \text{ and } \mathbf{y} = \cos^{-1} \frac{\langle \mathbf{x}, \mathbf{y} \rangle}{\|\mathbf{x}\| \cdot \|\mathbf{y}\|} \tag{24}$$

Indeed if $\mathbf{x}$ and $\mathbf{y}$ are orthogonal, $\langle \mathbf{x}, \mathbf{y} \rangle = 0$ and so their angle is 90°, which is the geometric value for right angles. So everything is perfectly consistent.

A.11.4 Formulas

If we consider the column vectors

$$\mathbf{x} = \begin{bmatrix} x_1 \\ x_2 \\ \vdots \\ x_n \end{bmatrix} \qquad \mathbf{y} = \begin{bmatrix} y_1 \\ y_2 \\ \vdots \\ y_n \end{bmatrix} \tag{25}$$

we can interpret the x_i and y_i as defining the coordinates of vectors $\mathbf{x}$ and $\mathbf{y}$ in R_n with respect to the natural basis. Hence, it should not come as a surprise that we define the scalar product of two column n-vectors in the following way.

Let $\mathbf{x}$ and $\mathbf{y}$ denote two column n-vectors as in Eq. (25). Then their scalar product $\langle \mathbf{x}, \mathbf{y} \rangle$ is

$$\langle \mathbf{x}, \mathbf{y} \rangle = \sum_{i=1}^{n} x_i y_i = \mathbf{x}' \mathbf{y} = \mathbf{y}' \mathbf{x} \tag{26}$$

One important property of the scalar product involving column vectors and matrices is as follows. Suppose that $\mathbf{x}$ is a column n-vector with components $x_1, x_2, \ldots, x_n$; suppose that $\mathbf{y}$ is a column m-vector with components $y_1, y_2, \ldots, y_m$. Let $\mathbf{A}$ be an $n \times m$ matrix with elements a_{ij} ($i = 1, 2, \ldots, n; j = 1, 2, \ldots, m$). Then

$$\langle \mathbf{x}, \mathbf{A}\mathbf{y} \rangle = \sum_{i=1}^{n} \sum_{j=1}^{m} a_{ij} x_i y_j = \mathbf{x}' \mathbf{A}\mathbf{y} \tag{27}$$

and

$$\langle \mathbf{x}, \mathbf{A}\mathbf{y} \rangle = \langle \mathbf{A}'\mathbf{x}, \mathbf{y} \rangle = \mathbf{y}' \mathbf{A}' \mathbf{x} = \sum_{j=1}^{m} \sum_{i=1}^{n} y_j a_{ij} x_i = \mathbf{x}' \mathbf{A}\mathbf{y} \tag{28}$$

A.12 QUADRATIC FORMS AND DEFINITE MATRICES

This section introduces certain definitions and concepts which are of fundamental importance in the study of networks and systems.

Suppose that $\mathbf{x}$ is a column n-vector with components x_i and that $\mathbf{A}$ is a real $n \times n$ symmetric matrix, that is, $\mathbf{A} = \mathbf{A}'$, with elements a_{ij}. Let us consider the scalar-valued function $f(\mathbf{x})$ defined by the scalar product

$$f(\mathbf{x}) = \langle \mathbf{x}, \mathbf{A}\mathbf{x} \rangle = \sum_{i=1}^{n} \sum_{j=1}^{n} a_{ij} x_i x_j \tag{1}$$

This is called a *quadratic form* because it involves multiplication by pairs of the elements x_i of $\mathbf{x}$. For example, if

$$\mathbf{x} = \begin{bmatrix} x_1 \\ x_2 \end{bmatrix} \qquad \mathbf{A} = \begin{bmatrix} 1 & 2 \\ 2 & 3 \end{bmatrix}$$

then

$$f(\mathbf{x}) = \langle \mathbf{x}, \mathbf{A}\mathbf{x} \rangle = x_1^2 + 4x_1 x_2 + 3x_2^2$$

which involves terms in the square of the components of $\mathbf{x}$ and their cross products.

We now offer certain definitions.

Definition 1 *If for all $\mathbf{x} \neq \mathbf{0}$*

(a) $f(\mathbf{x}) = \langle \mathbf{x}, \mathbf{A}\mathbf{x} \rangle \geq 0$ (2)
$f(\mathbf{x})$ *is called a positive semidefinite form and $\mathbf{A}$ is called a positive semidefinite matrix.*
(b) $f(\mathbf{x}) = \langle \mathbf{x}, \mathbf{A}\mathbf{x} \rangle > 0$ (3)
$f(\mathbf{x})$ *is called a positive definite form and $\mathbf{A}$ is called a positive definite matrix.*
(c) $f(\mathbf{x}) = \langle \mathbf{x}, \mathbf{A}\mathbf{x} \rangle \leq 0$ (4)
$f(\mathbf{x})$ *is called a negative semidefinite form and $\mathbf{A}$ is called a negative semidefinite matrix.*
(d) $f(\mathbf{x}) = \langle \mathbf{x}, \mathbf{A}\mathbf{x} \rangle < 0$ (5)
$f(\mathbf{x})$ *is called a negative definite form and $\mathbf{A}$ is called a negative definite matrix.*

We now give a procedure for testing whether a given matrix is positive definite. The basic technique is summarized in the following theorem.

Theorem 2 *Suppose that $\mathbf{A}$ is the real symmetric $n \times n$ matrix*

$$\mathbf{A} = \begin{bmatrix} a_{11} & a_{12} & \cdots & a_{1n} \\ a_{12} & a_{22} & \cdots & a_{2n} \\ \vdots & \vdots & \vdots & \vdots \\ a_{1n} & a_{2n} & \cdots & a_{nn} \end{bmatrix} \tag{6}$$

Let $\mathbf{A}_k$ be the $k \times k$ matrix, defined in terms of $\mathbf{A}$, for $k = 1, 2, \ldots, n$, by

$$\mathbf{A}_k = \begin{bmatrix} a_{11} & a_{12} & \cdots & a_{1k} \\ a_{12} & a_{22} & \cdots & a_{2k} \\ \vdots & \vdots & \vdots & \vdots \\ a_{1k} & a_{2k} & \cdots & a_{kk} \end{bmatrix} \tag{7}$$

Then $\mathbf{A}$ is positive definite if and only if

$$\det \mathbf{A}_k > 0 \tag{8}$$

for each $k = 1, 2, \ldots, n$.

   $\mathbf{A}$ is negative definite if and only if

$$\det \mathbf{A}_k < 0 \tag{9}$$

for each $k = 1, 2, \ldots, n$.

   There is a host of additional properties of definite and semidefinite symmetric matrices which we give below as theorems. Some of the proofs are easy, but others are very difficult.

Theorem 2 $\mathbf{A}$ is a positive definite matrix if and only if all its eigenvalues are positive. $\mathbf{A}$ is a negative definite matrix if and only if all its eigenvalues are negative. In either case, the eigenvectors of $\mathbf{A}$ are real and mutually orthogonal.

Theorem 3   If $\mathbf{A}$ is a positive semidefinite or negative semidefinite matrix, then at least one of its eigenvalues must be zero. If $\mathbf{A}$ is positive (negative) definite, then $\mathbf{A}^{-1}$ is positive (negative) definite.

Theorem 4   If both $\mathbf{A}$ and $\mathbf{B}$ are positive (negative) definite, and if $\mathbf{A} - \mathbf{B}$ is also positive (negative) definite, then $\mathbf{B}^{-1} - \mathbf{A}^{-1}$ is positive (negative) definite.

A.13   TIME-VARYING VECTORS AND MATRICES

A time-varying column vector $\mathbf{x}(t)$ is defined as a column vector whose components are themselves functions of time, i.e.,

$$\mathbf{x}(t) = \begin{bmatrix} x_1(t) \\ x_2(t) \\ \vdots \\ x_n(t) \end{bmatrix} \tag{1}$$

while a time-varying matrix $\mathbf{A}(t)$ is defined as a matrix whose elements are time functions, i.e.,

$$\mathbf{A}(t) = \begin{bmatrix} a_{11}(t) & a_{12}(t) & \cdots & a_{1m}(t) \\ a_{21}(t) & a_{22}(t) & \cdots & a_{2m}(t) \\ \vdots & \vdots & \vdots & \vdots \\ a_{n1}(t) & a_{n2}(t) & \cdots & a_{nm}(t) \end{bmatrix} \tag{2}$$

The addition of time-varying vectors and matrices, their multiplication, and the scalar-product operations are defined as before.

A.13.1 Time Derivatives

The time derivative of the vector $\mathbf{x}(t)$ is denoted by $d/dt\, \mathbf{x}(t)$ or $\dot{\mathbf{x}}(t)$ and is defined by

$$\frac{d}{dt}\mathbf{x}(t) \triangleq \dot{\mathbf{x}}(t) \triangleq \begin{bmatrix} \dot{x}_1(t) \\ \dot{x}_2(t) \\ \vdots \\ \dot{x}_n(t) \end{bmatrix} \tag{3}$$

The time derivative of the matrix $\mathbf{A}(t)$ is denoted by $d/dt\, \mathbf{A}(t)$ or $\dot{\mathbf{A}}(t)$ and is defined by

$$\frac{d}{dt}\mathbf{A}(t) \triangleq \dot{\mathbf{A}}(t) \triangleq \begin{bmatrix} \dot{a}_{11}(t) & \dot{a}_{12}(t) & \cdots & \dot{a}_{1m}(t) \\ \dot{a}_{21}(t) & \dot{a}_{22}(t) & \cdots & \dot{a}_{2m}(t) \\ \vdots & \vdots & \vdots & \vdots \\ \dot{a}_{n1}(t) & \dot{a}_{n2}(t) & \cdots & \dot{a}_{nm}(t) \end{bmatrix} \tag{4}$$

Of course, in order for $\dot{\mathbf{x}}(t)$ or $\dot{\mathbf{A}}(t)$ to make sense the derivatives $\dot{x}_i(t)$ and $\dot{a}_{ij}(t)$ must exist.

As a consequence of these definitions, the following properties hold:

$$\frac{d}{dt}[\mathbf{x}(t) + \mathbf{y}(t)] = \dot{\mathbf{x}}(t) + \dot{\mathbf{y}}(t) \tag{5}$$

$$\frac{d}{dt}[\alpha(t)\mathbf{x}(t)] = \dot{\alpha}(t)\mathbf{x}(t) + \alpha(t)\dot{\mathbf{x}}(t) \tag{6}$$

$$\frac{d}{dt}\langle \mathbf{x}(t), \mathbf{y}(t) \rangle = \langle \dot{\mathbf{x}}(t), \mathbf{y}(t) \rangle + \langle \mathbf{x}(t), \dot{\mathbf{y}}(t) \rangle \tag{7}$$

$$\frac{d}{dt}[\mathbf{A}(t) + \mathbf{B}(t)] = \dot{\mathbf{A}}(t) + \dot{\mathbf{B}}(t) \tag{8}$$

$$\frac{d}{dt}[\alpha(t)\mathbf{A}(t)] = \dot{\alpha}(t)\mathbf{A}(t) + \alpha(t)\dot{\mathbf{A}}(t) \tag{9}$$

$$\frac{d}{dt}[\mathbf{A}(t)\mathbf{B}(t)] = \dot{\mathbf{A}}(t)\mathbf{B}(t) + \mathbf{A}(t)\dot{\mathbf{B}}(t) \tag{10}$$

$$\frac{d}{dt}[\mathbf{A}(t)\mathbf{x}(t)] = \dot{\mathbf{A}}(t)\mathbf{x}(t) + \mathbf{A}(t)\dot{\mathbf{x}}(t) \tag{11}$$

$$\frac{d}{dt}\langle \mathbf{x}(t), \mathbf{A}(t)\mathbf{y}(t) \rangle = \langle \dot{\mathbf{x}}(t), \mathbf{A}(t)\mathbf{y}(t) \rangle + \langle \mathbf{x}(t), \dot{\mathbf{A}}(t)\mathbf{y}(t) \rangle + \langle \mathbf{x}(t), \mathbf{A}(t)\dot{\mathbf{y}}(t) \rangle \tag{12}$$

To illustrate these definitions we present the following two examples.

EXAMPLE 1 Consider the time-varying column vector

$$\mathbf{x}(t) = \begin{bmatrix} \sin t \\ t^2 \\ e^t + 2 \end{bmatrix}$$

Then

$$\dot{\mathbf{x}}(t) = \begin{bmatrix} \cos t \\ 2t \\ e^t \end{bmatrix}$$

EXAMPLE 2 Consider the time-varying matrix

$$\mathbf{A}(t) = \begin{bmatrix} 2t^3 & 5 & 4t \\ \cos t & t & t^2 \end{bmatrix}$$

Then

$$\dot{\mathbf{A}}(t) = \begin{bmatrix} 6t^2 & 0 & 4 \\ -\sin t & 1 & 2t \end{bmatrix}$$

A.13.2 Integration

We can define the integrals of vectors and matrices in a similar manner. Thus,

$$\int_{t_0}^{t_1} \mathbf{x}(t)\,dt \triangleq \begin{bmatrix} \int_{t_0}^{t_1} x_1(t)\,dt \\ \int_{t_0}^{t_1} x_2(t)\,dt \\ \vdots \\ \int_{t_0}^{t_1} x_n(t)\,dt \end{bmatrix} \tag{13}$$

$$\int_{t_0}^{t_1} \mathbf{A}(t)\,dt = \begin{bmatrix} \int_{t_0}^{t_1} a_{11}(t)\,dt & \cdots & \int_{t_0}^{t_1} a_{1m}(t)\,dt \\ \vdots & \vdots & \vdots \\ \int_{t_0}^{t_1} a_{n1}(t)\,dt & \cdots & \int_{t_0}^{t_1} a_{nm}(t)\,dt \end{bmatrix} \tag{14}$$

As a direct consequence of the definitions and of the properties of integration we have

$$\int_{t_0}^{t_1} [\mathbf{x}(t) + \mathbf{y}(t)]\,dt = \int_{t_0}^{t_1} \mathbf{x}(t)\,dt + \int_{t_0}^{t_1} \mathbf{y}(t)\,dt \tag{15}$$

$$\int_{t_0}^{t_1} [\mathbf{A}(t) + \mathbf{B}(t)]\,dt = \int_{t_0}^{t_1} \mathbf{A}(t)\,dt + \int_{t_0}^{t_1} \mathbf{B}(t)\,dt \tag{16}$$

A.14 GRADIENT VECTORS AND JACOBIAN MATRICES

Let us suppose that $x_1, x_2, \ldots, x_n$ are real scalars which are the components of the column n-vector $\mathbf{x}$

$$\mathbf{x} = \begin{bmatrix} x_1 \\ x_2 \\ \vdots \\ x_n \end{bmatrix} \tag{1}$$

Now consider a scalar-valued function of the x_i

$$f(x_1, x_2, \ldots, x_n) \triangleq f(\mathbf{x}) \tag{2}$$

Clearly, f is a function mapping n-dimensional vectors to scalars

$$f: R_n \to R \tag{3}$$

Definition 1 *The gradient of f with respect to the column n-vector $\mathbf{x}$ is denoted by $\partial f(\mathbf{x})/\partial \mathbf{x}$ and is defined by*

$$\frac{\partial f}{\partial \mathbf{x}} \triangleq \frac{\partial}{\partial \mathbf{x}} f(\mathbf{x}) \triangleq \begin{bmatrix} \dfrac{\partial f}{\partial x_1} \\ \dfrac{\partial f}{\partial x_2} \\ \vdots \\ \dfrac{\partial f}{\partial x_n} \end{bmatrix} \tag{4}$$

so that the gradient is also a column n-vector.

EXAMPLE 1 Suppose $f: R_3 \to R$ and is defined by

$$f(\mathbf{x}) = f(x_1, x_2, x_3) = x_1^2 x_2 e^{-x_3}$$

Then

$$\frac{\partial f}{\partial \mathbf{x}} = \begin{bmatrix} 2x_1 x_2 e^{-x_3} \\ x_1^2 e^{-x_3} \\ -x_1^2 x_2 e^{-x_3} \end{bmatrix}$$

Again let us suppose that $\mathbf{x} \in R_n$. Let us consider a function $\mathbf{g}$,

$$\mathbf{g}: R_n \to R_m \tag{5}$$

such that

$$\mathbf{y} = \mathbf{g}(\mathbf{x}) \qquad \mathbf{x} \in R_n \tag{6}$$
$$\mathbf{y} \in R_m$$

By this we mean

$$y_1 = g_1(x_1, x_2, \ldots, x_n) = g_1(\mathbf{x})$$
$$y_2 = g_2(x_1, x_2, \ldots, x_n) = g_2(\mathbf{x})$$
$$\vdots \tag{7}$$
$$y_m = g_m(x_1, x_2, \ldots, x_n) = g_m(\mathbf{x})$$

Definition 2   *The Jacobian matrix of* $\mathbf{g}$ *with respect to* $\mathbf{x}$ *is denoted by* $\partial\mathbf{g}(\mathbf{x})/\partial\mathbf{x}$ *and is defined as*

$$\frac{\partial\mathbf{g}(\mathbf{x})}{\partial\mathbf{x}} \triangleq \begin{bmatrix} \dfrac{\partial g_1}{\partial x_1} & \dfrac{\partial g_1}{\partial x_2} & \cdots & \dfrac{\partial g_1}{\partial x_n} \\ \dfrac{\partial g_2}{\partial x_1} & \dfrac{\partial g_2}{\partial x_2} & \cdots & \dfrac{\partial g_2}{\partial x_n} \\ \vdots & \vdots & \vdots & \vdots \\ \dfrac{\partial g_m}{\partial x_1} & \dfrac{\partial g_m}{\partial x_2} & \cdots & \dfrac{\partial g_m}{\partial x_n} \end{bmatrix} \tag{8}$$

Thus, if $\mathbf{g}: R_n \to R_m$, *its Jacobian matrix is an* $m \times n$ *matrix.*

As an immediate consequence of the definition of a gradient vector we have

$$\frac{\partial}{\partial\mathbf{x}}\langle\mathbf{x},\mathbf{y}\rangle = \mathbf{y} \tag{9}$$

$$\frac{\partial}{\partial\mathbf{x}}\langle\mathbf{x},\mathbf{A}\mathbf{y}\rangle = \mathbf{A}\mathbf{y} \tag{10}$$

$$\frac{\partial}{\partial\mathbf{x}}\langle\mathbf{A}\mathbf{x},\mathbf{y}\rangle = \frac{\partial}{\partial\mathbf{x}}\langle\mathbf{x},\mathbf{A}'\mathbf{y}\rangle = \mathbf{A}'\mathbf{y} \tag{11}$$

The definition of a Jacobian matrix yields the relation

$$\frac{\partial}{\partial\mathbf{x}}\mathbf{A}\mathbf{x} = \mathbf{A} \tag{12}$$

Now suppose that $\mathbf{x}(t)$ is a time-varying vector and that $f(\mathbf{x})$ is a scalar-valued function of $\mathbf{x}$. Then by the chain rule

$$\frac{d}{dt}f(\mathbf{x}) = \frac{\partial f}{\partial x_1}\dot{x}_1 + \frac{\partial f}{\partial x_2}\dot{x}_2 + \cdots + \frac{\partial f}{\partial x_n}\dot{x}_n = \sum_{i=1}^{n}\frac{\partial f}{\partial x_i}\dot{x}_i \tag{13}$$

which yields

$$\frac{d}{dt}f(\mathbf{x}(t)) = \left\langle\frac{\partial f}{\partial\mathbf{x}},\dot{\mathbf{x}}(t)\right\rangle \tag{14}$$

Similarly, if $\mathbf{g}: R_n \to R_m$, and if $\mathbf{x}(t)$ is a time-varying column vector, then

$$\frac{d}{dt}\mathbf{g}(\mathbf{x}) \triangleq \begin{bmatrix} \dfrac{d}{dt}g_1(\mathbf{x}) \\[4pt] \dfrac{d}{dt}g_2(\mathbf{x}) \\[4pt] \vdots \\[4pt] \dfrac{d}{dt}g_m(\mathbf{x}) \end{bmatrix} \tag{15}$$

$$= \begin{bmatrix} \left\langle \dfrac{\partial g_1}{\partial \mathbf{x}}, \dot{\mathbf{x}}(t) \right\rangle \\[6pt] \left\langle \dfrac{\partial g_2}{\partial \mathbf{x}}, \dot{\mathbf{x}}(t) \right\rangle \\[6pt] \vdots \\[6pt] \left\langle \dfrac{\partial g_m}{\partial \mathbf{x}}, \dot{\mathbf{x}}(t) \right\rangle \end{bmatrix} \tag{16}$$

$$= \left(\frac{\partial \mathbf{g}}{\partial \mathbf{x}} \right) \dot{\mathbf{x}}(t) \tag{17}$$

It should be clear that gradient vectors and matrices can be used to compute mixed time and partial derivatives.

A.15   VECTOR AND MATRIX NORMS

We conclude our brief introduction to column vectors, matrices, and their operations by discussing the concept of the norm of a column vector and the norm of a matrix. The norm is a generalization of the familiar magnitude of euclidean length of a vector. Thus, the norm is used to decide how large a vector is and also how large a matrix is; in this manner, it is used to attach a scalar magnitude to such multivariable quantities as vectors and matrices.

A.15.1   Norms for Column Vectors

The reader should recall that we have already discussed a particular vector norm in Sec. A.11. This norm is often called the euclidean norm. The particular properties of this norm yield the natural properties of more general norms.

Let us consider a column n-vector $\mathbf{x}$

$$\mathbf{x} = \begin{bmatrix} x_1 \\ x_2 \\ \vdots \\ x_n \end{bmatrix} \tag{1}$$

The euclidean norm of $\mathbf{x}$, denoted by $\|\mathbf{x}\|_2$, is simply defined by

$$\|\mathbf{x}\|_2 = (x_1^2 + x_2^2 + \cdots + x_n^2)^{1/2} = \sqrt{\langle \mathbf{x}, \mathbf{x} \rangle} \tag{2}$$

It should be clear that the value of $\|\mathbf{x}\|_2$ provides us with an idea of how big $\mathbf{x}$ is. We recall that the euclidean norm of a column n-vector satisfies the following conditions:

$$\|\mathbf{x}\|_2 \geq 0 \quad \text{and} \quad \|\mathbf{x}\|_2 = 0 \quad \text{if and only if } \mathbf{x} = \mathbf{0} \tag{3}$$

$$\|\alpha\mathbf{x}\|_2 = |\alpha| \cdot \|\mathbf{x}\|_2 \quad \text{for all scalars } \alpha \tag{4}$$

$$\|\mathbf{x} + \mathbf{y}\|_2 \leq \|\mathbf{x}\|_2 + \|\mathbf{y}\|_2 \quad \text{the triangle inequality} \tag{5}$$

For many applications, the euclidean norm is not the most convenient to use in algebraic manipulations, although it has the most natural geometric interpretation. For this reason, one can generalize the notion of a norm in the following way.

Definition 1   *Let $\mathbf{x}$ and $\mathbf{y}$ be column n-vectors. Then a scalar-valued function of $\mathbf{x}$ qualifies as a norm $\|\mathbf{x}\|$ of $\mathbf{x}$ provided that the following three properties hold:*

(a) $$\|\mathbf{x}\| > 0 \qquad \text{for all } \mathbf{x} \neq \mathbf{0} \tag{6}$$

(b) $$\|\alpha\mathbf{x}\| = |\alpha| \cdot \|\mathbf{x}\| \qquad \text{for all } \alpha \in R \tag{7}$$

(c) $$\|\mathbf{x} + \mathbf{y}\| \leq \|\mathbf{x}\| + \|\mathbf{y}\| \qquad \text{for all } \mathbf{x}, \mathbf{y} \tag{8}$$

The reader should note that Eqs. (6) to (8) represent a consistent generalization of the properties (3) to (5) of the euclidean norm.

In addition to the euclidean norm there are two other common norms

$$\|\mathbf{x}\|_1 = \sum_{i=1}^{n} |x_i| \tag{9}$$

$$\|\mathbf{x}\|_\infty = \max_i |x_i| \tag{10}$$

We encourage the reader to verify that the norms defined by (9) and (10) indeed satisfy properties (6) to (8).

EXAMPLE 1   Suppose that $\mathbf{x}$ is the column vector

$$\mathbf{x} = \begin{bmatrix} 2 \\ -1 \\ 3 \end{bmatrix}$$

Then $\|\mathbf{x}\|_1 = |2| + |-1| + |3| = 6$, $\|\mathbf{x}\|_2 = (4 + 1 + 9)^{1/2} = \sqrt{14}$, $\|\mathbf{x}\|_\infty = \max\{|2|, |-1|, |3|\} = 3$.

A.15.2   Matrix Norms

Next we turn our attention to the concept of a norm of a matrix. To motivate the definition we simply note that a column n-vector can also be viewed as an $n \times 1$ matrix. Thus, if we are to extend the properties of vector norms to those of the matrix norms, they should be consistent. For this reason, we have a definition.

> **Definition 2**   *Let* A *and* B *be real* $n \times m$ *matrices with elements* a_{ij} *and* b_{ij} ($i = 1, 2, \ldots, n; j = 1, 2, \ldots, m$). *Then the scalar-valued function* $\|A\|$ *of* A *qualifies as the norm of* A *if the following properties hold*:
>
> (a) $$\|A\| > 0 \qquad\qquad provided\ not\ all\ a_{ij} = 0 \qquad (11)$$
>
> (b) $$\|\beta A\| = |\beta| \cdot \|A\| \qquad for\ any\ scalar\ \ \beta \qquad (12)$$
>
> (c) $$\|A + B\| \leq \|A\| + \|B\| \qquad\qquad\qquad (13)$$

As with vector norms, there are many convenient matrix norms, e.g.,

$$\|A\|_1 = \sum_{i=1}^{n} \sum_{j=1}^{m} |a_{ij}| \qquad (14)$$

$$\|A\|_2 = \left(\sum_{i=1}^{n} \sum_{j=1}^{m} a_{ij}^2 \right)^{1/2} \qquad (15)$$

$$\|A\|_\infty = \max_{i} \sum_{j=1}^{m} |a_{ij}| \qquad (16)$$

Once more we encourage the reader to prove that these matrix norms do indeed satisfy the defining properties (11) to (13).

A.15.3   Properties

Two important properties that hold between norms which involve multiplication of a matrix with a vector and multiplication of two matrices are summarized in the following two theorems.

> **Theorem 1**   *Let* A *be an* $n \times m$ *matrix with real elements* a_{ij} ($i = 1, 2, \ldots, n$; $j = 1, 2, \ldots, m$). *Let* x *be a column* m-vector with elements x_j ($j = 1, 2, \ldots, m$). *Then*
>
> $$\|Ax\| \leq \|A\| \cdot \|x\| \qquad (17)$$
>
> *in the sense that*
>
> (a) $$\|Ax\|_1 \leq \|A\|_1 \cdot \|x\|_1 \qquad (18)$$
>
> (b) $$\|Ax\|_2 \leq \|A\|_2 \cdot \|x\|_2 \qquad (19)$$
>
> (c) $$\|Ax\|_\infty \leq \|A\|_\infty \cdot \|x\|_\infty \qquad (20)$$

PROOF Let $\mathbf{y} = \mathbf{Ax}$; then $\mathbf{y}$ is a column vector with n-components $y_1, y_2, \ldots,$ y_n.

(*a*)

$$\|\mathbf{Ax}\|_1 = \|\mathbf{y}\|_1 \triangleq \sum_{i=1}^{n} |y_i| = \sum_{i=1}^{n} \left| \sum_{j=1}^{m} a_{ij} x_j \right|$$

$$\leq \sum_{i=1}^{n} \sum_{j=1}^{m} |a_{ij} x_j| = \sum_{i=1}^{n} \sum_{j=1}^{m} |a_{ij}| |x_j|$$

$$\leq \sum_{i=1}^{n} \sum_{j=1}^{m} |a_{ij}| \cdot \|\mathbf{x}\|_1 \qquad \text{since } \|\mathbf{x}\|_1 \geq |x_j|$$

$$= \left(\sum_{i=1}^{n} \sum_{j=1}^{m} |a_{ij}| \right) \|\mathbf{x}\|_1 = \|\mathbf{A}\|_1 \cdot \|\mathbf{x}\|_1$$

(*b*)

$$\|\mathbf{Ax}\|_2 = \|\mathbf{y}\|_2 = \left(\sum_{i=1}^{n} y_i^2 \right)^{1/2} = \left[\sum_{i=1}^{n} \left(\sum_{j=1}^{m} a_{ij} x_j \right)^2 \right]^{1/2}$$

$$\leq \left[\sum_{i=1}^{n} \left(\sum_{j=1}^{m} a_{ij}^2 \right) \left(\sum_{j=1}^{m} x_j^2 \right) \right]^{1/2} \qquad \text{by the Schwartz inequality}$$

$$= \left[\sum_{i=1}^{n} \left(\sum_{j=1}^{m} a_{ij}^2 \right) \|\mathbf{x}\|_2^2 \right]^{1/2} = \left(\sum_{i=1}^{n} \sum_{j=1}^{m} a_{ij}^2 \right)^{1/2} \|\mathbf{x}\|_2$$

$$= \|\mathbf{A}\|_2 \cdot \|\mathbf{x}\|_2$$

(*c*)

$$\|\mathbf{Ax}\|_\infty = \|\mathbf{y}\|_\infty = \max_i |y_i| = \max_i \left| \sum_{j=1}^{m} a_{ij} x_j \right|$$

$$\leq \max_i \left(\sum_{j=1}^{m} |a_{ij} x_j| \right) = \max_i \left(\sum_{j=1}^{m} |a_{ij}| |x_j| \right)$$

$$\leq \max_i \left(\sum_{j=1}^{m} |a_{ij}| \cdot \|\mathbf{x}\|_\infty \right) \qquad \text{because } \|\mathbf{x}\|_\infty \geq |x_j|$$

$$= \max_i \left(\sum_{j=1}^{m} |a_{ij}| \right) \|\mathbf{x}\|_\infty = \|\mathbf{A}\|_\infty \|\mathbf{x}\|_\infty$$

We shall leave it to the reader to verify the following theorem by imitating the proofs of Theorem 1.

Theorem 2 *Let $\mathbf{A}$ be a real $n \times m$ matrix and let $\mathbf{B}$ be a real $m \times q$ matrix. Then*

$$\|\mathbf{AB}\| \leq \|\mathbf{A}\| \cdot \|\mathbf{B}\| \tag{21}$$

in the sense that

(*a*) $$\|\mathbf{AB}\|_1 \leq \|\mathbf{A}\|_1 \cdot \|\mathbf{B}\|_1 \tag{22}$$

(*b*) $$\|\mathbf{AB}\|_2 \leq \|\mathbf{A}\|_2 \cdot \|\mathbf{B}\|_2 \tag{23}$$

(*c*) $$\|\mathbf{AB}\|_\infty \leq \|\mathbf{A}\|_\infty \cdot \|\mathbf{B}\|_\infty \tag{24}$$

A multitude of additional results concerning the properties of norms are available.

A.15.4   The Spectral Norm

A very useful norm, called the *spectral norm* of a matrix, is denoted by $\|\mathbf{A}\|_s$. Let $\mathbf{A}$ be a real $n \times m$ matrix. Then $\mathbf{A}'$ is an $m \times n$ matrix, and the product matrix $\mathbf{A}'\mathbf{A}$ is an $m \times m$ real matrix. Let us compute the eigenvalues of $\mathbf{A}'\mathbf{A}$, denoted by $\lambda_i(\mathbf{A}'\mathbf{A})$, $i = 1, 2, \ldots, m$. Since the matrix $\mathbf{A}'\mathbf{A}$ is symmetric and positive semidefinite, it has real nonnegative eigenvalues, i.e.,

$$\lambda_i(\mathbf{A}'\mathbf{A}) \geq 0 \qquad i = 1, 2, \ldots, m \tag{25}$$

Then the spectral norm of $\mathbf{A}$ is defined by

$$\|\mathbf{A}\|_s = \max[\lambda_i(\mathbf{A}'\mathbf{A})]^{1/2} \tag{26}$$

i.e., it is the square root of the maximum eigenvalue of $\mathbf{A}'\mathbf{A}$.

REFERENCES

There are many excellent books on linear algebra, varying from the very abstract to the very applied. The list below is a representative sample.

1 BELLMAN, R.: "Introduction to Matrix Analysis," McGraw-Hill, New York, 1960.

2 TURNBULL, H. W., and A. C. AITKEN: "An Introduction to the Theory of Canonical Matrices," Blackie, London, 1932.

3 HALMOS, P. R.: "Finite Dimensional Vector Spaces," Van Nostrand, Princeton, N.J., 1958.

4 BECKENBACK, E. F., and R. BELLMAN: "Inequalities," Springer, Berlin, 1961.

5 BIRKHOFF, G., and S. MACLANE: "A Survey of Modern Algebra," Macmillan, New York, 1958.

6 HOFFMAN, K., and R. KUNZE: "Linear Algebra," Prentice-Hall, Englewood Cliffs, N.J., 1961.

7 GANTMACHER, R. F.: "Matrix Theory," 2 vols., Chelsea, London, 1959.

8 FADDEEVA, D. K., and V. N. FADDEEVA: "Computation Methods of Linear Algebra," Freeman, San Francisco, 1963.

The use of Taylor series is extremely common in system analysis problems, since it can be used to approximate the behavior of a general nonlinear system about an operating point either by a linear system or by a nonlinear system of specific nonlinear complexity, e.g., quadratic, cubic, or quartic.

In this appendix we summarize the formulas of interest for the following cases:

1 Scalar-valued function of a single scalar variable
2 Scalar-valued function of n variables
3 Vector-valued function of n variables

B.1 SCALAR-VALUED FUNCTION OF A SINGLE SCALAR VARIABLE

Let x be a real scalar variable, and let $f(x)$ be a scalar-valued function of x. More precisely, we have

$$x \in R \qquad f(.): R \to R \tag{1}$$

We assume that $f(x)$ is sufficiently differentiable.

Let x^* be a specific value of x. Then the Taylor series of $f(x)$ about x^* is the infinite series

$$f(x) = f(x^*) + \frac{\partial f}{\partial x}\bigg|_{x=x^*}(x - x^*) + \frac{1}{2!}\frac{\partial^2 f}{\partial x^2}\bigg|_{x=x^*}(x - x^*)^2$$

$$+ \cdots + \frac{1}{k!}\frac{\partial^k f}{\partial x^k}\bigg|_{x=x^*}(x - x^*)^k + \cdots \tag{2}$$

where

$$\left.\frac{\partial^k f}{\partial x^k}\right|_{x=x^*}$$

means that the kth derivative, which depends on x, is evaluated at the value $x = x^*$.

The Taylor series contains an infinite number of terms. However, as $|x - x^*|$ becomes smaller and smaller, terms of the form $(x - x^*)^{n+1}$ become negligible with respect to terms of the form $(x - x^*)^n$, and this allows us to obtain good approximations to the function $f(x)$ in the immediate vicinity of x^*. Of particular interest are *linear approximation*:

$$f(x) \approx f(x^*) + \left.\frac{\partial f}{\partial x}\right|_{x=x^*}(x - x^*) \tag{3}$$

and *quadratic approximation*:

$$f(x) \approx f(x^*) + \left.\frac{\partial f}{\partial x}\right|_{x=x^*}(x - x^*) + \frac{1}{2}\left.\frac{\partial^2 f}{\partial x^2}\right|_{x=x^*}(x - x^*)^2 \tag{4}$$

The validity of these approximations can be estimated by *Taylor's theorem*.

Taylor's theorem states, roughly speaking, that we can truncate a Taylor series at an arbitrary derivative and get an exact expression, provided that the last derivative is evaluated not at x^* but at a point $\hat{x}$ which is in the interval between x and x^*, that is,

$$x^* \leq \hat{x} \leq x \qquad \text{or} \qquad x \leq \hat{x} \leq x^* \tag{5}$$

In other words, the following expressions are true:

$$f(x) = f(x^*) + \left.\frac{\partial f}{\partial x}\right|_{x=\hat{x}}(x - x^*) \tag{6}$$

$$f(x) = f(x^*) + \left.\frac{\partial f}{\partial x}\right|_{x=x^*}(x - x^*) + \frac{1}{2}\left.\frac{\partial^2 f}{\partial x^2}\right|_{x=\hat{x}}(x - x^*)^2 \tag{7}$$

$$\vdots$$

$$f(x) = f(x^*) + \sum_{k=1}^{m} \frac{1}{k!}\left.\frac{\partial^k f}{\partial x^k}\right|_{x=x^*}(x - x^*)^k + \frac{1}{(m+1)!}\left.\frac{\partial^{m+1} f}{\partial x^{m+1}}\right|_{x=\hat{x}}(x - x^*)^{m+1} \tag{8}$$

B.2   SCALAR-VALUED FUNCTION OF n VARIABLES

Let

$$f(\mathbf{x}) = f(x_1, x_2, \ldots, x_n) \tag{9}$$

denote a scalar-valued function of the n-vector $\mathbf{x}$. More precisely

$$\mathbf{x} \in R_n \qquad f(.): R_n \to R \tag{10}$$

Before proceeding with the Taylor series expansion, let us recall:

1 The gradient vector $\mathbf{g}(\mathbf{x})$ of $f(\mathbf{x})$ is an *n*-vector defined by

$$\mathbf{g}(\mathbf{x}) \triangleq \frac{\partial f}{\partial \mathbf{x}} \triangleq \begin{bmatrix} \dfrac{\partial f}{\partial x_1} \\ \dfrac{\partial f}{\partial x_2} \\ \vdots \\ \dfrac{\partial f}{\partial x_n} \end{bmatrix} \tag{11}$$

2 The second derivative (or Hessian) matrix $\mathbf{H}(\mathbf{x})$ is an $n \times n$ matrix given by

$$\mathbf{H}(\mathbf{x}) \triangleq \frac{\partial^2 f}{\partial \mathbf{x}^2} \triangleq \begin{bmatrix} \dfrac{\partial^2 f}{\partial x_1{}^2} & \dfrac{\partial^2 f}{\partial x_1 \partial x_2} & \cdots & \dfrac{\partial^2 f}{\partial x_1 \partial x_n} \\ \dfrac{\partial^2 f}{\partial x_2 \partial x_1} & \dfrac{\partial^2 f}{\partial x_2{}^2} & \cdots & \dfrac{\partial^2 f}{\partial x_2 \partial x_n} \\ \vdots & \vdots & \vdots & \vdots \\ \dfrac{\partial^2 f}{\partial x_n \partial x_1} & \dfrac{\partial^2 f}{\partial x_n \partial x_2} & \cdots & \dfrac{\partial^2 f}{\partial x_n{}^2} \end{bmatrix} \tag{12}$$

When the above terminology is used, the Taylor series expansion of $f(\mathbf{x})$ about a vector $\mathbf{x}^*$ takes the form

$$f(\mathbf{x}) = f(\mathbf{x}^*) + \mathbf{g}'(\mathbf{x}^*)(\mathbf{x} - \mathbf{x}^*) + \tfrac{1}{2}(\mathbf{x} - \mathbf{x}^*)'\mathbf{H}(\mathbf{x}^*)(\mathbf{x} - \mathbf{x}^*) + \cdots \tag{13}$$

where

$$\mathbf{g}(\mathbf{x}^*) \triangleq \frac{\partial f}{\partial \mathbf{x}}\bigg|_{\mathbf{x}=\mathbf{x}^*} \tag{14}$$

and

$$\mathbf{H}(\mathbf{x}^*) \triangleq \frac{\partial^2 f}{\partial \mathbf{x}^2}\bigg|_{\mathbf{x}=\mathbf{x}^*} \tag{15}$$

Let us illustrate the above by an example. Let $\mathbf{x} \in R_2$, and let

$$f(\mathbf{x}) = f(x_1, x_2) = x_1{}^3 x_2{}^4 \tag{16}$$

Then

$$\mathbf{g}(\mathbf{x}) = \begin{bmatrix} 3x_1{}^2 x_2{}^4 \\ 4x_1{}^3 x_2{}^3 \end{bmatrix} \tag{17}$$

$$\mathbf{H}(\mathbf{x}) = \begin{bmatrix} 6x_1 x_2{}^4 & 12x_1{}^2 x_2{}^3 \\ 12x_1{}^2 x_2{}^3 & 12x_1{}^3 x_2{}^2 \end{bmatrix} \tag{18}$$

Suppose we want to expand $f(\mathbf{x})$ of (16) about the 2-vector

$$\mathbf{x}^* = \begin{bmatrix} 1 \\ 2 \end{bmatrix} \tag{19}$$

Then

$$f(\mathbf{x}^*) = 16 \tag{20}$$

$$\mathbf{g}(\mathbf{x}^*) = \begin{bmatrix} 48 \\ 32 \end{bmatrix} \tag{21}$$

$$\mathbf{H}(\mathbf{x}^*) = \begin{bmatrix} 96 & 96 \\ 96 & 48 \end{bmatrix} \tag{22}$$

Hence the Taylor series (13) takes the form

$$f(x_1, x_2) = 16 + [48 \quad 32] \begin{bmatrix} x_1 - 1 \\ x_2 - 2 \end{bmatrix}$$

$$+ \tfrac{1}{2}[x_1 - 1 \quad x_2 - 2] \begin{bmatrix} 96 & 96 \\ 96 & 48 \end{bmatrix} \begin{bmatrix} x_1 - 1 \\ x_2 - 2 \end{bmatrix} + \cdots \tag{23}$$

B.3 VECTOR-VALUED FUNCTION OF n VARIABLES

Let

$$\mathbf{f}(\mathbf{x}) = \begin{bmatrix} f_1(x_1, x_2, \ldots, x_n) \\ f_2(x_1, x_2, \ldots, x_n) \\ \vdots \\ f_m(x_1, x_2, \ldots, x_n) \end{bmatrix} \tag{24}$$

be an m-vector-valued function of the n-vector $\mathbf{x}$. More precisely,

$$\mathbf{x} \in R_n \qquad \mathbf{f}(.): R_n \to R_m \tag{25}$$

Before writing the Taylor series let us define the $m \times n$ Jacobian matrix (matrix of first partial derivatives) $\mathbf{G}(\mathbf{x})$ as follows (see also Appendix A):

$$\mathbf{G}(\mathbf{x}) \triangleq \frac{\partial \mathbf{f}}{\partial \mathbf{x}} = \begin{bmatrix} \dfrac{\partial f_1}{\partial x_1} & \dfrac{\partial f_1}{\partial x_2} & \cdots & \dfrac{\partial f_1}{\partial x_n} \\[2mm] \dfrac{\partial f_2}{\partial x_1} & \dfrac{\partial f_2}{\partial x_2} & \cdots & \dfrac{\partial f_2}{\partial x_n} \\[2mm] \vdots & \vdots & \vdots & \vdots \\[2mm] \dfrac{\partial f_m}{\partial x_1} & \dfrac{\partial f_m}{\partial x_2} & \cdots & \dfrac{\partial f_m}{\partial x_n} \end{bmatrix} \tag{26}$$

Also we define the m second-derivative (Hessian) $n \times n$ matrices $\mathbf{H}_1(\mathbf{x})$, $\mathbf{H}_2(\mathbf{x})$, ..., $\mathbf{H}_m(\mathbf{x})$ by

$$\mathbf{H}_i(\mathbf{x}) = \frac{\partial^2 f_i}{\partial \mathbf{x}^2} = \begin{bmatrix} \dfrac{\partial^2 f_i}{\partial x_1^2} & \dfrac{\partial^2 f_i}{\partial x_1 \partial x_2} & \cdots & \dfrac{\partial^2 f_i}{\partial x_1 \partial x_n} \\ \vdots & \vdots & \vdots & \vdots \\ \dfrac{\partial^2 f_i}{\partial x_n \partial x_1} & \dfrac{\partial^2 f_i}{\partial x_n \partial x_2} & \cdots & \dfrac{\partial^2 f_i}{\partial x_n^2} \end{bmatrix} \tag{27}$$

In this case, the Taylor series expansion of $\mathbf{f}(\mathbf{x})$ about the n-vector $\mathbf{x}^*$ is given by

$$\mathbf{f}(\mathbf{x}) = \mathbf{f}(\mathbf{x}^*) + \mathbf{G}(\mathbf{x}^*)(\mathbf{x} - \mathbf{x}^*)$$

$$+ \frac{1}{2} \begin{bmatrix} 1 \\ 0 \\ \vdots \\ 0 \end{bmatrix} (\mathbf{x} - \mathbf{x}^*)' \mathbf{H}_1(\mathbf{x}^*)(\mathbf{x} - \mathbf{x}^*)$$

$$+ \frac{1}{2} \begin{bmatrix} 0 \\ 1 \\ \vdots \\ 0 \end{bmatrix} (\mathbf{x} - \mathbf{x}^*)' \mathbf{H}_2(\mathbf{x}^*)(\mathbf{x} - \mathbf{x}^*) + \cdots \tag{28}$$

$$+ \frac{1}{2} \begin{bmatrix} 0 \\ 0 \\ \vdots \\ 1 \end{bmatrix} (\mathbf{x} - \mathbf{x}^*)' \mathbf{H}_m(\mathbf{x}^*)(\mathbf{x} - \mathbf{x}^*) + \cdots$$

Let us define by $\mathbf{e}_1$, $\mathbf{e}_2$, ..., $\mathbf{e}_m$ the natural-basis vectors in R_m, that is,

$$\mathbf{e}_i = \begin{bmatrix} 0 \\ 0 \\ \vdots \\ 1 \\ \vdots \\ 0 \\ 0 \end{bmatrix} \leftarrow i\text{th component} \tag{29}$$

Then, Eq. (28) can be written in the shorthand form

$$\mathbf{f}(\mathbf{x}) = \underbrace{\mathbf{f}(\mathbf{x}^*)}_{\substack{\text{Constant} \\ \text{term}}} + \underbrace{\mathbf{G}(\mathbf{x}^*)(\mathbf{x} - \mathbf{x}^*)}_{\substack{\text{Linear} \\ \text{term}}} + \underbrace{\frac{1}{2} \sum_{i=1}^{m} \mathbf{e}_i(\mathbf{x} - \mathbf{x}^*)' \mathbf{H}_i(\mathbf{x}^*)(\mathbf{x} - \mathbf{x}^*)}_{\substack{\text{Quadratic} \\ \text{term}}} + \cdots \tag{30}$$

Taylor's theorem carries through to this general vector case. Thus the following expressions are exact:

$$\mathbf{f(x)} = \mathbf{f(x^*)} + \mathbf{G(\hat{x})(x - x^*)} \tag{31}$$

$$\mathbf{f(x)} = \mathbf{f(x^*)} + \mathbf{G(x^*)(x - x^*)} + \frac{1}{2} \sum_{i=1}^{m} \mathbf{e}_i \mathbf{(x - x^*)'H}_i\mathbf{(\hat{x})(x - x^*)} \tag{32}$$

where $\hat{\mathbf{x}}$ is a vector in the line segment joining $\mathbf{x}$ and $\mathbf{x^*}$; that is, $\hat{\mathbf{x}}$ is defined by some real scalar a, and

$$\hat{\mathbf{x}} = a\mathbf{x} + (1 - a)\mathbf{x^*} \qquad 0 \le a \le 1 \tag{33}$$

APPENDIX C
A BRIEF REVIEW OF
LAPLACE TRANSFORMS

SUMMARY

This appendix provides notational conventions for Laplace transforms for scalar- and vector-valued functions and lists the main properties needed to carry out certain developments in the main body of this text. This appendix is *not* intended to serve as an introduction to Laplace transforms for readers not previously exposed to this concept. The use of Laplace transform ideas in the main body of the text is minimal, and that material can be omitted with no loss in continuity.

C.1   DEFINITIONS: SCALAR CASE

Throughout this appendix we shall deal with time functions defined for values of time in the time interval

$$0 \leq t < \infty \tag{1}$$

Let $x(t)$ denote a scalar-valued function. Its Laplace transform is denoted by $\hat{x}(s)$. We often write

$$\hat{x}(s) = \mathcal{L}\{x(t)\} \tag{2}$$

To recover $x(t)$ from its Laplace transform $\hat{x}(s)$ we write, symbolically,

$$x(t) = \mathcal{L}^{-1}\{\hat{x}(s)\} \tag{3}$$

Under suitable assumptions on $x(t)$, its Laplace transform $\hat{x}(s)$ is computed according to

$$\hat{x}(s) = \int_0^\infty x(t)e^{-st}\,dt \tag{4}$$

where s is a complex scalar (often called the complex frequency). We remark that as long as the integral in (4) exists, i.e., it does not blow up, we can define the Laplace transform of $x(t)$.

For all well-behaved time functions encountered in engineering, the Laplace transform $\hat{x}(s)$ of $x(t)$ (if it exists) is unique. Similarly, the inverse Laplace transform is unique. Thus,

$$x(t) = \mathcal{L}^{-1}\{\mathcal{L}\{x(t)\}\} \tag{5}$$

and

$$\hat{x}(s) = \mathcal{L}\{\mathcal{L}^{-1}\{\hat{x}(s)\}\} \tag{6}$$

C.2 DEFINITIONS: THE VECTOR AND MATRIX CASE

Let $\mathbf{x}(t)$ be an n-dimensional time-varying vector

$$\mathbf{x}(t) = \begin{bmatrix} x_1(t) \\ x_2(t) \\ \vdots \\ x_n(t) \end{bmatrix} \tag{7}$$

The Laplace transform $\hat{\mathbf{x}}(s)$ of $\mathbf{x}(t)$

$$\hat{\mathbf{x}}(s) = \mathcal{L}\{\mathbf{x}(t)\} \tag{8}$$

is also an n-dimensional vector

$$\hat{\mathbf{x}}(s) = \begin{bmatrix} \hat{x}_1(s) \\ \hat{x}_2(s) \\ \vdots \\ \hat{x}_n(s) \end{bmatrix} \tag{9}$$

The computation of $\hat{\mathbf{x}}(s)$ is component by component, i.e.,

$$\hat{x}_i(s) \triangleq \mathcal{L}\{x_i(t)\} \qquad i = 1, 2, \ldots, n \tag{10}$$

The inverse Laplace transform of a vector

$$\mathbf{x}(t) = \mathcal{L}^{-1}\{\hat{\mathbf{x}}(s)\} \tag{11}$$

is also defined component by component, i.e.,

$$x_i(t) \triangleq \mathcal{L}^{-1}\{\hat{x}_i(s)\} \tag{12}$$

In the matrix case, the definitions are element by element. Suppose that $\mathbf{F}(t)$ is an $n \times m$ time-varying matrix with elements $f_{ij}(t)$, $i = 1, 2, \ldots, n$ and $j = 1, 2, \ldots, m$. The Laplace transform matrix $\hat{\mathbf{F}}(s)$ of $\mathbf{F}(t)$, denoted by

$$\hat{\mathbf{F}}(s) = \mathcal{L}\{\mathbf{F}(t)\} \tag{13}$$

is also an $n \times m$ matrix. If we denote by $\hat{f}_{ij}(s)$ the ijth element of $\hat{\mathbf{F}}(s)$, then

$$\hat{f}_{ij}(s) \triangleq \mathcal{L}\{f_{ij}(t)\} \qquad \begin{aligned} i &= 1, 2, \ldots, n \\ j &= 1, 2, \ldots, m \end{aligned} \tag{14}$$

The same element-by-element procedure is used to find the inverse Laplace transform matrix

$$\mathbf{F}(t) = \mathcal{L}^{-1}\{\hat{\mathbf{F}}(s)\} \tag{15}$$

C.3 IMPORTANT PROPERTIES OF LAPLACE TRANSFORMS

C.3.1 Linearity

If $\mathbf{A}$ and $\mathbf{B}$ are constant matrices,

$$\mathcal{L}\{\mathbf{A}\mathbf{x}(t) + \mathbf{B}\mathbf{y}(t)\} = \mathbf{A}\mathcal{L}\{\mathbf{x}(t)\} + \mathbf{B}\mathcal{L}\{\mathbf{y}(t)\} \tag{16}$$

Similarly

$$\mathcal{L}^{-1}\{\mathbf{A}\hat{\mathbf{x}}(s) + \mathbf{B}\hat{\mathbf{y}}(s)\} = \mathbf{A}\mathcal{L}^{-1}\{\hat{\mathbf{x}}(s)\} + \mathbf{B}\mathcal{L}^{-1}\{\hat{\mathbf{y}}(s)\} \tag{17}$$

If $\mathbf{A}, \mathbf{B}, \mathbf{C}, \mathbf{D}$ are constant matrices,

$$\mathcal{L}\{\mathbf{A}\mathbf{F}(t)\mathbf{B} + \mathbf{C}\mathbf{G}(t)\mathbf{D}\} = \mathbf{A}\mathcal{L}\{\mathbf{F}(t)\}\mathbf{B} + \mathbf{C}\mathcal{L}\{\mathbf{G}(t)\}\mathbf{D} \tag{18}$$

$$\mathcal{L}^{-1}\{\mathbf{A}\hat{\mathbf{F}}(s)\mathbf{B} + \mathbf{C}\hat{\mathbf{G}}(s)\mathbf{D}\} = \mathbf{A}\mathcal{L}^{-1}\{\hat{\mathbf{F}}(s)\}\mathbf{B} + \mathbf{C}\mathcal{L}^{-1}\{\hat{\mathbf{G}}(s)\}\mathbf{D} \tag{19}$$

C.3.2 Differentiation

If $x(t)$ is a scalar function, then

$$\mathcal{L}\{\dot{x}(t)\} = s\hat{x}(s) - x(0^+) \tag{20}$$

where $x(0^+)$ is the value of $x(t)$ at time $t = 0^+$. In the case of a vector we have

$$\mathcal{L}\{\dot{\mathbf{x}}(t)\} = s\hat{\mathbf{x}}(s) - \mathbf{x}(0^+) \tag{21}$$

and in the case of a matrix

$$\mathcal{L}\{\dot{\mathbf{F}}(t)\} = s\hat{\mathbf{F}}(s) - \mathbf{F}(0^+) \tag{22}$$

If $x(t)$ is a scalar, we have the differentiation rule

$$\mathcal{L}\left\{\frac{d^k x(t)}{dt^k}\right\} = s^k \hat{x}(s) - s^{k-1}x(0^+) - s^{k-2}\frac{dx}{dt}(0^+) - \cdots - s\frac{d^{k-2}x}{dt}(0^+) - \frac{d^{k-1}x}{dt}(0^+) \tag{23}$$

This formula for higher-order derivatives can be extended to the vector and matrix case.

Table C.1   BRIEF TABLE OF LAPLACE TRANSFORMS

Time function $x(t)$	Laplace transform $\hat{x}(s)$
$\delta(t)$ = unit impulse at $t = 0$	1
$\mu(t)$ = unit step at $t = 0$	$\dfrac{1}{s}$
$\dfrac{t^k}{k!}$	$\dfrac{1}{s^{k+1}}$
be^{-at}	$\dfrac{b}{s+a}$
$b\dfrac{t^k}{k!}a^{-at}$	$\dfrac{b}{(s+a)^{k+1}}$
$a \cos wt$	$\dfrac{as}{s^2+w^2}$
$a \sin wt$	$\dfrac{aw}{s^2+w^2}$
$be^{-at}\cos wt$	$b\dfrac{s+a}{(s+a)^2+w^2}$
$be^{-at}\sin wt$	$b\dfrac{w}{(s+a)^2+w^2}$

Index